T0180869

Lecture Notes in Computer Science 13426

Founding Editors

Gerhard Goos
Karlsruhe Institute of Technology, Karlsruhe, Germany

Juris Hartmanis
Cornell University, Ithaca, NY, USA

Editorial Board Members

Elisa Bertino
Purdue University, West Lafayette, IN, USA

Wen Gao
Peking University, Beijing, China

Bernhard Steffen
TU Dortmund University, Dortmund, Germany

Moti Yung
Columbia University, New York, NY, USA

More information about this series at https://link.springer.com/bookseries/558

Christine Strauss · Alfredo Cuzzocrea ·
Gabriele Kotsis · A Min Tjoa ·
Ismail Khalil (Eds.)

Database and Expert Systems Applications

33rd International Conference, DEXA 2022
Vienna, Austria, August 22–24, 2022
Proceedings, Part I

 Springer

Editors
Christine Strauss
University of Vienna
Vienna, Austria

Alfredo Cuzzocrea
University of Calabria
Rende, Italy

Gabriele Kotsis
Johannes Kepler University of Linz
Linz, Austria

A Min Tjoa (iD)
Vienna University of Technology
Vienna, Austria

Ismail Khalil
Johannes Kepler University of Linz
Linz, Austria

ISSN 0302-9743   ISSN 1611-3349 (electronic)
Lecture Notes in Computer Science
ISBN 978-3-031-12422-8   ISBN 978-3-031-12423-5 (eBook)
https://doi.org/10.1007/978-3-031-12423-5

© The Editor(s) (if applicable) and The Author(s), under exclusive license
to Springer Nature Switzerland AG 2022
This work is subject to copyright. All rights are reserved by the Publisher, whether the whole or part of the material is concerned, specifically the rights of translation, reprinting, reuse of illustrations, recitation, broadcasting, reproduction on microfilms or in any other physical way, and transmission or information storage and retrieval, electronic adaptation, computer software, or by similar or dissimilar methodology now known or hereafter developed.
The use of general descriptive names, registered names, trademarks, service marks, etc. in this publication does not imply, even in the absence of a specific statement, that such names are exempt from the relevant protective laws and regulations and therefore free for general use.
The publisher, the authors, and the editors are safe to assume that the advice and information in this book are believed to be true and accurate at the date of publication. Neither the publisher nor the authors or the editors give a warranty, expressed or implied, with respect to the material contained herein or for any errors or omissions that may have been made. The publisher remains neutral with regard to jurisdictional claims in published maps and institutional affiliations.

This Springer imprint is published by the registered company Springer Nature Switzerland AG
The registered company address is: Gewerbestrasse 11, 6330 Cham, Switzerland

Preface

Welcome to the two-volume edition of the proceedings of the 33rd International Conference on Database and Expert Systems Applications (DEXA 2022). After a break of two years due to the COVID-19 pandemic situation, which forced us to use online formats, we were happy that we could finally meet in person in Vienna, Austria, during August 22–24, 2022. The wide variety of the topics, as well as the depth of the presented research, revealed, that sound research in the field of database and expert systems applications was not at all shut down by the pandemic. The papers accepted and presented at DEXA 2022, which are collated in these two volumes of proceedings, are an impressive collection of the research and development performed during the challenging recent times.

This year, the DEXA Program Committee accepted 43 full papers and 20 short papers, leading to an acceptance rate of 35%. The total number of submissions was comparable with recent DEXA editions, and we are proud to see again that the DEXA community is global as we received contributions from all around the world (Europe, America, Asia, Africa, Oceania). Our Program Committee performed more than 500 reviews, which not only serve the purpose of quality control for the conference but also contained valuable feedback and insights for the authors. We would like to sincerely thank our Program Committee members for their rigorous and critical, and at the same time motivating, reviews of DEXA 2022 submissions.

As is the tradition of DEXA conference series, all accepted papers were published in Lecture Notes in Computer Science (LNCS) and made available by Springer. Authors of selected papers presented at the conference will be invited to submit substantially extended versions of their conference papers for publication in special issues of two international journals: Knowledge and Information Systems (KAIS) and Transactions of Large Scale Data and Knowledge Centered Systems (TLDKS), both published by Springer. The submitted extended versions will undergo a further review process.

DEXA 2022 covered a wide range of relevant topics: (i) big data management and analytics, (ii) consistency, integrity, and quality of data, (iii) constraint modeling and processing, (iv) database federation and integration, interoperability, and multi-databases, (v) data and information semantics, (vi) data integration, metadata management, and interoperability, (vii) data structures and data management algorithms, (viii) graph databases, (ix) incomplete and uncertain data, (x) information retrieval, (xi) statistical and scientific databases, (xii) temporal, spatial, and high dimensional databases, (xiii) query processing and transaction management, (xiv) visual data analytics, data mining, and knowledge discovery, (xv) WWW and databases, as well as web services.

We would like to express our gratitude to the distinguished keynote speakers for their presented leading edge topics:

- Ricardo Baeza-Yates, Institute for Experiential AI, Northeastern University, USA
- Sabrina Kirrane, Institute for Information Systems and New Media, Vienna University of Economics and Business, Austria
- Philippe Cudré-Mauroux, University of Fribourg, Switzerland

DEXA 2022 also featured six international workshops that focused the attention on a variety of specific topics:

- The 2nd International Workshop on AI System Engineering: Math, Modelling and Software (AISys 2022);
- The 1st International Workshop on Applied Research, Technology Transfer and Knowledge Exchange in Software and Data Science (ARTE 2022);
- The 1st International Workshop on Distributed Ledgers and Related Technologies (DLRT 2022);
- The 6th International Workshop on Cyber-Security and Functional Safety in Cyber-Physical Systems (IWCFS 2022);
- The 4th International Workshop on Machine Learning and Knowledge Graphs (MLKgraphs 2022);
- The 2nd International Workshop on Time Ordered Data (ProTime 2022).

Like the success of every conference, DEXA's success is also built on the continuous and generous support of its participants and contributors and their perpetual and sustained efforts. Our sincere thanks go to the loyal and dedicated authors, distinguished Program Committee members, session chairs, organizing and steering committee members, and student volunteers who worked hard to ensure the continuity and the high quality of DEXA 2022.

We would also like to express our thanks to all institutions actively supporting this event, namely

- Software Competence Center Hagenberg (SCCH), Austria;
- Institute of Telecooperation, Johannes Kepler University Linz (JKU), Austria;
- Web Applications Society (@WAS);
- Austria Society for Artificial Intelligence (ASAI), Austria;
- Vienna University of Economics and Business (WU), Austria; and
- Austrian Blockchain Center (ABC Research), Austria.

We hope you enjoyed the DEXA 2022 conference: not only as an opportunity to present your own work to the DEXA community but also as an opportunity to meet new peers and foster and enlarge your network. We are looking forward to seeing you again next year!

August 2022   Christine Strauss
                                                                     Alfredo Cuzzocrea

Organization

Program Committee Chairs

Christine Strauss   University of Vienna, Austria
Alfredo Cuzzocrea   University of Calabria, Italy

Steering Committee

Gabriele Kotsis   Johannes Kepler University Linz, Austria
A Min Tjoa   Vienna University of Technology, Austria
Robert Wille   Software Competence Center Hagenberg, Austria
Bernhard Moser   Software Competence Center Hagenberg, Austria
Alfred Taudes   Vienna University of Economics and Business
     and Austrian Blockchain Center, Austria
Ismail Khalil   Johannes Kepler University Linz, Austria

Program Committee

Sonali Agarwal   IIIT Allahabad, India
Riccardo Albertoni   IMATI-CNR, Italy
Toshiyuki Amagasa   University of Tsukuba, Japan
Idir Amine Amarouche   USTHB, Algeria
Rachid Anane   Coventry University, UK
Mustafa Atay   Winston-Salem State University, USA
Ladjel Bellatreche   LIAS, ISAE-ENSMA, France
Nadia Bennani   LIRIS, INSA de Lyon, France
Karim Benouaret   Université Claude Bernard Lyon 1, France
Djamal Benslimane   Université Claude Bernard Lyon 1 , France
Vasudha Bhatnagar   University of Delhi, India
Andreas Both   DATEV eG, Germany
Athman Bouguettaya   University of Sydney, Australia
Omar Boussaid   ERIC, Université Lumiere Lyon 2, France
Stephane Bressan   National University of Singapore, Singapore
Pablo Garcia Bringas   University of Deusto, Spain
Barbara Catania   Università degli Studi di Genova, Italy
Ruzanna Chitchyan   University of Bristol, UK
Soon Chun   City University of New York, USA
Deborah Dahl   Conversational Technologies, USA

Jérôme Darmont	Université Lumiere Lyon 2, France
Soumyava Das	Imply Data, USA
Vincenzo Deufemia	University of Salerno, Italy
Dejing Dou	University of Oregon, USA
Cedric Du Mouza	Cnam, France
Johann Eder	University of Klagenfurt, Austria
Andreas Ekelhart	Secure Business Austria, Austria
Markus Endres	University of Passau, Germany
Noura Faci	Université Claude Bernard Lyon 1, France
Bettina Fazzinga	University of Calabria, Italy
Flavio Ferrarotti	Software Competence Centre Hagenberg, Austria
Flavius Frasincar	Erasmus University Rotterdam, The Netherlands
Bernhard Freudenthaler	Software Competence Centre Hagenberg, Austria
Steven Furnell	University of Nottingham, UK
Manolis Gergatsoulis	Ionian University, Greece
Vikram Goyal	IIIT-Delhi, India
Sven Groppe	University of Lübeck, Germany
Wilfried Grossmann	University Vienna, Austria
Francesco Guerra	Università di Modena e Reggio Emilia, Italy
Giovanna Guerrini	University of Genoa, Italy
Allel Hadjali	LIAS, ISAE-ENSMA, France
Abdelkader Hameurlain	IRIT, Paul Sabatier University, France
Sven Hartmann	Clausthal University of Technology, Germany
Manfred Hauswirth	Technical University Berlin, Germany
Ionut Iacob	Georgia Southern University, USA
Hamidah Ibrahim	Universiti Putra Malaysia, Malaysia
Sergio Ilarri	University of Zaragoza, Spain
Abdessamad Imine	Loria, France
Ivan Izonin	Lviv Polytechnic National University, Ukraine
Stephane Jean	LIAS, ISAE-ENSMA and University of Poitiers, France
Peiquan Jin	University of Science and Technology of China, China
Anne Kayem	Hasso Plattner Institute, University of Potsdam, Germany
Elmar Kiesling	Vienna University of Economics and Business, Austria
Uday Kiran	University of Tokyo, Japan
Carsten Kleiner	Hannover University of Applied Science and Arts, Germany
Henning Koehler	Massey University, New Zealand
Michal Kratky	VSB-Technical University of Ostrava, Czech Republic

Petr Kremen	Biomax Informatics AG and Czech Technical University in Prague, Czech Republic
Josef Küng	Johannes Kepler Universitaet Linz, Austria
Lenka Lhotska	Czech Technical University in Prague, Czech Republic
Chuan-Ming Liu	National Taipei University of Technology, Taiwan
Jorge Lloret	University of Zaragoza, Spain
Hui Ma	Victoria University of Wellington, New Zealand
Qiang Ma	Kyoto University, Japan
Elio Masciari	University of Naples Federico II, Italy
Jun Miyazaki	Tokyo Institute of Technology, Japan
Lars Moench	University of Hagen, Germany
Riad Mokadem	Pryamide, Paul Sabatier University, France
Anirban Mondal	University of Tokyo, Japan
Yang-Sae Moon	Kangwon National University. South Korea
Franck Morvan	IRIT, Paul Sabatier University, France
Philippe Mulhem	LIG-CNRS, France
Enzo Mumolo	University of Trieste, Italy
Francesco D. Muñoz-Escoí	Universitat Politècnica de València, Spain
Ismael Navas-Delgado	University of Málaga, Spain
Javier Nieves	Azterlan, Spain
Brahim Ouhbi	Université Moulay Ismail, Morocco
Marcin Paprzycki	Systems Research Institute, Polish Academy of Sciences, Poland
Louise Parkin	LIAS, ISAE-ENSMA, France
Iker Pastor-Lovez	University of Deusto, Spain
Dhaval Patel	IBM, USA
Clara Pizzuti	ICAR-CNR, Italy
Elaheh Pourabbas	IASI-CNR, Italy
Simone Raponi	Hamad Bin Khalifa University, Qatar
Claudia Roncancio	Grenoble Alpes University, France
Massimo Ruffolo	ICAR-CNR, Italy
Marinette Savonnet	LE2I, University of Burgundy, France
Florence Sedes	IRIT, Paul Sabatier University, France
Hossain Shahriar	Kennesaw State University, USA
Michael Sheng	Macquarie University, Australia
Patrick Siarry	Université Paris-Est Créteil, France
Tarique Siddiqui	Microsoft Research, USA
Gheorghe Cosmin Silaghi	Babes-Bolyai University, Romania
Srinath Srinivasa	International Institute of Information Technology, Bangalore, India
Bala Srinivasan	Monash University, Australia

Christian Stummer — Bielefeld University, Germany
Olivier Teste — IRIT, Université Toulouse Jean Jaurès, France
Jean-Marc Thevenin — Université Toulouse 1 Capitole, France
A Min Tjoa — Vienna University of Technology, Austria
Vicenc Torra — University of Skövde, Sweden
Traian Marius Truta — Northern Kentucky University, USA
Krishnamurthy Vidyasankar — Memorial University, Canada
Piotr Wisniewski — Nicolaus Copernicus University, Poland
Ming Hour Yang — Chung Yuan Chritian University, Taiwan
Haruo Yokota — Tokyo Institute of Technology, Japan
Yan Zhu — Southwest Jiaotong University, China
Qiang Zhu — University of Michigan - Dearborn, USA
Ester Zumpano — University of Calabria, Italy

External Reviewers

Amani Abusafia, Australia
Ahoud Alhazmi, Australia
Abdulwahab Aljubairy, Australia
Balsam Alkouz, Australia
Radim Baca, Czech Republic
Saidi Boumediene, France
Bernardo Breve, Italy
Taotao Cai, Australia
Luciano Caroprese, Italy
Dipankar Chaki, Australia
Peter Chovanec, Czech Republic
Gaetano Cimino, Italy
Stefano Cirillo, Italy
Matthew Damigos, Greece
Kaushik Das Sharma, India
Yves Denneulin, France
Kong Diison, China
Nabil El Malki, France
Marco Franceschetti, Austria
Myeong-Seon Gil, South Korea
Ramon Hermoso, Spain
Akm Tauhidul Islam, USA
Eleftherios Kalogeros, Greece
Sharanjit Kaur, Malaysia
Julius Koepke, Austria
Abdallah Lakhdari, Australia

Hieu Hanh Le, Japan
Ji Liu, China
Qiuhao Lu, USA
Josef Lubas, Austria
Petr Lukas, Czech Republic
Jorge Martinez-Gil, Austria
Amin Mesmoudi, France
Gabriele Oligeri, Qatar
Shaowen Peng, Japan
Gang Qian, USA
Savio Sciancalepore, The Netherlands
Babar Shahzaad, Australia
Zheng Song, USA
Piotr Sowinski, Poland
Junjie Sun, Japan
Raquel Trillo-Lado, Spain
Zheni Utic, USA
Eugenio Vocaturo, Italy
Alexander Voelz, Austria
Kai Wang, China
Yi-Hung Wu, Taiwan
Chengyang Ye, Japan
Chih-Chang Yu, Taiwan
Feng George Yu, USA
Xiao Zhang, China

Organizers

Abstracts of Keynote Talks

Responsible AI

Ricardo Baeza-Yates

Institute for Experiential AI @ Northeastern University

Abstract. In the first part we cover five current specific problems that motivate the needs of responsible AI: (1) discrimination (e.g., facial recognition, justice, sharing economy, language models); (2) phrenology (e.g., biometric based predictions); (3) unfair digital commerce (e.g., exposure and popularity bias); (4) stupid models (e.g., minimal adversarial AI) and (5) indiscriminate use of computing resources (e.g., large language models). These examples do have a personal bias but set the context for the second part where we address four challenges: (1) too many principles (e.g., principles vs. techniques), (2) cultural differences; (3) regulation and (4) our cognitive biases. We finish discussing what we can do to address these challenges in the near future to be able to develop responsible AI.

Following the Rules: From Policies to Norms

Sabrina Kirrane

Institute for Information Systems and New Media @ Vienna University
of Economics and Business

Abstract. Since its inception, the world wide web has evolved from a medium for information dissemination, to a general information and communication technology that supports economic and societal interaction and collaboration across the globe. Existing web-based applications range from e-commerce and e-government services, to various media and social networking platforms, many of whom incorporate software agents, such as bots and digital assistants. However, the original semantic web vision, whereby machine-readable web data could be automatically actioned upon by intelligent software web agents, has yet to be realized. In this talk, we will show how rules, in the form of policies and norms, can be used to specify a variety of data usage constraints (access policies, licenses, privacy preferences, regulatory constraints), in a manner that supports automated enforcement or compliance checking. Additionally, we discuss how, when taken together, policies, preferences, and norms can be used to afford humans more control and transparency with respect to individual and collaborating agents. Finally, we will highlight several open challenges and opportunities.

Contents – Part I

Warehousing Methodologies

Contents – Part II

Neural Networks

Efficient Data Processing Techniques

Advanced Analytics Methodologies and Methods

Knowledge Graphs

Jointly Learning Propagating Features on the Knowledge Graph for Movie Recommendation

Yun Liu(ID), Jun Miyazaki(✉)(ID), and Qiong Chang(ID)

Tokyo Institute of Technology, 2-12-1 Ookayama, Meguro-ku, Tokyo, Japan
liu@lsc.c.titech.ac.jp, {miyazaki,q.chang}@c.titech.ac.jp

Abstract. Knowledge graphs are widely used as auxiliary information to improve the performance in recommender systems. This enables items to be aligned with knowledge entities and provides additional item attributes to facilitate learning interactions between users and items. However, the lack of user connections in the knowledge graph may degrade the profiling of user preferences, especially for explicit user behaviors. Furthermore, learning knowledge graph embeddings is not entirely consistent with recommendation tasks due to different objectives. To solve the aforementioned problems, we extract knowledge entities from users' explicit reviews and propose a multi-task framework to jointly learn propagating features on the knowledge graph for movie recommendations. The review-based heterogeneous graph can provide substantial information for learning user preferences. In the proposed framework, we use an attention-based multi-hop propagation mechanism to take users and movies as center nodes and extend their attributes along with the connections of the knowledge graph by recursively calculating the different contributions of their neighbors. We use two real-world datasets to show the effectiveness of our proposed model in comparison with state-of-the-art baselines. Additionally, we investigate two aspects of the proposed model in extended ablation studies.

Keywords: Multi-task learning · Knowledge graph · Review-based recommendation · Personalized recommender systems

1 Introduction

Knowledge graphs (KGs) contain a large number of item attributes, which are widely used as auxiliary information to improve recommendation performance. One of the most commonly-used practices is aligning items with knowledge entities in a KG, which enables to explore item attributions along with the connections of the entities [1–5,22].

The key point of KG-based recommender systems (RSs) is how to profile user preferences on the basis of the KG. Existing works profile user preferences by first integrating user behaviors into the graph and then designing an effective method to learn user preferences along with the connections in the graph [6–8].

© The Author(s), under exclusive license to Springer Nature Switzerland AG 2022
C. Strauss et al. (Eds.): DEXA 2022, LNCS 13426, pp. 3–16, 2022.
https://doi.org/10.1007/978-3-031-12423-5_1

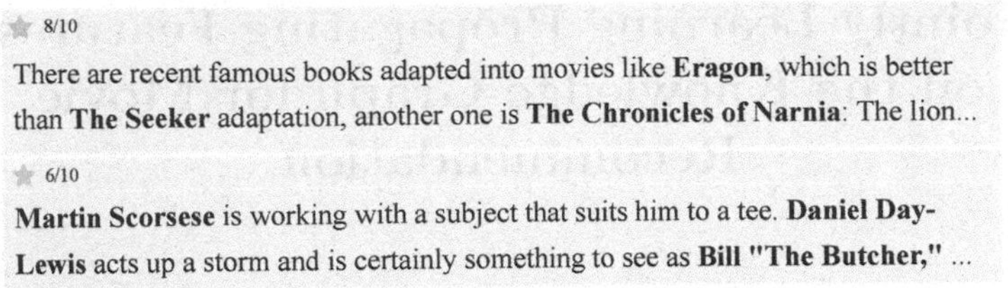

There are recent famous books adapted into movies like **Eragon**, which is better
than **The Seeker** adaptation, another one is **The Chronicles of Narnia**: The lion...

★ 6/10

Martin Scorsese is working with a subject that suits him to a tee. **Daniel Day-
Lewis** acts up a storm and is certainly something to see as **Bill "The Butcher,"** ...

Fig. 1. Illustration of movie reviews with knowledge mentions. The bold words are
knowledge mentions aligned with entities in the KG.

To handle the interactions between users and items, they treat the interactions
as KG edges, and define the built heterogeneous graph as a collaborative KG
[3]. Existing KG-based recommendation methods are roughly classified into two
types: path-based and embedding-based.

Path-based methods explore paths between users and items to learn the
multiple hops information as user preferences for enhancing recommendation
performances. They usually treat KG-based recommendation tasks as multi-
hop reasoning problems [2,9] or define meta-paths to extract specific patterns
between users and items to improve recommendation accuracy [10]. Embedding-
based methods represent users and items as entity embeddings by using current
KG embedding (KGE) algorithms, such as TransE [12] and TransR [13]. The
user preferences of these works are depicted by the linked neighbors of users in
the graph.

Although these methods can improve the corresponding recommendation
performances, they also have several deficiencies. First, they usually integrate
the user-item implicit interactions (e.g., clicks and browses) directly into the
graph, which is unsuitable for explicit user behaviors (e.g., ratings and reviews).
Second, all user neighbors in the graph are items, which is insufficient to profile user preferences based on explicit behaviors. Third, although KGs have their
benefits in learning user preferences on the basis of the connections on the graph,
directly using entity embeddings for recommendation tasks results unnecessary
losses in accuracy.

User reviews are widely used as auxiliary information in RSs and have been
successfully applied to improve recommendation performance [23,24]. Existing
review-based RSs usually extract topics or semantic embeddings from reviews
as features to profile user preferences for recommendation [14,24]. However none
consider the substantial knowledge information contained in reviews [23–25].
Figure 1 shows two movie reviews from users with knowledge mentions aligned
with KGs. We can see that movie reviews contain substantial knowledge mentions corresponding to knowledge entities.

To address the limitations represented by the current KG-based works, and
inspired by the success applying reviews to RSs, we propose a novel recommendation framework, jointly learning propagation features on the KG (JPKG),
which can learn multi-hop propagation features as user preferences on the basis

of explicit review behaviors of users for movies. The review entities extracted from reviews based on KGs can be considered as neighbors of users/movies in the graph to assist in profiling user preferences and movie properties. On the basis of the review entities, we first construct a review-based heterogeneous KG, as shown in Fig. 2. To fully exploit user preferences on the graph, we then introduce an attention-based multi-hop propagation mechanism that updates a node embedding of a user/movie on the basis of the different contributions of its neighbors. To bridge the differences between the knowledge embedding learning and recommendation, we adopt a multi-task learning framework to jointly learn the propagation feature on the KG to predict movie ratings.

The contributions of our work are summarized as follows:

- We built a review-based heterogeneous KG to address the lack of user connections, which considers the movie-related entities and contains users' connections to their review entities.
- We designed a multi-task framework to jointly learn multi-hop features of user/movies, which can recursively learn the different contributions of neighbors to users/movies.
- We conducted experiments on two public datasets, demonstrating the effectiveness of JPKG, especially on sparse datasets.

2 Methodology

2.1 Problem Formulation

In this paper, we focus on generating propagating links through the jointly learning of a recommendation task and KG linking task to recommend a movie to a user. Let $\mathcal{U} = \{u_1, u_2, ..., u_{|U|}\}$ and $\mathcal{M} = \{m_1, m_2, ..., m_{|M|}\}$ denote the user set containing $|U|$ users and the movie set containing $|M|$ movies, respectively. The user-movie rating matrix $Y \in \mathbb{R}^{|U| \times |M|}$ is defined in accordance with the rating behaviors from users to movies, and the element y_{u_i, m_j} is a rating value given from user u_i to movie m_j. In addition, the heterogeneous graph $\mathcal{G} = (\mathcal{V}, \mathcal{E})$ is comprised of heterogeneous nodes and undirected edges, where $\mathcal{V}$ consists of users, movies, and review entities, and $\mathcal{E}$ is the set of edges connecting users/movies and review entities. Here, we use $S = \{s_1, s_2, ...\}$ to represent the set of review entities in the graph $\mathcal{G}$, and $\mathcal{V} = \mathcal{U} \cup \mathcal{M} \cup S$. We use A to denote the adjacency matrix of the graph $\mathcal{G}$, where $A_{i,j} = 1$ if $(i, j) \in \mathcal{E}$ and $A_{i,j} = 0$ otherwise.

Given the user-movie rating matrix Y and the heterogeneous graph $\mathcal{G}$, we aim to predict the ratings between users and movies that have not interacted before.

2.2 Heterogeneous Graph Construction

We construct a heterogeneous graph containing users, movies, and their corresponding review entities. For review entities, we adopt the entity linking method

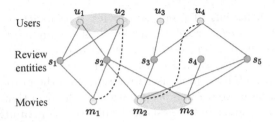

Fig. 2. Example of a review-based heterogeneous KG. Solid lines denote the real connections in the build graph, and dashed lines denote the interactions between users and movies. Grey circles denote similar users and movies discovered through connections to review entities.

[11] to find entities of reviews and each entity as a node in the heterogeneous graph. In review-based RSs, users and items can be represented by their corresponding reviews information [14,15]. Therefore, both users and movies can be linked with their review entities, as shown in Fig. 2. In the figure, we can see that since both u_1 and u_2 are linked to s_1 and s_2, and u_2 has watched movie m_1, we can recommend m_1 to user u_1 on the basis of the similar preferences of u_1 and u_2. Moreover, the multi-hop propagation mechanism can capture the connectivity lines $u_1 \rightarrow s_1 \rightarrow m_1$ and $u_1 \rightarrow s_2 \rightarrow m_1$ in the graph, and the lines reflect the relationship between user u_1 and movie m_1. Similarly, we can also recommend m_2 to u_3 because of the similar properties of m_2 and m_3 and the link propagation $u_3 \rightarrow s_3 \rightarrow m_2$.

2.3   The Proposed Framework

We designed a multi-task learning framework as shown in Fig. 3, which jointly learns the graph link prediction task and rating prediction task to predict the accurate ratings. The proposed framework consists of a graph attention learning module, multi-hop propagation module, and mutual learning module. The graph attention learning module computes the weights of edges in the graph by considering the contributions of review entities to their connected users/movies. We use lines with different thicknesses to represent different attention values, and the larger the value, the thicker the line. The multi-hop propagation module recursively propagates the node embeddings from their neighbors on the basis of the weighted KG. The mutual learning module seamlessly combines the graph link prediction task and the recommendation task to provide accurate ratings.

Graph Attention Learning Module. Given the heterogeneous graph $\mathcal{G}$, we represent the nodes in the graph as vectors by using a graph embedding layer. For a node of user u_i in the graph, the corresponding d-dimensional embedding can be represented by $\mathbf{e}_{u_i} \in \mathbb{R}^d$. Similarly, we use $\mathbf{e}_{m_j} \in \mathbb{R}^d$ to represent the embedding of a movie m_j in the graph. For a review entity s_r, we use $\mathbf{s}_r \in \mathbb{R}^d$ to represent its embedding vector.

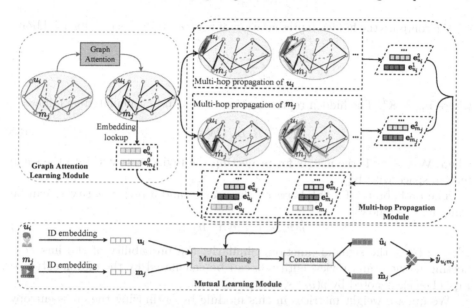

Fig. 3. Proposed JPKG framework.

We adopt the attention mechanism to learn the contributions of review entities to users/movies in the heterogeneous graph. The input of this module is the graph embeddings generated by mapping one-hot vectors through a fully-connected neural network. Given an embedding vector $\mathbf{e}_{u_i}$ of user u_i, and the embedding vector $\mathbf{s}_r$ of the r-th review entity linked with user u_i, the attention values between the user node and its neighbor can be calculated through this module. Specifically, the query vector of $\mathbf{e}_{u_i}$ can be formulated as follows:

$$\mathbf{q}_{u_i} = ReLU(\mathbf{W}_q \mathbf{e}_{u_i}),\qquad(1)$$

where $\mathbf{W}_q \in \mathbb{R}^{l \times d}$ is a matrix to project the user node from the d-dimension entity space into the l-dimension query space, and $ReLU(\cdot)$ [18] is a rectified linear unit.

$$\mathbf{k}_{s_r} = ReLU(\mathbf{W}_k \mathbf{s}_r),\qquad(2)$$

where $\mathbf{W}_k \in \mathbb{R}^{l \times d}$ is a matrix to transform the review entity into the key-space.

On the basis of the two aforementioned equations, we compute the attention score between user u_i and its linked review entity s_r as follows:

$$a(u_i, s_r) = \mathbf{q}_{u_i}^T \mathbf{k}_{s_r}.\qquad(3)$$

We normalize the attention scores of all the neighbors corresponding to the user u_i by using the softmax function:

$$a(u_i, s_r) = \frac{exp(a(u_i, s_r))}{\sum_{s_{r'} \in \mathcal{N}_{u_i}} exp(a(u_i, s_{r'}))},\qquad(4)$$

where $\mathcal{N}_{u_i}$ is the set of review entities linked to user u_i.

We compute the hidden representation of user u_i on the basis of its neighbors:

$$\mathbf{h}_{u_i} = \sum_{s_r \in N_{u_i}} a(u_i, s_r)\mathbf{s}_r, \tag{5}$$

where $\mathbf{h}_{u_i} \in \mathbb{R}^d$. The hidden representation of s_r can be calculated as follows:

$$\mathbf{h}_{s_r} = ReLU(\mathbf{W}\mathbf{s}_r), \tag{6}$$

where $\mathbf{W} \in \mathbb{R}^{l \times d}$ is the matrix for projecting the review entity $\mathbf{s}_r$ into the same hidden space with $\mathbf{h}_{u_i}$.

The probability of a link between the user u_i and a review entity s_r can be computed as follows:

$$p(u_i, s_r) = \sigma(\mathbf{h}_{u_i}^T \mathbf{h}_{s_r}), \tag{7}$$

where $\sigma(\cdot)$ is the sigmoid function. Similarly, the probability of the link connecting movie j and review entity s_r can be calculated by the aforementioned Eqs. (1)–(7), denoted by $p(m_j, s_r)$.

We update weight matrices in this module by optimizing the cross-entropy loss function as follows:

$$\mathcal{L}_G = \mathcal{L}_{GU} + \mathcal{L}_{GM}, \tag{8}$$

where $\mathcal{L}_{GU}$ and $\mathcal{L}_{GM}$ are the loss functions for user-centric and movie-centric link prediction, respectively, and each of them can be formulated as:

$$\begin{aligned} \mathcal{L}_{GU} &= -\sum_{(u_i,s_r) \in \mathcal{G}} A_{u_i,s_r} \log p(u_i, s_r) + (1 - A_{u_i,s_r}) \log(1 - p(u_i, s_r)) \\ \mathcal{L}_{GM} &= -\sum_{(m_j,s_r) \in \mathcal{G}} A_{m_j,s_r} \log p(m_j, s_r) + (1 - A_{m_j,s_r}) \log(1 - p(m_j, s_r)) \end{aligned}, \tag{9}$$

where the symbol $A_{.,.}$ denotes a value in the adjacency matrix.

Multi-hop Propagation Module. To compute the effect of multi-hop neighbors on a user/movie, we recursively propagate the embeddings along the connecting lines centered on the user/movie. Taking $m_1 \rightarrow s_1 \rightarrow u_1$ and $m_1 \rightarrow s_2 \rightarrow u_1$ as an example, in the one-hop propagation, movie m_1 and user u_1 take s_1 and s_2 as their attributes to enrich the representations, and in the two-hop propagation, m_1 and u_1 use the embedding information of each other to further enrich their feature representations.

Considering a user u_i in the graph, we use $\mathcal{N}_{u_i}$ to denote a set of neighbors centered around user u_i. The neighbor embeddings of user u_i can be represented by $\mathbf{e}_{\mathcal{N}_{u_i}}$, and

$$\mathbf{e}_{\mathcal{N}_{u_i}} = \sum_{s_r \in \mathcal{N}_{u_i}} a(u_i, s_r)\mathbf{s}_r, \tag{10}$$

where $a(u_i, s_r)$ denotes the attention weights from a review entity s_r linked to user u_i, indicating the contribution from s_r to u_i.

We leverage the method proposed in [3] to aggregate the embeddings of users/movies and their neighbor embeddings. Given the embedding $\mathbf{e}_{u_i}$ of user u_i

and its neighbor embeddings $e_{\mathcal{N}_{u_i}}$, the aggregation operation can be formulated as:

$$f = LeakyReLU(\mathbf{W}_1(e_{u_i} + e_{\mathcal{N}_{u_i}})) + LeakyReLU(\mathbf{W}_2(e_{u_i} \odot e_{\mathcal{N}_{u_i}})), \quad (11)$$

where $\mathbf{W}_1, \mathbf{W}_2 \in \mathbb{R}^{d \times d}$ are the trainable matrices, and $\odot$ indicates the element-wise product.

For the multi-hop propagation, we recursively propagate information from multi-hop distances to users/movies by stacking multiple aggregation layers. In the t-th aggregation layer, the embedding of u_i can be defined as:

$$e_{u_i}^t = f(e_{u_i}^{t-1}, e_{\mathcal{N}_{u_i}}^{t-1}), \quad (12)$$

where the embedding of $\mathcal{N}_{u_i}$ in the $(t-1)$-th aggregation layer is calculated as follows,

$$e_{\mathcal{N}_{u_i}}^{t-1} = \sum_{s_r \in \mathcal{N}_{u_i}} a(u_i, s_r) s_r^{t-1}, \quad (13)$$

where s_r^{t-1} is the embedding of review entity s_r generated from the previous propagation layers. Similarly, the multi-hop propagation embedding of m_j is represented as $e_{m_j}^t$. Note that when $t = 0$, vectors $e_{u_i}^0 = \mathbf{h}_{u_i}$ and $e_{m_j}^0 = \mathbf{h}_{m_j}$.

For user u_i and movie m_j, the corresponding outputs generated by $(t+1)$ aggregation layers can be gathered by $\{e_{u_i}^0, e_{u_i}^1, ..., e_{u_i}^t\}$ and $\{e_{m_j}^0, e_{m_j}^1, ..., e_{m_j}^t\}$, respectively.

Mutual Learning Module. In this module, we jointly learn the propagation embedding and the corresponding ID embedding of each user and movie to complete the information exchange from two different kinds of latent features.

We describe the mutual learning operation by introducing multiple interaction layers between the ID embedding $\mathbf{u}_i \in \mathbb{R}^d$ of user u_i and the corresponding t-hop propagation embeddings $\{e_{u_i}^0, e_{u_i}^1, ..., e_{u_i}^t\}$. In the n-th mutual learning layer, we build $d \times d$ pairwise interactions between them as follows:

$$\mathbf{C}^n = \mathbf{u}_i \left(e_{u_i}^n\right)^\top = \begin{bmatrix} u_{i1}e_{u_i1}^n & \cdots & u_{id}^1 e_{u_i1}^n \\ \cdots & & \cdots \\ u_{i1}e_{u_id}^n & \cdots & u_{id}e_{u_id}^n \end{bmatrix}, \quad (14)$$

where $\mathbf{C}^n \in \mathbb{R}^{d \times d}$ is the interaction matrix of $\mathbf{u}_i$ and $e_{u_i}^n$ in the n-th layer, and $n \le (t+1)$. The ID embedding of u_i in the n-th layer is generated as follows:

$$\mathbf{u}_i^n = \mathbf{C}^n \mathbf{w}_{ue} + (\mathbf{C}^n)^\top \mathbf{w}_{eu} + \mathbf{b} \quad (15)$$

where the vectors $\mathbf{w}_{ue} \in \mathbb{R}^d$ and $\mathbf{w}_{eu} \in \mathbb{R}^d$ denote the trainable projection weights for mapping $\mathbf{C}^n$ to the ID embedding space, and $\mathbf{b} \in \mathbb{R}^d$ is the trainable bias.

We concatenate $(t+1)$ ID embeddings corresponding to u_i as one vector, and then compute the final representation of u_i by using a linear projection:

$$\hat{\mathbf{u}}_i = \mathbf{W}' concatenate(\mathbf{u}_i^1, \mathbf{u}_i^2, ..., \mathbf{u}_i^n, ...), \quad (16)$$

where $\mathbf{W}' \in \mathbb{R}^{d \times (t+1)*d}$ is the trainable projection matrix. Similarly, the embedding of movie m_j can be represented as $\hat{\mathbf{m}}_j$. The final ratings of user u_i to movie m_j is calculated as:

$$\hat{y}_{u_i m_j} = \hat{\mathbf{u}}_i^\top \hat{\mathbf{m}}_j. \tag{17}$$

Optimization. To optimize the proposed model, the entire loss function is defined as follows:

$$\begin{aligned} \mathcal{L} &= \mathcal{L}_G + \mathcal{L}_{RS} + \mathcal{L}_{REG} \\ &= \lambda_1 \mathcal{L}_G + \sum_{u_i \in U, m_j \in M} \mathcal{J}\left(\hat{y}_{u_i, m_j}, y_{u_i, m_j}\right) + \lambda_2 \|\mathbf{W}\|_2^2, \end{aligned} \tag{18}$$

where $\mathcal{L}_G$ is the loss function of the graph link prediction task defined in Eq. 8, $\mathcal{L}_{RS}$ is the loss function of the rating prediction task, and $\mathcal{L}_{REG}$ is the regularization term. The symbol $\mathcal{J}(*)$ denotes the mean square error (MSE) function. We use λ_1 and λ_2 as the learning rate parameters to balance the loss.

3 Experiments

3.1 Datasets

We evaluated our model on two publicly available real-world movie datasets: IMDb and Amazon-movie.

- **IMDb** dataset. The dataset was published by a related work JMARS [16], which includes ratings and reviews information from users to movies, and the ratings are in the range of $[0, 10]$.
- **Amazon-movie** dataset. This dataset belongs to the "Amazon product data"[1], which has been widely used to evaluate review rating prediction works [14,17]. The ratings from users to movies are in the range of $[0, 5]$.

To analyze the impacts of different sparse data on recommendation performances, we filtered each dataset into eight different core versions ranging from 3-core to 10-core on the basis of the minimum number of reviews from users. For example, 3-core means each user has at least three reviews in the dataset. We removed the duplicate edges in each graph. The statistics of datasets are illustrated in Table 1.

3.2 Experimental Settings

Baselines. To evaluate the effectiveness of the proposed model, we chose three highly-relevant state-of-the-art works: rating-based matrix factorization methods, review-based neural networks, and knowledge-based mutual learning methods as our baselines.

---

[1] http://jmcauley.ucsd.edu/data/amazon/.

Table 1. Statistics of the two datasets with different sparsities

Dataset		3-core	4-core	5-core	6-core	7-core	8-core	9-core	10-core
IMDb	# users	1,833	1,648	1,504	1,393	1,318	1,237	1,160	1,095
	# movies	4,663	4,663	4,663	4,663	4,663	4,663	4,663	4,663
	# ratings	126K	125K	125K	124K	124K	123K	123K	122K
	# review entities	69K	69K	68K	68K	68K	68K	68K	68K
	# nodes	76K	75K	75K	75K	74K	74K	74K	74K
	# edges	1,008K	1,004K	1,000K	999K	995K	992K	988K	984K
Amazon-movie	# users	158K	123K	93K	68K	53K	43K	35K	29K
	# movies	59K	59K	58K	58K	58K	57K	57K	56K
	# ratings	1,448K	1,343K	1,223K	1,101K	1,009K	936K	876K	825K
	#review entities	190K	189K	187K	185K	182K	180K	178K	176K
	# nodes	408K	371K	340K	312K	294K	281K	271K	263K
	# edges	6,248K	6,058K	5,810K	5,471K	5,211K	4,992K	4,802K	4,641K

- Probabilistic matrix factorization (PMF). PMF is a matrix factorization model that learns the latent representations of users and items from a rating matrix to provide accurate recommendations [19].
- Generalized matrix factorization (GMF). GMF is a generalized version of matrix factorization (MF) [20] that uses a nonlinear layer to project the latent vectors of users and items into the same space, and models the interactions between users and items on the basis of their projected vectors [21].
- Multi-task feature learning for KG enhanced recommendation (MKR). This method treats items as head entities of the KG and learns latent vectors of items by mutual learning between an RS task and KGE task [22].
- Deep cooperative neural networks (DeepCoNN). DeepCoNN is a review-based neural network that adopts a convolution-based parallel structure framework to extract the latent representations of users and items from their corresponding reviews [23].
- Transformational neural networks (TransNets). This method is also a review-based neural network inspired by DeepCoNN that introduces a transform layer in a parallel neural network to transform reviews of users and items into the same representation space for recommendation [24].

Evaluation Metric. To measure the performances of all the tested models, we adopt root-mean-square error (RMSE) as the evaluation metric. Given a ground truth rating y_{u_i,m_j} rated by user u_i for movie m_j and its corresponding predicted rating $\widehat{y}_{u_i,m_j}$, the RMSE is calculated as:

$$RMSE = \sqrt{\frac{1}{N} \sum_{u_i,m_j} \left(\widehat{y}_{u_i,m_j} - y_{u_i,m_j} \right)^2}, \qquad (19)$$

where N indicates the number of user ratings for movies.

Table 2. Overall performance comparison. Best results are highlighted in bold.

		3-core	4-core	5-core	6-core	7-core	8-core	9-core	10-core
IMDb	PMF	1.837	1.812	1.812	1.84	1.767	1.789	1.77	1.758
	GMF	1.848	1.826	1.826	1.862	1.782	1.809	1.789	1.771
	MKR	1.827	1.819	1.804	1.818	1.806	1.806	1.806	1.802
	DeepCoNN	1.813	1.815	1.778	1.809	1.774	1.751	1.787	1.776
	TransNets	1.814	1.816	**1.763**	1.793	1.777	1.75	1.775	1.766
	JPKG	**1.772**	**1.775**	1.773	**1.762**	**1.763**	**1.746**	**1.736**	**1.748**
Amazon-movie	PMF	1.131	1.088	1.081	1.072	1.084	1.085	1.092	1.097
	GMF	1.175	1.172	1.172	1.167	1.160	1.160	1.154	1.148
	MKR	1.100	1.097	1.088	1.082	1.072	1.069	1.066	1.060
	DeepCoNN	1.045	1.034	1.024	1.026	1.018	1.017	1.020	1.016
	TransNets	1.047	1.042	1.030	1.041	1.022	1.023	1.014	1.021
	JPKG	**1.031**	**1.029**	**1.021**	**1.018**	**1.011**	**1.006**	**1.001**	**0.997**

Parameter Settings. We randomly selected 80%, 10%, and 10% of samples as the training, validation, and test sets, respectively. We set the learning rates of the recommendation task and graph linking prediction task to 2.0×10^{-4} and 8.0×10^{-6}, respectively. The values of λ_1 and λ_2 were fixed to 0.04 and 1.0×10^{-6}, respectively. The number of propagation layers in the multi-hop propagation module was set to 3. The dimensions of both ID embeddings and graph node embeddings were set to 16. The batch size in the training processes for the recommendation task and graph link prediction task were set to 64 and 1024, respectively. The training interval was set to 4, which means that we repeatedly train recommendation task 4 times before training the graph link prediction task once in each epoch.

3.3   Experimental Results

We report the experimental results of our proposed model and those of the baselines datasets with various sparsities in Table 2. We can see that the proposed JPKG outperforms the other models in most cases. In particular, it achieves the best performance on the Amazon-movie dataset with all the sparsities and on the IMDb dataset except the 5-core sparsity. In general, review-based methods perform better than rating-based methods, indicating that review information can reflect user preferences and item properties that do not exist in ratings. Moreover, as the data becomes denser, the improvement of the review-based method becomes smaller. However, the improvements of RMSEs for our method on both the IMDb and Amazon-movie datasets with various sparsities remains at about 2% and 9%, respectively, which demonstrates that our proposed method is effective for sparse datasets and maintains its effectiveness consistently as the datasets become denser.

Table 3. RMSE results of ablation study.

	IMDb		Amazon-movie	
	3-core	10-core	3-core	10-core
JPKG-ML	1.811	1.783	1.042	1.018
JPKG-PF	1.787	1.769	1.040	1.017
JPKG-attn	1.777	1.753	1.037	1.004
JPKG	**1.772**	**1.748**	**1.031**	**0.997**

3.4 Ablation Study

To investigate the effectiveness of the three modules in our work, we report the experimental results from two perspectives based on ablation studies: recommendation accuracy and convergence.

For the ablation methods, we first disabled the mutual learning layers and aggregated the multi-hop propagation features directly as the final representations of users/movies, termed JPKG-ML. We then disabled the multi-hop propagation module and jointly learned the attention-based node representations and ID embeddings to predict ratings, termed JPKG-PF. Finally, we disabled the attention mechanism on the graph and treated the contributions of all neighbors of a node as the same, termed JPKG-attn.

Recommendation Accuracy. Table 3 shows the RMSE results of the ablation methods and JPKG on the IMDb and Amazon-movie datasets with the 3-core and 10-core sparsities, respectively. We can see that disabling any of the three key modules degrades the performance of the model. We can also see that JPKG-ML underperforms other methods, which indicates the mutual learning module plays a more important role than the other two modules. This finding also reveals an empirical fact that directly using graph embeddings for recommendation may introduce noise and mislead the final recommendation. Furthermore, JPKG-attn performs better than JPKG-PF, which verifies that removing the multi-hop propagation module can have a more significant effect than removing the attention module on recommendation results. One possible reason is that learning multi-hop propagation features can substantially improve the quality of representation learning.

Convergence. We investigated the influences of the key modules on our model by observing the convergence of ablation methods on the IMDb and Amazon-movie datasets, and the results are presented in Fig. 4. We reported the RMSE results on the validation data by varying the training epochs to illustrate the convergence. Note that we adopted the early-stopping strategy to obtain the final experimental results. We can see that the convergence speed of JPKG is faster than those of JPKG-attn, JPKG-ML, and JPKG-PF. Moreover, JPKG can

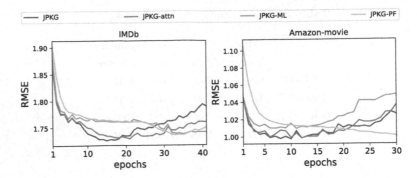

Fig. 4. Convergence comparisons on the two datasets among three ablation models and JPKG.

reach a smaller value than the other three ablation methods on the two datasets. Note that JPKG-PF needs more epochs for the convergence, which means that adopting multi-hop propagation can enable us to speed up the convergence. We can also see that JPKG-ML cannot converge to a relatively small loss on the two datasets. The aforementioned results illustrate the necessity of the three key modules in our model.

4   Conclusion

In this paper, we proposed JPKG, a multi-task framework that jointly learns multi-hop propagation features on a KG for movie recommendations. JPKG overcomes the limitation of insufficient user connections in current KG-based recommendations by integrating review entities, users, and movies into a heterogeneous graph. The attention learning module and multi-hop propagation module of JPKG achieve attention-based multi-hop propagation feature learning by recursively calculating the different contributions of neighbors on the graph. The mutual learning module of JPKG combines the entity embeddings learned from the two aforementioned modules to help provide more accurate recommendations. The experimental results on two real-world datasets demonstrate the effectiveness of our proposed model.

For future work, we will focus on providing explainable recommendations on the basis of the current work. Furthermore, we will explore other methods that can enhance the user preference mining ability on KGs.

Acknowledgements. This work was supported in part by Japan Society for the Promotion of Science (JSPS) KAKENHI Grant Numbers JP18H03242, JP18H03342, JP19H01138, and JP21K17868.

References

1. Wang, H., Zhang, F., Wang, J., et al.: RippleNet: propagating user preferences on the knowledge graph for recommender systems. In: Proceedings of the 27th ACM International Conference on Information and Knowledge Management, pp. 417–426 (2018)
2. Tai, C.Y., Huang, L.Y., Huang, C.K., et al.: User-centric path reasoning towards explainable recommendation. In: Proceedings of the 44th International ACM SIGIR Conference on Research and Development in Information Retrieval, pp. 879–889 (2021)
3. Wang, X., He, X., Cao, Y., et al.: KGAT: knowledge graph attention network for recommendation. In: Proceedings of the 25th ACM SIGKDD International Conference on Knowledge Discovery & Data Mining, pp. 950–958 (2019)
4. Zhang, F., Yuan, N.J., Lian, D., et al.: Collaborative knowledge base embedding for recommender systems. In: Proceedings of the 22nd ACM SIGKDD International Conference on Knowledge Discovery and Data Mining, pp. 353–362 (2016)
5. Wang, H., Zhang, F., Xie, X., et al.: DKN: deep knowledge-aware network for news recommendation. In: Proceedings of the 2018 World Wide Web Conference, pp. 1835–1844 (2018)
6. Ai, Q., Azizi, V., Chen, X., et al.: Learning heterogeneous knowledge base embeddings for explainable recommendation. Algorithms 11(9), 137 (2018)
7. Fu, Z., Xian, Y., Gao, R., et al.: Fairness-aware explainable recommendation over knowledge graphs. In: Proceedings of the 43rd International ACM SIGIR Conference on Research and Development in Information Retrieval, pp. 69–78 (2020)
8. Brams, A.H., Jakobsen, A.L., Jendal, T.E., et al.: MindReader: recommendation over knowledge graph entities with explicit user ratings. In: Proceedings of the 29th ACM International Conference on Information & Knowledge Management, pp. 2975–2982 (2020)
9. Xian, Y., Fu, Z., Muthukrishnan, S., et al.: Reinforcement knowledge graph reasoning for explainable recommendation. In: Proceedings of the 42nd International ACM SIGIR Conference on Research and Development in Information Retrieval, pp. 285–294 (2019)
10. Yu, X., Ren, X., Sun, Y., et al.: Personalized entity recommendation: a heterogeneous information network approach. In: Proceedings of the 7th ACM International Conference on Web Search and Data Mining, pp. 283–292 (2014)
11. Ratinov, L., Roth, D., Downey, D., et al.: Local and global algorithms for disambiguation to wikipedia. In: Proceedings of the 49th Annual Meeting of the Association for Computational Linguistics: Human Language Technologies, pp. 1375–1384 (2011)
12. Bordes, A., Usunier, N., Garcia-Duran, A., et al.: Translating embeddings for modeling multi-relational data. In: Advances in Neural Information Processing Systems, vol. 26 (2013)
13. Lin, Y., Liu, Z., Sun, M., et al.: Learning entity and relation embeddings for knowledge graph completion. In: 29th AAAI Conference on Artificial Intelligence (2015)
14. McAuley, J., Leskovec, J.: Hidden factors and hidden topics: understanding rating dimensions with review text. In: Proceedings of the 7th ACM Conference on Recommender Systems, pp. 165–172 (2013)
15. Liu, D., Li, J., Du, B., et al.: DAML: dual attention mutual learning between ratings and reviews for item recommendation. In: Proceedings of the 25th ACM SIGKDD International Conference on Knowledge Discovery & Data Mining, pp. 344–352 (2019)

16. Diao, Q., Qiu, M., Wu, C.Y., et al.: Jointly modeling aspects, ratings and sentiments for movie recommendation (JMARS). In: Proceedings of the 20th ACM SIGKDD International Conference on Knowledge Discovery and Data Mining, pp. 193–202 (2014)
17. Kim, D., Park, C., Oh, J., et al.: Convolutional matrix factorization for document context-aware recommendation. In: Proceedings of the 10th ACM Conference on Recommender Systems, pp. 233–240 (2016)
18. Nair, V., Hinton, G.E.: Rectified linear units improve restricted Boltzmann machines. In: ICML (2010)
19. Mnih, A., Salakhutdinov, R.R.: Probabilistic matrix factorization. In: Advances in Neural Information Processing Systems, vol. 20 (2007)
20. Koren, Y., Bell, R., Volinsky, C.: Matrix factorization techniques for recommender systems. Computer **42**(8), 30–37 (2009)
21. He, X., Liao, L., Zhang, H., et al.: Neural collaborative filtering. In: Proceedings of the 26th International Conference on World Wide Web, pp. 173–182 (2017)
22. Wang, H., Zhang, F., Zhao, M., et al.: Multi-task feature learning for knowledge graph enhanced recommendation. In: The World Wide Web Conference, pp. 2000–2010 (2019)
23. Zheng, L., Noroozi, V., Yu, P.S.: Joint deep modeling of users and items using reviews for recommendation. In: Proceedings of the 10th ACM International Conference on Web Search and Data Mining, pp. 425–434 (2017)
24. Catherine, R., Cohen, W.: TransNets: learning to transform for recommendation. In: Proceedings of the 11th ACM Conference on Recommender Systems, pp. 288–296 (2017)
25. Chen, C., Zhang, M., Liu, Y., et al.: Neural attentional rating regression with review-level explanations. In: Proceedings of the World Wide Web Conference, pp. 1583–1592 (2018)

Syntax-Informed Question Answering
with Heterogeneous Graph Transformer

Fangyi Zhu[⊠], Lok You Tan, See-Kiong Ng, and Stéphane Bressan

National University of Singapore, Singapore, Singapore
{fyzhu,seekiong,steph}@nus.edu.sg, lok.t@u.nus.edu

Abstract. Large neural language models are steadily contributing state-of-the-art performance to question answering and other natural language and information processing tasks. These models are expensive to train. We propose to evaluate whether such pre-trained models can benefit from the addition of explicit linguistics information without requiring retraining from scratch.

We present a linguistics-informed question answering approach that extends and fine-tunes a pre-trained transformer-based neural language model with symbolic knowledge encoded with a heterogeneous graph transformer. We illustrate the approach by the addition of syntactic information in the form of dependency and constituency graphic structures connecting tokens and virtual vertices.

A comparative empirical performance evaluation with BERT as its baseline and with Stanford Question Answering Dataset demonstrates the competitiveness of the proposed approach. We argue, in conclusion and in the light of further results of preliminary experiments, that the approach is extensible to further linguistics information including semantics and pragmatics.

Keywords: Question answering · Transformer · Graph neural network

1 Introduction

Question answering [2] is a field within natural language processing [16] that studies the design and implementation of algorithms, tools, and systems for the automatic answering of questions in natural language. Among the many types of question answering [1], this work focuses on extractive question answering. Extractive question answering refers to the task of, given a question and a passage, selecting from the passage a text span corresponding to the answer to the question.

Large language models such as Bidirectional Encoder Representations from Transformers (BERT) [4] brought competitive performance to many natural language processing tasks, including question answering [20]. Although these models are obviously able to learn relevant linguistic information [7], Kuncoro et al. show that BERT benefits from the addition of syntactic information for various structured prediction tasks [19].

Consider the question *"What kind of economy did northern California start to grow in the 2000s?"* from Stanford Question Answering Dataset (SQuAD) [26]. The part of the passage in which the answer is located reads *"[...] due to a stronger tech-oriented economy"*. The answer, according to SQuAD, is *"tech-oriented"* (it could also be *"a tech-oriented economy"* or *"a stronger tech-oriented economy"*). However, BERT is

© The Author(s), under exclusive license to Springer Nature Switzerland AG 2022

C. Strauss et al. (Eds.): DEXA 2022, LNCS 13426, pp. 17–31, 2022.
https://doi.org/10.1007/978-3-031-12423-5_2

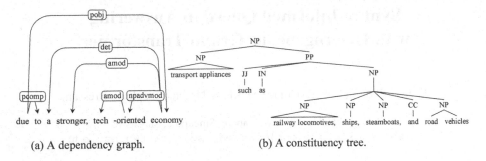

(a) A dependency graph.   (b) A constituency tree.

Fig. 1. Examples of syntactic graphic structures.

unable to find an answer. A dependency analysis of the sentence, represented by the dependency graph in Fig. 1a, reveals that the word "*tech-oriented*" is an adverb modifier of "*economy*". This dependency is relevant to the question of the form "*What kind of [...]*". Dependencies encode important information specifying grammatical functions between a dependent, here (*tech-oriented*), and its head, here (*economy*) [15].

Consider the other question "*Along with road vehicles, locomotives and ships, on what vehicles were steam engines used during the Industrial Revolution?*", also from SQuAD. The part of the passage in which the answer is located reads "*[...] Steam engines can be said to have been the moving force behind the Industrial Revolution and saw widespread commercial use driving machinery in factories, mills and mines; powering pumping stations; and propelling transport appliances such as railway locomotives, ships, steamboats and road vehicles. [...]*". The answer is "*steamboats*". BERT finds a redundant answer "*propelling transport appliances such as railway locomotives, ships, steamboats*". However, a constituency analysis, represented by the constituency tree in Fig. 1b, clearly indicates "*railway locomotives*", "*ships*", "*streamboats*", and "*road vehicles*" are coordinated noun phrases.

Generally, the integration of statistical machine learning with symbolic knowledge and reasoning "opens relevant possibilities towards richer intelligent systems" remark the authors of [6] arguing for a principled integration of machine learning and reasoning. Nevertheless, most existing neural language models are still plundering the benefits of statistical learning before they attempt to explicitly exploit old-fashion symbolic knowledge of the linguistic structures.

While the success of transformer-based neural language models is attributed to the self-attention mechanism [33] that they implement, the question arises whether the adjunction of a focused attention mechanism guided by structures representing symbolic linguistic information [21], such as dependency graphs and constituency trees, can further improve the performance of neural language models.

Therefore we devise, present and evaluate a linguistics-informed question answering approach that extends a pre-trained transformer-based neural language model with linguistic graphic structures encoded with a heterogeneous graph transformer [13]. The integration is relatively seamless because both models work in the space of embeddings, albeit not necessarily just embeddings of tokens but also of words and other linguistic units. The transformer-based neural language model is fine-tuned and the heterogeneous graph neural network is trained to compute and aggregate the embeddings under the constraints of the graphic structures [40].

We instantiate and evaluate the approach for the cases of the addition of syntactic information, in the form of dependency and constituency graphic structures connecting tokens and virtual vertices, for extractive question answering.

We refer to the resulting model as syntax-informed neural network with heterogeneous graph transformer (SyHGT), for dependencies (SyHGT-D) and constituencies (SyHGT-C). For the sake of simplicity, SyHGT is presented, discussed and evaluated here for extractive question answering.

Overall, there are three main contributions in this work: (1) present a syntax-informed approach via heterogeneous graph transformer for question answering; (2) propose to integrate virtual vertices that can be any linguistic symbolic for incorporating prior linguistic knowledge; (3) empirically evaluate our approach on SQuAD2.0 and it gains 1.22 and 0.98 improvement over the baseline in terms of EM and F1 metrices.

2   Related Work

Early question answering systems used syntactic analysis and rule-based approaches [8]. Later systems utilised heavy feature engineering [29]. Advancements in computer hardware then paved the way for neural models which require little feature engineering.

Language models learn the probability of a sequence of words. Neural language models are often used as encoders to obtain word embeddings. Since BERT, a neural language model, successfully executed on 11 natural language processing tasks, question answering has been dominated by large models built upon it [20].

Jawahar et al. probed BERT's layers, and found that lower layers captured surface features, middle layers captured syntactic features, and upper layers captured semantic features [14]. The upper layers were found to model the long-distance dependencies, making them crucial to performance in downstream tasks. However, it was also found that syntactic information is diluted in these upper layers. Kuncoro et al. extended BERT to take into account syntactic information by modifying its pre-training objective [19]. Using another syntactic language model as a learning signal, they added what they termed 'syntactic bias' to BERT.

Vashishth et al. used dependency trees and graph convolutional networks to learn syntax-based embeddings that encode functional similarity instead of traditional topical similarity [31]. The syntax-based embeddings were found to encode information complementary to ELMo [25] embeddings that only relied on sequential context. Zhang et al. proposed syntax-guided network (SG-Net), a question answering model that used dependency trees as explicit syntactic constraints for a self-attention layer [39]. SG-NET was effective especially with longer questions as it could select vital parts. The syntax-guided attention considered syntactic information that is complementary to traditional attention mechanisms. Syntax guidance provided more accurate attentive signals and reduced the impact of noise in long sentences.

For question answering, graph neural networks have found success in multi-hop reasoning [3, 30] on the WikiHop data set [35]. Graph neural networks operate directly on graphs and can capture dependencies between vertices.

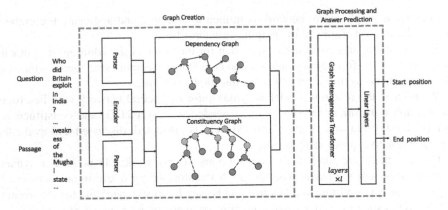

Fig. 2. The diagram of our proposed approach. There are two major components, graph creation, as well as graph processing and answer prediction. The green vertices are token vertices, including the dashed green vertices (virtual lexeme vertices). The dashed blue vertices in the constituency graph are virtual constituent vertices. The dashed arrows are morphology edges. The straight arrows are dependency edges or constituency edges. The dotted edges in the constituency graph are part-of-speech edges. (Color figure online)

This work shows the utility of syntax and graph neural networks in learning better representations. In our approach, we bolster the pre-trained BERT model with additional syntactic information. In the same vein as the approach by Mao et al. [22], we bridge old rule-based systems and new neural models by integrating symbolic knowledge into statistical machine learning. This is done by explicitly incorporating the syntactic information, namely constituencies and dependencies, into a question answering model, which is made possible by inserting a heterogeneous graph transformer into the question answering pipeline. This keeps our approach rooted linguistically, instead of solely relying on pre-trained language models that are not explainable. Unlike vanilla graph neural networks, a heterogeneous graph transformer can deal with a heterogeneous graph where multiple types of vertices associated with different relations exist. To the best of our knowledge, integrating the syntax information and heterogeneous graph transformers for extractive question answering has not yet been explored.

3   Methodology

In a standard neural language model applied to extractive question answering, the question and passage are encoded together, then passed through a linear layer that outputs the probabilities for each token to be the start and end of the answer span.

We propose SyHGT, a linguistics-informed architecture. We need to create, represent, and process linguistic graphic structures connecting the language model embeddings of the tokens of the question and passage. In the cases of syntactic information about dependencies and constituencies, the graphic structures are created by a parser. The result for a question-passage pair is a graph of embedded vectors of the tokens. This non-Euclidean graph structure cannot be used directly by the neural language model. However, the insertion of a heterogeneous graph transformer layer to the question answering pipeline allows us a relatively straightforward implementation combining both the statistical and symbolic information. By inserting the graph neural network between the neural language model and the output layer, we can process the graph before making a prediction. Figure 2 depicts, SyHGT, the proposed approach, with its two main components: graph creation and graph processing and answer predication.

3.1 Graph Creation Module

Both passage and question are parsed and encoded. The encoder produces embeddings for each token that correspond to the graph vertices. The syntax graphic structures define the graph vertices and edges. Respectively, we create a constituency graph and a dependency tree. The obtained graphs are the input to the following heterogeneous graph transformer. Note that the tokeniser of the neural language model may not align with the tokeniser of the parser, be it for dependencies or for constituencies, which most likely considers lexemes rather than morphemes. The graph heterogeneous graph neural network model easily alleviates this issue by the introduction of intermediary vertices aggregating tokens into lexemes, where needed.

Encoder. SyHGT requires a neural language model as its initial encoder which produces embedding vectors for the text. The Bidirectional Encoder Representations from Transformers (BERT), see Vaswani et al. [32] and Devlin et al. [4], is used for implementation and performance evaluation in this paper.

The question q and passage p are concatenated with the appropriate BERT-specific special tokens to form the sequence: [CLS] q [SEP] p [SEP]. The sequence is fed into BERT to obtain the token embeddings $T = t_1, ..., t_n$, which are the hidden states of the input sequence at the last layer, where n is the number of tokens.

Dependency Graph. Dependency is the notion that linguistic units, lexemes, e.g., words, are connected to each other by directed links. In a dependency structure, every lexical vertex is dependent on at one other lexical vertex or is the head of a dependency. The structure is therefore a directed graph, with vertices representing lexical elements and edges representing dependency relations [24]. Dependency parsing produces a dependency graph of a sentence. The dependency graphs are then processed to obtain the dependency relations. Each sentence in the question and passage is parsed individually.

Since most BERT implementations leverage the WordPiece tokenizer [36], which may split words into sub-words, i.e. ad hoc morphemes, we add, in such a case, a virtual lexeme vertex on top of the sub-word tokens to represent the original word, so that the graph construction happens at the correct level.

The edges in the dependency graph are grouped into two categories morphology edges and dependency edges. Morphology edges connect token vertices corresponding to sub-words to virtual lexeme vertices. Dependency edges connect the head vertex and the dependent vertex of a recognized dependency relation. There is one type of edge for each type of dependency relation.

Constituency Graph. Constituency analysis iteratively decomposes sentences into constituent or sub-phrases, which are clauses, phrases, and words. These constituents belong to one of several categories such as noun phrase (NP), verb phrase (VP) as well as parts of speech. Explicitly, given an input sentence, constituency analysis builds a tree, in which leaves or terminal vertices correspond to input words and the internal or non-terminal vertices are constituents.

The vertices in the constituency tree are grouped into three categories, token vertices, lexeme vertices, and constituent vertices. Token vertices correspond to common tokens. Lexeme vertices represent lexemes that need to be recomposed from the token vertices of their sub-words. Constituent vertices represent constituents. Inner nodes and the root of the constituency tree are virtual, lexeme or constituency, vertices.

The edges in the constituency tree are grouped into three categories, morphology edges, part-of-speech edges, and constituent edges. Morphology edges connect token vertices corresponding to subwords to lexeme vertices. Part-of-speech edges connect the part-of-speech vertices to lexemes vertices. Constituency edges connect low-level constituents to high-level constituents.

3.2   Graph Processing and Answer Prediction

The heterogeneous graph transformer takes the constructed graphs as input and passes its outputs to the linear layer. The output from the linear layer consists of two numbers for each vertex; one number denotes the probability of the vertex being the start of the answer span, and the other of the vertex being the end. The final predicted start position and end position of span is determined by the respective maximum scores.

Heterogeneous Graph Transformer. Graph neural networks, proposed by the authors of [28], are neural models that capture the dependence of graphs via message passing following the edges between the vertices in a graph [37,40]. Specifically, the target for a graph neural network layer is to yield a contextualized representations for each vertex via aggregating the information from its surrounding vertices. By stacking multiple layers, the obtained representations of the vertices can be fed into downstream tasks, such as vertex classification, graph classification, link prediction, etc.

Recent years have witnessed the emerging success of graph neural networks (GNNs) for modeling structured data. However, most GNNs are designed for homogeneous graphs, in which all vertices and edges belong to the same types, making them infeasible to represent heterogeneous structures [13]. Relational Graph Convolutional Network (R-GCN) first proposed relation-specific transformation in the message passing steps to deal with various relations [23]. Subsequently, several works focused on dealing with the heterogeneous graph [34,38]. Inspired by the architecture design of Transformer [33], Hu et al. [13] presented the Heterogeneous Graph Transformer that incorporates the self-attention mechanism in a general graph neural network structure that can deal with a heterogeneous graph.

Given a heterogeneous graph $G = (V, E)$, each vertex $v \in V$ and each edge $e \in E$ are associated with their type $c \in C$ and $r \in R$. The process in one heterogeneous graph transformer layer can be decomposed into three steps: heterogeneous mutual attention calculation, heterogeneous message passing, as well as target-specific aggregation.

Heterogeneous Mutual Attention Calculation. For a source vertex s of type c_s and a target vertex t of type c_t connected by an edge $e = (s, t)$ of type r_e, we first calculate a query vector Q_t and a key vector K_s, with the output from previous heterogeneous graph transformer layer, by two vertex type-specific linear projections $W_{c_t}^Q$ and $W_{c_s}^K$,

$$Q_t = W_{c_t}^Q h_t^{(l-1)}, \tag{1}$$

$$K_s = W_{c_s}^K h_s^{(l-1)}, \tag{2}$$

here $h_s^{(l-1)}$ and $h_t^{(l-1)}$ denote the representations of vertex s and vertex t by the $(l-1)$-th heterogeneous transformer layer, separately.

Then, we calculate a similarity score by taking the dot product of Q_t with K_s as shown in Equation (3). An edge type-specific linear projection $W_{r_e}^A$ is utilised in case that there are multiple types of edges between a same vertex type pair, while μ is a predefined vector indicating the general significance of each edge type. The obtained score is normalised by the square root of the dimension of key vector d_{K_s}. After the scores for all neighbors of t have been computed, a softmax function is applied to yield the normalised attention weights A_t for neighbor aggregation,

$$A_t = \underset{\forall s \in N_t}{softmax}(\frac{\mu K_s W_{r_e}^A Q_t^T}{\sqrt{d_{K_s}}}). \tag{3}$$

Heterogeneous Message Passing. Parallel to the mutual attention calculation, the representation of source vertex s from previous heterogeneous graph transformer layer $h_s^{(l-1)}$, is fed into another linear projection $W_{c_s}^M$ to produce a message vector M_s,

$$M_s = W_{c_s}^M h_s^{(l-1)} W_{r_e}^M, \tag{4}$$

here we add another projection $W_{r_e}^M$ to incorporate the edge dependency.

Target-Specific Aggregation. With the attention weights A_t and message vector M_s yielded by previous steps, we aggregate the information from all the neighbors to t,

$$h_t^{(l)} = \sigma(W_{c_t}^C \sum_{s \in N_t} A_t M_s) + h_t^{(l-1)}, \tag{5}$$

where $W_{c_t}^C$ is another linear projection mapping the aggregated representation back to t's type-specific feature space, followed by a non-linear activation operation. By conducting the residual connection operation [11], a highly contextualized representation $h_t^{(l)}$ for the target vertex t by the current l-th heterogeneous graph transformer layer is produced that can be fed into the following module for downstream tasks.

Answer Prediction. After propagation by the heterogeneous graph transformer layers, the produced representations for the vertices corresponding to common tokens h are passed to the linear layers with the learned parameters W_s, W_e and bias b_s, b_e,

$$y_s = softmax(W_s h + b_s), \tag{6}$$

$$y_e = softmax(W_e h + b_e). \tag{7}$$

The two probability distributions y_s and y_e indicate the probability of each vertex being the start or end of the answer span separately.

We compute the cross entropy loss as our training objective,

$$\mathcal{L} = -(y_s' \log y_s + y_e' \log y_e), \tag{8}$$

where y_s' and y_e' are the ground truth start position and end position of the answer.

4   Experiments and Discussions

We empirically evaluate the effectiveness of our proposed approach with version 2.0 of the Stanford Question Answering Dataset (SQuAD 2.0) [27].

4.1   Setup

The base encoder is a pre-trained BERT language model, in its public Pytorch implementation from the *Transformers*[1] library. We keep their default settings with a maximum input length of 384. We initialise the weights with the saved models available from *Hugging Face*[2]. We then fine-tune the weights during training. We use the standard BERT base model (cased), also known as *bert-base-cased*, model. To build the heterogeneous graph transformer, we use the *pytorch-geometric* library.[3]

In the dependency graph, the embeddings of the initial tokens are obtained from the pre-trained language model. The embeddings for the virtual lexeme tokens are initialised with the mean of the embeddings of their corresponding sub-words. The dependency graph edges are obtained from dependency parsing with the method of [5]. Their embeddings are initialised randomly according to the type of dependency.

In the constituency tree, similarly to the dependency tree, the embeddings of the initial tokens are obtained from the pre-trained language model and the embeddings for the virtual lexeme tokens are initialised with the mean of the embeddings of their corresponding sub-words. The constituent vertices are obtained from constituency parsing with the method of [18]. The embeddings of the virtual vertices and of the edges are initialised randomly according to their category.

The training uses AdamW optimizer [17] and a learning rate of 2e−5. We stack 2 heterogeneous graph transformer layers with 4 attention heads in each. The model is trained with a mini-batch size of 32 for 7 epochs. The code will be available on Github.

Training and testing use SQuAD 2.0, a data set of questions collected on a set of Wikipedia articles. The answer to every question is a text span or the question might be unanswerable. It contains around 130k training and 12k development examples.

4.2   Evaluation

Metrics. We use the following two metrics for the performance evaluation. F1 measures the normalised average overlap between the prediction and ground-truth answer. Exact match (EM) evaluates whether the prediction exactly matches the ground-truth.

Overall Experimental Results. The overall experimental results are shown in Table 1. We compare the performance of the pre-trained BERT alone, of SyHGT with a dependency graph and BERT, and of SyHGT with a constituency tree and BERT. The results are presented in Table 1, in which the three models are refered to as BERT, SyGHT-D (BERT), and SyGHT-C (BERT), respectively. We observe a slight improvement of 0.77 EM and 0.46 F1 of SyGHT-C over the BERT baseline and a more significant improvement of 1.22 EM and 0.97 F1 of SyGHT-D over the BERT baseline.

---

[1] github.com/huggingface/transformers.

[2] huggingface.co/.

[3] github.com/rusty1s/pytorch_geometric.

Table 1. Overall comparative empirical performance with SQuAD2.0.

Method	F1	EM
BERT	75.41	71.78
SyHGT-C (BERT)	75.87	72.55
SyHGT-D (BERT)	**76.38**	**73.00**

Microanalysis

Dependency Graph. We examine the dependencies[4] involved in question-answer pairs for which SyGHT-D and BERT alone find different answers. Each example shows the question (Q), the paragraph (P) as well as the BERT and SyHGT-D answers.

Examples 1 and 2, along with the corresponding dependency graphs, showcase the inferred utility of dependencies.

Example 1

Q: Colonial rule would be considered what type of imperialism?

P: ... Formal imperialism is defined as physical control or full-fledged colonial rule ...

BERT: Formal imperialism is defined as physical control or full-fledged

SyHGT-D: Formal

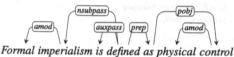

Formal imperialism is defined as physical control

Example 2

Q: What is the other country the Rhine separates Switzerland to?

P: The Alpine Rhine begins ... and later forms the border between Switzerland to the West and Liechtenstein ...

BERT: the West and Liechtenstein

SyHGT-D: Liechtenstein

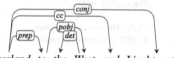

Switzerland to the West and Liechtenstein

In Example 1, BERT predicts a long and incorrect span whereas SyHGT-D, informed by the dependency graph, recognises that 'formal', as an adjectival modifier (*amod*) of 'imperialism', is the correct answer. In Example 2, the dependency tree shows that the phrase 'to the West' is connected to 'Switzerland' through a preposition (*prep*) as an object of preposition (*pobj*), while 'Liechtenstein' is a conjunct (*conj*). SyHGT-D uses the dependencies to correctly identify that 'to the West' is not a separate element from 'Switzerland', and that 'Liechtenstein' is the answer.

Overall, we report that several specific dependencies, in particular *prep, pobj, dobj, nsubj, conj, cc* seem to allow SyHGT-D to predict corresponding linguistically sound answers, albeit sometimes at the expense of the more general answer. SyHGT-D seems to be parsimonious.

Constituency Graph. We examine the constituencies[5] involved in question-answer pairs for which SyGHT-C and BERT alone find different answers. Each example shows the question (Q), the paragraph (P) as well as the BERT and SyHGT-C answers.

---

[4] The descriptions of the dependencies can be found in downloads.cs.stanford.edu/nlp/software/dependencies_manual.pdf.

[5] The descriptions of the dependencies can be found in http://surdeanu.cs.arizona.edu//mihai/teaching/ista555-fall13/readings/PennTreebankConstituents.html.

Examples 3 to 6, along with the corresponding constituency trees, showcase the inferred utility of constituencies.

Example 3

Q: *What is one example of what a clinical pharmacist's duties entail?*

P: *... The clinical pharmacist's role involves creating a comprehensive drug therapy plan for patient-specific problems, identifying goals of therapy, and reviewing ...*

BERT: *creating a comprehensive drug therapy plan for patient-specific problems, identifying goals of therapy, and reviewing all prescribed medications*

SyHGT-C: *creating a comprehensive drug therapy plan for patient-specific problems*

Example 4

Q: *Where did Kublai extend the Grand Canal to?*

P: *... Kublai expanded the Grand Canal from southern China to Daidu in the north. ...*

BERT: *southern China to Daidu in the north*
SyHGT-C: *Daidu in the north*

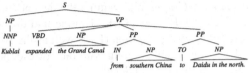

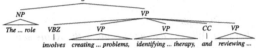

Example 5

Q: *Who published the State of the Planet 2008-2009 report?*

P: *Michael Oppenheimer, a long-time participant in the IPCC and coordinating lead author of the Fifth Assessment Report conceded in Science Magazine's State of the Planet 2008-2009 ...*

BERT: *Michael Oppenheimer*
SyHGT-C: *Science Magazine*

Example 6

Q: *What type of architecture is represented in the majestic churches?*

P: *Gothic architecture is represented in the majestic churches but also at the burgher houses and fortifications. ...*

BERT: *Gothic architecture*
SyHGT-C: *Gothic*

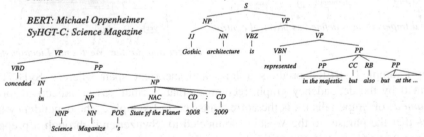

In Example 3, the question asks for one clinical pharmacist's duty where the answer is one of 'creating a comprehensive drug therapy plan for patient-specific problems', 'identifying goals of therapy', 'reviewing ...'. BERT is not able to distinguish different duties based on the syntax structure and thus gives all the duties in the passage. However, the constituency tree helps distinguish the three duties. Hence, with the constituencies, SyHGT-C can provide a correct answer. In Example 4, BERT predicts 'southern China to Daidu in the north. BERT confuses the coordinated prepositional phrases (PP) 'from southern China' and 'to Daidu in the north'. With the constituency tree, SyHGT-C can understand that 'Kublai expanded' from the start location 'southern China' to the end location 'Daidu in the north' and predict the correct answer. The passage in Example 5 is difficult to understand as it contains long and complex clauses. BERT fails to understand this sentence. The constituency tree clarifies that 'State of the Planet' belongs to 'Science Magazine' leading to the correct answer. Example 6 further illustrates that SyHGT-C can give more accurate answers that exactly match the ground-truth benefited from integrating the constituency information.

4.3 LingHGT, SemHGT, PragHGT: Towards Linguistics-Informed Language Models

Linguistics structures are numerous and, more often than not, amenable to a graph representation. Such structures exist not only for many aspects of syntax but also for semantics and pragmatics. The architecture we discussed applies to other linguistic graphic structures. Preliminary experiments with BERT and SQuAD seem to confirm the versatility and effectiveness of the model and its realisation for linguistics in general, LingHGT, and for semantics, SemHGT, and pragmatics, PragHGT, in particular.

We are conducting preliminary experiments of the utilisation of an entity-relationship graph SQuAD for semantics- and pragmatics-informed question answering. We used the spaCy library [12][6] for named entity recognition and the Open-NRE [9][7] for relationship extraction. The model is pre-trained on the the Wiki80 data set that is derived from FewRel and covers 80 relations [10][8]. We extract entities and relationships from the questions and passages for semantic information and we augment questions with contextual information about the questioner for pragmatic information. In the following, we look at examples where BERT incorporating semantics information can answer correctly while the original BERT cannot. In Example 7, the relevant relation identified is 'child' between the entity 'Lupe Mayorga' (*PERSON*), and the entity 'Aken' (*PERSON*). In Example 8 the relevant relation identified is 'member of political party' between the entity 'Annabel Goldie' (*PERSON*), and the entity 'Conservatives' (*NORP*).

Example 7
Q: Who was Bill Aiken's adopted mother?
P: Aken, adopted by Mexican movie actress
Lupe Mayorga
BERT: (no answer)
SemHGT: Lupe Mayorga

Example 8
Q: Who announced she would step down as leader of the Conservatives?
P: ...leader Annabel Goldie claiming that their support had held firm. Nevertheless, she too announced she would step down as leader of the party...
BERT: (no answer)
SemHGT: Annabel Goldie

We are exploring the opportunity and the applications of a pragmatics-informed language model. The following example is simulated in order to illustrate the targetted behaviour of a pragmatics-informed version of the proposed model.

Example 9
P: The IPCC receives funding through the IPCC Trust Fund, established in 1989 by the United Nations Environment Programme (UNEP) and the World Meteorological Organization

---

[6] https://spacy.io/.
[7] https://github.com/thunlp/OpenNRE.
[8] https://github.com/thunlp/FewRel.

(WMO), Costs of the Secretary and of housing the secretariat are provided by the WMO, while UNEP meets the cost of the Depute Secretary.
    Q (asked by the Secretary): Who funds my secretariat?
    PraHGT: the World Meteorological Organization
    Q (asked by the Deputy Secretary): Who funds my secretariat?
    PraHGT: the United Nations Environment Programme

In Example 9, PraHGT should leverage its knowledge of the classes of the entities 'The United Nations Environment Programme' and 'the World Meteorological Organization', namely 'ORG - organization', 'Secretary' and 'Deputy Secretary', 'PER - person', and the relationships connecting them directly or via other entities to the questioner to produce correct answers. The spaCy named-entity recogniser does not have a class 'Job Title' for 'Secretary' and 'Deputy Secretary'. The custom class needs to be added. The reader notices that a morphology informed tokenisation is also needed in order to guarantee the proper association of 'secretariat' with 'Secretary', and to understand the typographical error (original to SQuaD2.0) in 'Depute' (instead of 'Deputy'.)

In general we believe that LingHGT is a blueprint for the implementation of linguistics-informed models on top of the existing powerful pre-trained neural language models, wherever the linguistics information can be represented as a graph.

5 Conclusion

This paper presented a syntax-informed question answering model. The approach combines the statistical knowledge of neural language model with the symbolic information contained in linguistic graphic structures such as dependencies graphs and constituency trees. The seamless integration is realised by the means of a heterogeneous graph transformer added to a pre-trained transformer-based neural language model. The models therefore combines the self-attention mechanism of the transformer-based neural language model with a focused attention guided by graphic structures representing linguistics information in a heterogeneous graph transformer model.

An empirical performance evaluation of the proposed approach in comparison to the neural language model alone for question answering with SQuAD2.0 shows improvement. An initial microanalysis of the results suggest that the proposed model makes more focused predictions thanks to its awareness of syntax. Several examples, for which the proposed approach does not find the correct answer, even suggest that a better syntax parser could be key to addressing the shortcomings.

Preliminary results for LingHGT, SemHGT, and PragHGT confirm the versatility and effectiveness of linguistics-informed language models and give a blueprint for the implementation of incorporating the linguistics information as a graph into the powerful transformed-based language models.

Acknowledgement. This research is supported by the National Research Foundation, Singapore under its Industry Alignment Fund - Pre-positioning (IAF-PP) Funding Initiative. Any opinions, findings and conclusions or recommendations expressed in this material are those of the authors and do not reflect the views of National Research Foundation, Singapore.

References

1. Calijorne Soares, M.A., Parreiras, F.S.: A literature review on question answering techniques, paradigms and systems. J. King Saud Univ. Comput. Inf. Sci. **32**(6), 635–646 (2020)
2. Cimiano, P., Unger, C., McCrae, J.: Ontology-based interpretation of natural language. Synth. Lect. Hum. Lang. Technol. **7**(2), 1–178 (2014)
3. De Cao, N., Aziz, W., Titov, I.: Question answering by reasoning across documents with graph convolutional networks. arXiv arXiv:1808.09920 [cs, stat] (April 2019)
4. Devlin, J., Chang, M.W., Lee, K., Toutanova, K.: BERT: pre-training of deep bidirectional transformers for language understanding (2019)
5. Dozat, T., Manning, C.D.: Deep biaffine attention for neural dependency parsing. In: 5th International Conference on Learning Representations, ICLR 2017, Toulon, France, 24–26 April 2017, Conference Track Proceedings. OpenReview.net (2017)
6. d'Avila Garcez, A., Gori, M., Lamb, L.C., Serafini, L., Spranger, M., Tran, S.N.: Neural-symbolic computing: an effective methodology for principled integration of machine learning and reasoning (2019)
7. Goldberg, Y.: Assessing BERT's syntactic abilities. arXiv arXiv:1901.05287 [cs] (January 2019)
8. Green, B.F., Wolf, A.K., Chomsky, C., Laughery, K.: Baseball: an automatic question-answerer. In: Papers Presented at the May 9–11, 1961, Western Joint IRE-AIEE-ACM Computer Conference, IRE-AIEE-ACM 1961 (Western), pp. 219–224. Association for Computing Machinery, New York, NY, USA (1961)
9. Han, X., Gao, T., Yao, Y., Ye, D., Liu, Z., Sun, M.: OpenNRE: an open and extensible toolkit for neural relation extraction. In: Proceedings of the 2019 Conference on Empirical Methods in Natural Language Processing and the 9th International Joint Conference on Natural Language Processing (EMNLP-IJCNLP): System Demonstrations, Hong Kong, China, pp. 169–174. Association for Computational Linguistics (November 2019). https://doi.org/10.18653/v1/D19-3029. https://aclanthology.org/D19-3029
10. Han, X., et al.: FewRel: a large-scale supervised few-shot relation classification dataset with state-of-the-art evaluation. In: Proceedings of the 2018 Conference on Empirical Methods in Natural Language Processing, Brussels, Belgium, pp. 4803–4809. Association for Computational Linguistics (October–November 2018). https://doi.org/10.18653/v1/D18-1514. https://aclanthology.org/D18-1514
11. He, K., Zhang, X., Ren, S., Sun, J.: Deep residual learning for image recognition. In: 2016 IEEE Conference on Computer Vision and Pattern Recognition (CVPR), pp. 770–778 (2016). https://doi.org/10.1109/CVPR.2016.90
12. Honnibal, M., Montani, I., Van Landeghem, S., Boyd, A.: spaCy: industrial-strength natural language processing in Python (2020). https://doi.org/10.5281/zenodo.1212303
13. Hu, Z., Dong, Y., Wang, K., Sun, Y.: Heterogeneous graph transformer. In: Proceedings of the Web Conference 2020, WWW 2020, New York, NY, USA, pp. 2704–2710. Association for Computing Machinery (2020)
14. Jawahar, G., Sagot, B., Seddah, D.: What does BERT learn about the structure of language? In: Proceedings of the 57th Annual Meeting of the Association for Computational Linguistics, Florence, Italy, pp. 3651–3657. Association for Computational Linguistics (2019)
15. Jurafsky, D., Martin, J.H.: Dependency parsing. In: Speech and Language Processing, 3rd edn. (2020)
16. Jurafsky, D., Martin, J.H.: Speech and Language Processing: An Introduction to Natural Language Processing, Computational Linguistics, and Speech Recognition, 1st edn. Prentice Hall, USA (2000)

17. Kingma, D.P., Ba, J.: Adam: A method for stochastic optimization. In: Bengio, Y., LeCun, Y. (eds.) 3rd International Conference on Learning Representations, ICLR 2015, San Diego, CA, USA, 7–9 May 2015, Conference Track Proceedings (2015)

18. Kitaev, N., Klein, D.: Constituency parsing with a self-attentive encoder. In: Proceedings of the 56th Annual Meeting of the Association for Computational Linguistics (Volume 1: Long Papers), Melbourne, Australia, pp. 2676–2686. Association for Computational Linguistics (July 2018). https://doi.org/10.18653/v1/P18-1249. https://aclanthology.org/P18-1249

19. Kuncoro, A., et al.: Syntactic structure distillation pretraining for bidirectional encoders. arXiv arXiv:2005.13482 [cs] (May 2020)

20. Lan, Z., Chen, M., Goodman, S., Gimpel, K., Sharma, P., Soricut, R.: ALBERT: a lite BERT for self-supervised learning of language representations. In: 8th International Conference on Learning Representations, ICLR 2020, Addis Ababa, Ethiopia, 26–30 April 2020. OpenReview.net (2020). https://openreview.net/forum?id=H1eA7AEtvS

21. Crabtree, M., Powers, J.: Language Files: Materials for an Introduction to Language, 5th edn. Ohio State University Press, Columbus (1991)

22. Mao, J., Gan, C., Kohli, P., Tenenbaum, J.B., Wu, J.: The neuro-symbolic concept learner: interpreting scenes, words, and sentences from natural supervision. CoRR abs/1904.12584 (2019). http://arxiv.org/abs/1904.12584

23. Schlichtkrull, M., Kipf, T.N., Bloem, P., van den Berg, R., Titov, I., Welling, M.: Modeling relational data with graph convolutional networks. In: Gangemi, A., et al. (eds.) ESWC 2018. LNCS, vol. 10843, pp. 593–607. Springer, Cham (2018). https://doi.org/10.1007/978-3-319-93417-4_38

24. Nivre, J.: An efficient algorithm for projective dependency parsing. In: Proceedings of the 8th International Conference on Parsing Technologies, Nancy, France, pp. 149–160 (April 2003). https://aclanthology.org/W03-3017

25. Peters, M., et al.: Deep contextualized word representations. In: Proceedings of the 2018 Conference of the North American Chapter of the Association for Computational Linguistics: Human Language Technologies, Volume 1 (Long Papers), New Orleans, Louisiana, pp. 2227–2237. Association for Computational Linguistics (June 2018). https://doi.org/10.18653/v1/N18-1202. https://www.aclweb.org/anthology/N18-1202

26. Rajpurkar, P., Jia, R., Liang, P.: Know what you don't know: unanswerable questions for SQuAD. arXiv:1806.03822 [cs] (June 2018)

27. Rajpurkar, P., Jia, R., Liang, P.: Know what you don't know: unanswerable questions for SQuAD. In: Proceedings of the 56th Annual Meeting of the Association for Computational Linguistics (Volume 2: Short Papers), Melbourne, Australia, pp. 784–789. Association for Computational Linguistics (July 2018). https://doi.org/10.18653/v1/P18-2124. https://aclanthology.org/P18-2124

28. Scarselli, F., Gori, M., Tsoi, A.C., Hagenbuchner, M., Monfardini, G.: The graph neural network model. Trans. Neur. Netw. 20(1), 61–80 (2009). https://doi.org/10.1109/TNN.2008.2005605

29. Shen, D., Klakow, D.: Exploring correlation of dependency relation paths for answer extraction. In: Proceedings of the 21st International Conference on Computational Linguistics and 44th Annual Meeting of the Association for Computational Linguistics, Sydney, Australia, pp. 889–896. Association for Computational Linguistics (July 2006). https://doi.org/10.3115/1220175.1220287. https://aclanthology.org/P06-1112

30. Tu, M., Wang, G., Huang, J., Tang, Y., He, X., Zhou, B.: Multi-hop reading comprehension across multiple documents by reasoning over heterogeneous graphs. In: Proceedings of the 57th Annual Meeting of the Association for Computational Linguistics, Florence, Italy, pp. 2704–2713. Association for Computational Linguistics (2019). https://doi.org/10.18653/v1/P19-1260. https://www.aclweb.org/anthology/P19-1260

31. Vashishth, S., Bhandari, M., Yadav, P., Rai, P., Bhattacharyya, C., Talukdar, P.: Incorporating syntactic and semantic information in word embeddings using graph convolutional networks. arXiv arXiv:1809.04283 [cs] (July 2019)
32. Vaswani, A., et al.: Attention is all you need. In: Guyon, I., et al. (eds.) Advances in Neural Information Processing Systems, vol. 30. Curran Associates, Inc. (2017)
33. Vaswani, A., et al.: Attention is all you need. In: Advances in Neural Information Processing Systems, pp. 5998–6008 (2017)
34. Wang, X., et al.: Heterogeneous graph attention network. In: The World Wide Web Conference, New York, NY, USA, pp. 2022–2032. Association for Computing Machinery (2019)
35. Welbl, J., Stenetorp, P., Riedel, S.: Constructing datasets for multi-hop reading comprehension across documents. Trans. Assoc. Comput. Linguist. **6**, 287–302 (2018). https://aclanthology.org/Q18-1021
36. Wu, Y., et al.: Google's neural machine translation system: bridging the gap between human and machine translation. arXiv preprint arXiv:1609.08144 (2016)
37. Wu, Z., Pan, S., Chen, F., Long, G., Zhang, C., Philip, S.Y.: A comprehensive survey on graph neural networks. IEEE Trans. Neural Netw. Learn. Syst. **32**(1), 4–24 (2020)
38. Zhang, C., Song, D., Huang, C., Swami, A., Chawla, N.V.: Heterogeneous graph neural network. In: Proceedings of the 25th ACM SIGKDD International Conference on Knowledge Discovery and Data Mining, KDD 2019, pp. 793–803. Association for Computing Machinery, New York (2019). https://doi.org/10.1145/3292500.3330961
39. Zhang, Z., Wu, Y., Zhou, J., Duan, S., Zhao, H., Wang, R.: SG-Net: syntax-guided machine reading comprehension. arXiv arXiv:1908.05147 [cs] (November 2019)
40. Zhou, J., et al.: Graph neural networks: a review of methods and applications. AI Open **1**, 57–81 (2020)

Automated Process Knowledge Graph Construction from BPMN Models

Stefan Bachhofner[1]([📧])[ID], Elmar Kiesling[1,2][ID], Kate Revoredo[1][ID],
Philipp Waibel[1][ID], and Axel Polleres[1,3][ID]

[1] Institute for Data, Process and Knowledge Management,
Vienna University of Economics and Business, Vienna, Austria
{Stefan.Bachhofner,Elmar.Kiesling,Kate.Revoredo,Philipp.Waibel,
Axel.Polleres}@wu.ac.at
[2] Austrian Center for Digital Production, Vienna, Austria
[3] Complexity Science Hub Vienna, Vienna, Austria

Abstract. Enterprise knowledge graphs are increasingly adopted in industrial settings to integrate heterogeneous systems and data landscapes. Manufacturing systems can benefit from knowledge graphs as they contribute towards implementing visions of interconnected, decentralized and flexible smart manufacturing systems. Process knowledge is a key perspective which has so far attracted limited attention in this context, despite its usefulness for capturing the context in which data are generated. Such knowledge is commonly expressed in diagrammatic languages and the resulting models can not readily be used in knowledge graph construction. We propose BPMN2KG to address this problem. BPMN2KG is a transformation tool from BPMN2.0 process models into knowledge graphs. Thereby BPMN2KG creates a frame for process-centric data integration and analysis with this transformation. We motivate and evaluate our transformation tool with a real-world industrial use case focused on quality management in plastic injection molding for the automotive sector. We use BPMN2KG for process-centric integration of dispersed production systems data that results in an integrated knowledge graph that can be queried using SPARQL, a standardized graph-pattern based query language. By means of several example queries, we illustrate how this knowledge graph benefits data contextualization and integrated analysis. In a broader context, we contribute towards the vision of a process-centric enterprise Knowledge Graph (KG). BPMN2KG is available at https://short.wu.ac.at/BPMN2KG, and the sample queries and results at https://short.wu.ac.at/DEXA2022.

Keywords: Business process management · BPMN 2.0 · Semantic web · Knowledge graph · Model transformation · Industry 4.0

This research has received funding from the Teaming.AI project, which is part of the European Union's Horizon 2020 research and innovation program under grant agreement No 957402.

© The Author(s), under exclusive license to Springer Nature Switzerland AG 2022
C. Strauss et al. (Eds.): DEXA 2022, LNCS 13426, pp. 32–47, 2022.
https://doi.org/10.1007/978-3-031-12423-5_3

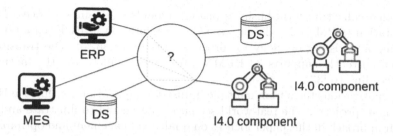

Fig. 1. Motivation: Dispersed data and unknown dependencies across the automation hierarchy (ERP, MES), data stores (DS), and production system components (I4.0 component).

1 Introduction

Relating domain and process knowledge to disparate and heterogeneous data is a challenge in most enterprise settings, which is particularly pronounced in the data-rich context of Cyber-physical Production Systems (CPPSs). Such systems currently drive a paradigm shift in manufacturing that is alluded to as the fourth industrial revolution and associated with the term Industry 4.0 (I4.0) [20]. This fundamental shift in industry is inherently driven by data [22] and characterized by requirements for flexible, networked, and self-configurable processes [32]. Consequently, data and process landscapes are expanding rapidly in smart manufacturing, but they typically remain disparate and fragmented (i) across information systems and data stores [15], (ii) between office and shop floor environments, and (iii) across business functions. Figure 1 illustrates this disconnect between various information systems across the automation hierarchy, data stores, and production system components. This disconnect raises the following challenges,

C1 *Integration* across multiple organizational, functional, and temporal levels of granularity,
C2 *Contextualization* of raw sensor data with higher-level operational information and quality requirements, and
C3 *Linking and aggregation* of decisions and goals on the production and operational levels to higher-level business goals.

KGs – which are characterized by a flexible schema, decentralized architecture, and ability to support data and knowledge integration – provide a promising foundation for such challenges. To integrate the fragmented process and data landscape through KGs, however, it is necessary to consider the process context. To address these challenges we therefore propose a combination of (i) Business Process Modeling, which was proposed as a method to tackle fragmentation challenges in manufacturing [1] and (ii) KG modeling based on Semantic Web (SW) standards, which have recently shown promising results in I4.0 applications [7,26,29]. We specifically propose BPMN2KG[1] as a tool

---

[1] BPMN2KG is available at https://short.wu.ac.at/BPMN2KG.

to automatically transform business process models in Business Process Model and Notation (BPMN) 2.0 [25] into a KG representation in RDF, based on and extending an existing ontology for process representation [5]. The transformation tool currently supports 33 BPMN 2.0 elements, including the most used ones according to [33].

Our motivation stems from multiple industrial applications within the H2020 Teaming.ai project[2]. The real-world scenario we selected to illustrate and validate our approach in this paper focuses on quality management and optimization of injection molding processes in the automotive industry. This use case illustrates how the proposed approach combines heterogeneous manufacturing data and process landscapes by integrating domain-specific and semantic abstraction models. BPMN2KG contributes towards using untapped process knowledge for integration initiatives using process graph modeling. It facilitates integrated querying of manufacturing data and process knowledge with SW methods and tools. This creates the ability to easily link data across sources and with manufacturing domain knowledge and provides a foundation for process-centric enterprise KG construction in I4.0 and beyond.

The remainder of the paper is structured as follows: Sect. 2 provides an introduction to process knowledge representation and knowledge graphs; Sect. 3 introduces the problem by means of a real-world quality management focused use case in an industrial setting; Sect. 4 introduces BPMN2KG and covers requirements, architecture, and implementation details; and Sect. 5 shows the results of the questions raised by our quality management focused use-case. Section 6 details related work. Finally, the paper concludes with remarks on BPMN2KG for our use case in Sect. 7.

2 Background

Process knowledge representation in Industry 4.0. A business process is a sequence of events, activities, and decision points that involve a number of actors and objects and leads to an outcome that is of value to at least one customer. It is typically represented in graphical models [19]. In recent years, BPMN 2.0 [25] has become a de-facto standard for modeling business processes, and it has also attracted increasing attention in the manufacturing domain [2,4]. It provides a wide range of graphical syntax elements that allow to describe process aspects in semantically well-defined terms, in various complexities, and for different use cases. eXtensible Markup Language (XML) is commonly used as a data format for BPMN. Due to space constraints, we do not discuss individual BPMN elements here; the full specification can be found in [25].

Knowledge Graphs. A KG is *"a graph of data intended to accumulate and convey knowledge of the real world, whose nodes represent entities of interest and whose edges represent relations between these entities."* [18]. Initially developed in the

---

[2] http://teamingai-project.eu.

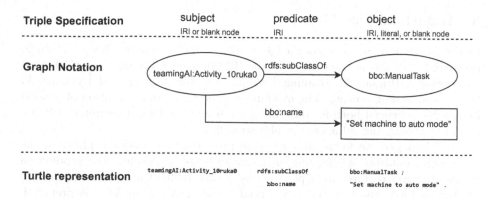

Fig. 2. RDF triple [13] (top) example in graph notation (middle) and turtle representation (bottom).

context of the Semantic Web (SW), KGs have today seen widespread adoption in web technology companies such as Microsoft, Google, Facebook, IBM [24], and Apple [23], where they provide an infrastructure to support services such as search, recommendations, and automation. KGs rely on graph data models such as labeled property graphs or directed edge-labelled graphs [18].

For modeling KGs, Resource Description Framework (RDF) [13] is a widely used language recommended by the World Wide Web Consortium (W3C). KGs in RDF are formed from triples, each of which consists of a subject, a predicate, and an object. Figure 2 illustrates two triples in RDFs from our motivating use case - the two triples belong to the first task of our mass production process. The first triple encodes the statement *"Activity 10ruka0 is a subclass of bbo:ManualTask"*, and the second *"Activity 10ruka0 has the bbo name of Set machine to auto mode"*. In graph notation, IRIs and blank nodes are represented with an ellipse and literals with a rectangle.

BPMN-based Ontology (BBO) [5] is an ontology to represent business processes modeled in BPMN 2.0 in a KG. An ontology is necessary as it for example allows us to sub-class it's concepts, or use their properties. We see this in the example above from Fig. 2, where we the define the thing that is identified by the Universal Resource Identifier *teamingAI:Activity_10ruka0* as a sub-class of *bbo:ManualTask*, and use the property *bbo:name* to give the Universal Resource Identifier (URI) a name. In addition to standard BPMN elements, BBO also provides some non-standard elements, such as a description in which manufacturing facility the process should be executed. We will use BBO as a basis and extend it with additional BPMN elements that are not covered in BBO.

3   Industrial Use Case

Our research into process knowledge graphs is motivated by three real-world industrial use cases in the context of the H2020 project Teaming.ai, which aims to develop a Human-AI Teaming Platform for Maintaining and Evolving AI Systems in Manufacturing. The machine-interpretable representation of process knowledge is crucial both for data integration and the human-centered collaboration of human and AI agents in I4.0 scenarios.

In this paper, we focus on a use case provided by a major supplier in the automotive industry specializing in plastic injection moulding. The production of plastic parts requires multiple processes, each defined in a separate process model: (i) First, the production material – the plastic granules – is prepared, which involves inspecting its quality. If quality is approved, the granules are fed to the manufacturing machine, otherwise, the material supplier is informed. (ii) Next, the manufacturing machine is configured by setting various machine parameters. These settings are then tested by producing a trial part and inspecting its quality. If the quality of the part is not satisfactory, the machine parameters are further readjusted, and another trial part is produced. This is continued until the quality meets the requirements. (iii) Finally, mass production starts with the determined machine parameter settings.

These processes are linked directly and indirectly through shared objects and data flows. They are also linked to other processes not considered in our use case scenario, such as mold engineering and setup, logistics processes, inventory handling, or order handling.

Figure 3 depicts the mass production process in BPMN 2.0. Figure 3a defines the start of the mass production and Fig. 3b illustrates the production process itself, together with the quality inspection of the produced part, as a sub-process. This sub-process is repeated for each unit produced until a stopping event is received or an error occurs.

The mass production process starts with the production of a part using an injection molding machine. During this step, a wealth of machine log data are generated and stored in a database. In our use case scenario, this log data will be used to populate a KG. Next, the quality of the part is checked by means of an automatic Visual Quality Inspections (VQIs) system. If this VQI system is not confident about its result (determined by a confidence threshold), a human-based manual inspection takes place. The result of both checks are again stored in a database. The quality of the part then determines the next activity. If the quality is *ok*, additional information about the part, e.g., part id, is persisted, and the part is handed over to packaging. If the quality is *not ok*, the next step depends on whether or not it is a recurring defect. In case of a recurring defect, mass production stops and a reconfiguration of the machine parameters is requested. For non-recurring defects, the part is scrapped and the next part is produced.

To produce in high volume, the company uses several manufacturing machines. Not all of these machines are of the same type, produce the same quality, and do not have the same capabilities or parameters. Moreover, some

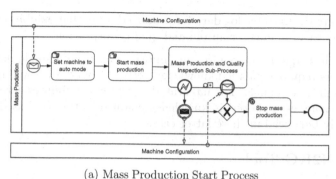

(a) Mass Production Start Process

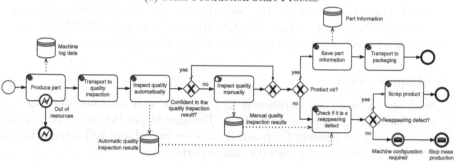

(b) Mass Production and Quality Inspection Sub-Process

Fig. 3. Mass production process of plastic parts industrial use case in BPMN 2.0.

production parts require unique treatments, for example a special finishing, and for some parts, automated VQI is not feasible.

Thus, a wide range of specialized variants of the discussed processes are used by the company, resulting in an extensive process landscape with many different process models. In addition, the execution of these processes creates vast amounts of data, including parameter and sensor data from the injection machines, energy and water sensor data, quality inspection data, and part information for each part produced. These persisted data are used for different purposes, for example in the design of new production process models and products/molds, to optimize the production process by analyzing the quality inspection results, or when performing the machine parameter setup.

A key challenge in this context is the fragmented nature of the data produced, which are not contextualized or linked to process knowledge. This makes it difficult to answer common questions such as:

(Q1) Across process models, which activities store data in or consume data from the various data stores?

(Q2) Which production processes include a quality control activity (of any kind)?

(Q3) What are the observed defect rates per defect type, across all production variants, for manual versus automatic visual quality inspection?

(Q4) What are the machine log data for produced units that exhibit a particular type of defect in manual or automated quality inspection?

Some of these questions require only information from a single data source, whereas others require combined knowledge derived from data stored in multiple systems and process models. The integrated querying capabilities across process models, data sources, and domain knowledge enabled by the KG-based approach is particularly beneficial in these latter cases.

4   BPMN2KG Tool

In the following, we summarize the requirements we elicited from our industrial use cases (Sect. 4.1), outline the KG construction with BPMN2KG on a schematic level (Sect. 4.2), and finally discuss implementation aspects (Sect. 4.3).

4.1   Requirements

Informed by the use case introduced in Sect. 3 as well as other use cases in the manufacturing domain as part of the Teaming.ai project, we collected the following set of requirements for business process model and KG integration in several rounds of workshops with domain experts:

(R1) Flexible semantic data model and schema: Supporting integration of process knowledge with domain knowledge and data requires a flexible model that can express relations between resource, process, and data elements. The tool shall not extend existing process modeling tools to encode explicit semantics into BPMN models, but rather impose semantics on the schema level through KG construction, curation, and completion techniques.

(R2) Automated model transformation: BPMN2KG shall automatically transform any valid BPMN model into a KG representation. All core as well as the most widely used other BPMN elements shall be supported. We will base the choice of these elements on studies such as [33], which found that only 20% of BPMN syntax elements are regularly used in their sample. In particular, the syntax elements to be supported are, in descending order of popularity, task, sequence flow, start event, end event, gateway, parallel gateway, data-based eXclusive OR (XOR) gateway, pool, and lane.

(R3) Rich process-oriented querying across functional areas, heterogeneous data sources, model and instance data and the process hierarchy. This necessitates both navigational queries, e.g., to express precedence patterns along the sequence flows, and graph-pattern based queries, e.g., to search for specific matches such as the use of particular data across administrative, support, and production-level processes (cf. the motivating questions Q1-Q4).

(R4) Modularity and extensibility: Whereas the prototype shall support basic transformation of the process structure of any valid BPMN 2.0 model, it should be modular and extensible through custom mappings. Due to this extensibility, the tool shall also be universally applicable beyond the manufacturing domain.

4.2   Process Knowledge Graph Construction

BPMN2KG constructs a KG from BPMN models, transforming multiple isolated models into a uniform representation that can be queried, linked to background knowledge, and integrated with instance data in a single, integrated graph. We call this KG a *process knowledge graph*, as it contains explicit process knowledge. The tool thereby makes this knowledge accessible for other systems via widely used standards. Figure 4 illustrates this concept by means of our mass production process (top). The red and green rectangles relate the graphical elements to their XML (middle) representation and show how these elements – after the transformation with BPMN2KG – are represented in RDF (bottom). Let us consider the first task of the process (green rectangle), which is the manual task *"Set machine to auto mode"* marked with a hand in the upper left corner. This task is represented in XML with the *bpmn:manualTask* tag, and has the two attributes *id* and *name* with values *"Activity_1e93nvu"* and *"Set machine to auto mode"*. BPMN2KG transforms these two attributes into the two triples (**tai:Activity_1e93nvu rdfs:subClassOf bbo:ManualTask**) and (**Activity_1e93nvu bbo:name "Set machine to auto mode"@en**). We use unit-tests to verify the correctness of such transformations.

To transform BPMN models into a knowledge graph representation, we use RDF Mapping Language (RML) as a declarative mapping language. Declarative software languages enable a higher level of abstraction – consider for example the Structured Query Language (SQL), another declarative language, where we define data structures without worrying about their physical realization, and queries without worrying about their procedural execution. RML allows for a similar abstraction for the relationship of heterogeneous data structures to RDF. RML by definition is a *"a generic mapping language, based on and extending"* the Relational data base to Resource description framework Mapping Language (R2RML) standard [12]. R2RML is a W3C recommendation [14], but is specialized for *"relational databases to RDF datasets"* [14]. To transform BPMN into a KG representation, we use RML to define a relation from XML to RDF for each XML element. RML uses XPath [10] to create logical sources that are mapped to one or more RDF triples. For the mapping to BBO, this results in 23 *rr:TriplesMap* definitions[3].

Supported BPMN Elements. A study that analyzed 120 BPMN diagrams found that only 20% of BPMN syntax elements are regularly used in their sample [33]. These syntax elements are, in descending order of popularity, task, sequence flow, start event, end event, gateway, parallel gateway, data-based XOR gateway, pool, and lane. Based on this observation, we decided that the first version of the tool has to support these syntactic elements. Unfortunately, BBO [5] does not include classes for pools and lanes. For this reason, we extended BBO into Business process model and notation Based Ontology Extension (BBOExt). In addition to pools and lanes, this extension supports message flows, association,

---

[3] https://short.wu.ac.at/BPMN2BBO, https://short.wu.ac.at/BPMN2BBOExt, and https://short.wu.ac.at/BPMN2BBOExtANDBBO.

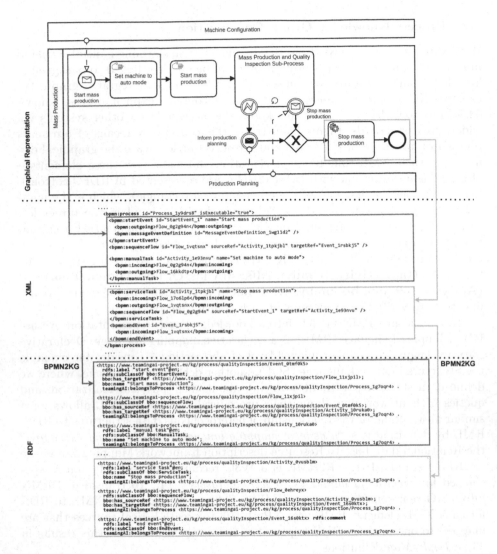

Fig. 4. Mapping example for excerpts of mass production process.

text annotation, data object, data object reference, data output association, data input association, and data store reference.

4.3 Implementation

Command Line Tool. Business Process Model and Notation to Knowledge Graph (BPMN2KG) (See footnote 1) is implemented as a command line tool in Python[4] with five arguments, two of them are required: --bpmn-input

---

[4] https://www.python.org/.

(the BPMN file or a directory) and --kg-output (the RDF file that the KG should be saved to or a directory where files should be saved to) that point BPMN2KG to the two files needed for the transformation. The other three arguments let the user choose the ontology (--ontology), the subject's URI template (--uri-template), and the serialization format of the output file (--serialization-format).

Benefits. The software layer on top of the RML adds a number of convenient features. First, the user can specify a folder instead of only a single file at a time for the transformation. Second, the URI templates need to be set only once and do not need to be manually exchanged each time. Third, it allows the user to easily change the target ontology. And finally, it encapsulates the complexity of RML into simple command line calls. As an engine for the RML transformations, we decided to use RMLMapper[5] as it offers a command line as well as a library interface which enables us to change to a different architecture in the future without changing the technology behind the transformation. We can for example change from Python to Java without replacing the engine. And finally, BPMN2KG can be integrated with any BPMN2.0 (which we follow) compliant software.

5   Application Scenarios

In this section, we focus on the use case introduced in Sect. 3, which tackles quality management and analytic challenges in plastic injection moulding. Specifically, we transform the graphical knowledge on the processes involved in production setup, execution, and quality control – which are captured in several BPMN models – into a KG representation. We enrich the KG with information about the manufacturing machines used by the various process activities and the involved resources and validate the capability to contextualize data in a KG with the transformed process knowledge.

Next, we illustrate how the resulting KG supports sensor data contextualization and analysis in quality management - addressing the previously identified requirements in Sect. 4.1 via the examples queries introduced in Sect. 3. See Appendix A for the queries pertaining *(Q1)* to *(Q3)*, their results, and the link to all queries - including *(Q4)* and the full syntax.

(Q1) Data Flow Analysis. A common challenge in complex production systems – as well as information systems more generally – is the proliferation of heterogeneous systems and dispersed data stores. The lack of a (at least high-level) overview of data flows makes it difficult to trace data provenance as well as to understand the complex interdependencies that exist between various process activities, systems, and data stores. Modeling data flows in BPMN using

---

[5] RMLMapper: https://github.com/RMLio/rmlmapper-java with commit *54bf875*.

data store and *data association* elements is helpful to document such relationships, but the resulting process models can not be readily queried and connections across process models are not visible. The integrated KG produced by BPMN2KG can help to untangle data flows and dependencies in large process landscape.

Specifically, example query (Q1) illustrates how the KG can support integrated querying of data flows from and to data stores. The query retrieves all data associations between activities and data stores (cf. Fig. 5) and indicates the direction of the flow. The result shows, for instance, that the activity *"Check if it is a reappearing defect"* consumes data from the data store *"Manual quality checking result"*, while the activity *"Check quality manually"* writes to it. The graph-based structure also provides a foundation for more complex object-centric analyses of data flows across process models and highlights how the KG can contribute towards mapping the process and data landscapes.

(Q2) Cross-Model Activity Querying. Process knowledge becomes even more useful once the model elements are associated with semantic concepts. For instance, abstraction hierarchies across activities allow for efficient querying. In our use case, for instance, automated and manual quality inspection activities are all subclasses of *tai:QualityManagementActivity*. Therefore, it is possible to use the domain knowledge captured in the activity model in the queries. (Q2) selects all activities that are sub classes of *tai:QualityManagementActivity* and return their IDs, the name of the activity, and the ids of their respective processes (cf. ??). The result shows four activities related to quality management in three different processes.

(Q3) Comparing Detection Rates of Manual and Automated Quality Inspection. Beyond interlinking process models and associating them with domain knowledge, the process KG can also link process models to instance data such as quality inspection results. In our use case, which focuses on quality management, this can be used to investigate observed defects by defect type, for different types of quality inspection activities, and across process variants. The query in ??, for example, aggregates the observed cases for different defect types across processes and groups them by task type and defect type. Note here how we use BPMN2KG to contextualize quality management data with respect to process knowledge.

(Q4) Retrieving Machine Log Data for Defects. As a final illustrative application scenario, the process KG references instance-level machine log data and makes it available for process-oriented querying[6]. This makes it possible to contextualize and retrieve sensor readings when diagnosing quality issues. For instance, the SPARQL query and result for (Q4) retrieves the machine log data for produced units with a particular type of defect in manual or automated quality inspection. The query in particular retrieves all defect parts with their product ID, product name, part ID, and stroke measurements for cushion, plasticisation, and transfer.

---

[6] You can find the query *(Q4)* at https://short.wu.ac.at/DEXA2022-Q4.

6   Related Work

Various KG applications within business process management have been developed in the literature, including KGs as a means to support process modeling [11], process model querying [27], and event log generation [8]. In this paper, we present a transformation tool for process models into a KG representation, with a focus on the integration with raw data and domain knowledge.

Reference [9] proposes a modeling language that combines process and domain knowledge. The approach set up a hybrid knowledge base derived from diagrammatic models, semantically lifted legacy data and open geospatial data. The specification is done manually. The proposed vision is similar to BPMN2KG, as it also aims to create an integrated semantic data fabric for process model contents and contextual data. However, the focus is not on any particular business process modeling language and the authors do not provide a mechanism to transform these models into a KG representation. They instead propose to integrate the semantic model into a customised BPMN front end.

Motivated by cost reduction through the reuse of data from legacy systems, [21] propose an approach for the transformation of BPMN models into OWL2 ontologies. Similar to [5], this work does not provide automatic integration of process and domain knowledge in a single representation.

In a similar domain as the one tackled in this paper, [30] models industrial business processes for querying and retrieval using OWL and SWRL. This also results in a semantic representation of business process models; key differences are the more limited set of transformed BPMN elements (they do not include data sources) and the use of OWL as a representation formalism. Furthermore, the paper does not address the integration of production systems data.

Reference [28] aimed to semantically annotate process models at design time. This is accomplished in a four step process. Similarly to our work, they also map, for example, an activity in a process model to an entity in a KG. However, this approach is based on Event-driven Process Chains (EPCs) [31] rather than BPMN. Another major difference is the execution of the four-step process at *design-time*. Our approach does not focus on assistance during modeling, but transforms models for integration and contextualization at *execution time*. Hence, their work is complementary to ours.

Similar to our work, [16] construct a KG for CPPS. They focus on KG construction from multiple design perspectives to achieve integration among these. However, this introduces uncertainty as different design perspectives might model the same construct differently, or leave it out completely. This is different to our work since we have no uncertainty as we have one perspective, the process perspective. Their work is hence complementary to ours.

7   Conclusions

In this paper, we introduce BPMN2KG to integrate process knowledge, domain knowledge, and dispersed data into a KG representation. We motivate the need for the approach by challenges that arise in the context of an I4.0 use case – which requires flexible processes, has a large process variety, and has to cope with increased "datafication" of the shop floor. BPMN2KG eases (i) the integration across multiple views and granularity levels using data stores (**C1**), (ii) the contextualization by adding process context to data (**C2**), and (iii) the linking and aggregation of production and operational levels (**C3**) - which we illustrate by the example of a plastic injection molding and quality management use case.

Our automatic transformation can further replace a manual semantic annotation of process models, which is generally not feasible in the face of large process landscapes [11]. Additionally, we provide the means to answer complex questions that require the combined knowledge of lower-level shop floor data and higher-level process information. Our work also contributes towards the vision of a process-centric, or at least a process knowledge informed enterprise KG. This is linked to the concept of layered KGs for CPPS presented in [6], where a KG has different domain views (for example process engineering and quality control) and layers based on Reference Architectural Model Industrie 4.0 (RAMI 4.0), which are decoupled I4.0 layers. And finally, as a minor contribution, we open source the RML rules which map BPMN models to the ontologies in our public repository, which means they can be used freely by anyone.

In future work, we plan to build upon and extend the current transformation tool. First, in the present paper we assume that the process logs and the machine logs are available in triple format. For the former, we are indeed already working on an accompanying transformation tool for XES[7]. This software tool will be used alongside BPMN2KG in a software system called Teaming.AI [17]. Beyond, we evaluate the usefulness of other target process ontologies, such as BPMN [3], and an extension that allows for a transformation from the KG to a BPMN XML model.

Acknowledgement. This work has also received funding from the Teaming.AI project in the European Union's Horizon 2020 research and innovation program under grant agreement No 95740.

A SPARQL Queries and Results

Due to space constraints, we deleted all prefix statements in the following queries and completely exclude *(Q2), (Q3), and (Q4)* – you can find all queries with the full syntax and the results at https://short.wu.ac.at/DEXA2022.

---

[7] https://git.ai.wu.ac.at/teaming-ai/extensible-event-stream-to-knowledge-graph.

```
SELECT ?dataAssociation ?dataStoreReferenceName ?predicateToActivity ?activityName
WHERE {
    ?dataAssociation rdfs:subClassOf ?o .
    FILTER ( ?o = bboe:DataInputAssociation || ?o = bboe:DataOutputAssociation ) .
    ?dataAssociation ?predicateReferences ?reference .
    FILTER ( ?predicateReferences = bbo:has_sourceRef ||
            ?predicateReferences = bbo:has_targetRef ) .
    ?reference rdfs:subClassOf bboe:DataStoreReference .
    ?dataAssociation ?predicateToActivity ?activity .
    FILTER ( ?predicateToActivity = bboe:is_dataInputFor ||
            ?predicateToActivity = bboe:is_dataOutputFrom ) .
    ?activity bbo:name ?activityName .
    ?reference bbo:name ?dataStoreReferenceName .
}
```

dataAssociation	dataStoreReferenceName	predicateToActivity	activityName
tai:DataOutputAssociation_12b2glg	Machine Log Data	bboe:is_dataOutputFrom	Produce part
tai:DataOutputAssociation_17xegjx	Automatic quality checking result	bboe:is_dataOutputFrom	Check quality automatically
tai:DataOutputAssociation_1spmd24	Part information	bbe:is_dataOutputFrom	Save part information
tai:DataInputAssociation_0blyrx1	Automatic quality checking result	bboe:is_dataInputFor	Check if it is a reappearing defect
tai:DataInputAssociation_10dfmxn	Manual quality checking result	bboe:is_dataInputFor	Check if it is a reappearing defect
tai:DataOutputAssociation_0iuiqz3	Manual quality checking result	bboe:is_dataOutputFrom	Check quality manually
tai:DataOutputAssociation_12tyfc6	Machine parameters	bboe:is_dataOutputFrom	Produce a test part
tai:DataOutputAssociation_1tg1wtf	Quality result	bboe:is_dataOutputFrom	Check quality

Fig. 5. SPARQL query and result for (Q1) showing data flows between activities and data stores.

References

1. Erasmus, J., Vanderfeesten, I., Traganos, K., Grefen, P.: Using business process models for the specification of manufacturing operations. Comput. Ind. **123**, 103297 (2020)
2. Abouzid, I., Saidi, R.: Proposal of BPMN extensions for modelling manufacturing processes. In: 5th International Conference on Optimization and Applications (ICOA), pp. 1–6. IEEE (2019)
3. Abramowicz, W., Filipowska, A., Kaczmarek, M., Kaczmarek, T.: Semantically enhanced business process modeling notation. In: Semantic Technologies for Business and Information Systems Engineering: Concepts and Applications, pp. 259–275. IGI Global (2012)
4. Ahn, H., Chang, T.-W.: Measuring similarity for manufacturing process models. In: Moon, I., Lee, G.M., Park, J., Kiritsis, D., von Cieminski, G. (eds.) APMS 2018. IAICT, vol. 536, pp. 223–231. Springer, Cham (2018). https://doi.org/10.1007/978-3-319-99707-0_28
5. Annane, A., Aussenac-Gilles, N., Kamel, M.: BBO: BPMN 2.0 based ontology for business process representation. In: 20th European Conference on Knowledge Management (ECKM 2019), vol. 1, pp. 49–59, Lisbon, Portugal, September 2019
6. Bachhofner, S., Kiesling, E., Kabul, K., Sallinger, E., Waibel, P.: Knowledge graph modularization for cyber-physical production systems. In: International Semantic Web Conference (Poster). Virtual Conference, October 2021
7. Buchgeher, G., Gabauer, D., Martinez-Gil, J., Ehrlinger, L.: Knowledge graphs in manufacturing and production: a systematic literature review. IEEE Access **9**, 55537–55554 (2021)

8. Calvanese, D., Kalayci, T.E., Montali, M., Tinella, S.: Ontology-based data access for extracting event logs from legacy data: the onprom tool and methodology. In: Abramowicz, W. (ed.) BIS 2017. LNBIP, vol. 288, pp. 220–236. Springer, Cham (2017). https://doi.org/10.1007/978-3-319-59336-4_16

9. Cinpoeru, M., Ghiran, A.-M., Harkai, A., Buchmann, R.A., Karagiannis, D.: Model-driven context configuration in business process management systems: an approach based on knowledge graphs. In: Pańkowska, M., Sandkuhl, K. (eds.) BIR 2019. LNBIP, vol. 365, pp. 189–203. Springer, Cham (2019). https://doi.org/10.1007/978-3-030-31143-8_14

10. Clark, J., DeRose, S.: XML path language (XPath) version 1.0. W3C recommendation, W3C, November 1999. https://www.w3.org/TR/1999/REC-xpath-19991116/

11. Corea, C., Fellmann, M., Delfmann, P.: Ontology-based process modelling - will we live to see it? In: Ghose, A., Horkoff, J., Silva Souza, V.E., Parsons, J., Evermann, J. (eds.) ER 2021. LNCS, vol. 13011, pp. 36–46. Springer, Cham (2021). https://doi.org/10.1007/978-3-030-89022-3_4

12. Cyganiak, R., Sundara, S., Das, S.: R2RML: RDB to RDF mapping language. W3C recommendation, W3C, September 2012. https://www.w3.org/TR/2012/REC-r2rml-20120927/

13. Cyganiak, R., Wood, D., Lanthaler, M.: RDF 1.1 Concepts and Abstract Syntax. W3c recommendation, World Wide Web Consortium, 25 February 2014. https://www.w3.org/TR/2014/REC-rdf11-concepts-20140225/

14. Das, S., Sundara, S., Cyganiak, R.: R2RML: RDB to RDF mapping language. W3C recommendation, W3C, September 2012. https://www.w3.org/TR/2012/REC-r2rml-20120927/

15. Erasmus, J., Vanderfeesten, I., Traganos, K., Grefen, P.: The case for unified process management in smart manufacturing. In: 2018 IEEE 22nd International Enterprise Distributed Object Computing Conference (EDOC), pp. 218–227 (2018)

16. Grangel-González, I., Halilaj, L., Vidal, M.-E., Rana, O., Lohmann, S., Auer, S., Müller, A.W.: Knowledge graphs for semantically integrating cyber-physical systems. In: Hartmann, S., Ma, H., Hameurlain, A., Pernul, G., Wagner, R.R. (eds.) DEXA 2018. LNCS, vol. 11029, pp. 184–199. Springer, Cham (2018). https://doi.org/10.1007/978-3-319-98809-2_12

17. Hoch, T., et al.: Teaming.AI: enabling human-AI teaming intelligence in manufacturing. In: Proceedings of Interoperability for Enterprise Systems and Applications Workshops: AI Beyond Efficiency: Interoperability towards Industry 5.0. Springer, Valencia (2022)

18. Hogan, A., et al.: Knowledge graphs. ACM Comput. Surv. (CSUR) **54**(4), 1–37 (2021)

19. Indulska, M., Recker, J., Rosemann, M., Green, P.: Business process modeling: current issues and future challenges. In: van Eck, P., Gordijn, J., Wieringa, R. (eds.) CAiSE 2009. LNCS, vol. 5565, pp. 501–514. Springer, Heidelberg (2009). https://doi.org/10.1007/978-3-642-02144-2_39

20. Kagermann, H., Wahlster, W., Helbig, J., et al.: Recommendations for implementing the strategic initiative Industrie 4.0: final report of the Industrie 4.0 working group. Technical report, Berlin, Germany (2013)

21. Kchaou, M., Khlif, W., Gargouri, F., Mahfoudh, M.: Transformation of BPMN model into an OWL2 ontology. In: International Conference on Evaluation of Novel Approaches to Software Engineering, pp. 380–388. Virtual Event, April 2021

22. Klingenberg, C.O., Borges, M.A.V., Antunes Jr., J.A.V.: Industry 4.0 as a data-driven paradigm: a systematic literature review on technologies. J. Manuf. Technol. Manag. (2019)
23. Malyshev, S., Krötzsch, M., González, L., Gonsior, J., Bielefeldt, A.: Getting the most out of Wikidata: semantic technology usage in Wikipedia's knowledge graph. In: International Semantic Web Conference, pp. 376–394, Monterey, California, USA, October 2018
24. Noy, N., Gao, Y., Jain, A., Narayanan, A., Patterson, A., Taylor, J.: Industry-scale knowledge graphs: lessons and challenges. Commun. ACM **62**(8), 36–43 (2019)
25. Business Process Model and Notation (BPMN) 2.0 specification (2011). https://www.omg.org/spec/BPMN/2.0/PDF, version 2
26. Patel, P., Ali, M.I., Sheth, A.: From raw data to smart manufacturing: AI and semantic web of things for industry 4.0. IEEE Intell. Syst. **33**(4), 79–86 (2018)
27. Polyvyanyy, A., Pika, A., ter Hofstede, A.H.: Scenario-based process querying for compliance, reuse, and standardization. Inf. Syst. **93**, 101563 (2020)
28. Riehle, D.M., Jannaber, S., Delfmann, P., Thomas, O., Becker, J.: Automatically annotating business process models with ontology concepts at design-time. In: de Cesare, S., Frank, U. (eds.) ER 2017. LNCS, vol. 10651, pp. 177–186. Springer, Cham (2017). https://doi.org/10.1007/978-3-319-70625-2_17
29. Rivas, A., Grangel-González, I., Collarana, D., Lehmann, J., Vidal, M.-E.: Unveiling relations in the Industry 4.0 standards landscape based on knowledge graph embeddings. In: Hartmann, S., Küng, J., Kotsis, G., Tjoa, A.M., Khalil, I. (eds.) DEXA 2020. LNCS, vol. 12392, pp. 179–194. Springer, Cham (2020). https://doi.org/10.1007/978-3-030-59051-2_12
30. Roy, S., Dayan, G.S., Devaraja Holla, V.: Modeling industrial business processes for querying and retrieving using OWL+SWRL. In: Panetto, H., Debruyne, C., Proper, H.A., Ardagna, C.A., Roman, D., Meersman, R. (eds.) OTM 2018. LNCS, vol. 11230, pp. 516–536. Springer, Cham (2018). https://doi.org/10.1007/978-3-030-02671-4_31
31. Scheer, A.W., Thomas, O., Adam, O.: Process Modeling using Event-Driven Process Chains, Chap. 6, pp. 119–145. Wiley, New York (2005)
32. Schneider, P.: Managerial challenges of industry 4.0: an empirically backed research agenda for a nascent field. Rev. Manag. Sci. **12**(3), 803–848 (2018)
33. Muehlen, M., Recker, J.: How much language is enough? Theoretical and practical use of the business process modeling notation. In: Seminal Contributions to Information Systems Engineering. LNCS, pp. 429–443. Springer, Heidelberg (2013). https://doi.org/10.1007/978-3-642-36926-1_35

CAKE: A Context-Aware Knowledge Embedding Model of Knowledge Graph

Jiadong Chen[1], Hua Ke[2], Haijian Mo[3], Xiaofeng Gao[1(✉)], and Guihai Chen[1]

[1] MoE Key Lab of Artificial Intelligence, Department of Computer Science
and Engineering, Shanghai Jiao Tong University, Shanghai, China
chenjiadong998@sjtu.edu.cn, {gao-xf,gchen}@cs.sjtu.edu.cn
[2] Tongji University, Shanghai, China
hke@tongji.edu.cn
[3] Electronics Technology Group Corporation, Beijing, China

Abstract. Recently, knowledge embedding on knowledge Graph (KG) has drawn increasing attention from both academia and industry for its concise rationale and promising prospects. However, performances of existing knowledge embedding methods are mostly either far from satisfactory, or exhibits weakness for generalization. In this work, a context-aware knowledge embedding model (CAKE) has been proposed for applications like knowledge completion and link prediction. We model the generative process of KG formation based on latent Dirichlet allocation and hierarchical Dirichlet process, where the latent semantic structure of knowledge elements is learned as contexts. Contextual information, i.e. the context-specific probability distribution over elements, is thereafter leveraged in a translation-based embedding model. Essentially, we develop loss function in a probabilistic style to approximately realize the "attention" mechanism in our model. In this work, the learned embeddings of entities and relations are applied to link prediction and triple classification in experiments and our model shows the best performance compared with multiple baselines.

Keywords: Knowledge graph embedding · Information system · Latent Dirichlet allocation · Hierarchical Dirichlet process

1 Introduction

Knowledge graph (KG), proven to be a powerful tool leading to intelligent applications at semantic-level, organizes facts in real world into an highly interactive, machine-readable and triplet-based network [5]. Knowledge graph regularizes a generalizable paradigm, or rather, a standard protocol, to unambiguously describe facts extracted from the real world. In this sense, construction of ontology performs an objective depiction of cognitions towards world and law of causation.

Knowledge embedding, a genre of methods that show the most promising prospects in solving knowledge sparsity and grit of a specific domain in KG, which is also known as other methodological branches that applies the idea of "vectorization" to make machine-readable the data carrying semantic information. Similar applications include word embeddings [9,12] and graph (network)

© The Author(s), under exclusive license to Springer Nature Switzerland AG 2022
C. Strauss et al. (Eds.): DEXA 2022, LNCS 13426, pp. 48–62, 2022.
https://doi.org/10.1007/978-3-031-12423-5_4

embeddings [3,4,16]. A variety of tactics have been presented to learn robust knowledge embedding, among which attention mechanism shows good potentials to improve the embedding accuracy: only part of the attributes of an entity are usable information for inferring a new knowledge. TransR [10] initially proposed such observation and hypothesis, while TransA [13] first explicitly formulated attention mechanism in the scenario of knowledge graph, using a ontology-level prejudgement as well as a relation-specific filtering skill to select only usable components in the latent space.

However, several drawbacks require our further attention, some of which may come from the existing attention mechanism. Existing methods that allow for attention mechanism tend to solely model relation-level attention, hypothesizing only relations carry multiple semantic components that discriminatively support inference in face of different entity pairs. But entities also carry such indicative information (e.g., entity *RogerFederer* works as a hint to place higher possibility to the relations related to *Tennis*) and can help with knowledge inference. TransG [20] uses Chinese Restaurant Process (CRP) to model infinitely many semantic components carried by relations, thus regarding relation semantics a probability distribution over latent components. In this work, we parallel extend the intuition to entities, assuming that entities and relations in knowledge graph probabilistically associate with a set of semantic components shared by all knowledge elements. We deem that the semantic components are consistent with factual domains for logically sound knowledge graphs; in case of ambiguity, we refer to such objects as contexts in this work, since they essentially interact with observable elements in role of contextual information. On the other hand, relation-specific attention mechanism overlooks meso-level constraints that should have been exerted on elements' embedding learning. In other words, relations exhibit internal structures w.r.t. their attention properties: compared with the relation *assassinate*, *citeThePaper* is more likely to co-occur with the relation *coauthorWith* and help with relevant inference.

In this paper, we utilize a hierarchical feedback structure to exert attentions on elements. We do not explicitly set parameters for attention learning but depend on contextual projections (Fig. 1) as well as probability-based penalty, so that trivial parameters are avoided while attention mechanism is aware under context and element. We formulate this work as a knowledge embedding model for applications like knowledge completion and link prediction, the latter of which is an experimental metrics in this paper. Overall framework of our work is shown in Fig. 1. We use latent Dirichlet allocation and hierarchical Dirichlet process, respectively, to model the generative process of a general knowledge graph, through which the latent semantic structure of knowledge elements is learned as contexts. Contextual information, i.e. the context-specific probability distribution over elements is thereafter leveraged in a translation-based embedding model. The learned embeddings of entities and relations are applied to link prediction and triple classification.

The contributions of this work are presented as follows: (1) We establish a context-aware model that develops loss function in a probabilistic style to approximately realize the "attention" layer. We claim that although the method belongs to no rigorously defined attention mechanism, it equivalently serves the model by discriminatively treating semantic information carried by knowledge elements. (2) The model automatically determines the probability of a triplet showing up in a specific context and thereby decides whether to utilize the corresponding contextual information for embedding learning and inference. (3) The proposed knowledge embedding model learns knowledge representations based on contextual information since contexts are learned a priori and the joint probability of a given triplet automatically adjust the loss distribution over different contexts. Our method shows the best performance in extensive experiments compared with multiple baselines.

The rest of this paper is organized as follows. In Sect. 2, related work is enumerated. Section 3 systematically introduces our proposed context-aware knowledge embedding (CAKE) model from two aspects: context-learning and embedding-learning. In Sect. 4, we conduct extensive experiments, showing and analyzing the results in link prediction and triple classification. We also discusses necessary details in engineering and cast insights into the embedding problem in this section. In Sect. 5, we conclude our work and future direction.

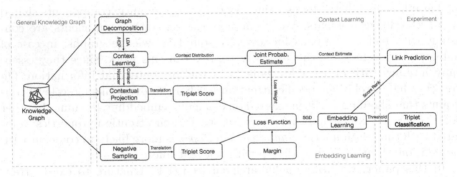

Fig. 1. Overall framework to develop context-aware knowledge embedding model from general knowledge graph

2 Related Work

Existing methods on knowledge embedding are chronologically enumerated and analyzed in this section. Since the proposed CAKE model is a translation-based embedding model (also the most mainstream model family), we mainly review previous studies along the branch while introduce other methods selectively.

TransE [1]. TransE is a canonical and easy-to-train model, modeling relationships by interpreting them as translations operating on the low-dimensional embeddings of the entities. In **TransH** [19] model, Wang et al. presented improvement solutions aiming at the aforementioned weakness of TransE. A relation-specific hyperplane is introduced so that the translation rule is relaxed to allow for a geometric flexibility. **TransR** [10] defines entity embeddings in entity space but projects them into relation space and exerts translation rules afterwards. Relation-specific mapping matrix M_r is introduced to the model so that entity pairs are projected first before any further translation. The loss function in TransR inherit the basic translation rule formulated by TransE. **CTransR** [10] holds the view that a single embedding for each relation can be underrepresentative in response to various entity pairs that exhibit heterogeneity. For each relation r, entity pairs (h, t) are clustered so that the pairs in the same group emphasize similar semantic components of r. Ji et al. proposed **TranSparse** [6] based on the archetype of TransR, and the novelty of TranSparse consists in that a sparsity-aware transfer matrix $M_r(\theta_r)$ is defined for each relation to map entities into the semantic space w.r.t. relations $e_h^r = M_r(\theta_r)e_h, e_t^r = M_r(\theta_r)e_t$, where θ_r is calculated to measure the degree of sparsity w.r.t. relation r; only a hyperparameter θ_{min} is required in the sparsity calculation. **TransA** [7]. Jia et al. presented this work essentially to improve the robustness and generalization ability of translation-based knowledge embedding models. TransA aims to find the optimal loss function so that the model is locally and structurally adaptive to multiple (or open-domain) knowledge bases. **TransAt** [13] initially formulates attention mechanism in the scenario of knowledge graph. It emphasizes the observation that only part of an entity's attribute information is factually utilized in knowledge inference, and thus such information should discriminatively contribute to the embedding learning of knowledge elements. **RDF2Vec** [15] is an unsupervised technique that can create task-agnostic numerical representations of the nodes in a knowledge graph by extending successful language modelling techniques.

Some recently emerging methods make attempts to develop neural network frameworks based on general embedding models to solve ranking problems induced by knowledge completion. **HOLE** [11] deploys circular correlation to develop associative memory to create compositional representations of knowledge elements. **ProjE** [14] formulates the task into a ranking problem and develops the framework conjugating a combination layer and projection layer. Loss function is defined in point-wise and list-wise style respectively. **CrossE** [21] explicitly simulates crossover interactions between knowledge elements and learns interaction-specific embeddings for entities and relations in the model. Another genre of studies seek to gain information from temporal and spatial factors based on the observation that facts only holds true within a timespan (or took place at a specific moment if the relation is instant). Jiang et al. [8] take into account the temporal priority between two facts w.r.t. a head entity and assumes that embedding of a previous relation associated to the entity can transform into a temporally subsequent one via a transition matrix. They first consider temporal factor in knowledge embedding, but failed to explicitly leverage temporal knowledge. **HyTE** [2] points out the drawback and proposes a novel model where discretized timestamps are depicted by hyperplanes in the semantic space. Projections of (embeddings of) head, relation and tail onto the hyperplane are trained to obey the translation rule.

3   Context-Aware Knowledge Embedding (CAKE)

This section introduces our context-aware knowledge embedding models and the methods our models based on. First we introduce LDA-based and HDP-based context learning methods. Then we develop the context-aware embedding models by deriving its loss function and negative sampling algorithms. We will give the optimization in the end.

Algorithm 1: Generative Process of LDA-based Context Learning

---

1 **foreach** *context z_k* **do**
2 $\quad$ draw parameter vector θ_k^ϵ from Dirichlet prior with parameter $\alpha^\epsilon \in \mathbb{R}^{|\epsilon|}$;
3 $\quad$ draw parameter vector θ_k^R from Dirichlet prior with parameter $\alpha^R \in \mathbb{R}^{|R|}$;

4 **foreach** *subgraph G_i, draw a context distribution $\phi_i \sim Dir(\beta)$* **do**
5 $\quad$ **foreach** *entity blank (denoted by e_{ij}) in G_i* **do**
6 $\quad\quad$ draw a context w.r.t. the entity as $z_{ij} \sim Multi(\phi_i)$;
7 $\quad\quad$ draw an entity $e \sim Multi(\theta_{z_{ij}}^\epsilon)$;

8 $\quad$ **foreach** *relation blank (denoted by r_{ij}) in G_i* **do**
9 $\quad\quad$ draw a context w.r.t. the relation as $z_{ij} \sim Multi(\phi_i)$;
10 $\quad\quad$ draw an relation $r \sim Multi(\theta_{z_{ij}}^\epsilon)$;

3.1   LDA-Based Context Learning

Latent Dirichlet Allocation, known as a generative probabilistic model for collections of discrete data, leverages a three-layer hierarchical Bayesian framework to model the formation of internally-structured data such as corpus and human populations (with haplotype). Teh et al. [17] refers to the corresponding inference problem as Grouped Clustering Problems (GCP). The problem is formulated as: given a fully observable knowledge graph $\mathcal{G}_i$, with a specified context number K and the prior distribution parameters α and β, we aim to learn the parameters of the multinomial distributions and w.r.t. both entities and relations.

A reasonable generative process of an arbitrary subgraph $\mathcal{G}_i$ from knowledge graph entails estimate of conditional probability such as $p(r|h)$ and $p(t|h,r)$. However, the introduction of such probability term breaks the conjugate structure between the Dirichlet prior and the multinomial distribution in both the scenarios of subgraph-context and context-element (i.e., relations and entities); without an intractable posterior, we can not carry out Gibbs sampling and update parameters. Therefore, we relax the conditions of generative process and treat entities and relations respectively, with two separate sets of LDA frameworks. The relaxation follows the "bag of word" simplification in LDA, HDP and other models in family. We deem the generative process of an arbitrary bunch of subgraphs $\mathcal{G}_i$ from knowledge graph $\mathcal{G}$ in Algorithm 1.

3.2 HDP-Based Context Learning

Dirichlet process (DP) is in a sense an infinitely dimensional generalization of Dirichlet distribution, normally specified by a base distribution G and a concentration parameter α. The formal mathematical definition of DP is as follows. Given a measurable set Ω, a base probability distribution G_0 and a positive real number α, if for any measurable finite partition of Ω (denoted by $(A_i, A_2, \ldots, A_n)$) it holds that $(G(A_1), \ldots, G(A_n)) \sim Dir(\alpha G_0(A_1), \ldots, \alpha G_0(A_n))$ Specifically, $Dir(\cdot)$ represents Dirichlet distribution. Then, $G \sim DP(G_0, \alpha)$, that is, G is subject to the Dirichlet process with base distribution G_0 and concentration parameter α.

Dirichlet Process theoretically suffices to formulate a mixture model with latently structured data. However, since base probability distribution G_0 can only be explicitly defined by a continuous distribution (e.g., Gaussian distribution) and $G|G_0 \sim DP(G_0, \alpha)$ is, on the other hand, discrete due to its definition over a finite partition of the measurable set. It causes the probability of each draw sharing probability atoms to be zero; in other words, the generative process following the DP with continuous base distribution G_0 do construct grouped data, but data cannot be shared across groups. G_0 is supposed to be discrete, so that probability atoms can be shared across all groups. To obtain a discrete G_0, another priori layer is added and we draw G_0 from DP (G, α) where G can be a continuous distribution. In this way, G_0 remains the base distribution shared by all groups, with finite, repeatable atoms to choose from. We use the term object to refer to the probability atom for convenience. Objects can be substantialized as topic and context. The whole generative process is then governed by the continuous base distribution G and a set of concentration parameters α_i^I. After a discrete base distribution G_0 is drawn, for each observation unit (e.g., a document), we draw the parameter θ_{ij} and further draw a realization according to the distribution characterized by θ_k. In terms of interpretation, HDP allows for the sharing of objects, and shows the property that "some objects are more likely to co-occur in an observation unit than other combinations". In the case with document generation, a topic related to *pop music* tends to co-occur with the topic related to *entertainment*, and it is hard to imagine it shows up with *computer architecture* in the same documents. Such *tendency* mathematically refers to a probability distribution, and HDP manages to capture the mechanism compared with a flat DP.

Generative Process of HDP. We use the application of HDP to topic learning for better description of its generative process. Note that a distribution G_i is drawn for each document and a probability distribution over topics is drawn via this step. Similar to the metaphor proposed for DP, a Chinese Restaurant Franchise (CRF) Process is also proposed to figuratively illuminate the generative process from the perspective of conditional probability. Interpretations based on conditional probability integrate out G, G_0 and G_i that cannot be explicitly represented and explained, so that the generative process is more understandable. CRF process assumes that there are multiple Chinese restaurants sharing the same menu. For each observation unit, we focus on the "enter" events within one corresponding restaurant.

HDP-Based Context Learning. Similar to LDA-based context learning model, HDP-based context learning problem can be formulated as: given a fully observable knowledge graph $\mathcal{G}$, a continuous base distribution G and hierarchical concentration parameters α and α_0, we aim to learn the context number activated by observed elements in $\mathcal{G}$ as well as the multinomial distribution parameters θ characterizing the aforementioned contexts. Algorithmic description has been summarized as Algorithm 2.

3.3   Context-Aware Knowledge Embedding

We define a semantic hyperplane for each context and characterize the cth context with its normal vector, denoted by ω_c. We inherit the form and notation of triplets as (h, r, t) where h, r and t represent the head entity, relation and the tail entity, respectively. To enable knowledge entities to own multiple interpretations in various contexts, we denote the projections of head h, relation r and tail t onto the hyperplane corresponding to context c by $P_c(e_h) = e_h - (\omega_c^\top e_h)\omega_c$, $P_c(e_r) = e_r - (\omega_c^\top e_r)\omega_c$, $P_c(e_t) = e_t - (\omega_c^\top e_t)\omega_c$. Similar to TransR, we deem that the projection operation provides a semantic reflection of the knowledge element, through which the distribution over latent semantic space is altered. We then exert the classic translation rule on the semantic reflections of the original embeddings. The scoring function takes the form as $d(e_h + e_r, e_t) = \|P_c(e_h) + P_c(e_r) - P_c(e_t)\|_{l_2}^2$ Note that such distance (or score) not only measures the validity of a given triplet, but also describes conformity of the given triple belonging to a specified context. Although the mechanism is not rigorously defined as attention, this layer equivalently serves model by selectively filtering information. The model automatically determine the probability of a triplet showing up in a specific context and thereby decide whether utilize the corresponding contextual information for embedding learning and inference. Contexts are learned a priori and the joint probability of a given triplet automatically adjust the loss distribution over different contexts. To effectively leverage the associations between knowledge elements and contexts, we add probabilistic components into the modified loss function.

The result of LDA-and-HDP-based context learning is the probability distribution of contexts over all knowledge elements. For context c, we denote its probability distribution over entity as $\mathcal{F}_c^\varepsilon(\epsilon)$, where the superscript ε indicates the element type. Similarly, we denote the probability distribution over relation as $\mathcal{F}_c^R(r)$. As contexts are learned with a "word-of-bag" model, we deem that

---

Algorithm 2: Generative Process of HDP-based Context Learning

---

1   Generate parameters $\theta_{ij}^{\mathcal{R}}$ and $\theta_{ij}^{\varepsilon}$ (probability distribution over relations and entities, respectively) for infinitely many courses on the franchise menu;

2   **foreach** i-th *restaurant* **do**

3   **foreach** *the j-th entity-customer enters and chooses k-th table subject to the distribution*

$$t_{ij}|t_{i1}, \ldots, t_{ij-1}, \alpha_0 \sim \sum_k \frac{n_{ik}^{\varepsilon} + n_{ik}^{\mathcal{R}}}{\sum_{t'} n_{jz'}^{\varepsilon} + n_{jk'}^{\mathcal{R}} + \alpha_0} \delta_k + \frac{\alpha_0}{\sum_{t'} n_{jz'}^{\varepsilon} + n_{jk'}^{\mathcal{R}} + \alpha_0} \delta_{k^{new}} \textbf{ do}$$

4   **if** *k-th table has been chosen by previous customers* **then**

5   adopts the context z_{ij} assigned to it;

6   **if** *a new table is chosen* **then**

7   choose a course z_{ij} from the menu subject to the distribution $z_{ij}|t_{11}, \ldots, z_{ij-1}, \alpha \sim$

$$\sum_k \frac{m_z^{\varepsilon} + m_z^{\mathcal{R}}}{\sum_{z'} m_{jz'}^{\varepsilon} + m_{jk'}^{\mathcal{R}} + \alpha} \delta_z + \frac{\alpha}{\sum_{z'} m_{jz'}^{\varepsilon} + m_{jk'}^{\mathcal{R}} + \alpha_0} \delta_{z^{new}};$$

8   Draw an entity subject to the multinomial distribution with parameter $\theta_{ij}^{\varepsilon}$, that is $e|\theta_{ij}^{\varepsilon} \sim Multi(\theta_{ij}^{\varepsilon})$;

9   **foreach** *the j-th relation-customer enters and chooses kth table subject to the distribution*

$$t_{ij}|t_{i1}, \ldots, t_{ij-1}, \alpha_0 \sim \sum_k \frac{n_{ik}^{\varepsilon} + n_{ik}^{\mathcal{R}}}{\sum_{t'} n_{jz'}^{\varepsilon} + n_{jk'}^{\mathcal{R}} + \alpha_0} \delta_k + \frac{\alpha_0}{\sum_{t'} n_{jz'}^{\varepsilon} + n_{jk'}^{\mathcal{R}} + \alpha_0} \delta_{k^{new}} \textbf{ do}$$

10   **if** *k-th table has been chosen by previous customers* **then**

11   adopts the context z_{ij} assigned to it;

12   **if** *a new table is chosen* **then**

13   choose a course z_{ij} from the menu subject to the distribution $z_{ij}|t_{11}, \ldots, z_{ij-1}, \alpha \sim$

$$\sum_k \frac{m_z^{\varepsilon} + m_z^{\mathcal{R}}}{\sum_{z'} m_{jz'}^{\varepsilon} + m_{jk'}^{\mathcal{R}} + \alpha} \delta_z + \frac{\alpha}{\sum_{z'} m_{jz'}^{\varepsilon} + m_{jk'}^{\mathcal{R}} + \alpha_0} \delta_{z^{new}};$$

14   Draw an entity subject to the multinomial distribution with parameter $\theta_{ij}^{\mathcal{R}}$, that is $r|\theta_{ij}^{\mathcal{R}} \sim Multi(\theta_{ij}^{\mathcal{R}})$;

---

the probability of an entity belonging to a context and that of a relation is independent. Thus, the joint probability of a triplet (h, r, t) emerging in context c can be calculated by

$$P(h, r, t|c) = P(h|c) \times P(r|c) \times P(t|c)$$
$$= F_c^{\varepsilon}(h) \times F_c^{\mathcal{R}}(r) \times F_c^{\varepsilon}(t)$$

For convenience, we use the notation $p_c(h, r, t)$ to represent the joint probability calculated above. We now discuss the interpretation of the joint probability of a given triplet with regard to context c. If $p_c(h, r, t)$ is relatively large, then entity h, t and relation r are highly likely to co-occur in context c; this intuitively indicates that the fact determined by this triplet is more likely to make sense and have practical interpretations.

We use the joint probability as a weight parameter for positive triplets. In this way, the triplets that are more semantically reasonable can be highlighted in the training process, enhancing the coherence of the corresponding context. For a negative triplet (h, r, t'), a large $p_c(h, r, t')$ value indicates that the three elements are semantically related in the context c but the fact does not hold true. In contrast, a small $p_c(h, r, t')$ means neither the elements are semantically linked, nor the fact is real. Since the latter case deserves a greater penalty, we adopt $(\kappa - p_c(h, r, t'))$ as the penalty weight of negative triplets. With a constant margin γ inherited from classic frameworks, the loss function can be developed into

$$
\mathcal{L} = \sum_{c \in C} \sum_{(h,r,t) \in D^+} \sum_{(h',r,t') \in D^-} [p_c(h, r, t) \times d(e_h + e_r, e_t)
$$
$$
- (\kappa - p_c(h', r', t')) \times d(e_{h'} + e_{r'}, e_{t'}) + \gamma]_+ \ ,
$$

where $[\cdot]_+$ represents the positive part of the inner terms and κ is a hyperparameter to scale the penalty degree. D^+ and D^- represent two respective sample sets which given in following field. The loss function preferentially encourages the positive facts that are semantically coherent in some context, and strongly punishes the negative triplets that neither semantically make sense nor represent valid facts.

For a valid triplet (h, r, t), researchers expect $e_h + e_r$ gets as close to et as possible [1]. It is also reasonable to expect that for invalid (or corrupted) triplets (h, r, t') we have $e_h + e_r$ gets as far as possible from $e_{t'}$. Hereby, a margin-based loss is developed in the form

$$
\mathcal{L} = \sum_{(h,r,t) \in \mathcal{D}^+} \sum_{(h',r,t') \in \mathcal{D}^-} [\gamma + d(e_h + e_r, e_t) - d(e_{h'} + e_r, e_{t'})]_+ \ ,
$$

where D^+ and D^- represent positive and negative (corrupted) sample set respectively; the former includes all valid facts provided by the knowledge graph, while the latter is an artificially constructed invalid sample set where fake triplets are manually created by corrupting and then knocking together elements from valid ones. Negative sampling algorithms are thereby developed according to the construction process. Then, we will mathematically represent the components of corrupted set D^- as following.

We use margin-based methods to construct a negative (invalid) sample set corresponding to the positive (valid) ones so that the training results are more satisfactory. In this case, negative samples are drawn parallel to positive ones so that the translation can be exclusively accurate, that is, $P_c(e_h) + P_c(e_r)$ gets close to $P_c(e_t)$ and stays far away from other tail entities when (h, r, t) is a valid triplet. We adopt the same way of construction for negative sample set as in TransE:

$$
\mathcal{D}^- = \{(h, r, t')|t' \in \varepsilon, (h, r, t) \in \mathcal{D}^+, (h, r, t') \notin \mathcal{D}^+\}
$$
$$
\cup \{(h', r, t)|h' \in \varepsilon, (h, r, t) \in \mathcal{D}^+, (h', r, t) \notin \mathcal{D}^+\}
$$

Considering optimization, the loss function of CAKE can be optimized with stochastic gradient descent (SGD) by calculating the gradients and selecting a proper learning rate. Entity embedding, relation embedding and normal vectors of each hyperplane associated to contexts need to be learned through the training process. We manually calculate the gradient formula for reproductability. The detailed update functions are shown in Appendix.

4   Experimental Evaluation

This section describes the experimental protocol, metrics and experimental results of our method (CAKE) as well as four baselines (TransE, TransH, TransR, TransAt). We conduct extensive experiments for two tasks, link prediction and triplet classification, on two datasets, Freebase and AceKG. We show the experimental results in the last subsection of this part, and present sufficient analysis based on observations and complementary tests.

4.1   Experiment Protocol

Following [13], we evaluate the performance of proposed CAKE on two tasks, link prediction and triplet classification.

We conducted experiments on both the Freebase and the AceKG subset for link prediction with metrics Hit@10 and Mean Rank and for triplet classification with metrics F1-score. For all the 5 methods, we experiment with margin value as $\{1, 2, 3, 4, 5, 6\}$ and the learning rate of $\{0.001, 0.005, 0.01\}$ respectively. For our method CAKE, we apply the hyperparameter κ as values from $\{1, 1.5, 2, 3\}$, and set context number as $\{10, 20, 30\}$ for LDA-based CAKE. Latent dimension number was set as $\{50, 100, 200\}$ respectively. We uniformly used batches with the size of 150 and trained each model for 2000 epoch for convergence.

4.2   Experiment Settings

Datasets. Freebase and AceKG [18] datasets are utilized in experiments, the former of which is known as an endeavor of general knowledge graph while the latter is an academy- oriented one, containing authors, papers, citation relationships and academic fields. We briefly introduce them as follows. For experiments, We extracted a tractable subset of Freebase with 24, 624 entities, 351 relations and 194, 328 facts; and a small subset of AceKG containing 30, 752 entities, 7 relations and 146, 917 facts.

Baselines. We use representative graph embedding models, **TransE** [1], **TransH** [20], **TransR** [10] and **TransAt** [13] as baselines in this work to testify the effectiveness of our proposed model CAKE.

Metrics. For three link prediction settings, we use **Hits@10** and **Mean Rank**, and for triplet classification, we use **F1-score** to validate models' effectiveness.

4.3 Experimental Results

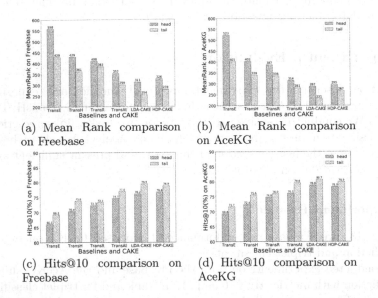

(a) Mean Rank comparison on Freebase

(b) Mean Rank comparison on AceKG

(c) Hits@10 comparison on Freebase

(d) Hits@10 comparison on AceKG

Fig. 2. Mean rank and Hits@10 comparison in task of link prediction

Link Prediction. We summarize the experimental results of link prediction in Table 1 and Fig. 2. As is shown, LDA-based CAKE model and HDP-based CAKE model exhibit the best performances among all tested methods, reaching the Mean Rank of 311 and 254 for the tasks of head and tail prediction respectively on Freebase dataset; the metrics are even better for AceKG, reaching 287 and 231 respectively. The most classic but basic method, TransE, shows the worst performance. TransH exhibits better performance than TransE due to the compatibility of special relation properties such as one-to-many, many-to-one and reflexive relations. TransR receives better performance compared with TransH due to its more flexible mapping rule: independent semantic spaces are set for entities and relations respectively, and thus semantic information can be inter-dimensionally manipulated in the training process. The consideration of attention layer over knowledge graph effectively improves the learning result.

Performances of the experimented methods show a similar tendency on Hits@10. HDP-based CAKE reaches nearly as good performance as LDA-based CAKE model on this metric in general; specifically, the former exceeds the latter on head prediction Hits@10 on Freebase dataset, since HDP framework learns latent contexts in a more faithful way to practice for general knowledge graph,

Table 1. Mean Rank and Hits@10 comparison in link prediction

Dataset	Freebase				AceKG			
Metric	Mean Rank		Hits@10(%)		Mean Rank		Hits@10(%)	
	Head	Tail	Head	Tail	Head	Tail	Head	Tail
TransE [1]	558	429	66.1	69.3	523	401	69.4	71.7
TransH [20]	429	361	70.4	73.8	402	339	72.4	75.6
TransR [10]	408	383	72.3	73.3	387	336	74.8	76.0
TransAt [13]	352	299	74.7	77.0	314	281	76.1	79.6
LDA-CAKE	311	254	76.2	79.5	287	231	78.8	80.7
HDP-CAKE	326	278	76.8	78.9	295	267	78.3	79.9

and the coherence better assists the inverse-translation inference in head entity prediction. For AceKG, however, contextual structure is more easy to a priori estimated by human, and thus LDA-based framework suffices to excellently capture the semantic coherence, making the LDA-based CAKE model the champion on this metric with AceKG data.

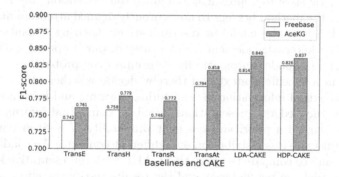

Fig. 3. F1-score comparison in task of triplet classification

Triplet Classification. The experimental for triplet classification is shown in Table 2 and Fig. 3. As is explained, we use F1-score as metric of bi-classification task. Generally, classification results on AceKG surpass those on Freebase since data of the latter exhibits less heterogentity compared to a general knowledge graph; we say the latter dataset is more semantically dense (only 7 relations are involved in the subset we use). As is shown by Table 2 and Fig. 3,

CAKE model exhibits the optimal performance among all compared methods. Specifically, HDP-based CAKE model exceeds LDA-based one on the dataset of Freebase due to more severe semantic heterogeneity of general knowledge graphs; with regard to AceKG, the results are quite the opposite. TransR, however, exhibits relatively worse performance compared with TransH in the task of triplet classification; inter-dimensional mappings of semantic components provide flexibility at the expense of larger parameter set and more uncertainty, which makes the bi-classification threshold (θ_r) learning a relatively hard task due to the non-linearity.

Table 2. F1-score Comparison in Triplet Classification

Dataset	Freebase	AceKG
TransE [1]	0.742	0.761
TransH [20]	0.758	0.779
TransR [10]	0.746	0.772
TransAt [1]	0.794	0.818
LDA-CAKE	0.814	0.840
HDP-CAKE	0.826	0.837

5   Conclusion

In this work, we establish a context-aware model that develops loss function in a probabilistic style to approximately realize the "attention" layer. We claim that although the method belongs to no rigorously defined attention mechanism, it equivalently serves the model by discriminatively treating semantic information carried by knowledge elements. Superiority of our proposed model should be noted: (1) The model automatically determines the probability of a triplet showing up in a specific context and thereby decide whether utilize the corresponding contextual information for embedding learning and inference. (2) The model learns knowledge representations based on contextual information since contexts are learned a priori and the joint probability of a given triplet automatically adjust the loss distribution over different contexts. We conduct extensive experiments to compare the proposed CAKE with representative knowledge embedding models on two datasets and the results testify the effectiveness and superiority of our proposed model. In future, We aim to capture more real world semantics and in turn serve the relevant applications.

Acknowledgement. This work was supported by the National Key R&D Program of China [2018YFB1004700]; the National Natural Science Foundation of China [61872238, 61972254]; and the State Key Laboratory of Air Traffic Management System and Technology [SKLATM20180X].

A   Appendix: Optimization of CAKE

For entity ϵ, the gradient of its embedding is

$$\frac{1}{2}\frac{\partial \mathcal{L}}{\partial e_\epsilon^l} = \sum_{c\in\mathcal{C}} \sum_{(h,r,t)\in\mathcal{D}^+} \sum_{(h',r',t')\in\mathcal{D}^-} \sum_{l'=1}^{L} (w_c^l w_c^{l'} - 1)\{p_c(h,r,t)$$
$$\times [e_h^{l'} - e_t^{l'} - \omega_c^\top(e_h - e_t)\omega_c^{l'} + P_c^{l'}(e_r)] \times (I_{t=\epsilon}$$
$$- I_{h=\epsilon}) - (\kappa - p_c(h',r',t')) \times [e_{h'}^{l'} - e_{t'}^{l'}$$
$$- \omega_c^\top(e_{h'} - e_{l'})\omega_c^{l'} + P_c^{l'}(e_{r'})] \times (I_{t'=\epsilon-I_{h'=\epsilon}})\}$$

For relation π, the gradient of its embedding is

$$\frac{1}{2}\frac{\partial \mathcal{L}}{\partial e_\pi^l} = \sum_{c\in\mathcal{C}} \sum_{(h,r,t)\in\mathcal{D}^+} \sum_{(h',r',t')\in\mathcal{D}^-} \sum_{l'=1}^{L} w_c^l w_c^{l'} \times \{I_{r=\pi}$$
$$\times p_c(h,r,t) \times [P_c^{l'}(e_h) + e_r^{l'} - (\omega_c^\top e_r)\omega_r^{l'}$$
$$- P_c^{l'}(e_t)] - I_{r'=\pi} \times (\kappa - p_c(h',r',t'))$$
$$\times [P_c^{l'}(e_{h'} + e_{r'}^{l'} - (\omega_c^\top e_{r'})\omega_{r'}^{l'} - P_c^{l'})]\}$$

For context δ, the gradient of its normal vector is

$$\frac{1}{2}\frac{\partial \mathcal{L}}{\partial e_\pi^l} = \sum_{c\in\mathcal{C}} \sum_{(h,r,t)\in\mathcal{D}^+} \sum_{(h',r',t')\in\mathcal{D}^-} \sum_{l'=1}^{L} (e_t^l - e_h^l - e_r^l) \times \{I_{c=\delta}$$
$$\times p_c(h,r,t) \times (\omega_c^{l'} + I_{l=l'} \times \omega_c^l) \times [e_h^{l'} + e_r^{l'} - e_t^{l'}$$
$$- \omega_c^\top(e_h + e_r + e_t)\omega_c^{l'}] - I_{c'=\epsilon} \times (\kappa - p_c'(h',r',t'))$$
$$\times (\omega_{c'}^{l'} + I_{l=l'} \times \omega_{c'}^l) \times [e_h^{l'} + e_r^{l'} - e_t^{l'}$$
$$- \omega_{c'}^\top(e_h + e_r + e_t)\omega_{c'}^{l'}]\}$$

References

1. Bordes, A., Usunier, N., Garcia-Duran, A., Weston, J., Yakhnenko, O.: Translating embeddings for modeling multi-relational data. In: Conference on Neural Information Processing Systems (NeurIPS), pp. 2787–2795 (2013)
2. Dasgupta, S.S., Ray, S.N., Talukdar, P.: HyTE: hyperplane-based temporally aware knowledge graph embedding. In: Conference on Empirical Methods in Natural Language Processing (EMNLP), pp. 2001–2011 (2018)
3. Du, L., Lu, Z., Wang, Y., Song, G., Wang, Y., Chen, W.: Galaxy network embedding: a hierarchical community structure preserving approach. In: International Joint Conferences on Artificial Intelligence (IJCAI), pp. 2079–2085 (2018)
4. Du, L., Wang, Y., Song, G., Lu, Z., Wang, J.: Dynamic network embedding: an extended approach for skip-gram based network embedding. In: International Joint Conferences on Artificial Intelligence (IJCAI), pp. 2086–2092 (2018)

5. Hoffart, J., Suchanek, F.M., Berberich, K., Weikum, G.: YAGO2: a spatially and temporally enhanced knowledge base from Wikipedia. Artif. Intell. (AI) **194**, 28–61 (2013)
6. Ji, G., Liu, K., He, S., Zhao, J.: Knowledge graph completion with adaptive sparse transfer matrix. In: AAAI Conference on Artificial Intelligence (AAAI), pp. 985–991 (2016)
7. Jia, Y., Wang, Y., Lin, H., Jin, X., Cheng, X.: Locally adaptive translation for knowledge graph embedding. In: AAAI Conference on Artificial Intelligence (AAAI), pp. 992–998 (2016)
8. Jiang, T., et al.: Encoding temporal information for time-aware link prediction. In: Conference on Empirical Methods in Natural Language Processing (EMNLP), pp. 2350–2354 (2016)
9. Levy, O., Goldberg, Y.: Neural word embedding as implicit matrix factorization. In: Conference on Neural Information Processing Systems (NeurIPS), pp. 2177–2185 (2014)
10. Lin, Y., Liu, Z., Sun, M., Liu, Y., Zhu, X.: Learning entity and relation embeddings for knowledge graph completion. In: AAAI Conference on Artificial Intelligence (AAAI), pp. 2181–2187 (2015)
11. Nickel, M., Rosasco, L., Poggio, T.: Holographic embeddings of knowledge graphs. In: AAAI Conference on Artificial Intelligence (AAAI), pp. 1955–1961 (2016)
12. Pennington, J., Socher, R., Manning, C.: Glove: global vectors for word representation. In: Conference on Empirical Methods in Natural Language Processing (EMNLP), pp. 1532–1543 (2014)
13. Qian, W., Fu, C., Zhu, Y., Cai, D., He, X.: Translating embeddings for knowledge graph completion with relation attention mechanism. In: International Joint Conferences on Artificial Intelligence (IJCAI), pp. 4286–4292 (2018)
14. Shi, B., Weninger, T.: ProjE: embedding projection for knowledge graph completion. In: AAAI Conference on Artificial Intelligence (AAAI), pp. 1236–1242 (2017)
15. Steenwinckel, B., et al.: Walk extraction strategies for node embeddings with RDF2Vec in knowledge graphs. In: Kotsis, G., et al. (eds.) DEXA 2021. CCIS, vol. 1479, pp. 70–80. Springer, Cham (2021). https://doi.org/10.1007/978-3-030-87101-7_8
16. Tang, J., Qu, M., Wang, M., Zhang, M., Yan, J., Mei, Q.: Line: large-scale information network embedding. In: International World Wide Web Conference (WWW), pp. 1067–1077 (2015)
17. Teh, Y.W., Jordan, M.I., Beal, M.J., Blei, D.M.: Sharing clusters among related groups: hierarchical Dirichlet processes. In: Conference on Neural Information Processing Systems (NeurIPS), pp. 1385–1392 (2005)
18. Wang, R., et al.: AceKG: a large-scale knowledge graph for academic data mining. In: ACM International Conference on Information and Knowledge Management (CIKM), pp. 1487–1490 (2018)
19. Wang, Z., Zhang, J., Feng, J., Chen, Z.: Knowledge graph embedding by translating on hyperplanes. In: AAAI Conference on Artificial Intelligence (AAAI), pp. 1112–1119 (2014)
20. Xiao, H., Huang, M., Zhu, X.: TransG: a generative model for knowledge graph embedding. In: Annual Meeting of the Association for Computational Linguistics (ACL), pp. 2316–2325 (2016)
21. Zhang, W., Paudel, B., Zhang, W., Bernstein, A., Chen, H.: Interaction embeddings for prediction and explanation in knowledge graphs. In: ACM International Conference on Web Search and Data Mining (WSDM), pp. 96–104 (2019)

The Digitalization of Bioassays in the Open Research Knowledge Graph

Jennifer D'Souza[1]([✉]) [iD], Anita Monteverdi[2], Muhammad Haris[3] [iD],
Marco Anteghini[4] [iD], Kheir Eddine Farfar[1] [iD], Markus Stocker[1] [iD],
Vitor A.P. Martins dos Santos[4] [iD], and Sören Auer[1,3] [iD]

[1] TIB Leibniz Information Centre for Science and Technology, Hannover, Germany
{jennifer.dsouza,kheir.farfar,markus.stocker,auer}@tib.eu
[2] Brain Connectivity Center, IRCCS Mondino Foundation, 27100 Pavia, Italy
anita.monteverdi01@universitadipavia.it
[3] L3S Research Center, Leibniz University Hannover, Hanover, Germany
haris@l3s.de
[4] Lifeglimmer GmbH, Markelstr. 38, 12163 Berlin, Germany
{anteghini,vds}@lifeglimmer.com

Abstract. *Background:* Recent years are seeing a growing impetus in the semantification of scholarly knowledge at the fine-grained level of scientific entities in knowledge graphs. The Open Research Knowledge Graph (ORKG, orkg.org) represents an important step in this direction, with thousands of *scholarly contributions* as structured, fine-grained, machine-readable data. There is a need, however, to engender change in traditional community practices of recording contributions as unstructured, non-machine-readable text. For this in turn, there is a strong need for AI tools designed for scientists that permit easy and accurate semantification of their scholarly contributions. We present one such tool, ORKG-ASSAYS. *Implementation:* ORKG-ASSAYS is a freely available AI micro-service in ORKG written in Python designed to assist scientists obtain semantified bioassays as a set of triples. It uses an AI-based clustering algorithm which on gold-standard evaluations over 900 bioassays with 5,514 unique property-value pairs for 103 predicates shows competitive performance. *Results and Discussion:* As a result, semantified assay collections can be surveyed on the ORKG platform via tabulation or chart-based visualizations of key property values of the chemicals and compounds offering smart knowledge access to biochemists and pharmaceutical researchers in the advancement of drug development.

Keywords: Open research knowledge graph · Scholarly digital library · Bioassays · K-means clustering · Artificial intelligence

Supported by TIB Leibniz Information Centre for Science and Technology, the EU H2020 ERC project ScienceGraph (GA ID: 819536) and the ITN PERICO (GA ID: 812968).

© The Author(s), under exclusive license to Springer Nature Switzerland AG 2022
C. Strauss et al. (Eds.): DEXA 2022, LNCS 13426, pp. 63–68, 2022.
https://doi.org/10.1007/978-3-031-12423-5_5

1   Introduction

The Open Research Knowledge Graph (ORKG) [3] digital library addresses scholarly content digitalization as a distributed, decentralized, and collaborative scholarly knowledge creation process that can be powered with automated semantification modules via a continuous, ongoing development cycle of autonomously maintained AI micro-services. To this end, this paper presents ORKG-ASSAYS an AI-based semantification micro-service trained on structured data based on the Bioassay ontology (BAO), and fitted in the ORKG for the rapid assimiliation of digitalized biological assays (bioassays). While ORKG-ASSAYS will be the first Life Science domain supported by an automated semantification micro-service in the ORKG, to our knowledge, it fosters the development of the first end-to-end bioassay digitalization workflow in the overall scholarly community as well.

The ORKG-ASSAYS micro-service workflow involves four steps. **1)** Querying a bioassay depositor for their unstructured or semi-structured assays. Commonly, bioassays raw data are obtained via the PubChem depository [12] – a major depositor of bioassays from various research institutes. **2)** Semantifying the assay via the ORKG-ASSAYS AI clustering model. **3)** Linking the depositor-provided assay cross-references to their scientific articles. And, **4)** integrating the bioassay semantic graph in the ORKG. Programmed in Python, ORKG-ASSAYS provides web-based and programmatic tools for semantifying bioassay texts. The semantified bioassay once entered in the ORKG is *editable* via user-friendly frontend interfaces, is *surveyable* via tabulations [11] or 2-D chart visualizations, and is *queryable* for various scientific semantic ORKG relationships. The ORKG-ASSAYS AI clustering method demonstrates high semantification performance F1 scores above 80% and has been chosen after diverse methodological tests including the state-of-the-art, bidirectional transformer-based SciBERT model discussed in prior work [1].

Summing up, ORKG-ASSAYS offers a highly accurate and pragmatic semantification model alleviating unrealistic expectations on scientists to semantify their bioassays from scratch, by instead offering them a mere curatorial role of the automatic annotations. The pace with which novel bioassays are being submitted suggests that we have only begun to explore the scope of possible assay formats and technologies to interrogate complex biological systems. Thus this data domain, specifically, promises long-standing future application discovery many of which remain potentially untapped. Furthermore, inspired by the method we demonstrate, by drastically reducing the time required to semantify data for other scholarly domains as well, digitalization can be realistically advocated to become a standard part of the publication process.

2   Bioassay Digitalization in the ORKG

ORKG-ASSAYS will now be discussed as its implementation w.r.t. the KG Lifecycle requirements consisting of the graph creation, hosting, curation, and deployment modules. The ORKG-ASSAYS micro-service belongs in an early stage of graph creation, i.e. when generating the graph itself. Thus, while the graph creation module handling the normalization of variously formatted graph data is beyond the scope of ORKG-ASSAYS, it addresses extracting the assay texts from heterogeneous bioassay depositories each with different file formats, generating a BAO-based structured graph. The end-to-end ORKG-ASSAYS semantification pipeline in a micro-service is discussed below.

Data Preparation. This step relies on public access availability to an assay depository's querying mechanism. PubChem, reported to have over 1 million assays [8], is queryable via its public REST API for its bioassays where some assays have depositor-provided cross-references to scientific articles in PubMed. Depending on the depositor, the data could be returned in JSON, XML, or CSV. We implemented a specific pipeline for "The Scripps Research Molecular Screening Center" which returned JSON query responses. It reported nearly 1,600 bioassays. However, to prepare the data, the bioassay description-specific sections had to be located in its JSON response file and the text then extracted. The text was merged from two separate parts, viz. assay overview and assay protocol summary. We noted that this parsing heuristic can be applied to most depositor responses, although there maybe some exceptions.

Automated Clustering-Based Semantification. Traditionally, AI-based scholarly KG construction is addressed by the recognition of entities and relations in scientific articles as sequence labeling and classification objectives [5–7,9,10]. We instead address the problem of bioassay semantification with a clustering objective. We choose clustering from our corpus observations that bioassays with similar text descriptions are semantified with similar sets of logical statements. Thus, the bioassays could be clustered based on their text descriptions and each cluster could be collectively semantified by the labels of the trained cluster. Indeed while entity and relation classification are sound strategies, they would be unnecessarily more complex and time-consuming methods for the problem at hand. We refer the reader to our prior work [2] which contrasts a classification versus a clustering objective for bioassay semantification.

The final semantification function in ORKG-ASSAYS was arrived at by an experimental process. This entailed testing two different vectorizations, i.e. TF-IDF and SciBERT [4], for the bioassay text to find the optimal representation for clustering by K-means with the elbow optimization strategy to find the best K value. While the TF-IDF vector is fitted on a training collection of assays, the SciBERT embeddings are directly queried for their pretrained 768 dimensional vectors. The results are shown in Table 1. We see that the direct TF-IDF vectorization on bioassay text outperforms the scholarly-articles-based pretrained SciBERT at 0.83 $F1$ vs. 0.77 $F1$ with fewer clusters (450 vs. 550).

Building the Knowledge Graph. We leverage the ORKG to convert our structured annotations to a KG. The assay's article's PubMED metadata is first fetched, following which the digitalized bioassay is added in the form of research contributions of the paper via the ORKG KG building functions.

Table 1. Semantification results by K-means clustering of vectorized bioassays

# Clusters (K)	TF-IDF			SciBERT		
	P	R	$F1$	P	R	$F1$
400	0.80	0.85	0.82	0.72	0.79	0.75
450	0.81	0.85	**0.83**	0.74	0.79	0.76
500	0.82	0.85	0.83	0.75	0.78	0.76
550	0.82	0.84	0.83	0.75	0.78	**0.77**
600	0.83	0.84	0.83	0.77	0.78	0.77

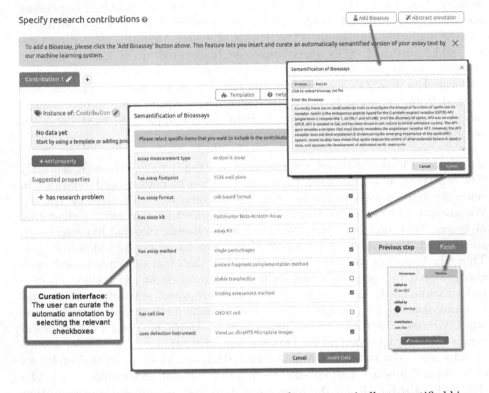

Fig. 1. ORKG frontend screens for user curation of an automatically semantified bioassay.

Data Workflows. **1. Add Paper Wizard.** In the ORKG Frontend, as shown in Fig. 1, the user can add an assay by clicking the 'Add Bioassay' button. The assay gets automatically semantified with the result on a screen with checkboxes enabling accept or reject user interactions. On clicking 'Insert Data,' all selected statements and the user provenance form the ORKG. **2. Bulk Import via REST API.** To ingest the data in bulk, iterative calls to the ORKG REST API with article metadata and structured bioassay as contributions encapsulated in a JSON object can be made. This process is depicted in Figs. 2 and 3.

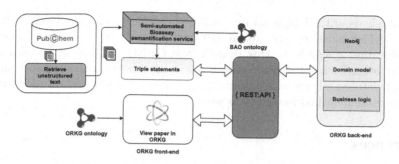

Fig. 2. End-to-end ORKG-ASSAYS semantification pipeline which practically realizes the digitalization of digitized data involving data sources, data retrieval, an annotation service, and resulting triple statements.

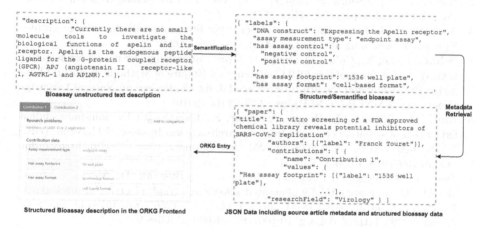

Fig. 3. Conversion of an unstructured Bioassay to its equivalent digitalized representation and finally presented in the ORKG frontend (https://tinyurl.com/orkg-assay).

3   Conclusion

We presented ORKG-ASSAYS—an end-to-end digitalization workflow of unstructured descriptions of bioassays within a next-generation digital library, the ORKG. Its supplementary information is released online https://github.com/jd-coderepos/bioassays-ie. The hybrid design of ORKG-ASSAYS complementarily integrates automated and manual semantification methods since pure machine learning on its own tends to be insufficiently accurate and expecting scientists to find the time to semantify their assays from scratch is unrealistic.

Bioassays being highly diverse are clearly a non-trivial semantification domain posing challenges to standardizing and integrating the data with the goal to maximize their scientific and ultimately their public health impact as the assay screening results are carried forward into drug development programs with intelligent machine assistance. The current coronavirus pandemic situation sheds critical light on advancing the drug development research life-cycle for which bioassays are crucial, offering credence to our domain choice for semantification research. In this respect, the ORKG will not serve as a mere mirror of other Bioassay depositories, but will itself be a unique application of a highly-structured science-wide knowledge graph of scholarly contributions which incoporates semantified bioassays as well.

References

1. Anteghini, M., D'Souza, J., Dos Santos, V.A.M., Auer, S.: Scibert-based seman-tification of bioassays in the open research knowledge graph. In: EKAW-PD 2020, pp. 22–30 (2020)
2. Anteghini, M., D'Souza, J., Santos, V.A., Auer, S.: Easy semantification of bioas-says (2021). arXiv preprint arXiv:2111.15182
3. Auer, S., et al.: Improving access to scientific literature with knowledge graphs. Bibliothek Forschung und Praxis **44**(3), 516–529 (2020)
4. Beltagy, I., Lo, K., Cohan, A.: Scibert: A pretrained language model for scientific text. In: Proceedings of the 2019 Conference on Empirical Methods in Natural Lan-guage Processing and the 9th International Joint Conference on Natural Language Processing (EMNLP-IJCNLP). pp. 3606–3611 (2019)
5. Brack, A., D'Souza, J., Hoppe, A., Auer, S., Ewerth, R.: Domain-independent extraction of scientific concepts from research articles. In: Jose, J.M., et al. (eds.) ECIR 2020. LNCS, vol. 12035, pp. 251–266. Springer, Cham (2020). https://doi.org/10.1007/978-3-030-45439-5_17
6. Dessì, D., Osborne, F., Reforgiato Recupero, D., Buscaldi, D., Motta, E., Sack, H.: AI-KG: an automatically generated knowledge graph of artificial intelligence. In: Pan, J.Z., et al. (eds.) ISWC 2020. LNCS, vol. 12507, pp. 127–143. Springer, Cham (2020). https://doi.org/10.1007/978-3-030-62466-8_9
7. D'Souza, J., Auer, S., Pedersen, T.: SemEval-2021 Task 11: NLPContribution-Graph - structuring scholarly nlp contributions for a research knowledge graph. In: Proceedings of the 15th SemEval-2021, pp. 364–376. ACL, August 2021
8. Kim, S., et al.: Literature information in pubchem: associations between pubchem records and scientific articles. J. Cheminformatics **8**(1), 1–15 (2016)
9. Liu, H., Sarol, M.J., Kilicoglu, H.: UIUC_BioNLP at SemEval-2021 task 11: A cas-cade of neural models for structuring scholarly NLP contributions. In: Proceedings of the 15th SemEval-2021, pp. 377–386. ACL, August 2021
10. Luan, Y., He, L., Ostendorf, M., Hajishirzi, H.: Multi-task identification of entities, relations, and coreference for scientific knowledge graph construction. In: Proceed-ings of the 2018 EMNLP, pp. 3219–3232. ACL, October–November 2018
11. Oelen, A., Jaradeh, M.Y., Farfar, K.E., Stocker, M., Auer, S.: Comparing research contributions in a scholarly knowledge graph. In: CEUR Workshop Proceedings, vol. 2526, pp. 21–26. RWTH, Aachen (2019)
12. Wang, Y., et al.: Pubchem's bioassay database. Nucleic Acids Res. **40**(D1), D400–D412 (2012)

Privacy-Preservation Approaches

Privacy Issues in Smart Grid Data: From Energy Disaggregation to Disclosure Risk

Kayode Sakariyah Adewole[1,2]([✉]) [ID] and Vicenç Torra[1] [ID]

[1] Department of Computing Science, Umeå University, Umeå, Sweden
{kadewole,vtorra}@cs.umu.se, adewole.ks@unilorin.edu.ng
[2] Department of Computer Science, University of Ilorin, Ilorin, Nigeria

Abstract. The advancement in artificial intelligence (AI) techniques has given rise to the success rate recorded in the field of Non-Intrusive Load Monitoring (NILM). The development of robust AI and machine learning algorithms based on deep learning architecture has enabled accurate extraction of individual appliance load signature from aggregated energy data. However, the success rate of NILM algorithm in disaggregating individual appliance load signature in smart grid data violates the privacy of the individual household lifestyle. This paper investigates the performance of Sequence-to-Sequence (Seq2Seq) deep learning NILM algorithm in predicting the load signature of appliances. Furthermore, we define a new notion of disclosure risk to understand the risk associated with individual appliances in aggregated signals. Two publicly available energy disaggregation datasets have been considered. We simulate three inference attack scenarios to better ascertain the risk of publishing raw energy data. In addition, we investigate three activation extraction methods for appliance event detection. The results show that the disclosure risk associated with releasing smart grid data in their original form is on the high side. Therefore, future privacy protection mechanisms should devise efficient methods to reduce this risk.

Keywords: Smart grid data · Non-intrusive load monitoring · Energy disaggregation · Data privacy · Disclosure risk

1 Introduction

The significant development in artificial intelligence, Internet-of-things, smart meter and smart grid solutions have contributed to the realization of smart sustainable cities [1]. Part of the goals of sustainable cities include appropriate use of available resources, energy conservation and improving the well-being of the societies [2]. Energy conservation focuses on efficient use of energy resources to achieve sustainability and self-reliance in energy management. Energy conservation requires monitoring and controlling of energy usage with the aim of optimizing energy demand to reduce energy consumption [3]. There have been immense research efforts in developing methodologies to address energy demands

© The Author(s), under exclusive license to Springer Nature Switzerland AG 2022
C. Strauss et al. (Eds.): DEXA 2022, LNCS 13426, pp. 71–84, 2022.
https://doi.org/10.1007/978-3-031-12423-5_6

[2, 4]. Hence, communicating with consumers to provide demand-response services based on their fine-grained energy consumption can help in reducing energy wastage. One methodology to achieve this goal is Non-intrusive load monitoring (NILM).

NILM or energy disaggregation is the task of separating a building's aggregated load consumption into constituent energy demands by the individual appliances [5]. NILM provides a smart solution to the problem of electrical energy monitoring of households at the appliance level. This method guarantees a real-time feedback by analyzing the aggregated power data measured by smart meter and extracting the information about the consumption of individual electrical devices. The goal of NILM is to provide real-time feedback about energy consumption to consumers, to detect faults and events, and to encourage energy-saving behaviours [6].

Majorly, NILM research domain involves the development of classification and regression algorithms to predict appliance state and load consumption respectively. The classification results of the NILM algorithms help to ascertain when a particular appliance is in use during the day. Several techniques have been studied to extract appliance activations from appliance data for the purpose of developing classification models. For instance, Laviron et al. [7] proposed three activation extraction techniques: Cartesio, ValmA and SimBA, to extract appliance signatures. Kelly et al. [8] proposed activation time extraction (ATE) algorithm which was specifically tuned on UK-DALE NILM dataset. Desai et al. [9] proposed Variance-Sensitive Thresholding (VST), which aims to improve over the Middle-Point Thresholding (MPT) method [6]. Event data obtained after applying a particular activation extraction method can be used to build a machine learning classifier. Regression models on the other hand take an aggregated household consumption data and produce the individual appliance load signatures [5, 8, 10]. Both classification and regression NILM models have shown significant performance over the years.

However, despite the benefits offered by NILM system, the ability to infer individual load signature from an aggregated consumption has posed a privacy issue in the domain of smart grid data publishing. Fine-grained electricity consumption data has been characterized with privacy-sensitive consumer behaviours, which are capable of revealing general habits and lifestyles of households [11, 12]. Information obtained through appliance-level inferencing and analysis is useful to third parties like marketers, law enforcement, and criminals. For instance, cases of attacks on smart grid infrastructure which lead to electricity blackout have been reported in Ukraine in 2015, 2016 and January 2017 where hackers were able to shutdown the energy system that supply heat and light to millions of households. This may have occurred as a result of privacy violation and security breach that may emanate from a specific household [13, 14].

In addition, most of the existing smart grid datasets in NILM domain are accompanied with meta-data, which provide some background information about the data collection procedures. These information can be explored by attackers as an external knowledge and along with the knowledge inferred from NILM to reveal the identity of individual households. Consequently, sharing of fine-grained electricity usage data in its original form has higher probability of revealing individual household lifestyles as attackers may want to deduce the type of appliance that is in use at any given time. Therefore, it is important to hide the individual appliance signatures from the protected datasets to be published by the utility company. This requirement creates a major challenge to existing privacy protection mechanisms for smart grid data. Existing protection mechanisms for smart grid data, such as data anonymization [13,15–19] and differential privacy [18,20] pay less attention to obfuscation of individual appliance signature in the aggregated masked data.

In this paper, we evaluate the disclosure risk associated with publishing raw smart grid data. We adapt Seq2Seq deep learning NILM algorithm to detect the signature of individual appliance in the aggregated signal. Additionally, we subject the individual predictive appliance load to three activation methods: ATE, MPT and VST to confirm the efficacy of the algorithm in correctly predicting the state of each appliance from their load signatures. Furthermore, we simulate three inference attack scenarios and empirically compute the disclosure risk probability for individual appliances. This helps us to concretize a new risk measure that is termed disaggregation risk. Therefore, our paper makes the first attempt to reveal this type of disclosure risk. Two publicly available NILM datasets for energy disaggregation have been considered. The findings from this study provide new insight for future privacy preserving mechanisms in the domain of smart grid data publishing.

The remaining parts of this paper are organized as follows: Sect. 2 presents related work in energy disaggregation and privacy preserving mechanisms for smart grid data. Section 3 focuses on the proposed approach for disclosure risk assessment. Section 4 highlights the experimental settings for the different attack scenarios. Section 5 discusses the results obtained from the different experiments and finally Sect. 6 concludes the paper and highlights feature research directions.

2 Related Work

2.1 Non-intrusive Load Monitoring

The field of NILM started with the noticeable work of [21], which centered on clustering analysis for appliance identification and load disaggregation based on the steady state and transient state features extracted from the aggregated energy. The goal of NILM is to monitor events and load consumed at appliance level from aggregated energy using a NILM device and a single smart meter. Due to the challenges of sub-metering every appliances in a building for load monitoring and demand-response services, NILM device can process the measured aggregated energy from the smart meter for event detection and energy disaggregation of individual appliances.

Recently, deep learning architectures have been proposed which allows an automatic extraction of salient features from the aggregated power signal. Deep learning algorithms have shown comparable performance over the algorithms that relied on the conventional feature extraction methods [8, 22]. Kelly et al. [8] adapted deep learning architectures such as Recurrent Neural Networks (RNNs) and Denoising Autoencoders (dAE) to extract appliance level energy consumption from aggregated signal. dAE was used to reconstruct the target appliance signal while treating the aggregated signal as noisy input. These algorithms have been shown to outperform the earlier NILM algorithms that relied on the conventional feature extraction methods such as Combinatorial Optimization (CO) and Factorial Hidden Markov Model (FHMM) on significant number of appliances. Long short term memory (LSTM) deep learning architecture have also been experimented for appliance signal reconstruction [2]. Sequence-to-Sequence (Seq2Seq) and Sequence-to-point (Seq2Point) deep learning architectures have been studied in [8, 10]. There have been extensive progress in the field of deep learning for NILM domain. The reader is referred to a recent review published in [2]. Nevertheless, despite the progress recorded in NILM domain, the ability to disaggregate the signature of individual appliances from the aggregated signal has been characterized with privacy issues. This reveals the lifestyle of the individual households in the smart grid data.

2.2   Privacy Preserving Data Publishing in Smart Grid

In the domain of privacy preserving smart grid data publishing, several techniques have been studied ranging from data anonymization using Battery-based Load Hiding (BLH) [15], data anonymization using k-anonymity [16, 17, 19], Generative Adversarial Network (GAN) and additive correlated noise [23, 24] and differential privacy (DP) [18, 20] among others.

BLH aims to install a rechargeable battery at the consumer end, which can be charged or discharged to make the electricity meter incapable of precisely obtaining the consumption data of electric appliances while also hiding the appliance actual energy consumption. This masking method is mainly theoretic and its empirical validation for real-world application is still a major concern [15]. K-anonymity [25] is a condition that protected data need to satisfy to guarantee the privacy of the individual in the masked data. The goal of k-anonymity is to ensure that each individual in a protected data cannot be identified within a set of k individuals. [16, 17] adopted k-ward microaggregation algorithm, which is one of the algorithms that satisfied k-anonymity to protect smart grid and building occupancy data. Thouvenot et al. [19] investigated the performance of microaggregation algorithm for time series data. Other privacy protection methods, such as random noise, data permutation, data transformation, time slicing, differential privacy and scope aggregation were discussed.

One of the benefits of GAN is its ability to model the uncertainties of original data and based on this model new data are generated. Two deep neural networks are usually trained; one to capture the distribution of the data and the other to estimate the probability that the input originates from the real data. This approach is promising to protect energy consumption data, however, its capability to prevent disclosure risk attacks is missing in the literature. Differential privacy has become a de-facto mechanism for privacy preserving which guaranteed $\epsilon-DP$ for every individual record in the protected data. [18,20] proposed DP mechanisms to protect smart grid data. However, none of the existing studies in the literature has attempted to reduce the disclosure risk that may occur due to inference attack on aggregated signals. These studies made an implicit assumption regarding the protection method used and therefore did not focus on testing the possibility of inferring useful knowledge which could assist attackers in linking individual households in the protected data.

3   Proposed Method for Disclosure Risk Assessment

Figure 1 shows the proposed framework for disclosure risk assessment. The framework consists of several stages with the goal of assessing the extent to which Seq2Seq disaggregation algorithm can predict the status and load signatures of individual appliances. Aggregated signal and individual appliance load data were used to train Seq2Seq algorithm. The trained model was applied to disaggregate unseen aggregated signal. The disaggregated individual appliance loads were subjected to three activation extraction methods. These methods were used to ascertain the efficacy of the NILM algorithm in detecting both the state (ON/OFF) and load signature of individual appliance. This approach provides three possible advantages. First, it allows us to investigate the performance of NILM algorithm for load disaggregation and event detection. Second, it helps us to quantify a disclosure risk measure which reveals appliance-level disaggregation risk particularly for smart grid data. Lastly, the disaggregated appliance loads can be used by future privacy preserving model to develop efficient mechanisms that can hide each appliance signature taking into consideration utility-privacy trade-off.

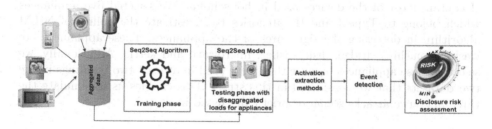

Fig. 1. Proposed framework for disclosure risk assessment

NILM system takes an aggregated energy consumption and predicts the load of each appliance. The aggregated power P_t at time t is the sum over all appliances loads as given in Eq. (1):

$$P_t = \sum_{\ell=1}^{L} P_t^{(\ell)} + \epsilon_t \qquad (1)$$

where L is the total number of appliances in the building, $P_t^{(\ell)}$ is the power of appliance ℓ at time t, and ϵ_t is the unidentified residual load. This residual load is usually characterized as noise. There are two types of noise in this case: structured and unstructured noise. The noise introduced at the appliance sub-meter is called structured noise which comes from appliances that were not sub-metered. The noise at the mains meter (i.e. aggregate meter) is called unstructured noise. To predict appliance loads from the aggregated signal, this study selected Seq2Seq deep learning algorithm based on its performance as reported in the literature [5,10]. The deep learning model has the tendency of extracting internal representational features of the signature of each appliance from the aggregated data.

To simulate the disclosure risk attacks, this paper considered three inference attack scenarios. The first scenario is an inference attack simulation on the same household in the same dataset. Second scenario is an inference attack on different households in the same dataset and lastly, the third scenario is an inference attack on different households in different datasets.

3.1 Appliance Selection

This study focused on two categories of appliances: Type I and Type II. An appliance with two states of operation (ON/OFF) is categorized as Type I. These include appliances such as kettle, toaster, light bulb, lamps, microwave etc. They consume energy only during the ON state. Multi-state appliances or finite state machines are Type II appliances that have finite number of operating states which may be executed repeatedly. State transitions can be detected using rising/falling edges of power consumption over a period of time. Appliances such as refrigerator, stove burner, dish washer and washing machine are the popular examples for multi-state/Type II appliances. The categories Type I and Type II contain most of the devices used in households. We selected five appliances which belong to Type I and II categories to investigate the ability of NILM algorithm in detecting the signatures of the appliances. These appliances are washing machine, fridge, dish washer, microwave and kettle. Additionally, we selected these appliances because they were used in at least two buildings in the two datasets that we considered in this study. This enables us to simulate the three inference attacks scenarios briefly discussed in Sect. 3.

3.2 Seq2Seq Disaggregation Algorithm

Seq2Seq NILM algorithm [8,10] is based on a deep learning architecture with different Convolutional Neural Networks (CNNs) layers. The algorithm maps aggregated input sequence to its corresponding target appliance sequence. Suppose F_s is a neural network that maps the input sequence with sliding windows $Y_{t:t+W-1}$ corresponding to the aggregated mains power to the corresponding windows $X_{t:t+W-1}$ of the target appliance power load sequence. The regression is then defined as $X_{t:t+W-1} = F_s(Y_{t:t+W-1}, \theta_s) + \epsilon$, where ϵ is W-dimensional

Gaussian random noise and θ_s are the parameters of the neural network F_s. Figure 2 shows the adapted Seq2Seq learning architecture in this study. It is important to mention that between CNN layer 4 and 5, there is a Dropout layer with dropout probability of 0.2. This was also used between CNN layer 5 and the fully connected layer and between the fully connected layer and the output layer. The output from CNN layer 5 was flatten before the fully connected layer was applied.

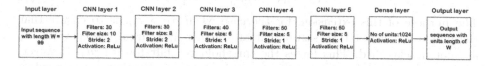

Fig. 2. Seq2Seq NILM deep learning architecture

3.3 Event Detection

After training a Seq2Seq algorithm to output the target appliance load, we subjected this result to event detection algorithms to extract the appliance state signature. A majority of event detection algorithms in the literature are threshold-based. They define threshold values to determine when a typical appliance is switched ON or OFF. In this study, we employed Middle-Point Thresholding (MPT), Variance-Sensitive Thresholding (VST) and Activation Time Extraction (ATE) for this purpose. Given the ON power threshold $\lambda^{(\ell)}$ for each appliance that is obtained from a particular threshold method, the ON state of the appliance ℓ at time t can be obtained as.

$$s_t^{(\ell)} = I(P_t^{(\ell)} \geq \lambda^{(\ell)}) \tag{2}$$

where $s_t^{(\ell)}$ is the ON state of appliance ℓ at time t if the condition is satisfied. I is a function that maps the boolean result to ON or OFF state of appliance ℓ. $P_t^{(\ell)}$ is the power of appliance ℓ at time t. The three threshold methods are briefly discussed in the subsequent sections.

3.4 Middle-Point Thresholding

Given the power consumption of the training data from individual appliance, the MPT method [6] applies a clustering technique to split the training data into two clusters and considers the centroid of each cluster. In this study, we applied K-means algorithm for this purpose. The two centroids from the clustering are denoted as $m_0^{(\ell)}$ for OFF state and $m_1^{(\ell)}$ for ON state. Therefore, the event detection threshold for MPT is fixed between these two values and it is given as,

$$\lambda^{(\ell)} = \frac{m_0^{(\ell)} + m_1^{(\ell)}}{2} \tag{3}$$

3.5   Variance-Sensitive Thresholding

Variance-Sensitive Thresholding (VST) extends MPT approach by incorporating standard deviation $\sigma_k^{(\ell)}$ for the data points in each cluster such that,

$$d = \frac{\sigma_0^{(\ell)}}{\sigma_0^{(\ell)} + \sigma_1^{(\ell)}}$$

$$\lambda^{(\ell)} = (1 - d)m_0^{(\ell)} + dm_1^{(\ell)} \tag{4}$$

The purpose of this extension, for the case when $\sigma_1 > \sigma_0$, is to ensure that the points in Class 1 (ON state) that are farther away from the centroid m_1 are not misclassified. This forces the threshold to shift toward m_0. However, when $\sigma_0 = \sigma_1$, VST becomes MPT approach. Since substituting σ_0 for σ_1 when calculating d will cause the value of d to be 0.5, hence, Eq. (4) becomes Eq. (3).

3.6   Activation Time Extraction

MPT and VST approaches only consider data from the distribution of power measurements to fix the threshold for a specific appliance. It often happens that due to noise in the smart meters, some measurements during short time intervals are either absent while the device is operating, or produce abnormal peaks during the OFF state (especially for multi-state appliances). Based on this, [8] proposed the ATE algorithm, which was tuned specifically for UK-DALE dataset. ATE considered both power threshold and time threshold as defined for each appliance in [8].

3.7   Disaggregation Risk

We define a new measure of disclosure risk which is termed *disaggregation risk*. This disclosure risk is particularly associated with smart grid data and we claim that privacy preserving mechanisms should minimize this risk for smart grid data publishing.

Definition 1 *(Disaggregation risk). We define disaggregation risk as the probability that NILM algorithm predicts the load signature of appliance ℓ and its corresponding ON events from the aggregated signal.*

We formalize *disaggregation risk* using Eq. (5).

$$DR^{(\ell)} = TP^{(\ell)}/(TP^{(\ell)} + FN^{(\ell)}) \tag{5}$$

where $TP^{(\ell)}$ is the number of correctly predicted ON state of appliance ℓ, $FN^{(\ell)}$ is the number of ON state of appliance ℓ that was mistakenly predicted as OFF state and $DR^{(\ell)}$ is the disaggregation risk for appliance ℓ which takes a value in the interval [0,1]. The higher the value of DR, the higher the disclosure risk.

Equation (5) fits conveniently in our case of disclosure risk as NILM energy disaggregation is characterized as being a highly class imbalance problem. Since we are interested in whether or not an attacker can successfully predict the operating period of the appliance of interest, we formalized this disclosure risk to evaluate the performance of Seq2Seq algorithm based on this requirement.

4   Experimental Setup

In this study, all experiments have been conducted using Python programming language. We used NILMTK and NILMTK-Contrib API [5] for energy disaggregation. Experiments were performed on a Dell Laptop with GeForce GTX 1050 Ti with Max-Q Design GPU, CUDA version 11.2 and Intel(R) Core^{TM} i9-8950H CPU @2.90 GHz 1TB HDD 32 GB RAM. Batch size of 32 and 50 epochs were used for Seq2Seq algorithm. Both aggregated and appliance data were resampled to a period of 1 min (60 s). Two widely used NILM datasets: UK-DALE and REFIT were investigated.

4.1   Datasets

As stated in Sect. 4, this study considered two publicly available and widely used NILM datasets: UK-DALE [26] and REFIT [27]. UK-DALE is made up of 5 households. Each appliance sub-meters in UK-DALE recorded aggregate apparent mains power sampled at every 6 s. Household 1, 2 and 5 also recorded both active and reactive mains power every second. REFIT dataset has a total of 20 households with both aggregate and appliance level data sampled at 8 s intervals. As stated in Sect. 4, both aggregated and appliance data were resampled to a period of 1min during training and testing of the model. We applied the same sampling frequency when learning the threshold values for MPT and VST methods from the appliance data. This is to ensure that both mains and appliance-level signals are properly aligned.

4.2   Training and Testing Period

To simulate the three inference attacks scenarios briefly discussed in Sect. 3, three different experimental settings have been adopted. For the first inference attack scenario, household 2 with active mains and appliance power were used. The training period is between 20/05/2013 and 20/09/2013 while the testing period is between 21/09/2013 and 10/10/2013. Similarly, household 2 in REFIT dataset was used based on active mains and appliance power. The training period is between 17/09/2013 and 17/01/2014 while the testing period is between 01/03/2014 and 01/04/2014. For the second inference attack scenario, household 2 was used to train the model and the model was tested on household 1. The training period is between 20/05/2013 and 20/09/2013 while the testing period is between 21/09/2013 and 10/10/2013 for UK-DALE. For REFIT dataset, household 2 was used for training the model and the model was tested on household 5. The training period is between 17/09/2013 and 17/01/2014 while the testing period is between 01/03/2014 and 01/04/2014. Lastly, for the third inference attack scenario, a model was trained with household 2 in UK-DALE dataset and tested on household 2 in REFIT dataset. The training period is between 20/05/2013 and 20/09/2013 while the testing period is between 01/03/2014 and 01/04/2014.

4.3   Threshold Computation

As stated in Sect. 3.3, ATE threshold was fixed by [8] for UK-DALE dataset based on a thorough investigation of appliance energy consumption rate. In this study, we retained these settings for both UK-DALE and REFIT for ATE method as we needed to test the efficacy of this approach across different datasets. Therefore, for MPT and VST methods, we computed the ON power threshold values from the individual appliance data. Table 1 shows the detail of these threshold for the two datasets.

Table 1. ON power threshold in Watt computed for the two datasets

Appliance	UK-DALE			REFIT		
	ATE	MPT	VST	ATE	MPT	VST
Washing machine	20.0	864.49	219.8	20.0	1028.20	592.29
Fridge	50.0	47.85	18.73	50.0	44.73	3.85
Dish washer	10.0	1054.8	146.22	10.0	1100.99	669.62
Microwave	200.0	562.56	72.30	200.0	555.67	67.51
Kettle	2000.0	1059.66	117.34	2000.0	1359.92	241.59

5   Results and Discussion

This section presents the disclosure risk results associated with each appliance. As stated in Sect. 3, three inference attack scenarios were considered in this study. The findings from this study will help future research in privacy preserving smart grid data publishing to prevent inference attack on smart grid data, which may reveal some background information about individual household lifestyles. The results presented show the efficacy of Seq2Seq disaggregation algorithm in predicting the appliance load signature and the status of each appliance at the different timestamps.

5.1   Inference Attack on the Same Building

Table 2 shows that the disclosure risk associated with individual appliance is on the high side as Seq2Seq algorithm was able to predict the signature of the devices. The disaggregation risk for each threshold method is very close, which shows that these methods are capable of revealing the ON power event of the individual appliance. However, the algorithms faced a challenge to accurately predict the positive event in the case of microwave on REFIT dataset for this scenario. Nevertheless, the probability of disaggregating individual appliance by NILM algorithm when attacker has a trained model on the same household is particularly on the high side.

Table 2. Disaggregation risk for each appliance in the two datasets based on attack scenario 1

Appliance	Disaggregation risk - UK-DALE			Disaggregation risk - REFIT		
	ATE	MPT	VST	ATE	MPT	VST
Washing machine	0.87	0.48	0.62	0.56	0.44	0.70
Fridge	0.96	0.96	0.99	0.75	0.80	0.99
Dish washer	0.98	0.98	1.00	0.93	0.64	0.81
Microwave	0.83	0.62	0.89	0.00	0.00	0.00
Kettle	0.90	0.95	0.99	0.41	0.55	0.72

5.2 Inference Attack on Different Buildings in the Same Dataset

The disclosure risk results of inference attack on different buildings in the same dataset are shown in Table 3. We observed that the disclosure risk of washing machine and microwave dropped when compared with the first scenario. This is as a result of the different sources of the training and testing data as regards the household consumption patterns. We observed that the results of the threshold methods are close. However, the probability of predicting the signature of the individual appliance is noticeable and this poses risk to the privacy of the individual households.

Table 3. Disaggregation risk for each appliance in the two datasets based on attack scenario 2

Appliance	Disaggregation risk - UK-DALE			Disaggregation risk - REFIT		
	ATE	MPT	VST	ATE	MPT	VST
Washing machine	0.26	0.13	0.15	0.41	0.01	0.10
Fridge	0.88	0.89	0.99	0.55	0.64	0.99
Dish washer	0.93	0.70	1.00	0.92	0.10	0.44
Microwave	0.40	0.19	0.57	0.00	0.00	0.03
Kettle	0.51	0.73	0.85	0.36	0.56	0.77

5.3 Inference Attack on Different Buildings from Different Datasets

Table 4 shows the disclosure risk associated with each appliance when Seq2Seq algorithm was trained on UK-DALE data and prediction was done with REFIT data. We observed high success rate (except in the case of washing machine) of the NILM algorithm in this scenario despite the difference in the datasets used for the training and testing. Surprisingly, microwave signature in REFIT was successfully disaggregated when a model trained on UK-DALE was used during the inference attack simulation. VST method successfully disaggregated microwave signature with 61% disclosure risk, which gives better performance than the other methods for this appliance. This behaviour shows how successful attacks can be launched by simply using a pre-trained model from one dataset to reveal the behaviour of appliances in another dataset. This result further confirms the ability of energy disaggregation algorithm in revealing the load consumption patterns of individual appliance. Again, this finding poses a privacy issue on smart grid data publishing. In this scenario, VST outperformed other threshold methods.

Table 4. Disaggregation risk for each appliance in the two datasets based on attack scenario 3

Appliance	Disaggregation risk - UK-DALE and REFIT		
	ATE	MPT	VST
Washing machine	0.12	0.01	0.04
Fridge	0.81	0.83	0.99
Dish washer	0.99	0.83	0.99
Microwave	0.44	0.22	0.61
Kettle	0.54	0.73	0.87

6   Conclusion

The results presented in this paper confirmed the ability of NILM algorithm to predict appliance loads signatures from an aggregated signal. The deep learning algorithm (Seq2Seq) and the threshold methods investigated have shown promising results in load disaggregation and event detection for the appliances. The success rates of the algorithms investigated in this study create challenges for future privacy preserving mechanism for smart grid data. The ability to hide appliance signatures from the aggregated signal would apparently improve the privacy of the smart grid data before publishing. Inference attack on energy data can be minimized by designing an effective privacy preserving method to counter energy disaggregation risk.

Acknowledgement. This work was partially supported by the Wallenberg AI, Autonomous Systems and Software Program (WASP) funded by the Knut and Alice Wallenberg Foundation. The first author is supported by the Kempe foundation.

References

1. Ibrahim, M., El-Zaart, A., Adams, C.: Smart sustainable cities roadmap: readiness for transformation towards urban sustainability. Sustain. Urban Areas **37**, 530–540 (2018)
2. Gopinath, R., Kumar, M., Joshua, C.P.C., Srinivas, K.: Energy management using non-intrusive load monitoring techniques-state-of-the-art and future research directions. Sustain. Urban Areas **62**(2020), 102411 (2020)
3. Janik, A., Ryszko, A., Szafraniec, M.: Scientific landscape of smart and sustainable cities literature: a bibliometric analysis. Sustainability **12**(3), 779 (2020)
4. Lin, X., Tian, Z., Lu, Y., Niu, J., Cao, Y.: An energy performance assessment method for district heating substations based on energy disaggregation. Energy Build. **255**, 111615 (2022)
5. Batra, N., et al.: Towards reproducible state-of-the-art energy disaggregation. In: Proceedings of the 6th ACM International Conference on Systems for Energy-Efficient Buildings, Cities, and Transportation, pp. 193–202. ACM (2019)
6. Precioso, D., Gomez-Ullate, D.: NILM as a regression versus classification problem: the importance of thresholding. arXiv preprint arXiv:2010.16050 (2020)
7. Laviron, P., Dai, X., Huquet, B., Palpanas, T.: Electricity demand activation extraction: from known to unknown signatures, using similarity search. In: Proceedings of the ACM International Conference on Future Energy Systems, e-Energy. ACM (2021)
8. Kelly, J., Knottenbelt, W.: Neural NILM: deep neural networks applied to energy disaggregation. In: Proceedings of the 2nd ACM International Conference on Embedded Systems for Energy-Efficient Built Environments, pp. 55–64. ACM (2015)
9. Desai, S., Alhadad, R., Mahmood, A., Chilamkurti, N., Rho, S.: Multi-state energy classifier to evaluate the performance of the NILM algorithm. Sensors **19**(23), 5236 (2019)
10. Zhang, C., Zhong, M., Wang, Z., Goddard, N., Sutton, C.: Sequence-to-point learning with neural networks for non-intrusive load monitoring. In: Proceedings of the AAAI Conference on Artificial Intelligence, vol. 32. AAAI (2018)
11. Mashima, D., Serikova, A., Cheng, Y., Chen, B.: Towards quantitative evaluation of privacy protection schemes for electricity usage data sharing. ICT Express **4**(1), 35–41 (2018)
12. Tudor, V., lmgren, M., Papatriantafilou, M.: A study on data de-pseudonymization in the smart grid. In: Proceedings of the Eighth European Workshop on System Security, pp. 1–6 (2015)
13. Armoogum, S., Bassoo, V.: Privacy of energy consumption data of a household in a smart grid. In: Yang, Q., Yang, T., Li, W. (eds.) Smart Power Distribution Systems, pp. 163–177. Academic Press (2019)
14. BBCNews, "Ukraine power cut 'was cyber-attack'" (2017). https://www.bbc.com/news/technology-38573074
15. Chin, J.-X., De Rubira, T.T., Hug, G.: Privacy-protecting energy management unit through model-distribution predictive control. IEEE Trans. Smart Grid **8**(6), 3084–3093 (2017)
16. Jia, R., Sangogboye, F.C., Hong, T., Spanos, C., Kjærgaard, M.B.: PAD: protecting anonymity in publishing building related datasets. In: Proceedings of the 4th ACM International Conference on Systems for Energy-Efficient Built Environments, pp. 1–10 (2017)

17. Sangogboye, F.C., Jia, R., Hong, T., Spanos, C., Kjærgaard, M.B.: A framework for privacy-preserving data publishing with enhanced utility for cyber-physical systems. ACM Trans. Sens. Netw. (TOSN) **14**(3–4), 1–22 (2018)
18. Soykan, E.U., Bilgin, Z., Ersoy, M.A., Tomur, E.: Differentially private deep learning for load forecasting on smart grid. In: 2019 IEEE Globecom Workshops (GC Wkshps), pp. 1–6. IEEE (2019)
19. Thouvenot, V., Nogues, D., Gouttas, C.: Data-driven anonymization process applied to time series. In: SIMBig, pp. 80–90 (2017)
20. Fioretto, F., Van Hentenryck, P.: Differential private stream processing of energy consumption. arXiv preprint arXiv: 1808.01949 (2018)
21. Hart, G.W., Kern Jr., E.C., Schweppe, F.C.: Non-intrusive appliance monitor apparatus, 15 August 1989. US Patent 4,858,141
22. Çimen, H., Bazmohammadi, N., Lashab, A., Terriche, Y., Vasquez, J.C., Guerrero, J.M.: An online energy management system for AC/DC residential microgrids supported by non-intrusive load monitoring. Appl. Energy **307**, 118136 (2022)
23. Feng, X., Lan, J., Peng, Z., Huang, Z., Guo, Q.: A novel privacy protection framework for power generation data based on generative adversarial networks. In: 2019 IEEE PES Asia-Pacific Power and Energy Engineering Conference (APPEEC), pp. 1–5. IEEE (2019)
24. Khwaja, A.S., Anpalagan, A., Naeem, M., Venkatesh, B.: Smart meter data obfuscation using correlated noise. IEEE Internet Things J. **7**(8), 7250–7264 (2020)
25. Samarati, P.: Protecting respondents identities in microdata release. IEEE Trans. Knowl. Data Eng. **13**(6), 1010–1027 (2001)
26. Kelly, J., Knottenbelt, W.: The UK-dale dataset, domestic appliance-level electricity demand and whole-house demand from five UK homes. Sci. Data **2**(1), 1–14 (2015)
27. Murray, D., Stankovic, L., Stankovic, V.: Refit: Electrical load measurements (cleaned) (2016). https://pureportal.strath.ac.uk/en/datasets/refit-electrical-load-measurements-cleaned

CoK: A Survey of Privacy Challenges in Relation to Data Meshes

Nikolai J. Podlesny$^{(\boxtimes)}$, Anne V. D. M. Kayem, and Christoph Meinel

Hasso Plattner Institute, University of Potsdam, Potsdam, Germany
{Nikolai.Podlesny,Anne.Kayem,Christoph.Meinel}@hpi.de

Abstract. The growing volumes of data that appear on multiple distributed platforms raise the question of how to compose data meshes that can be published and/or shared safely amongst multiple cooperating parties. Data meshes are composed of subsets (or whole sets) of data repositories that are owned by autonomous parties. This raises new challenges in terms of guaranteeing privacy across various data mesh compositions. In this paper, we present a survey of the issues that emerge in guaranteeing the privacy of distributed mesh data. We discuss the limitations of existing solutions in handling personal data privacy with respect to meshed data. Finally, we postulate that identifying personal data in such datasets must be handled with a performance efficient algorithm that can determine (on-the-fly), potential linkages across various data repositories, that could be exploited to subvert privacy.

1 Introduction

Dealing with mesh data from the privacy perspective is important in the IT industry. In fact, data meshes are in reality, a special case of distributed data repositories where the data exist in a flexible ecosystem but with clear user-ownership properties. Unlike standard relational database management systems, a central authority is absent and is instead replaced by separate authorities that co-exist in a "mutually exclusive and collectively exhaustive" environment. That is, data mesh instances can interact with each other and share data across different domains. For instance, an online marketing platform shares data with banking platforms and shopping regulatory services to validate a purchase request from a given customer. In essence, the goal is that there should be no centralised communication orchestrator required under this paradigm to guarantee data privacy across the different domains. While each database instance allows flexibility nuances, they adhere to overarching architecture principles and guarantees service level agreements to each other through data contracts (illustrated in Fig. 1). This paradigm of data meshes can be referred to as micro-service architecture in software engineering, where each service is encapsulated and isolated to allow more flexibility.

Problem Statement. Distributing private information and fragmenting their identifiers significantly impede their tracing and discovery. This may sound good in the first moment, but it exacerbates privacy work to protect the same. To

© The Author(s), under exclusive license to Springer Nature Switzerland AG 2022
C. Strauss et al. (Eds.): DEXA 2022, LNCS 13426, pp. 85–102, 2022.
https://doi.org/10.1007/978-3-031-12423-5_7

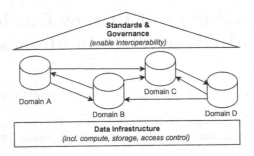

Fig. 1. Illustration of data meshes within an organisation

adhere to a high ethical standard and be compliant with most legislation like GDPR, HIPAA or CCPA, personally identifiable information (PII) even being distributed, must be protected, deleted upon request, and held secure. To do this, their existence and location must be known, even in a fragmented environment. Despite that individual data points might not initially be considered a privacy risk, their combination can be. Such attribute combinations are known as quasi-identifiers (QID). Traditional use cases of QID discovery imply static datasets with a standard relational database model, where standard metrics have to be addressed. In (data) mesh environments, there might be no, or only dynamically changing relational models. With the absence of any centralised layer that can identify, classify, label and alienate PII data records, differential privacy mechanisms by nature cannot help and a different solution is needed.

Contribution. In this work, we review, discuss and analyse the privacy implication of data mesh environments. We consolidate and systematise the state-of-the-art of related privacy work to do so. Based on this systematisation of knowledge (SoK), this work derives privacy fallacies in data mesh settings. Further, it discusses why practically the right of deletion, and other privacy actions are difficult to realise. We then offer experiments on implications for the search of privacy-compromising quasi-identifiers as vanishing points for de-anonymisation activities through comparing data mesh vs traditional RDBMS setups.

Outline. The rest of the paper is structured in the following manner: We assemble, consolidate and systematise latest related work in Sect. 2. This includes research on syntactic data anonymisation in Subsect. 2.3, semantic data anonymisation and differential privacy in Subsect. 2.4, unique column combinations in Subsect. 2.5, high-dimensional data anonymisation in Subsect. 2.6, quasi-identifier discovery in Subsect. 2.7, as well as data mesh databases in Subsect. 2.1 and privacy in data mesh environments in Subsect. 2.2. Section 3 then offers a characterisation of data meshes and quasi-identifiers in their context. Section 4 contributes experiments on discovering quasi-identifiers to avoid private data exposure in data mesh environments. Section 5 finally concludes our results and suggests avenues for future work.

2 State-of-the-Art

Data mesh databases are not a completely new research field, and have been addressed partially in the fields of peer-to-peer databases, distributed databases, data mesh topologies, syntactic-, semantic data anonymisation, high-dimensional data anonymisation and quasi-identifier discovery. The following subsections will summarise the most recent and extraordinary related work.

2.1 Data Mesh Databases

Back in 1997, Beall et al. reviewed systems for a general-purpose mesh database based on a hierarchy of topological entities [7]. Their hierarchical analysis for topology concluded that the hierarchic representation does not add a significant amount of extra storage to a mesh database. Rather, this representation can easily be extended to represent non-manifold models properly. In 2001, Gribble et al. published work on peer-to-peer systems and their behaviour towards the semantics of data [30]. Further, Gribble et al. highlight that P2P databases have unique challenges like the data placement problem where it is necessary to figure out how to distribute data and work so database queries can run at a low cost under resource and bandwidth constraints. As an outlook, new architectural designs are mentioned promising to help P2P databases to implement distributed query answering systems that are more scalable, reliable, and performant.

On a different venue, the mappings between peer-to-peer (P2P) databases are typically described to be local with no global schema accordingly to Bernstein et al. [57]. Also, the configurations and mappings between peers are highly dynamic that require semi-automatic solutions. In their work, Bernstein et al. presents Local Relational Model (LRM) as an architecture that can help resolve these issues for modern P2P databases. Franconi et al. [26] proposed a new model for P2P databases where nodes can request data from another node and use the third node for evaluation, but there can be no complex queries across the entire network. In contrast to standard first-order semantics, Franconi et al.'s new model captures the intended semantics of P2P systems. The model also halts the propagation of inconsistencies from node to node, so the database remains consistent, even if some of the nodes have inconsistent data. Remacle et al. [64] offered work on an Algorithm Oriented Mesh Database (AOMD) to manage mesh databases. Due to storage and algorithmic complexity, it is not possible to maintain complete graphs of data meshes according to Remacle et al. [64]. AOMD uses dynamic mesh representation to decrease computer memory use and increase algorithmic efficiency. It results in a light and efficient software implementation for mesh databases. Eunyoung Seegyoung Seol presented in his PhD thesis a mesh that is a piece-wise decomposition of the space/time domain where used by numerical simulation procedures [68]. Flexible distributed mesh database (FMDB) capable of shaping its representation based on the application's specific needs. FMDB embedded in SCOREC simulation packages effectively supporting automated adaptive analyses. Further, Seol et al. [67] published work on flexible distributed Mesh database (FMDB), that is a partition model and a distributed mesh management system. Seol et al. model has been used to efficiently support parallel automated adaptive analysis processes. The integration of mesh technology with the unified theory of acceptance and use of technology (UTAUT) can help businesses with analytics and technology adoption accordingly to Shirazi et al. [69]. Customised UTAUT models for mesh app, service, and conversational systems adoption that add motivation, innovation, privacy, and AI problem solving to traditional UTAUT can lead to intelligent mesh technology [69]. Rodríguez-Gianolli et al. [65] presented a hyperion prototype that demonstrates the possibility of using Peer-to-Peer (P2P) computing to share data. In their prototype, each peer includes a database with its own schema. The peers can join and leave the network independently. In a Hyperion P2P Database Network, the peer nodes share data by clustering into interest groups and pairing up using acquaintance links. The P2P Layer handles the peer-to-peer data sharing, while the Local Database Layer handles traditional database functions [65].

A P2P database system (PDBS) is a collection of autonomous databases that communicate with each other in a peer-to-peer fashion. Bonifati et al. [63] elaborated on how PDBS can borrow ideas from distributed database systems (DDBS) and multi-database systems (MDBS). For that purpose, Bonifati et al. compared past distributed database systems to PDBS, emphasising the database-centric and P2P-centric features of PDBS [63]. On the same note, Masud et al. investigated transaction processing in a peer-to-peer database network [47]. Their work looked into the problems around the consistent execution of concurrent transactions. Masud et al. also proposed solutions like Merged Transactions and OTM-based propagation to guarantee consistent performance [47].

Various venues broach the issue of data mesh environments, their technical realisation and implication towards distributed datasets. Yet, the fragmentation of data records into distributed databases and the consequences to overarching, traditional central tasks like security and privacy themes remains mostly unresolved.

2.2   Privacy in Mesh Networking

A few privacy questions have been discussed in the context of mesh networks and mesh structures. Wu et al. illustrated privacy attacks on mesh network based on the open medium property of wireless channel [77]. Traditional anonymous routing algorithm cannot be directly applied to Mesh network. In their paper, Wu et al. designed a private routing algorithm that used "Onion", i.e., layered encryption, to hide routing information [77]. Ganesh et al. proposed a strategy that applies self-organising maps (SOM) algorithm separately in each distributed dataset relative to database horizontal partitions [28]. In the sequence, these representative subsets are sent to a central site, which performs a fusion of partial results and applies K-means algorithms.

While research has been done on privacy in mesh networks, their findings and concepts are not easily transferable to data mesh environments. Data mesh is a special case of databases, while mesh networks originate from network topologies.

A similar paradigm but different application context. As open problems remain the question of how to find distributed describing attributes forming personally identifiable information (PII), how data deletion or data lineage can be realised in fragmented landscapes.

2.3 Syntactic Data Anonymisation

Randomisation [33,45], generalisation [27,72], suppression [27,72], and perturbation [45] are among the data transformation methods used in syntactic data anonymisation. Generalisation restructures the content of a dataset by changing its values according to a pre-defined term replacement taxonomy, whereas suppression simply erases data. As one travels up the ladder in a hierarchy-based taxonomy, each value gradually loses its uniqueness.

The k-anonymity Family. One of the first and best-known is k-anonymity, limiting distinguishability by classifying each tuple in the data set with at least $k - 1$ identical data records. Sweeney claims that the $k - 1$ closest neighbors are chosen based on similar descriptive features and enforced via generalisation and suppression [72]. The pattern of generalisation is to aggregate data values through a pre-defined hierarchy, such as combining the individual year 2021 into a year range of 2020–2025. Suppression, on the other hand, fully removes the selected data value. The generalisation toolset appears to be sensitive to attacks based on homogeneity and background knowledge [46]. To mitigate this, l-diversity takes the granularity of sensitive data representations into account, ensuring a factor of l diversity for each quasi-identifier within a particular equivalence class (usually a size of k). By evaluating the relative distributions of sensitive values in specific equivalence classes and throughout the entire dataset, t-closeness as an extension handles skewness and background knowledge attacks [42]. k-anonymity is also a privacy metric denoted k-map. If every combination of attribute values for quasi-identifiers appears at least k times in a dataset, it meets the k-map constraint [72]. To protect against symmetric assaults, Nergiz et al. [54] presented δ-presence, which builds on both k-anonymity and k-map. δ-min and δ-max are hidden in the δ-parameter. These two characteristics deal with the fact that no one is present.

Data Transformation Techniques. To support data transformation, the prior anonymisation techniques and their modifications used generalisation and suppression [27,49]. This is useful for theoretical demonstrations, but it quickly reaches its limits when dealing with larger datasets. Syntactic data anonymisation methods like k-anonymity [72], l-diversity [46], and t-closeness [42] are NP-hard, as Meyerson et al. [49] and Bayardo et al. [29] have shown. Because of their iterative and incremental character, the dependent *generalisation* methods are NP-hard in and of themselves. Applying generalisation and suppression to high dimensional data results in considerable information loss, rendering the data worthless for data analytics, according to Aggrawal et al. [4]. This is especially true because generalisation's runtime grows exponentially for several descriptive attributes, making it unfeasible. As a result, suppression persists and obliterates attribute values, resulting in significant information loss. Given the algorithm's complexity, all variations can only employ heuristics like k-optimise [6] to get improved approximations to perfect privacy, not perfect privacy [5].

Perturbation has been proposed as a viable alternative to generalisation [45]. The alternation of the real value to the nearest similar findable value is referred to as perturbation. This includes the effect of introducing an aggregated value or employing a close-by value so that just one value needs to be modified rather than numerous ones to form clusters. Finding such a value can take longer in certain cases due to iteratively rechecking the newly produced value(s), which negatively influences performance.

Optimal k-anonymity has been demonstrated to be an NP-hard task [5,49]. Due to their algorithmic nature, applying generalisation and suppression strategies to high-dimensional data results in a substantial level of information loss, leaving the data essentially unusable for data analytics. Tassa et al. [74] recommend using k-concealment to reduce the information loss caused by generalising database entries. However, in the case of high-dimensional application fields, both contributions degrade the NP-hardness. Fredj et al. [27] provided an in-depth review, categorisation, and advice for selecting generalisation algorithms.

The problem of ensuring k-anonymity with either optimal or holistic techniques to syntactic data anonymisation has been demonstrated to be an NP-hard task [49]. Heuristics can only be used to achieve better approximations to perfect privacy, not perfect privacy, in all types of k-anonymity algorithms [5]. As a result, scaling, particularly generalisation and perturbation in high-dimensional data, produces an impractical runtime [58,62] and a large level of information loss, rendering the data worthless for data analytics. With the help of GPU acceleration [61], it has been proven to shift the time complexity amplitude as runtime explosion from smaller $n < 20$ to larger $n < 150$ for 2^n, yet the nature of the growth remains.

2.4   Semantic Data Anonymisation and Differential Privacy

Semantic data anonymisation approaches sum up the statistical distributions of data values and the semantic meanings drawn from linking (defining patterns) between data points in an attempt to re-define privacy not just as a process of syntactically transforming datasets but also to consider both the statistical distributions of data values and the semantic meanings drawn from linking (defining patterns) between data points. The data veracity is tampered with by deleting significant ties between the data and an individual. Noise injection, permutation, or statistical shifting are commonly used to achieve this [19,33,44]. These algorithms are also known as *differential privacy*, and their statistical approaches are highly optimised for pre-defined use cases and mass data processing. In differential privacy, for example, this is accomplished by deciding how many noise injections to add to the output dataset at query runtime to assure anonymity in each situation [18]. Further, Dwork et al. extend their work with a vast introduction into the algorithmic foundations of differential privacy [22].

Individual contributions to differential privacy include the use of the exponential mechanism to expose statistical information about a dataset while concealing the private specifics of individual data items [48]. By applying controlled random distribution sensitive noise additions, the Laplace method for perturbation facilitates statistical shifting in differential privacy [20,38]. Because both sensitive attributes and quasi-identifiers are evaluated on a per-row basis during anonymisation [41], the discretised version [44] is known as a matrix mechanism. Because these anonymisation are done at runtime and on a case-by-case basis, the anonymisation processing is deferred until query runtime, increasing the risk of data leakage [36]. Leoni introduced "non-interactive" differential privacy [40] by performing statistical adjustments a priori to user searches. Another difficulty with differential privacy is that it is computationally infeasible to apply differential privacy to huge datasets (impractical). Dwork et al. shows that differential privacy is likewise NP-hard [21]. Experts are still debating whether approximation differential privacy algorithms provide adequate privacy assurances. An arbitrary family of attribute sets could be used to link a single data record back to its owner in certain conditions [24]. Abadi et al. offered the application of incorporating differential privacy into the deep learning context [1]. Even the US Census Bureau plans to adopt differential privacy accordingly to John Abowd [3]. But as Lee et al. have highlighted, the concept of differential privacy received considerable attention in the literature, yet little discussion is available on how to apply it in practice [39].

These revelations lead to an unsolved issue. Due to their complexity, anonymising a large dataset using either approximate procedures that may leave data inferences that can be exploited to de-anonymize people or precise counterparts results in exponentially growing runtime.

Randomisation techniques have gained increased attention as a result of the issues surrounding syntactic data anonymisation [33,45]. This semantic data anonymisation technique aims to re-define privacy as a process of considering both statistical distributions of data values and semantic meanings extracted from linking (defining patterns) between data points, rather than simply as a process of syntactically altering datasets. Dwork et al. [23] provide an in-depth survey of past work, in addition to the previous description of relatively recent contributions. Dankbar et al. have provided a comprehensive overview of the current literature on unequal privacy. They also pointed out some important general constraints, such as the theoretical character of the privacy parameter, which limits the ability to quantify the level of anonymity that would be guaranteed to patients [14]. Ji et al. explored the relationship between machine learning and differential privacy [34]. To illustrate both its strong guarantees and limitations, Li et al. focus on empirical accuracy performances of algorithms and semantic implications of differential privacy [43].

Semantic data anonymisation methods, such as differential privacy, have been demonstrated to be NP-hard for big datasets [21]. Given their runtime and use case-specific nature, they are computationally infeasible (impractical performance-wise) when applied to large high-dimensional data.

2.5   Unique Column Combinations

Unique column combinations (UCC) are attribute combinations that generate a unique identifier for the given dataset in data profiling (table). Discovering these unique column combinations (UCC) is a major scientific challenge.

Abedjan et al. [2] compiled and formalised the most recent breakthroughs in the finding of UCCs in their paper. Heise et al. built on their work by presenting a scalable discovery of unique column combinations based on parallelisation and the scale-out concept [32]. Feldmann has done the same thing [25]. Han et al. build on similar ideas [31] and use Hadoop with its MapReduce technology [15] to create a distributed computing environment. Papenbrock et al. [56] offered a comparison of alternative discovery strategies. Papenbrock et al., on the other hand, proposed a hybrid of quick approximation approaches and efficient validation procedures for UCCs [56]. Ruiz et al. published a patent recently that summarised several dataset profiling tools, techniques, and systems, including efficient UCC finding [66].

The search for UCC may be encapsulated in a cyclical dependence on the Hitting-Set issue as a family of W[2]-complete problems [9,17], according to Bläsius et al. [9]. In the worst-case scenario, this implies a super polynomial runtime, rendering its use to huge, high-dimensional data impracticable for the time being.

2.6   High-Dimensional Data

Given past advances in syntactic and semantic data anonymisation, more attention has shifted to hybrid systems that incorporate aspects from the initial syntactic and semantic data anonymisation approaches and provide abstractions from the raw dataset via aggregations or separations. For example, in attribute compartmentation [58,62], privacy is ensured by separating attributes that constitute quasi-identifiers using the notion of maximum partial unique column combinations (mpUCC) from the data profiling domain (mpUCCs). Quasi-identifiers are attribute value combinations that uniquely identify persons in a dataset (QID). By removing those QIDs, the re-identification attack of mixing QIDs with auxiliary data to draw inferences and extract private information is also prevented [76]. However, finding quasi-identifiers is difficult.

The enormous number of rows and columns distinguishes high-dimensional data. While the growing number of rows is seldom a problem, the growing number of columns can fast cause state-space explosions in enumeration issues [8]. The higher the dataset dimensions, the faster it reaches computational infeasibility. As can be seen from the preceding subsections, several disciplinary approaches for obtaining privacy, such as data profiling and mining, anonymisation processing, and differential privacy, eventually run into NP-hard difficulties.

In a few cases, high-dimensional data is being anonymized in great detail. Kohlmayer et al. proposed adaptations based on the Secure Multi-party Computing (SMC) protocol as a flexible approach on top of k-anonymity, l-diversity, and

t-closeness, as well as heuristic optimisation, to anonymize distributed and separated data silos in the medical field [37]. Mohammed et al. propose *LKC-privacy* to achieve privacy in both centralised and distributed scenarios [50], promising scalability for anonymising large datasets *LKC-privacy*, however, restricts the length of quasi-identifier tuples to a pre-determined number of characters that offers a practical approach but does not guarantee the entire absence of privacy-violating identifiers in high-dimensions. Other initiatives, such as Zhang et al. [80], employ a MapReduce approach based on the Hadoop distributed file system (HDFS) to increase compute capacity. On the other hand, the NP-hard nature swiftly beats the economic scalability options. Large numbers of entities defining characteristics (hundreds of attributes) must be handled in a performance-efficient and privacy-preserving way.

There are two reasons why discriminating between sensitive and non-sensitive properties is problematic, according to Manolis Terrovitis' study [75]. First, we can see that sensitive features are not the main reason for the success of de-anonymisation assaults (homogeneity, similarity, and background information). Second, creating an exhaustive collection of sensitive and non-sensitive qualities is problematic for high-dimensional datasets with distinct patterns that expand with the amount of data acquired on an individual. Podlesny et al. proposed modeling the attribute linkage problem for generating privacy-preserving data silos as a Bayesian network [59,60] to reduce the complexity of the compartmentation problem [58,62]. To train a Bayesian network, exact inference learning [53] and approximate inference learning [13] have the same NP-hardness. Recent contributions, however, show that using attribute linkage techniques to compress the network enables for performance-scalable data processing even on huge datasets [60]. Clifton et al. provided a balanced review of outstanding concerns in both syntactic and semantic data anonymisation methods, as well as its benefits, belongings, and summarised critiques [12]. Clifton et al. point out that the differences between different syntactic and semantic anonymisation origin models are less pronounced than previously supposed. Both archetypes, however, will have problems in large-scale data settings. Differential privacy is frequently the best empirical privacy for a fixed (empirical) utility level, however syntactic anonymity models may be preferred for more precise answers.

Regardless of where it came from, data anonymisation is yet to be applied to large-scale, multi-attribute, high-dimensional datasets in a reasonable amount of time and with limited resources. Each solution suffers from considerable complexity restrictions for huge quantities of descriptive characteristics (columns), resulting in massive information loss, calculation demands, and hence runtime, or privacy guarantees through approximation approaches.

2.7   Quasi-Identifier Discovery

Byun et al. addressed the lack of diversity through equivalence classes and their information-loss by transforming the k-anonymity problem to a k-member clustering problem [11], based on Sweeneys work on the family of k-anonymity techniques [72,73]. While Byun et al. technique uses distance and cost functions works for numeric and categorical data, it does not guarantee approximation factors. For clustering purposes, the projection of quasi-identifier similarity remains data-specific.

Xiao et al. published anatomy, a novel approach that immediately releases all quasi-identifiers and sensitive values in two independent tables [78]. This, in conjunction with grouping operations, should allow for the capture of correlation while minimising reconstruction error. Zhang et al. investigated the scalability benefits of horizontal scaling in cloud computing environments, as well as the use of a quasi-identifier index-based technique to speed up data querying on huge datasets [79]. Statistical de-anonymisation attacks on high-dimensional datasets were proven by Narayanan et al. for re-identifying people in the Netflix Prize dataset with tolerance for certain inaccuracies in the adversary's prior information [51]. Soria-Comas et al. summarised the topic of re-linkage using quasi-identifiers. They explored data governance issues like user permission, purpose limitation, transparency, individual rights of access, correction, and deletion. When deleting specified qualities against extra personally identifiable information (PII), Narayanan et al. expounded on the PII fallacy of the HIPAA privacy law [52], as the eradication of all quasi-identifiers is not assured. Soria-Comas et al. work also highlighted the need for new privacy models built from the ground up with big data requirements in mind, such as continuous and vast data collected from numerous source systems, resulting in multi-attribute and high-dimensional datasets [70]. Braghin et al. have submitted an optimised quasi-identifier strategy that uses parallelisation for efficient QID discovery [10], even though parallelisation is not a novel concept. Braghin et al. study can serve as a comparative baseline for our research due to its extensive description and encouraging outcomes.

The discovery of quasi-identifiers summarised as *Find-QID* problem [61] remains NP-hard and W[2]-complete [9,61]. Heuristic and greedy approach exist, they even weaken the exponential implication of the same *Find-QID* problem, yet particularly in high-dimensional spaces a lasting solution remains open unless the W-hierarchy collapses [9]. This assumes an already pre-compiled, static dataset. Adding now a distributed factor in, like in the case of data meshes, the search and identification of QIDs become even more complex.

In summary, the community has done a lot of research on peer-to-peer database, mesh network and anonymisation techniques individually. Yet, to the best of our understanding, the paradigm of data mesh in databases and its side effects with, against and towards privacy is largely unexplored. In particular, this includes the topics around data deletion, quasi-identifier discover and data lineage under the constraint of distributed, highly fragmented data records across multiple data mesh instances. To emphasis the underlying complexity, we demonstrate the differences of data mesh to more traditional database approaches in the experiments of the following Sect. 4.

3   Data Meshes

To recapitulate on essential terminologies, we briefly summarise the current understanding and state of data mesh in database and quasi-identifiers in the same domain. The concept of data mesh centers around the democratisation and decentralisation of development activities. Instead of a central and predominating database with strict governance, a distributed setup build the basis of data meshes. Each data repository is somehow coupled, can have upstream and downstream dependencies guaranteed through data contracts defining their usage, availability, quality and content. This structural paradigm offers flexibility in its configuration. Still, the same gained flexibility introduces looser governance challenges like the absence of data lineage, which we will describe in the following more profoundly. A similarity can be found in software engineering, where a trend from monolith- towards microservices as architecture patterns has been observed [35,55].

Characteristics of a Data Mesh. Given the decoupled nature of data meshes [16], different data records might be split or even duplicated across multiple data repositories. Traditionally, each data mesh instance is dedicated to a certain data domain, with a clear owned business entity and corresponding dependencies, inputs and outputs objective. While each data mesh instance is somehow autarkic, it may directly consume each other. Figure 1 illustrates this setup on a high level perspective. Data between each instance can be linked through identifiers, but this is not guaranteed. Such a fractured landscape brings value through its flexibility. Each data domain can act and scale independently, yet learnings from different sectors include that the same paradigm re-balances the weight against arbitrary governance structures. As seen in the healthcare domain, the archetype of various detached data repositories introduces a challenge for overarching topics like data privacy, common interfaces and standardisation.

In the case of a central place, the same overarching objectives can be easily monitored, traced and supported like in the case of the implementation of GDPRs *data deletion* right. A simple act like deleting personally identifiable information (PII) sounds trivial, but imagine there are hundreds of data mesh instances across hundreds of teams and each acts on its own. In various decoupled data repositories, tracking down distributed user attributes can only work with thoroughly conducting *data lineage* which requires a lot of dedication and documentation work for each development team as cross-linkages may be possible. Figure 2 depicts such perspective, where each domain holds a subset of user data. Each subset individually may not look concerning from a privacy perspective, but joining these through existing identifiers they can become concerning.

Quasi-identifiers in a Data Mesh. Quasi-identifiers (QID) are attribute combinations that jointly form identifiers while independently might seem unsuspicious. A quasi-identifier does not have to identify all individuals, but serves at least one individual to be exposed and cause harm to their privacy. Formally, QIDs are defined as

Definition 1. *Quasi-identifier*
Let $F = \{f_1, ..., f_n\}$ *be a set of all features and* $B := \mathcal{P}(F) = \{B_1, ..., B_k\}$
its power set, i.e. the set of all possible feature combinations. A set of selected
features $B_i \in B$ *is called a quasi-identifier, if* B_i *identifies at least one entity*
uniquely and all features $f_j \in B_i$ *are not standalone identifiers.*

To make this tangible, the readers attention is pointed towards Fig. 2 one more
time. Here, one can see that Domain A holds a *ZIP code* information, Domain B
age and *gender* and both are linked through the *Call Center ID*. Further, Domain
D holds analytical results like the *disease prediction* or *medical adherence*. When
following all identifiers, one can easily build a data profile including *age, gender,*
ZIP code, disease prediction and *medical adherence* without touching the Domain
C. Now, as Sweeney et al. showed that 87% of the entire population are identifiable
through the combination of age, gender and zip [71], an attacker may infer *disease*
prediction and *medical adherence* to those 87%.

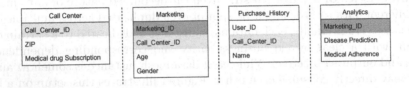

Fig. 2. Indirect linkage of quasi-identifiers in a data mesh

4   Experiments

To fortify the novolum that the data mesh paradigm creates towards data privacy topics, we will build on the prior knowledge and characteristic summary and
outline through a series of experiments the same theses and raised challenges.
For that purpose, we leverage a semi-synthetic dataset and state-of-the-art hardware to compare different database archetypes and their runtime implications
on finding PII compromising quasi-identifiers.

Hardware. Our examination runs on a GPU-accelerated high-performance compute cluster, housing 64 vCPU cores (E5-4650), 240 GB RAM, and 8x NVIDIA
GeForce 3060 with 3584 CUDA cores each and a combined Tensor performance
of 816 Tensor TFLOPs. GPU-related experiments' execution environment will
be restricted to one dedicated CPU core and a single, dedicated Tesla V100
GPU.

Dataset. For the purpose of evaluation, a semi-synthetic health dataset has been
compiled based on publicly available contributions, previous work and publications. The dataset consists of genomic data, fake but consistent names, addresses,
SSN, passwords and telephone numbers, as well as medical records randomly

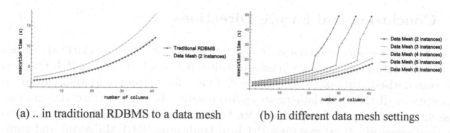

(a) .. in traditional RDBMS to a data mesh   (b) in different data mesh settings

Fig. 3. Projected runtime growth of discovered QID over increasing columns.

assigned but adhering to known statistical distributions. For transparency, the full dataset can be downloaded from github.com[1].

Evaluation. To demonstrate the differences in time complexity when different database archetypes are being introduced, these experiments build on Sweeney's k-anonymity approach of finding quasi-identifiers [73]. A GPU-accelerated search schema without heuristics purely based on *groupby* and *count* statements developed by Podlesny et al. [61] is being utilised in the following. Figure 3a delineates the runtime growth for discovering the quasi-identifiers. The Y-axis represents the execution time to find all QIDs in an exact manner (not heuristic) while the x-axis the increasing number of describing attributes being stored in the associated database archetype. The different database archetypes of traditional central RDBMS and data mesh are clearly visible. Both runtime portray an exponential increase, while the growth of the data mesh answers to a higher factor (see Fig. 3a). While both, a traditional central RDBMS and a data mesh can be scaled horizontally and vertically in number of nodes and hardware used, the data mesh suffers a fragmentation of describing data attributes that can form quasi-identifiers. This fragmentation needs to be first compensated which essentially answers to more network I/O and therefore longer processing time. The larger the fragmentation, the higher the network I/O and the longer the compute.

Following the same line of thoughts, Fig. 3b depicts the evolution of the same metrics over different data mesh sizes. The data mesh size answers to the number of instances involved with equally distributed data attributes, starting from two and increasing. Given the nature of the search, the complexity is exponential already. Yet, two things stand out. First, the more data mesh instances exist with equivalent data distribution, the sooner runtime increases due to the higher degree of fragmentation and therefore, more data shifting and joining is required. Second, the more data meshes exist, the earlier one experiences an uncontrolled explosion of execution time as, given the hardware constraint, the capacities of main memory and GPU memory are exceeded.

---

[1] https://github.com/jaSunny/synthetic_genome_data.

5   Conclusion and Future Directions

The previous sections offered a systematisation of knowledge and clarified characteristics of data mesh and how quasi-identifiers potentially exposing PII. Further, the summarised state-of-the-art delineates gaps for privacy and anonymisation concepts in distributed data mesh environments. To demonstrate the uniqueness and scalability of this problem, we have offered a variety of experiments to discover quasi-identifier exposing PII in a traditional RDBMS setup and compared these metrics against same algorithms running in a data mesh setup. The increase of complexity and runtime is clearly visible.

Based on this understanding, we formulate the open distributed Quasi-identifiers problem: To find usage of PII data within a data mesh, elements of one quasi-identifiers (QIDs) might be distributed and linked across more than one database instance. To find these distributed QIDs, all describing attribute combination of any length that can be cross-linked through arbitrary identifiers need to be considered. Due to its distributed nature, this represents a special case of the W[2]-complete *Find-QID* problem [61].

References

1. Abadi, M., et al.: Deep learning with differential privacy. In: Proceedings of the 2016 ACM SIGSAC Conference on Computer and Communications Security, pp. 308–318. ACM (2016)
2. Abedjan, Z., Naumann, F.: Advancing the discovery of unique column combinations. In: Proceedings of the 20th ACM International Conference on Information and Knowledge Management, pp. 1565–1570 (2011)
3. Abowd, J.M.: The US census bureau adopts differential privacy. In: Proceedings of the 24th ACM SIGKDD International Conference on Knowledge Discovery & Data Mining, p. 2867 (2018)
4. Aggarwal, C.C.: On k-anonymity and the curse of dimensionality. In: Proceedings of the 31st International Conference on Very Large Data Bases, VLDB 2005, VLDB Endowment, pp. 901–909 (2005)
5. Barth-Jones, D.: The 're-identification' of governor William Weld's medical information: a critical re-examination of health data identification risks and privacy protections, then and now (July 2012)
6. Bayardo, R.J., Agrawal, R.: Data privacy through optimal k-anonymization. In: 2005 Proceedings of the 21st International Conference on Data Engineering, ICDE 2005, pp. 217–228. IEEE (2005)
7. Beall, M.W., Shephard, M.S.: A general topology-based mesh data structure. Int. J. Numer. Meth. Eng. **40**(9), 1573–1596 (1997)
8. Birnick, J., Bläsius, T., Friedrich, T., Naumann, F., Papenbrock, T., Schirneck, M.: Hitting set enumeration with partial information for unique column combination discovery. Proc. VLDB Endow. **13**(12), 2270–2283 (2020)
9. Bläsius, T., Friedrich, T., Schirneck, M.: The parameterized complexity of dependency detection in relational databases. In: 11th International Symposium on Parameterized and Exact Computation, Dagstuhl, Germany, vol. 63, pp. 6:1–6:13 (2017)

10. Braghin, S., Gkoulalas-Divanis, A., Wurst, M.: Detecting quasi-identifiers in datasets (16 January 2018). US Patent 9,870,381
11. Byun, J.-W., Kamra, A., Bertino, E., Li, N.: Efficient k-anonymization using clustering techniques. In: Kotagiri, R., Krishna, P.R., Mohania, M., Nantajeewarawat, E. (eds.) DASFAA 2007. LNCS, vol. 4443, pp. 188–200. Springer, Heidelberg (2007). https://doi.org/10.1007/978-3-540-71703-4_18
12. Clifton, C., Tassa, T.: On syntactic anonymity and differential privacy. In: 2013 IEEE 29th International Conference on Data Engineering Workshops (ICDEW), pp. 88–93. IEEE (2013)
13. Dagum, P., Luby, M.: Approximating probabilistic inference in Bayesian belief networks is NP-hard. Artif. Intell. **60**(1), 141–153 (1993)
14. Dankar, F.K., El Emam, K.: Practicing differential privacy in health care: a review. Trans. Data Priv. **6**(1), 35–67 (2013)
15. Dean, J., Ghemawat, S.: MapReduce: simplified data processing on large clusters. Commun. ACM **51**(1), 107–113 (2008)
16. Dehghani, Z.: Data mesh principles and logical architecture. martinfowler.com (2020)
17. Downey, R.G., Fellows, M.R.: Fundamentals of Parameterized Complexity. TCS, vol. 4. Springer, London (2013). https://doi.org/10.1007/978-1-4471-5559-1
18. Dwork, C.: Differential privacy: a survey of results. In: Agrawal, M., Du, D., Duan, Z., Li, A. (eds.) TAMC 2008. LNCS, vol. 4978, pp. 1–19. Springer, Heidelberg (2008). https://doi.org/10.1007/978-3-540-79228-4_1
19. Dwork, C.: Differential privacy. In: van Tilborg, H.C.A., Jajodia, S. (eds.) Encyclopedia of Cryptography and Security, pp. 338–340. Springer, Boston (2011). https://doi.org/10.1007/978-1-4419-5906-5_752
20. Dwork, C., McSherry, F., Nissim, K., Smith, A.: Calibrating noise to sensitivity in private data analysis. In: Halevi, S., Rabin, T. (eds.) TCC 2006. LNCS, vol. 3876, pp. 265–284. Springer, Heidelberg (2006). https://doi.org/10.1007/11681878_14
21. Dwork, C., Naor, M., Reingold, O., Rothblum, G.N., Vadhan, S.: On the complexity of differentially private data release: efficient algorithms and hardness results. In: Proceedings of the 41st Annual ACM Symposium on Theory of Computing, STOC 2009, pp. 381–390. ACM, New York, NY, USA (2009)
22. Dwork, C., Roth, A.: The algorithmic foundations of differential privacy. Found. Trends® Theor. Compu. Sci. **9**(3–4), 211–407 (2013)
23. Dwork, C., Smith, A.: Differential privacy for statistics: what we know and what we want to learn. J. Priv. Confid. **1**(2), 135–154 (2010)
24. European Commission: Opinion 05/2014 on anonymisation techniques (April 2014)
25. Feldmann, B.: Distributed Unique Column Combinations Discovery. Hasso-Plattner-Institute, January 2020. https://hpi.de/fileadmin/user_upload/fachgebiete/friedrich/documents/Schirneck/Feldmann_masters_thesis.pdf
26. Franconi, E., Kuper, G., Lopatenko, A., Serafini, L.: A robust logical and computational characterisation of peer-to-peer database systems. In: Aberer, K., Koubarakis, M., Kalogeraki, V. (eds.) DBISP2P 2003. LNCS, vol. 2944, pp. 64–76. Springer, Heidelberg (2004). https://doi.org/10.1007/978-3-540-24629-9_6
27. Fredj, F.B., Lammari, N., Comyn-Wattiau, I.: Abstracting anonymization techniques: a prerequisite for selecting a generalization algorithm. Procedia Comput. Sci. **60**, 206–215 (2015)
28. Ganesh, P., KamalRaj, R., Karthik, S.: Protection of privacy in distributed databases using clustering. Int. J. Mod. Eng. Res. **2**, 1955–1957 (2012)

29. Ghinita, G., Karras, P., Kalnis, P., Mamoulis, N.: Fast data anonymization with low information loss. In: Proceedings of the 33rd International Conference on Very Large Data Bases, VLDB Endowment, pp. 758–769 (2007)

30. Gribble, S.D., Halevy, A.Y., Ives, Z.G., Rodrig, M., Suciu, D.: What can database do for peer-to-peer? In: WebDB, vol. 1, pp. 31–36 (2001)

31. Han, S., Cai, X., Wang, C., Zhang, H., Wen, Y.: Discovery of unique column combinations with hadoop. In: Chen, L., Jia, Y., Sellis, T., Liu, G. (eds.) APWeb 2014. LNCS, vol. 8709, pp. 533–541. Springer, Cham (2014). https://doi.org/10.1007/978-3-319-11116-2_49

32. Heise, A., Quiané-Ruiz, J.A., Abedjan, Z., Jentzsch, A., Naumann, F.: Scalable discovery of unique column combinations. Proc. VLDB Endow. **7**(4), 301–312 (2013)

33. Islam, M.Z., Brankovic, L.: Privacy preserving data mining: a noise addition framework using a novel clustering technique. Knowl. Based Syst. **24**(8), 1214–1223 (2011)

34. Ji, Z., Lipton, Z.C., Elkan, C.: Differential privacy and machine learning: a survey and review (2014)

35. Kalske, M., Mäkitalo, N., Mikkonen, T.: Challenges when moving from Monolith to microservice architecture. In: Garrigós, I., Wimmer, M. (eds.) ICWE 2017. LNCS, vol. 10544, pp. 32–47. Springer, Cham (2018). https://doi.org/10.1007/978-3-319-74433-9_3

36. Kifer, D., Machanavajjhala, A.: No free lunch in data privacy. In: Proceedings of the 2011 ACM SIGMOD International Conference on Management of Data, SIGMOD 2011, pp. 193–204. ACM, New York (2011)

37. Kohlmayer, F., Prasser, F., Eckert, C., Kuhn, K.A.: A flexible approach to distributed data anonymization. J. Biomed. Inform. **50**, 62–76 (2014)

38. Koufogiannis, F., Han, S., Pappas, G.J.: Optimality of the Laplace mechanism in differential privacy. arXiv preprint arXiv:1504.00065 (2015)

39. Lee, J., Clifton, C.: How much is enough? Choosing ε for differential privacy. In: Lai, X., Zhou, J., Li, H. (eds.) ISC 2011. LNCS, vol. 7001, pp. 325–340. Springer, Heidelberg (2011). https://doi.org/10.1007/978-3-642-24861-0_22

40. Leoni, D.: Non-interactive differential privacy: a survey. In: Proceedings of the 1st International Workshop on Open Data, pp. 40–52. ACM (2012)

41. Li, C., Miklau, G., Hay, M., McGregor, A., Rastogi, V.: The matrix mechanism: optimizing linear counting queries under differential privacy. VLDB J. **24**(6), 757–781 (2015). https://doi.org/10.1007/s00778-015-0398-x

42. Li, N., Li, T., Venkatasubramanian, S.: t-closeness: privacy beyond k-anonymity and l-diversity. In: 2007 IEEE 23rd International Conference on Data Engineering, pp. 106–115 (April 2007)

43. Li, N., Lyu, M., Su, D., Yang, W.: Differential privacy: from theory to practice. Synth. Lect. Inf. Secur. Priv. Trust **8**(4), 1–138 (2016)

44. Liu, F.: Generalized gaussian mechanism for differential privacy. arXiv preprint arXiv:1602.06028 (2016)

45. Liu, K., Kargupta, H., Ryan, J.: Random projection-based multiplicative data perturbation for privacy preserving distributed data mining. IEEE Trans. Knowl. Data Eng. **18**(1), 92–106 (2006)

46. Machanavajjhala, A., Kifer, D., Gehrke, J., Venkitasubramaniam, M.: L-diversity: privacy beyond k-anonymity. ACM Trans. Knowl. Discov. Data (TKDD) **1**(1), 3 (2007)

47. Masud, M., Kiringa, I.: Transaction processing in a peer to peer database network. Data Knowl. Eng. **70**(4), 307–334 (2011)

48. McSherry, F., Talwar, K.: Mechanism design via differential privacy. In: 2007 48th Annual IEEE Symposium on Foundations of Computer Science, pp. 94–103. IEEE (2007)
49. Meyerson, A., Williams, R.: On the complexity of optimal k-anonymity. In: Proceedings of the 23rd ACM SIGMOD-SIGACT-SIGART Symposium, pp. 223–228. ACM (2004)
50. Mohammed, N., Fung, B., Hung, P.C., Lee, C.K.: Centralized and distributed anonymization for high-dimensional healthcare data. ACM Trans. Knowl. Discov. Data (TKDD) **4**(4), 18 (2010)
51. Narayanan, A., Shmatikov, V.: Robust de-anonymization of large sparse datasets. In: 2008 IEEE Symposium on Security and Privacy, SP 2008, pp. 111–125. IEEE (2008)
52. Narayanan, A., Shmatikov, V.: Myths and fallacies of "personally identifiable information". Commun. ACM **53**(6), 24–26 (2010)
53. Neapolitan, R.E.: Probabilistic reasoning in expert systems: theory and algorithms. CreateSpace Independent Publishing Platform (2012)
54. Nergiz, M.E., Atzori, M., Clifton, C.: Hiding the presence of individuals from shared databases. In: Proceedings of the 2007 ACM SIGMOD International Conference on Management of Data, pp. 665–676 (2007)
55. Newman, S.: Monolith to Microservices: Evolutionary Patterns to Transform Your Monolith. O'Reilly Media (2019)
56. Papenbrock, T., Naumann, F.: A hybrid approach for efficient unique column combination discovery. Proc. der Fachtagung Business, Technologie und Web (BTW). GI, Bonn, Deutschland (accepted) Google Scholar (2017)
57. Phil, B., Giunchiglia, F., Kementsietsidis, A., Mylopoulos, J., Serafini, L., Zaihrayeu, I.: Data management for peer-to-peer computing: a vision. In: 5th International Workshop on the Web and Databases, WebDB 2002 (2002)
58. Podlesny, N.J., Kayem, A.V., Meinel, C.: Attribute compartmentation and greedy UCC discovery for high-dimensional data anonymization. In: Proceedings of the 9th ACM Conference on Data and Application Security and Privacy, pp. 109–119 (2019)
59. Podlesny, N.J., Kayem, A.V., Meinel, C.: Identifying data exposure across high-dimensional health data silos through Bayesian networks optimised by multigrid and manifold. In: 2019 IEEE 17th International Conference on Dependable, Autonomic and Secure Computing (DASC). IEEE (2019)
60. Podlesny, N.J., Kayem, A.V.D.M., Meinel, C.: Towards identifying de-anonymisation risks in distributed health data silos. In: Hartmann, S., Küng, J., Chakravarthy, S., Anderst-Kotsis, G., Tjoa, A.M., Khalil, I. (eds.) DEXA 2019. LNCS, vol. 11706, pp. 33–43. Springer, Cham (2019). https://doi.org/10.1007/978-3-030-27615-7_3
61. Podlesny, N.J., Kayem, A.V.D.M., Meinel, C.: A parallel quasi-identifier discovery scheme for dependable data anonymisation. In: Hameurlain, A., Tjoa, A.M. (eds.) Transactions on Large-Scale Data- and Knowledge-Centered Systems L. LNCS, vol. 12930, pp. 1–24. Springer, Heidelberg (2021). https://doi.org/10.1007/978-3-662-64553-6_1
62. Podlesny, N.J., Kayem, A.V.D.M., von Schorlemer, S., Uflacker, M.: Minimising information loss on anonymised high dimensional data with greedy in-memory processing. In: Hartmann, S., Ma, H., Hameurlain, A., Pernul, G., Wagner, R.R. (eds.) DEXA 2018. LNCS, vol. 11029, pp. 85–100. Springer, Cham (2018). https://doi.org/10.1007/978-3-319-98809-2_6

63. Record, A.S.: Distributed databases and peer-to-peer databases. SIGMOD Rec. **37**(1), 5 (2008)
64. Remacle, J.F., Shephard, M.S.: An algorithm oriented mesh database. Int. J. Numer. Meth. Eng. **58**(2), 349–374 (2003)
65. Rodríguez-Gianolli, P., et al.: Data sharing in the hyperion peer database system. In: Proceedings of the 31st International Conference on Very Large Data Bases, pp. 1291–1294. Citeseer (2005)
66. Ruiz, J.A.Q., Naumann, F., Abedjan, Z.: Datasets profiling tools, methods, and systems (11 June 2019). US Patent 10,318,388
67. Seol, E.S., Shephard, M.S.: Efficient distributed mesh data structure for parallel automated adaptive analysis. Eng. Comput. **22**(3–4), 197–213 (2006)
68. Seol, E.S.: FMDB: flexible distributed mesh database for parallel automated adaptive analysis. Rensselaer Polytechnic Institute Troy, NY (2005)
69. Shirazi, F., Keramati, A.: Intelligent digital mesh adoption for big data (2019)
70. Soria-Comas, J., Domingo-Ferrer, J.: Big data privacy: challenges to privacy principles and models. Data Sci. Eng. **1**(1), 21–28 (2016)
71. Sweeney, L.: Simple demographics often identify people uniquely. Health (San Francisco) **671**(2000), 1–34 (2000)
72. Sweeney, L.: Achieving k-anonymity privacy protection using generalization and suppression. Int. J. Uncertain. Fuzziness Knowl. Based Syst. **10**(05), 571–588 (2002)
73. Sweeney, L.: k-anonymity: a model for protecting privacy. Int. J. Uncertain. Fuzziness Knowl. Based Syst. **10**(05), 557–570 (2002)
74. Tassa, T., Mazza, A., Gionis, A.: k-concealment: an alternative model of k-type anonymity. Trans. Data Priv. **5**(1), 189–222 (2012)
75. Terrovitis, M., Mamoulis, N., Kalnis, P.: Privacy-preserving anonymization of set-valued data. Proc. VLDB Endow. **1**(1), 115–125 (2008)
76. Wong, R.C.-W., Fu, A.W.-C., Wang, K., Pei, J.: Anonymization-based attacks in privacy-preserving data publishing. ACM Trans. Database Syst. **34**(2), 1–46 (2009)
77. Wu, X., Li, N.: Achieving privacy in mesh networks. In: Proceedings of the 4th ACM Workshop on Security of Ad Hoc and Sensor Networks, pp. 13–22 (2006)
78. Xiao, X., Tao, Y.: Anatomy: simple and effective privacy preservation. In: Proceedings of the 32nd International Conference on Very Large Data Bases, pp. 139–150. VLDB Endowment (2006)
79. Zhang, X., Liu, C., Nepal, S., Chen, J.: An efficient quasi-identifier index based approach for privacy preservation over incremental data sets on cloud. J. Comput. Syst. Sci. **79**(5), 542–555 (2013)
80. Zhang, X., Yang, L.T., Liu, C., Chen, J.: A scalable two-phase top-down specialization approach for data anonymization using MapReduce on cloud. IEEE Trans. Parallel Distrib. Syst. **25**(2), 363–373 (2014)

Why- and How-Provenance
in Distributed Environments

Paulo Pintor[1]([envelope]) [iD], Rogério Luís de Carvalho Costa[2] [iD], and José Moreira[1] [iD]

[1] IEETA - University of Aveiro, 3810-193 Aveiro, Portugal
{paulopintor,jose.moreira}@ua.pt
[2] CIIC - Polytechnic of Leiria, 2411-901 Leiria, Portugal
rogerio.l.costa@ipleiria.pt

Abstract. With the emergence of new paradigms in data management
and processing like Cloud services, the Internet of Things (IoT), and
NoSQL, there is a growing trend for distributing data across multiple
platforms and using the technologies most suited for each case according
to criteria such as performance and cost. But it also raises new challenges
and needs, like understanding the sources, the transformations, and the
processes made on the data to infer their quality and reliability. Data
provenance becomes particularly relevant in such a context.

This paper presents a solution to deal with why- and how-provenance
queries on distributed data sources and different database paradigms.
The proposed solution does not require any change to the query execution
engine. It uses pure SQL with annotations and an algorithm to build data
provenance information from the result obtained by the query. We also
present experimental evaluation results obtained using an open-source
logical integration tool.

Keywords: Provenance · Data provenance · Databases · Distributed
systems

1 Introduction

Data management and processing have been changing over the past years. Several factors have made data increasingly distributed, including the emergence of
the Cloud, smart devices, and the Internet of Things (IoT). Also, the rise of open
data and data science attracted experts in several domains who became interested in data manipulation and processing, knowledge extraction, and results
sharing. This context leads to issues regarding the quality, veracity, completeness, and correctness of data sources, thus increasing the need to understand
where data comes from, whether the source is trustworthy, and the transformations made on data. Data provenance is metadata information (annotations)
about data origin and transformations made on data and helps solve such issues.

Although there exists a standard (PROV [15], a W3C recommendation) for
describing this information in terms of agents, entities, activities, and their relationships, an important research topic is how to disclose provenance information

© The Author(s), under exclusive license to Springer Nature Switzerland AG 2022
C. Strauss et al. (Eds.): DEXA 2022, LNCS 13426, pp. 103–115, 2022.
https://doi.org/10.1007/978-3-031-12423-5_8

in database queries, i.e., to know where query results come from and how they were computed.

Some solutions with different approaches allow obtaining data provenance information from database queries, but all are in centralized environments or specific database management systems. The data provenance issue becomes even more essential in distributed database environments in which several (and possibly heterogeneous) databases are accessed to answer a single user's query.

This paper discusses issues and challenges involving data provenance in distributed and heterogeneous databases. It presents a solution for *how-* and *why-provenance* that does not need to make changes to the database engine nor use system-specific functions and procedures. Hence, our solution can build provenance information for the results of a query independently of the data source type (e.g., file, and relational database and NoSQL databases), which is an important feature when dealing with distributed and heterogeneous data sources.

The following section presents some background and related work. Section 3 discusses data provenance in distributed environments and then describes the proposed solution. Then, Sect. 4 presents results from an experimental evaluation. Finally, Sect. 5 concludes the paper and describes future work.

2   Background and Related Work

This section presents some background on building the provenance of the results of database queries (data provenance) and the problems that arise when working with distributed databases. It also reviews existing related works.

2.1   Data Provenance

In [11], the authors proposed four types of provenance, divided hierarchically: provenance meta-data, information system provenance, workflow provenance, and data provenance. Data provenance aims to collect the provenance information from queries over a database. Due to the fact dealing with databases with specific schemas and because the provenance can be at the tuple level, this type of provenance has requirements that do not appear in other types of provenance.

The three most common types of data provenance are *why-*, *how-* and *where-provenance* [4,5,10]. With the increased interest and research in data provenance, other categories have been proposed, such as *Why-not-provenance* [2,11] and *Which-provenance* [10] and perhaps more might follow. The focus of this paper is on *why-* and *how-provenance*.

Why-Provenance – Collects all the inputs that contributed to a query result [3–5,10]. The technique to collect why-provenance information is called Witnesses basis. It is a set of tuples that contribute to a particular result. These tuples are called witnesses of the production of the resulting tuple.

Based on the definition in [3,5], given a database I, a query Q over I and a tuple t in $Q(I)$, an instance of $I' \subseteq I$ is a witness for t if $t \in Q(I')$. This can be denoted as: $Why(Q, I, t) = \{I' \subseteq I | t \in Q(I')\}$

Orderspt			
sname	dest	vehicle	provtoken
LisboaStore	Porto	Train	tk3
LisboaStore	Braga	Truck	tk4
PortoStore	Braga	Train	tk5
PortoStore	Madrid	Airplane	tk6

Fig. 1. An example of a table with orders.

For instance, consider the table in Fig. 1 representing orders. The field "sname" is the supplier name, "dest" is the destination, "vehicle" is the type of vehicle, and the tuple identifier is called "provtoken".

$$Q1 : \pi_{dest}\sigma_{dest=\text{"Braga"}}(\pi_{sname,dest}Orderspt \bowtie \pi_{vehicle,dest}Orderspt)$$

The *Why-provenance* is the set of tuples with all the possible combinations, without duplicates. The result of Q1 is displayed in Table 1 and shows that the witnesses of *"Braga"* are *tk4* and *tk5* alone or the conjugation of both.

Table 1. Result of Q1

dest	why	how
Braga	{tk4}, {tk4,tk5}, {tk5}	$(tk4 \otimes tk4) \oplus (tk4 \otimes tk5) \oplus (tk5 \otimes tk4) \oplus (tk5 \otimes tk5)$

How-Provenance – Explains how the inputs contributed to the result and is obtained using algebraic identities and polynomials (semirings) [3–5,9,10,16]. Each tuple must also have an annotation called prove token.

A semiring is defined as $(K, 0, 1, \oplus, \otimes)$ where K is a set of data elements that will be annotated using the constants 0 and 1. Given a query Q if the tuple t contributes to the output result is annotated with 1, otherwise is annotated with 0. The binary operators $\oplus, \otimes$ are used as alternative $\oplus$ and as joint $\otimes$.

Different types of semirings can be used to achieve different answers. For *how-provenance* the universal semiring or how-semiring $(N[X], 0, 1, \oplus, \otimes)$ is used. As stated in [9], *unions* are associative and commutative operations and are represented by $\oplus$. The *joins* also have those two properties, but they are also distributive over *unions* and they represented by $\otimes$. The *projections* and *selections* are also commutative among themselves.

Hence regarding the result present in Table 1 about *How-Provenance*, *"Braga"* is obtained by the conjugation of *tk4* with itself (*join*), or (*union*) by the conjugation of *tk5* with itself (*join*), or (*union*) by the conjugation of *tk4* and *tk5* (*join*) or (*union*) by the conjugation of *tk5* and *tk4*.

Regarding the *joins* properties, more specifically, the distributive property of the results in Table 1 can also be simplified to: $(tk4 \oplus tk5) \otimes (tk4 \oplus tk5)$. In [16] it is proposed to use m-semirings with the operator *monus* ($\ominus$) to be able to give the provenance for non-monotone queries.

2.2   Distributed Databases

While Multi-Model databases allow having different types of models (e.g., graph, key-value, and documents) in the same Database Management System (DBMS) [13], Polystore databases are built on the top of multiple storage engines that are integrated and enable to query multiple data sources using different models and paradigms [7].

Using distributed query engines (e.g., Presto [18]), users may query over distributed and heterogeneous databases using standard SQL language. The query engines act as mediators between the querying interface and the underlying systems, but they do not deal with distribution transparency, i.e., the location of each data structure (e.g., table) must be included in the query. This forces users to have deep knowledge about the different data sources and their schemas.

Distribution transparency can be achieved by logical data integration. It commonly comprises a high-level global model, i.e., a Global Conceptual Schema (GCS), and Local Conceptual Schemas (LCS), which represent the physically distributed data [21]. The GCS stores the information about how to link global and local entities. There are no Extract-Transform-Load (ETL) methods. Queries are written considering the global entities, thus hiding distribution complexity from the end-users. This approach is especially useful in scenarios where the users need data to always be up to date.

But the logical integration requires the mapping between global and local entities. One global entity may match a single entity of a specific data source (local entity). But a single logical global entity may map to two or more local entities (i.e., partitioning). In **horizontal partitioning**, a global entity maps to two or more local entities (i.e., partitions) storing distinct instances of conceptually related data. For example, a global entity representing customers' data can map to two local entities, one storing data about customers from Europe and another storing data about customers from America. Thus, the global entity is the union of the local partitions. In **vertical partitioning**, a global entity maps to two or more local entities (i.e., partitions), and each partition stores distinct features (attributes) of the global entity. Thus, to retrieve all the attributes of a global entity instance (e.g., tuple), one should join data from two or more local entities (i.e., vertical partitions). For example, a global entity representing customers' information can map to two local entities at distinct sources, one storing customers' mailing addresses and another storing customers' billing data.

Figure 2 exemplifies partitioning over the table *Orderspt* represented in Fig. 1. In Fig. 2, the table is split into two, one physically stored in Portugal and the other in Spain. The data in Portugal represent the stores located in Portugal and the same for Spain. Figure 2 also represents the Stores tables, which contain the store's name, localization, and e-mail.

In a distributed database scenario, one must obtain data provenance information considering all the data sources involved in the distributed query. Thus it is not possible to use plugins for a specific database, and in the case of the use of a mediator, it needs to deal with different types of databases.

Portugal			
Storespt			
name	city	email	provtoken
LisboaStore	Lisboa	ls@store.pt	t1
PortoStore	Porto	ps@store.pt	t2

Orderspt			
sname	dest	vehicle	provtoken
LisboaStore	Porto	Train	tk3
LisboaStore	Braga	Truck	tk4
PortoStore	Braga	Train	tk5
PortoStore	Madrid	Airplane	tk6

Spain			
Storesen			
name	city	email	provtoken
MadridStore	Madrid	ms@store.en	tk1
BarcelonaStore	Barcelona	bs@store.en	tk2

Ordersen			
sname	dest	vehicle	provtoken
BarcelonaStore	Madrid	Truck	tk7
BarcelonaStore	Braga	Train	tk8
MadridStore	Barcelona	Truck	tk9
MadridStore	Bilbao	Truck	tk10

Fig. 2. An example of a distributed environment for stores and orders.

2.3 Related Work

In the literature, there are several works with methods to apply W3C PROV, most of them in Workflows [12,20]. There are also works to describe Geospatial datasets in distributed environments [6].

In terms of data provenance there are examples such as ProvSQL [17], Perm [8], and GProM [1]. These are of solutions to visualize information about *where-*, *how-* and *why-provenance* and solutions for probabilistic query evaluation. ProvSQL is a lightweight extension for PostgreSQL that supports several relational database formalisms, including *where-provenance* and *how-provenance*. GProM approached it with a middleware solution for Oracle, SQLite, and PostgreSQL, but only in a centralized environment. Perm promotes rewriting the queries. However, extending these formalisms to distributed environments with different data sources (e.g., NoSQL and semi-structured) is an open issue.

The transparency in distributed environments integration helps the users to have a high-level model of the domain. Hence, they do not need to be concerned about how the data sources are connected and distributed or their heterogeneity. Nevertheless, users continue to need the information to infer the veracity and quality of the result, making the use of data provenance essential.

3   Provenance in Distributed Databases

This section shows how to obtain *how-* and *why-provenance* in a distributed databases environment using SQL. In [17], ProvSQL is an extension to PostgreSQL that changes the query execution engine. Our approach is non-intrusive and aims to work independently of the database and without changing the engine. This improves portability because our solution uses only standard SQL and not functions or stored procedures coded in languages that depend on the database management system. Furthermore, nowadays there are distributed query engines that can create an abstraction layer across data sources of different paradigms using SQL, our solution can also be used to build data provenance over distributed and heterogeneous databases (e.g., relational and NoSQL).

3.1   Architecture

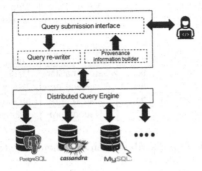

Fig. 3. Architecture and main components.

The architecture of the proposed solution is depicted in Fig. 3. The user submission interface allows users to write the queries to retrieve data from one or more databases. It is assumed that the mapping between global entities and local entities in source databases is known a priori, as discussed in the previous section. Then, the Query re-writer adds annotations to the query to obtain the provenance data and submits the request to the source databases through a distributed query engine. The latter transforms the query into sub-queries that are sent to be executed in the source databases. The distributed query execution engine gets query results containing provenance tokens from each data source and assembles a global query execution result. Then, the engine sends such a result to the Provenance Information Builder, which builds the provenance sentences and sends them to the user together with the user's query results.

For instance, considering that a user wants to execute query Q1 in a distributed environment using the data displayed in Fig. 2, the query would be as follows.

$$Q2 : \pi_{dest}\sigma_{dest=\text{``Braga''}}(\pi_{sname,dest}Orderspt \cup \pi_{sname,dest}Ordersen) \bowtie$$
$$(\pi_{vehicle,dest}Orderspt \cup \pi_{vehicle,dest}Ordersen)$$

Despite that the user might only see the global entities, the *unions* in the query are required to retrieve the data from the two local data sources. The provenance information resulting from Q2 is as follows.

Why-Provenance – {{p.orderspt:tk4,p.orderspt:tk5}, {p.orderspt:tk4,p.orderspt:tk8}, {p.orderspt:tk4}, {p.orderspt:tk5}, {p.orderspt:tk5,p.orderspt:tk8}, {p.orderspt:tk8}}

How-Provenance – (p.orderspt:tk4 $\otimes$ (p.orderspt:tk5 $\oplus$ c.ordersen:tk8))
$\oplus$ (p.orderspt:tk4 $\otimes$ p.orderspt:tk4) $\oplus$ (p.orderspt:tk5 $\otimes$ (p.orderspt:tk5 $\oplus$
c.ordersen:tk8)) $\oplus$ (p.orderspt:tk5 $\otimes$ p.orderspt:tk4) $\oplus$ (c.ordersen:tk8 $\otimes$
(p.orderspt:tk5 $\oplus$ c.ordersen:tk8)) $\oplus$ (c.ordersen:tk8 $\otimes$ p.orderspt:tk4)

Since we are in a distributed environment, and the data provenance information is given with tokens, we add additional information. The format of the provenance results has three parts separated by dot and colon characters: the first is the data source ("*p*" for PostgreSQL and "*c*" for Cassandra, in the example), the second is the table name (*orderspt* or *ordersen*) and the third is the provenance token.

3.2 Annotations

The solution proposed in this work has two premises. First, each data element (e.g., a token) in a data source must have a unique identifier as shown in [5, 10, 16]. The annotations can be seen as provenance tokens and they support the witness basis theory for *why-provenance* and the semiring theory for *how-provenance*.

As almost all databases have a function to create Universally Unique Identifiers (UUID), these are a natural choice to be used as provenance tokens. If the system does not provide UUIDs, it is needed to create a column with a unique identifier, e.g., a number or a string.

We also assume the existence of a distributed query engine (as shown in Fig. 3) that supports the standard SQL function *Listagg* [14] or a similar one. This function allows to aggregate/concatenate string values from a group of rows and separate them with a delimiter.

Our approach is to add annotations to user queries to retrieve provenance information from the data sources together with the query results themselves. The annotations depend on the operators in the query.

Distinct, Union and Group By – The annotation consists of adding columns to the user queries. In the case of a *distinct* clause, as a tuple t in a query result $Q(I)$ may have several witnesses ($I' \in I$), we use the function *listagg* to aggregate all the tokens of I' into a single value. The tokens are separated using the special character $\odot$. The *distinct* clause must be removed from the query because, as each tuple has a unique identifier (token), it would prevent the aggregation of the witnesses i' of t in a single tuple. The annotation for the operator *union* is similar to the *distinct* clause because there are also no duplicates in the result of a query, and in the case of a *group by*, we need to use a different separator, in this case $\oplus$. The different separators will help the algorithm to combine the tokens properly.

Join – In this case, it is not necessary to use the function *listaagg*, only add the tokens columns for the tables involved in the *join*. If the query is composed of sub-queries, it is required that the sub-queries have the tokens columns. For example, if we want to *join* two *unions*, we need to apply the *union* transformation explained above and add the *unions* token columns to the *join* projection.

The splitters and the columns for the *joins* will allow the built provenance information algorithm to interpret it by splitting and joining the columns and applying the *how-* and *why-provenance* methods as defined in the literature. All the columns with annotations have the name *"prov"*.

3.3   Build Provenance Information

Algorithm 1 demonstrates how the annotations are processed to obtain *how-* and *why-provenance*. Even though most of the times it is possible to derive *why-* from *how-provenance*, we opted for separate approaches in this solution. This option was based on [10], where it is demonstrated that the derivation is not straightforward. Also, if we utilize the m-semiring technique [17], the derivation becomes even more complex.

Before using the functions *HowProvenance* and *WhyProvenance*, the columns with the annotations are aggregated in an array of arrays, which will be the input parameter of both functions. As an example, the first column of *"prov"* can contain *"tk4;tk3|tk5"* and the second column *"tk8|tk9;tk10"*, and the array will have the final result *[["tk4;tk3", "tk8"], ["tk5", "tk9;tk10"]]*. This avoids the repeated looping through the annotations columns for each function.

The *HowProvenance* function starts looping through the input array and initiates variables *"temp"* and *"paren"*. In the second loop if the tokens has the character ";", it replaces the character by ⊕ because it means a *union* or a *distinct*. Between the replace function, it adds to the string *"temp"* the parenthesis and the ⊗ because the next token is part of a *join*.

If the character is not present, it adds the token and ⊗ to the string *"temp"* for the same reason as above, and the boolean *"paren"* helps place the parenthesis in the right place. In lines 12 and 13, the extra characters are removed from *"temp"* and added to an array since the second loop ended. The function will return a string that concatenates all the array positions with ⊕. This last step uses the ⊕ because the *"aggTokens"* array is created by splitting the *group by* character clause.

To obtain the *why-Provenance* we need to apply the distributive property to the *how-provenance's* result and apply the rules of witnesses basis. Thus, we need two nested loops again because the *WhyProvenance* input parameter is an array of arrays. In the first iteration of the second loop (lines 28 to 32), we populate the array *"why"* with a *set* for every token obtained from the split by the character ";".

In the subsequent iterations, we need to apply the distribution. If the array *"why"* length is higher than the length of the split array, for each *set* in *"why"* we add the tokens obtained from the split (lines 34 to 38). Else for each token in the split, we loop through the *"why"* array and copy the *"why"* to a temporary variable, add to this temporary variable the token in the split and add it to a temporary array. In the end, *"why"* will be equal to the temporary array. The return clause will return a string constructed by the function *CheckDoubles* that also removes the possible similar sets.

Algorithm 1. How- and Why-provenance algorithm

```
 1: function HOWPROVENANCE(aggTokens)
 2:     how ← []
 3:     for each agt ∈ aggTokens do
 4:         temp ←''; paren ← False;
 5:         for each t ∈ agt do
 6:             if t ⊂';' then
 7:                 temp ← temp +' ('+t.replace(';' ,' ⊕')+')⊗'
 8:             else
 9:                 if notparen then
10:                     temp ←' ('+temp + t +' ⊗'
11:                     paren ← True
12:                 else
13:                     temp ← temp + t+')⊗'
14:                     paren ← False
15:                 end if
16:             end if
17:         end for
18:         temp ←' ('+temp.RemoveExtraChars()+')'
19:         how.add(temp)
20:     end for
21:     return how.join(⊕)
22: end function
23:
24: function WHYPROVENANCE(aggTokens)
25:     why ← []
26:     for each agt ∈ aggTokens do
27:         for i = 0, 1, . . . length(agt) do
28:             if i == 0 then
29:                 for each t ∈ agt[i].split(';' ) do
30:                     tSet ← Set(); tSet.add(t); why.add(tSet)
31:                 end for
32:             else
33:                 if length(agt[i].split(';' )) < lenght(why) then
34:                     for each wt ∈ why do
35:                         for each t ∈ agt[i].split(';' ) do
36:                             wt.add(t)
37:                         end for
38:                     end for
39:                 else
40:                     copyWT ← []
41:                     for each t ∈ agt[i].split(';' ) do
42:                         for each wt ∈ why do
43:                             temp ← wt.copy(); temp.add(t); copyWT.add(temp)
44:                         end for
45:                     end for
46:                     why = copyWT
47:                 end if
48:             end if
49:         end for
50:     end for
51:     return CheckDoubles(why)
52: end function
```

4   Experimental Evaluation

As proof of concept for our solution, we use EasyBDI [19], an open-source prototype for logical integration of distributed databases that provides mapping functionalities between local and global schemas. EasyBDI has a graphical interface that allows the users to query over the global schemas without writing SQL commands. The interface provides different frames where the user can drag and drop entity columns and the operators (e.g., group by) to use on the query.

When the user executes the query, EasyBDI builds the SQL command based on the mapping between GCS and LCSs and submmits it to Trino, a distributed query execution engine. Since the software is open-source, we modified the query generation module to add the annotation columns when performing the query build. We also applied the proposed algorithm to the query execution result.

As dataset, we used the tables represented in Fig. 2. PostgreSQL stores the data about Portugal, and the ones about Spain is in a Cassandra database. EasyBDI allows the user to identify mapping types. In this case, there is a horizontal mapping (which means that the global entity is horizontally partitioned through two data sources), i.e., the global entity representing the orders maps to structures in Cassandra and PostgreSQL. The first example is a *distinct* query to obtain all the vehicles used in orders. The executed query is:

> **SELECT** vehicle, listagg(prov, ';') **WITHIN GROUP (ORDER BY vehicle) as prov FROM (SELECT** sname, dest, vehicle, listagg(provtoken, ';') **WITHIN GROUP (ORDER BY sname) as prov FROM(** SELECT sname, dest, vehicle, **provtoken** FROM postgresql.public.orderspt UNION SELECT sname, dest, vehicle, **provtoken** FROM cassandra.stkspace.ordersen) **GROUP BY sname, dest, vehicle) GROUP BY vehicle**

In the above query, the clauses in bold are the ones we added to the query. Starting with the sub-query, the local schemas' union is needed to obtain the global entity. Since we add *"provtoken"* to the tables and they might be different, the *union* result would be erroneous without the *group by* clause. Thus, we also add the *group by* clause and the *listagg* function. In the main query, the *distinct* clause has been removed, and we used a *group by* clause again with the column in the *distinct* and the *listagg* to aggregate the tokens. The result obtained is the following:

For *"Airplane"*, the provenance is simple. We have only one token as a witness for the *why-provenance* and the same token for *how-provenance*. For *"Train"* and *"Truck"* we have different witnesses, and we can also obtain each row using one of the tokens. Since in the *How-provenance* column the tokens are separated by ⊕, we can use one of the tokens only to obtain the rows.

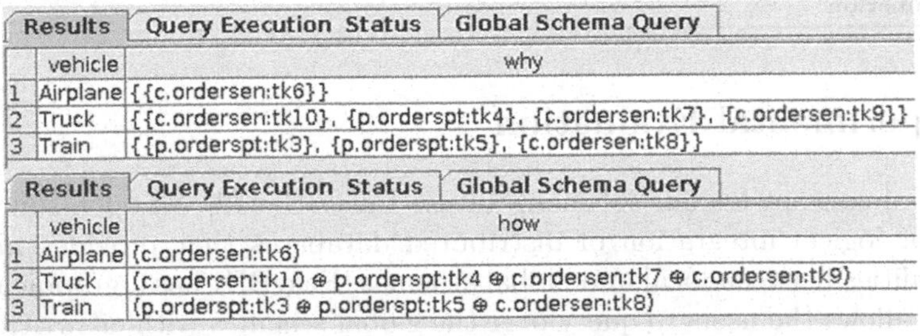

Results	Query Execution Status	Global Schema Query
vehicle		why
1 Airplane	{{c.ordersen:tk6}}	
2 Truck	{{c.ordersen:tk10}, {p.orderspt:tk4}, {c.ordersen:tk7}, {c.ordersen:tk9}}	
3 Train	{{p.orderspt:tk3}, {p.orderspt:tk5}, {c.ordersen:tk8}}	

Results	Query Execution Status	Global Schema Query
vehicle		how
1 Airplane	{c.ordersen:tk6}	
2 Truck	{c.ordersen:tk10 ⊕ p.orderspt:tk4 ⊕ c.ordersen:tk7 ⊕ c.ordersen:tk9}	
3 Train	{p.orderspt:tk3 ⊕ p.orderspt:tk5 ⊕ c.ordersen:tk8}	

Fig. 4. The result of the distinct query.

The following query is a *join* between the stores and orders to obtain the orders' destination and the stores' e-mail responsible for the orders. The *unions* are simplified, since they are equal to the last query, just now for the two tables.

SELECT s.email, o.dest, **s.prov as prov, o.prov as prov** FROM (– UNION STORES –) s, (– UNION ORDERS –) o WHERE s.name = o.sname

The *unions* are again applied to obtain the GCS. We added the annotations' columns for the tables/view/query involved in the join to the query projection. The result in Fig. 5 shows that all *why-provenance's* tokens are in pairs of witnesses: in order to obtain any row, we need both of the tokens. In contrast with the query of Fig. 4, now the tokens are separated by $\otimes$ in *how-provenance*, each means that we need a join between both tokens.

Results	Query Execution Status	Global Schema Query	
	email	dest	why	how
1	ls@store.pt	Porto	{{p.storespt:tk1 , p.orderspt:tk3}}	(p.storespt:tk1 ⊗ p.orderspt:tk3)
2	ls@store.pt	Braga	{{p.storespt:tk1 , p.orderspt:tk4}}	(p.storespt:tk1 ⊗ p.orderspt:tk4)
3	ps@store.pt	Madrid	{{p.storespt:tk2 , p.orderspt:tk6}}	(p.storespt:tk2 ⊗ p.orderspt:tk6)
4	ps@store.pt	Braga	{{p.storespt:tk2 , p.orderspt:tk5}}	(p.storespt:tk2 ⊗ p.orderspt:tk5)
5	ms@store.pt	Barcelona	{{c.storesen:tk1 , c.ordersen:tk9}}	(c.storesen:tk1 ⊗ c.ordersen:tk9)
6	ms@store.pt	Bilbao	{{c.storesen:tk1 , c.ordersen:tk10}}	(c.storesen:tk1 ⊗ c.ordersen:tk10)
7	bs@store.pt	Madrid	{{c.storesen:tk2 , c.ordersen:tk7}}	(c.storesen:tk2 ⊗ c.ordersen:tk7)
8	bs@store.pt	Braga	{{c.storesen:tk2 , c.ordersen:tk8}}	(c.storesen:tk2 ⊗ c.ordersen:tk8)

Fig. 5. The result of the query with join

The last query example is a *group by* the previous query applied to *"dest"*. Since it is a *group by*, we need to use the *listagg* function in the *joins'* columns.

SELECT o.dest, **listagg(s.prov, '|') WITHIN GROUP (ORDER BY o.dest) as prov, listagg(p.prov, '|') WITHIN GROUP (ORDER BY o.dest) as prov** FROM (– UNION STORES –) s, (– UNION ORDERS –) o WHERE s.name = o.sname GROUP BY o.dest

Results	Query Execution Status	Global Schema Query
	dest	why
1	Barcelona	{c.storesen:tk1 , c.ordersen:tk9}
2	Bilbao	{c.storesen:tk1 , c.ordersen:tk10}
3	Braga	{{c.storesen:tk2 , c.ordersen:tk8}, {p.storespt:tk1 , p.orderspt:tk4}, {p.storespt:tk2 , p.orderspt:tk5}}
4	Madrid	{{c.storesen:tk2 , c.ordersen:tk7}, {p.storespt:tk2 , p.orderspt:tk6}}
5	Porto	{p.storespt:tk1 , p.orderspt:tk3}

	dest	how
1	Barcelona	(c.storesen:tk1 ⊗ c.ordersen:tk9)
2	Bilbao	(c.storesen:tk1 ⊗ c.ordersen:tk10)
3	Braga	{{c.storesen:tk2 ⊗ c.ordersen:tk8} ⊕ (p.storespt:tk1 ⊗ p.orderspt:tk4) ⊕ (p.storespt:tk2 ⊗ p.orderspt:tk5)}
4	Madrid	{{c.storesen:tk2 ⊗ c.ordersen:tk7} ⊕ (p.storespt:tk2 ⊗ p.orderspt:tk6)}
5	Porto	(p.storespt:tk1 ⊗ p.orderspt:tk3)

Fig. 6. The result of the group by query

As demonstrated in Fig. 6, some rows now have more than one pair of witnesses for the *why-provenance*. *How-provenance* column shows that it is possible to *join* different tokens to obtain the rows. In the result of destinations *"Braga"* and *"Madrid"*, we can see that the result can be obtained from the two databases because both *"why"* and *"how"* have tokens from the two sources.

5   Conclusions and Future Work

This work discusses data provenance in distributed environments, which is essential to infer the data's veracity and quality.

We present a solution to generate *how-* and *why-provenance* using pure SQL queries with annotations and an algorithm to build the provenance information. It is a non-intrusive solution that does not require any change to the distributed query execution engine. Also, it is not specific to any database system or model. We also present an implementation of our proposals on EasyBDI. It is a logical database integration tool based on which users query entities from global schemas that abstract the data organization on each data source. There is no materialization. Distributed query processing and provenance data generation are done on the fly, without materializations.

In future work, we plan to study how to generate other types of provenance (e.g., *where-provenance*) following the same logic used here. Since we are working with distributed environments, another issue is how to generate provenance information in contexts where materializations are used for database integration and analytic processing.

Acknowledgments. Paulo Pintor has a research grant awarded by the Portuguese public agency for science, technology and innovation FCT - Foundation for Science and Technology - under the reference 2021.06773.BD. This work is partially funded by National Funds through the FCT under the Scientific Employment Stimulus - Institutional Call - CEECINST/00051/2018, and in the context of the projects UIDB/04524/2020 and UIDB/00127/2020.

References

1. Arab, B.S., Feng, S., Glavic, B., Lee, S., Niu, X., Zeng, Q.: GProM - a swiss army knife for your provenance needs. IEEE Data Eng. Bull. **41**(1), 51–62 (2018). http://sites.computer.org/debull/A18mar/p51.pdf
2. Bidoit, N., Herschel, M., Tzompanaki, A.: Efficient computation of polynomial explanations of why-Not questions. In: International Conference on Information and Knowledge Management, Proceedings 19–23 October 2015, pp. 713–722 (2015). https://doi.org/10.1145/2806416.2806426
3. Buneman, P., Khanna, S., Tan, W.C., Chiew, W.: Why and where: a characterization of data provenance. Comput. Sci. **1973**, 316–330 (2001)
4. Buneman, P., Tan, W.C.: Data provenance: what next? SIGMOD Rec. **47**(3), 5–16 (2018). https://doi.org/10.1145/3316416.3316418

5. Cheney, J., Chiticariu, L., Tan, W.C.: Provenance in databases: why, how, and where. Found. Trends Databases **1**, 379–474 (2007). https://doi.org/10.1561/1900000006

6. Closa, G., Masó, J., Proß, B., Pons, X.: W3C PROV to describe provenance at the dataset, feature and attribute levels in a distributed environment. Comput. Environ. Urban Syst. **64**, 103–117 (2017). https://doi.org/10.1016/j.compenvurbsys.2017.01.008, http://dx.doi.org/10.1016/j.compenvurbsys.2017.01.008

7. Duggan, J., et al.: The BigDAWG polystore system. SIGMOD Rec. **44**(2), 11–16 (2015). https://doi.org/10.1145/2814710.2814713

8. Glavic, B., Alonso, G.: Perm: processing provenance and data on the same data model through query rewriting. In: Proceedings of the International Conference on Data Engineering, pp. 174–185. IEEE, Shanghai, China (2009). https://doi.org/10.1109/ICDE.2009.15

9. Green, T.J., Karvounarakis, G., Tannen, V.: Provenance semirings. In: Proceedings of the Twenty-Sixth ACM SIGMOD-SIGACT-SIGART Symposium on Principles of Database Systems, pp. 31–40, PODS 2007. Association for Computing Machinery, New York, NY, USA (2007). https://doi.org/10.1145/1265530.1265535

10. Green, T.J., Tannen, V.: The semiring framework for database provenance. In: Proceedings of the 36th ACM SIGMOD-SIGACT-SIGAI Symposium on Principles of Database Systems, pp. 93–99, PODS 2017. Association for Computing Machinery, New York, NY, USA (2017). https://doi.org/10.1145/3034786.3056125

11. Herschel, M., Diestelkämper, R., Ben Lahmar, H.: A survey on provenance: What for? What form? What from? VLDB J. **26**(6), 881–906 (2017). https://doi.org/10.1007/s00778-017-0486-1

12. Kock-Schoppenhauer, A.K., Hartung, L., Ulrich, H., Duhm-Harbeck, P., Ingenerf, J.: Practical extension of provenance to healthcare data based on the W3C PROV standard. Stud. Health Technol. Inform. **253**, 28–32 (2018). https://doi.org/10.3233/978-1-61499-896-9-28

13. Lu, J., Holubová, I.: Multi-model databases: a new journey to handle the variety of data. ACM Comput. Surv. **52**, 1–38 (2019). https://doi.org/10.1145/3323214

14. Michels, J., et al.: The new and improved SQL: 2016 standard. SIGMOD Rec. **47**, 51–60 (2018). https://doi.org/10.1145/3299887.3299897

15. Moreau, L., Groth, P., Cheney, J., Lebo, T., Miles, S.: The rationale of PROV. J. Web Semant. **35**, 235–257 (2015). https://doi.org/10.1016/j.websem.2015.04.001

16. Senellart, P.: Provenance and probabilities in relational databases: from theory to practice. SIGMOD Rec. **46**, 5–15 (2017). https://doi.org/10.1145/3186549.3186551

17. Senellart, P., Jachiet, L., Maniu, S., Ramusat, Y.: ProvSQL: provenance and probability management in PostgreSQL. Proc. VLDB Endow. **11**(12), 2034–2037 (2018). https://doi.org/10.14778/3229863.3236253

18. Sethi, R., et al.: Presto: SQL on everything. In: 2019 IEEE 35th International Conference on Data Engineering (ICDE), pp. 1802–1813 (2019). https://doi.org/10.1109/ICDE.2019.00196

19. Silva, B., Moreira, J., de Costa, R.L.C.: EasyBDI: near real-time data analytics over heterogeneous data sources. In: Advances in Database Technology - EDBT 2021-March, pp. 702–705 (2021). https://doi.org/10.5441/002/edbt.2021.88

20. Zhang, M., Jiang, L., Zhao, J., Yue, P., Zhang, X.: Coupling OGC WPS and W3C PROV for provenance-aware geoprocessing workflows. Comput. Geosci. **138**, 104419 (2020). https://doi.org/10.1016/j.cageo.2020.104419

21. Özsu, M.T., Valduriez, P.: Principles of Distributed Database Systems. Springer, Cham (2020). https://doi.org/10.1007/978-3-030-26253-2

Provenance-Based SPARQL Query Formulation

Yael Amsterdamer[(✉)] and Yehuda Callen

Bar-Ilan University, Ramat-Gan, Israel
{yael.amsterdamer,calleny}@biu.ac.il

Abstract. We present in this paper a novel solution for assisting users in formulating SPARQL queries. The high-level idea is that users write "semi-formal SPARQL queries", namely, queries whose structure resembles SPARQL but are not necessarily grounded to the schema of the underlying Knowledge Base (KB) and require only basic familiarity with SPARQL. This means that the user-intended query over the KB may differ from the specified semi-formal query in its structure and query elements. We design a novel framework that systematically and gradually refines the query to obtain candidate formal queries that do match the KB. Crucially, we introduce a formal notion of provenance tracking this query refinement process, and use the tracked provenance to prompt the user for fine-grained feedback on parts of the candidate query, guiding our search. Experiments on a diverse query workload with respect to both DBpedia and YAGO show the usefulness of our approach.

1 Introduction

A huge body of information is stored in RDF knowledge Bases (KBs) such as YAGO [33], DBpedia [11] and others. Such KBs can be queried using SPARQL, the W3C standard query language [32]. Familiarity with the SPARQL syntax, however, is not sufficient for using it: writing a SPARQL query requires a deep understanding of the KB content and structure. In turn, in many useful KBs, this structure is highly complex and contains many irregularities. The KB structure often further evolves over time, and so even users who are initially familiar with it may struggle to adapt to such modifications.

To this end, our system SPARQLIt[1] (see Fig. 1 for a high-level architecture) allows users to write a "semi-formal" query (step 0 in the Figure), i.e., a query whose syntax follows that of SPARQL, but its contents – entities and properties – do not necessarily match the KB. For example, a user seeking graduates of Columbia University living in India may write a query with the selection criteria `?x livesIn India. ?x graduatedFrom columbia`. When evaluated over YAGO, the user may discover that the query returns little or no results. The reason is that in YAGO, `livesIn` is usually populated with cities rather than countries and that Columbia is represented as `<Columbia_University>` – but it is

---

[1] The code of SPARQLIt implementation is available at [10].

© The Author(s), under exclusive license to Springer Nature Switzerland AG 2022
C. Strauss et al. (Eds.): DEXA 2022, LNCS 13426, pp. 116–129, 2022.
https://doi.org/10.1007/978-3-031-12423-5_9

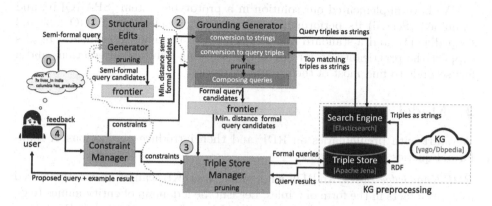

Fig. 1. Overall architecture of SPARQLIt

difficult for a user to discover this. To this end, our solution generates candidate formal queries, i.e., queries that match the KB. Semi-formal queries are transformed to candidate formal queries via sequences of operations of two flavors: *structural edits* (step 1 in Fig. 1), which are proposals for similar, alternative structures of semi-formal queries, e.g., replacing `?x livesIn India` by `?x livesIn ?y. ?y isLocatedIn India`; and *groundings* (step 2) where elements of semi-formal queries are replaced by elements of the KB, e.g., replacing `columbia` by `<Columbia_University>`. We detail in Sects. 3–5 the ways in which structural edits and groundings are generated to yield relevant candidate queries. Candidate queries are evaluated over the KB (step 3) so that query results may also serve to filter candidates (see below).

Crucially, we introduce a formal notion of *provenance tracking* throughout the process: we track the sequence of transformations that are performed, leading from the input semi-formal query to each candidate formal query, as well as the binding of variables of candidate queries to the KB. Provenance tracking serves as the basis of procuring *fine-grained feedback* from the user with respect to proposed formal queries. Namely, for a candidate formal query, we present not only the query itself and its example evaluation results, but also the *way* in which each element in the candidate query was obtained, i.e., from which element in the user's semi-formal query it has been (indirectly) transformed, if any. This is combined with the provenance of query evaluation (i.e. binding of formal query variables to KB elements). See bottom pane of Fig. 2 for an example of presented provenance, in the form of (semi-formal element, formal element, binding example). Feedback is then procured with respect to each such piece of provenance: the user may mark "must", "must not" or "don't care". The user feedback is converted by the *Constraints Manager* (step 5) to constraints on proposed queries, which are accumulated and used by the other modules to prevent the generation of queries that do not comply with user feedback. This is repeated until the user finds a proposed query satisfactory.

We have implemented our solution in a prototype system (SPARQLIt) and examined (Sect. 6) its performance over two large-scale KBs, YAGO [33] and DBpedia [11], and a standard query benchmark [34]. The experimental results support the practicality of our approach, requiring only few interactions and a few seconds to find most of the examined queries.

2   Model

We start with preliminaries on RDF and then introduce our notions of semi-formal queries and provenance.

RDF Knowledge Bases. An RDF Knowledge Base (KB) can be abstractly viewed as a set of facts in the form of triples. Let Ent be a domain of entity names (e.g., India, University) and Lit be a domain of literals (e.g., 2021). Let Pred be a domain of predicate names (e.g., livesIn). An *RDF knowledge base* is a set of triples of the form (s, p, o) where $s \in$ Ent is the subject, $p \in$ Pred is the predicate and $o \in$ Ent $\cup$ Lit is the object. We use *element* to uniformly refer to an entity, literal or predicate. We will sometimes represent multiple triples $\langle s, p, o \rangle, \langle s', p', o' \rangle$ by the *n3 notation* s p o. s' p' o'.

Formal SPARQL Selection Queries. To query RDF, we use the notion of Basic Graph Patterns (BGPs). Let Var $= \{?\text{x}, ?\text{y}, \dots \}$ be a set of variables. A *Triple Pattern* is a member of the set (Ent $\cup$ Var) $\times$ (Pred $\cup$ Var) $\times$ (Lit $\cup$ Pred $\cup$ Var). Namely, in a triple pattern, subjects, predicates and objects may be replaced by variables. A BGP is a set of triple patterns, and its graph view may be obtained in the same way RDF KBs are encoded as graphs. A *formal SPARQL selection query* $Q = (G_Q, V_Q)$ then consists of a BGP G_Q and a set of output variables V_Q, which is a subset of the variables occurring in G_Q.

Query Evaluation. Given a formal SPARQL selection query $Q = (G_Q, V_Q)$ and an RDF KB G, let φ be a mapping of all variables in G_Q to RDF terms in G. Denote by $\varphi(G_Q)$ the result of replacing in G_Q every variable v by $\varphi(v)$. If $\varphi(G_Q) \subseteq G$ (i.e. all obtained triples are in the KB G) then we say that φ is a *binding*. Each binding φ yields a *query answer* $A = \varphi|_{V_Q}$ (φ restricted to output variables) and the *query result* $Q(G)$ is the set of all such answers.

Semi-formal Queries. We now introduce the notion of *semi-formal queries*, that have a similar form to that of formal SPARQL queries, but their labels are not necessarily bound to the corresponding set of names in the KG. Additionally, they may include special temporary placeholder elements from $\text{Temp}^{sf} = \{??\text{X}, ??\text{Y}, \dots \}$ to be replaced by any KB term (entity/literal/predicate) when we transform the query into a formal one (see below). Formally, let $\text{Ent}^{sf} \supset \text{Ent}$, $\text{Lit}^{sf} \supset \text{Lit}$, $\text{Pred}^{sf} \supset \text{Pred}$ be extended sets of entity, literal and predicate names, abstractly capturing any element the user may write, including formal elements. A semi-formal BGP is then a BGP whose triple patterns are elements of $(\text{Ent}^{sf} \cup \text{Var} \cup \{\text{Temp}^{sf}\}) \times (\text{Pred}^{sf} \cup \text{Var} \cup \{\text{Temp}^{sf}\}) \times (\text{Ent}^{sf} \cup \text{Lit}^{sf} \cup \text{Var} \cup \{\text{Temp}^{sf}\})$. A *semi-formal query* $Q = (G_Q, V_Q)$ consists of a semi-formal BGP and a distinguished subset V_Q of output variables.

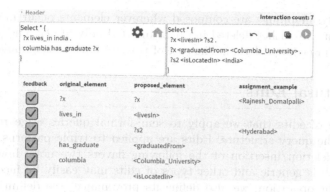

Fig. 2. SPARQLIt User interface

Example 1. Figure 2 is a screenshot of SPARQLIt, where the top-left panel displays a semi-formal query: its syntax follows that of SPARQL, yet some of its elements, e.g., `columbia`, `has_graduate`, do not occur in the queried KB (YAGO). The top-right panel displays a formal SPARQL query matching YAGO, for which an example binding is $\varphi(?x)$ =`Rajnesh_Domalpalli` and $\varphi(?s2)$ =`Hyderabad`.

Provenance Model. We introduce two types of provenance. The first is "standard": the provenance of a binding φ obtained by evaluating a formal SPARQL Query Q over a KB G, denoted $\mathrm{prov}(Q, G, \varphi)$, is represented as a set of variable-value pairs of the form (x, v). The provenance of a query answer $A \in Q(G)$, denoted $\mathrm{prov}(Q, G, A)$ is then a set of such provenance representations, for all the bindings of Q in G yielding A. The second type of provenance is novel, and is geared towards tracking the gradual refinement of queries, as follows.

Definition 1. *Given two (formal or semi-formal) BGPs Q and Q', a provenance expression for a transformation of Q to Q' is denoted by $\mathrm{prov}(Q, Q') = (P, C)$: P is a set of pairs (e, e') where e is either $\perp$ or an element of Q and e' is either $\perp$ or an element of Q', such that (1) $(\perp, \perp) \in P$ and (2) each element of Q and of Q' appears in exactly one pair. $C \in \mathbb{N}$ is the transformation cost.*

Intuitively, pairs encode "mappings" of individual elements in Q to elements in Q'; the notation $\perp$ is used to mark deletions/insertions of elements.

We will show in the sequel how to attach provenance to concrete transformations, yet we already note that an important property of provenance is composability: namely, need to be able to combine provenance expressions of a sequence of refinements to yield a provenance expression for the entire sequence.

Definition 2. *Let Q_0, Q_1, Q_2 be three queries, and let $\mathrm{prov}(Q_0, Q_1) = (P_0, C_0)$ and $\mathrm{prov}(Q_1, Q_2) = (P_1, C_1)$. We compose them to provenance $\mathrm{prov}(Q_0, Q_2) = (P_2, C_2)$ by $P_2 = \{(e_0, e_2) \mid \exists e_1 \neq \perp.(e_0, e_1) \in P_0 \wedge (e_1, e_2) \in P_1\} \cup \{(\perp, e) \mid (\perp, e) \in P_1\} \cup \{(e, \perp) \mid (e, \perp) \in P_0\}$. As for cost, $C_2 = C_0 + C_1$.*

Intuitively, mappings are composed wherever elements occur in Q_1; otherwise, elements are either deleted in Q_1 or inserted only in Q_2. The provenance records this insertion/deletion. The cost of transformations is cumulative.

3   Structural Edits

A first type of edits that we apply to semi-formal queries is geared towards modifying the query structure. Edits are applied to triple patterns, capturing reordering/deletion/insertion of the following flavors (we note, however, that our approach is generic and other types of edits may easily be incorporated). For each edit operation, we also define its provenance (see definition 1), but keep the costs abstract at this point and discuss concrete cost choices below.

Subject-Object Switching: switch the subject and object of $T = \langle s, p, o \rangle$, yielding $T' = \langle o, p, s \rangle$. The provenance captures an identity mapping, i.e., its set of pairs is $\{(e, e)\}$ for each element e of the (original and result) query.

Element Exclusion: replace the subject/object/predicate of $T = \langle s, p, o \rangle$ by a fresh variable ?x, yielding $T' = \langle ?x, p, o \rangle$ or $T' = \langle s, ?x, o \rangle$ or $T' = \langle s, p, ?x \rangle$. The fresh variable is not in the output and thus can be bound to any element of the KB. For $T' = \langle ?x, p, o \rangle$ the provenance will include the pairs $(s, \perp)$, $(\perp, ?x)$, (p, p) and (o, o) and (e, e) for every other element e; the provenance is similarly defined for $T' = \langle s, ?x, o \rangle$ or $T' = \langle s, p, ?x \rangle$. Note that the excluded element is mapped to $\perp$, which means we indeed stop tracking it. *Predicate Splitting:* replace $T = \langle s, p, o \rangle$ by two triples $T' = \langle s, p, ?x \rangle$, $T'' = \langle ?x, ??Y, o \rangle$ where ?x is a fresh variable and ??Y is a fresh placeholder. This stands for replacing a predicate by a path with two predicates. The provenance includes (s, s), (p, p), $(\perp, ?x)$, $(\perp, ??Y)$, (o, o), and (e, e) for every other element e.

Example 2. Consider the triple pattern ?x lives_In India. If an inverse predicate is used in the KB, a candidate query may be generated by Subject-Object Switching yielding India lives_In ?x. Alternatively, applying Element Exclusion could yield ?x ?p India or ?x lives_In ?y, generalizing the query by placing a variable that may be bound to any predicate/entity. The provenance includes a record of the newly added variable $((\perp, \{?p)$ or $(\perp, \{?y))$ and associates the other nodes and edges with their counterparts in the refined query. Last, the KB may include information about people living in cities rather than directly in countries. Applying Predicate Splitting on ?x lives_In India would yield the two triple patterns ?x lives_In ?s2. ?s2 ??P1 India. The placeholder $??P1 \in \text{Temp}^{sf}$ will be ultimately replaced in further edit steps by a predicate such as isLocatedIn.

Example 3. Reconsider the semi-formal query in Fig. 2, and consider the application of Predicate Splitting to ?x lives_In India to ?x lives_In ?s2. ?s2 ??P1 India, followed by Subject-Object Switching for columbia has_graduate ?x. Composing the two provenance expressions we obtain the pairs $(\perp, ?s2)$ and $(\perp, ??P1)$ along with pairs (e, e) for $e = $ India, ?x, etc.

Searching for Structural Edits. We store a *frontier* of semi-formal queries, initially including only the input one. Whenever prompted, the Generator applies to each semi-formal query currently in the frontier, the least costly possible edit. Two types of pruning are applied to the generation of semi-formal queries. First, multiple sequences of edit operations may result in the same semi-formal query, in which case we keep in the frontier only the minimum-cost representative. Second, the structural edits generator maintains a cache of semi-formal sub-queries for which no formal query exists, i.e., they were rejected by other modules (as described in the sequel). These and queries contained in them are ignored in subsequent steps. The overall minimal-cost candidate is passed on to the Grounding Generator with its provenance.

Cost. The assignment of cost for each operation may be viewed as a configuration choice. We have experimented with different cost assignments, and observed that it generally useful to render a single structural edit operation more costly than grounding the entire query (i.e. we prefer groundings that use the current structure, if exist). We thus set the structural edit costs to be greater than C, which is an upper bound on the grounding cost (see Sect. 4). Specifically, we have superior optimal results with costs $2C$, $100C$ and $30C$ for Object-Subject Switching, Element Exclusion and Predicate Splitting respectively.

4 Grounding Generator

The Grounding Generator gets as input a semi-formal query Q' and the KB G. It generates formal queries by replacing entities/predicates in Q' that do not occur in G by ones that do. We start by generating a ranked list of groundings for individual triple patterns in Q and then combine groundings that are consistent with each other to form a query. We next explain each of the two steps.

Triple Groundings. Given a semi-formal triple pattern t' we generate a ranked list of k formal triple patterns $t_1, ..., t_k$. These are generated as follows. First, we represent t as a string $s(t)$ by removing SPARQL syntax (including variables and placeholders) and performing tokenization. Then we feed $s(t)$ to a *black-box search engine* that indexes triples from the KB. The engine returns the top-k relevant KB triples along with their string representation. We augment these triples back to triple patterns, plugging into them any variable that has occurred in t'. Unlike variables, placeholders are not added back to the triple patterns, so that they are grounded by KB elements.

Example 4. Consider for example the triple pattern `?x lives_in India`. First, we remove SPARQL notations and perform an initial tokenization, which yields *"lives in India"*. The search engine results includes, e.g., the strings *"Aadya lives in India"* and *"Aarav lives in Indianapolis"*, attached to the KB triples `<Aadya> <livesIn> <India>`, `<Aarav> <livesIn> <Indianapolis>`. Since the original triple pattern has `?x` as subject, we replace the subject in the KB triples by `?x` and obtain the ranked list `?x <livesIn> <India>`, `?x <livesIn> <Indianapolis>`.

Provenance. We define the provenance for groundings in a similar way to that of refinements. For a semi-formal triple pattern $t' = (s', p', o')$ and a choice of formal triple pattern $t = (s, p, o)$ as grounding, we introduce the pairs $(s', s), (p', p), (o', o)$ to be stored in the provenance. We discuss costs below.

From Grounded Triple Patterns to Formal BGPs. We use the ranked lists of formal triple patterns obtained for each triple pattern t' in the semi-formal query, to yield candidate formal BGPs. We traverse these lists in order, each time choosing a single candidate triple pattern for each t' (i.e. we start with the set of all top-1 triple patterns). For each choice of triples, we check their provenance for consistency, namely, that no two pairs $(x, y), (x, z)$ such that $y \neq z$ appear in their provenance. If the set is consistent then the formal triple patterns are concatenated to form a BGP G_Q. Otherwise (or when the Grounding Generator is prompted for the next query), the process is repeated, with one of the triple patterns being replaced by the next-best one in the ranked list, and so forth.

Provenance Revisited. Recall that only triple patterns with consistent provenance expressions were combined. The overall provenance is then defined as follows: its pairs set is the union of pairs sets in the provenance of all triple patterns; the cost is the sum of costs stored in these provenance expressions.

From BGPs to Queries. So far, we have generated formal BGPs as candidates, mapping the semi-formal BGP to each of them. Recall that the semi-formal query Q' includes a distinguished subset $V_{Q'}$ of output variables. For each formal BGP G_Q with provenance $\text{prov}(G_{Q'}, G_Q) = (P, C)$, the set of output variables is defined as $V_Q = \{v \mid (v', v) \in P \land v' \in V_{Q'}\}$. The query $Q = (G_Q, V_Q)$ is the obtained candidate, with the carried provenance staying intact, i.e., $\text{prov}(Q', Q) = \text{prov}(G_{Q'}, G_Q)$. We next show how provenance of edits and groundings may be composed.

Example 5. Following Example 3, we generate candidate groundings for `lives_In`, `India`, `has_graduate`, `columbia` and the placeholder `??P1`. A formal query resulting from one such combination of groundings is shown on the top-right part of Fig. 2 with its provenance on the bottom: e.g., `has_graduate` has transformed to `<graduatedFrom>`, while `isLocatedIn` is newly added (so it has no counterpart).

Cost. The cost for grounding a triple pattern t to t' is set based on the string distance measures between their representative strings $s(t), s(t')$, generated by the search engine as explained above. Specifically, we use the Levenshtein edit distance. An exception is the grounding of placeholders, which has 0 cost by definition. To account for *semantic synonyms* that are represented by very different strings, we generate a set of synonyms (using https://www.datamuse.com/api/) for each string, and take the minimal edit distance between a synonym of $s(t)$ and a synonym of $s(t')$. We revisit this design choice in Sect. 6. Last, recall that our choice of structural edit costs relied on an upper bound C for the grounding cost; we set C to be the maximal string length of a representative string of an element in the KB, multiplied by the number of query terms.

5   Procuring Feedback

The Triple Store Manager receives as input a formal SPARQL query Q produced by the Grounding Generator, and executes it over a black-box triple store (we have used Apache Jena [4]). The query result may be empty: this is typically an indication that Q does not match the user intention, and we search for alternative queries (we revisit this assumption in Sect. 6). If the query result is non-empty, we choose an example result, and an example binding yielding it. To procure feedback, we combine the provenance accumulated throughout the process of generating the candidate query, with the provenance of the example query result.

Definition 3. *Let Q', Q be a semi-formal and formal query respectively and let* $\text{prov}(Q', Q) = (P, C)$. *Further let G be a KB, $A \in Q(G)$ an answer, φ a binding yielding A and* $\text{prov}(Q, G, \varphi)$ *its provenance. The provenance* $\text{prov}(Q', Q, G, \varphi) = (P', C)$ *where* $P' = \{(e, e', v) \mid (e, e') \in P \wedge e' \in Var \wedge (e', v) \in \text{prov}(Q, G, \varphi)\} \cup \{(e, e', \perp) \mid (e, e') \in P \wedge e' \notin Var \wedge (e, e') \neq (\perp, \perp)\}$.

The user is then prompted for feedback on each triplet in the provenance, and may choose one of the following responses for a given triplet (e, e', v). MUST: from now on, only formal queries Q for which $(e, e', v) \in \text{prov}(Q', Q, G, \varphi')$ for some binding φ' will be proposed. MUST NOT: only formal queries Q for which $(e, e', v) \notin \text{prov}(Q, Q'', G, \varphi')$ for all bindings φ' will be proposed. MAYBE: no restrictions are imposed on the triplet.

Example 6. The rightmost column on the bottom table of Fig. 2 shows variable bindings for a query output example, e.g., the binding of `?s2` to `Hyderabad`. This allows procuring informative feedback for each corresponding provenance triplet, through the checkboxes to the left. For instance, the user may confirm that `<livesin>` is a correct grounding for `lives_in`. This will lead to considering only queries with (`lives_in`, `<livesin>`, $\perp$). In contrast, they may e.g. convey that the assignment of `Rajnesh_Domalpalli` to `?x` is incorrect, leading to pruning any candidate that includes (`?x`, `?x`, `Rajnesh_Domalpalli`) (using our edit operations, every variable is always mapped to itself or $\perp$). Both positive and negative feedback significantly narrows the search space of possible queries.

6   Experiments

We have implemented our solution in a prototype called SPARQLIt. The prototype is implemented in .Net, using Blazor [8] for its front-end, Elasticsearch [17] for the Search Engine (used for groundings) and Apache Jena [4] for the Triple Store. All experiments were run on Intel i7-core processors with 32GB of RAM.

As Knowledge Bases, we have used YAGO [33] English facts (approx. 160M triples) and DBPedia [11]. Since the notion of semi-formal queries is novel, to our knowledge, no existing benchmarks are available. To this end, we have constructed benchmarks based on the first 50 NL questions in the training set of QALD-9 [34]. Their translation to gold formal queries w.r.t. DBpedia is given

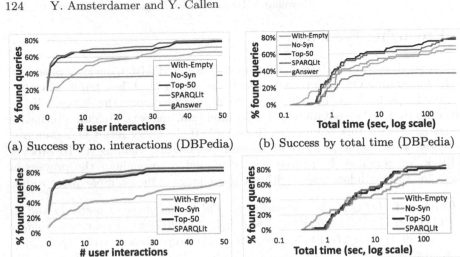

(a) Success by no. interactions (DBPedia)   (b) Success by total time (DBPedia)

(c) Success by no. interactions (YAGO)   (d) Success by total time (YAGO)

Fig. 3. Overall performance for DBPedia and YAGO

in [34], and we have formulated the gold queries w.r.t. YAGO; we have stripped aggregation to obtain selection queries, see [10]. We have then constructed two benchmarks of semi-formal queries: (1) *QALD Translated* is based on manual translation that is oblivious to the terminology used in the KBs, e.g., "Who is the tallest player of the Atlanta Falcons?" is translated to the semi-formal selection query `?x type AtlantaFalcons. ?x height ?y` (see [10]); (2) *QALD Cross-KB* uses the formal query for YAGO as a semi-formal query to be evaluated with respect to DBPedia, and vice versa. Intuitively, *QALD Translated* and *QALD Cross-KB* are used to simulate users who are unfamiliar with none of the KBs and users who are familiar with one KB and wish to use the other, respectively (user feedback is simulated using the underlying, hidden, formal query). Finally, we have generated a synthetic benchmark, where we have varied different aspects of the query structure, to examine their effect on our solutions.

As solution baselines we have used the NL-to-SPARQL query engine *gAnswer* [22], which achieved the best results in the QALD-9 Challenge [34], as well as multiple variants of our solution: (1) *With-Empty*. A variant that does not prune queries with empty results; (2) *No-Syn*. This variant does not use synonyms for distance computation, unlike our standard implementation (see Sect. 4); (3) *Top-50*. By default, we configure the Search Engine to return the top-100 results, whereas in this variant it is configured to return only 50.

Our evaluation metrics are the *number of user interactions*, i.e., the number of executions of step 4 in Fig. 1; and the *total response time*, i.e., the total computation time of proposed queries, throughout the interactive session.

Table 1. Percent of successfully found QALD queries

	SPARQLIt+ QALD Translated	SPARQLIt+ QALD Cross-KB	With-Empty+ QALD Translated	gAnswer+ QALD-9
DBpedia	78%	81%	64%	36%
YAGO	84%	81%	64%	–

(a) DBPedia   (b) YAGO

Fig. 4. Time segmentation for SPARQLIt

We next summarize our experimental results.

Overall Performance. In Figs. 3a and 3b, we show the cumulative percentage of formal queries successfully found by the different solutions, for the *QALD Translated* workload and DBpedia, with different bounds on the number of user interactions and total computation time. For both metrics, SPARQLIt exhibited the best results. In particular, interacting *once* with the user is already sufficient to outperform gAnswer, and to successfully find 52% of the queries; with up to 3 interactions this percentage increases to 62%; and with up to 10 interactions (and up to 23 s total time) this percentage increases to 68%. All restricted variants perform worse than SPARQLIt in terms of the number of interactions, showing the effect of our design choices. Figures 3c and 3d show results for the same experiment over the YAGO KB. We exclude gAnswer here since it is tailored for DBpedia. SPARQLIt achieves the best results in both metrics; With-Empty is significantly worse, indicating that the design choice of discarding queries that yield empty results is effective. We summarize the success rates of SPARQLIt using up to 50 interactions in Table 1 and contrast them with With-Empty and gAnswer (columns 1, 3 and 4 respectively).

We have executed the above experiments using the QALD Cross-KB workload (where YAGO queries are used as semi-formal queries over DBpedia and vice versa). The trends were similar: with one interaction, we have successfully found 44% of the queries in both KBs; using up to 3 interactions we found 53% (resp., 56%) of the queries in DBpedia (resp., YAGO); and using up to 10 interactions we found 67% (resp., 65%) of the DBpedia (resp., YAGO) queries. We summarize the success rates with up to 50 interactions in Table 1 (second column). *Overall, SPARQLIt had a high success rate and has succeeded in finding formal queries for the same semi-formal input over different KBs, as well as in finding the same formal query starting from different semi-formal queries.*

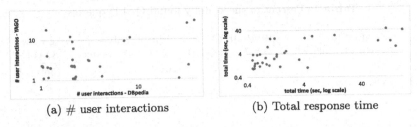

(a) # user interactions   (b) Total response time

Fig. 5. KBs correlation (each point denotes a QALD query)

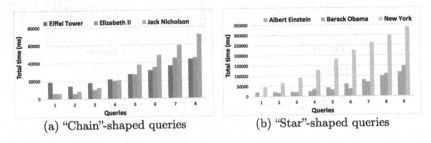

(a) "Chain"-shaped queries   (b) "Star"-shaped queries

Fig. 6. Response time vs. number of Subject-Object Switches.

Component Breakdown. Figure 4 shows a breakdown of the total computation time to (1) the Search Engine; (2) Triple Store and (3) all other components. The Triple Store and Search Engine are indeed responsible for a large fraction of the overall execution time (median 93% and 86% of the total time, respectively for YAGO and DBpedia); among the two, the time incurred by the Search Engine is typically higher: many query candidates are typically pruned and do not reach the Triple Store Manager. In contrast, when the overall response time is slower, we observe that it is mainly due to high latency Triple Store queries.

Effect of the KB. Figures 5a and 5b examine the effect of the KB on the difficulty of finding the target query. They show for each query the needed number of interactions (resp., total response time) for DBpedia (x-axis) vs. YAGO (y-axis). The graphs show relatively weak correlation (Pearson correlation coefficient is~0.6 for both graphs), given that DBpedia and YAGO have many common information sources (most notably, Wikipedia). This serves as evidence that the specifics of the KB structure are indeed essential when writing formal queries.

Synthetic Queries. Figure 6 shows the response time of SPARQLIt for representative synthetic queries. Queries include 8 triple patterns structured as chains (subsequent triple patterns share a single variable) or stars (all triple patterns share a single variable), and we vary the number of edits needed to obtain the correct formal query. The response time grows roughly linearly, although the space of relevant structures grows exponentially with the number of edits. This demonstrates the effectiveness of our approach in pruning irrelevant sub-queries.

7   Related Work

Many lines of research study solutions that assist users in query formulation. In particular, there is a large body of work on NL interfaces over KBs (e.g., [13,15,22,36,40]) or databases (e.g., [7,23,26,27,31]). Compared to the NL approach, our solution requires users to provide a more structured specification, yet leverages this structure for an improved interactive process (see Sect. 6). Another approach is that of Keyword Search over a KG (e.g., [20,25]); here again a major challenge is recovering from a situation where no suitable query was found, i.e., how to make the process interactive. Other works focus on autocompletion of SPARQL queries (e.g., [16,29,39]); these lines of work are complementary to ours: auto-completion tools may be adapted to our framework. Another relevant line of work focuses on similarity search, studying means of finding, for a given initial query, similar queries that return additional results (e.g., [18,28,30,37,41]. In particular, Zheng et al. [41] studied semantic similarity search for SPARQL, and introduced an edit distance notion for RDF graphs. The edit operations that we consider are different, since we do not require a formal query as input. Instead, we use measures based on syntactic similarity and string similarity. In Query-by-example, queries are reverse-engineered based on positive/negative result examples provided by the user (e.g., [1,2,5,9,12,14,24,38]). This method can be effective when the users search typed instances for which they can easily provide positive and negative examples, but is challenging to use when the query includes non-categorical predicates, which are typically very sparse and heterogeneous, and when users cannot provide sufficient examples. Finally, Faceted (navigational) Search enables users to refine their search options by navigating (drilling) down, and has been studied in the context of RDF querying (e.g., [6,19,21,35]). A challenge for interaction in this context arises when the browsed query parts may not match the other intended parts of the user query. If the user performs a sequence of drilling-down steps leading to a "dead-end", it is unclear which steps should be modified and how. Finally, the SPARQLIt system prototype was demonstrated in [3].

8   Conclusion

We have introduced a novel framework that assists users in querying RDF KBs. Users write queries that may not match the KB in contents and structure, and are given proposals for queries that do. Leveraging provenance, the framework procures fine-grained feedback on the proposed queries, guiding the translation. In future research, we will extend our Structural Edits operators as well as the fragment of SPARQL we have focused on, to account, e.g., for aggregation.

Acknowledgements. This work was partly funded by the Israel Science Foundation (grant No. 2015/21) and by the Israel Ministry of Science and Technology.

References

1. Abouzied, A., Angluin, D., Papadimitriou, C.H., Hellerstein, J.M., Silberschatz, A.: Learning and verifying quantified Boolean queries by example. In: PODS (2013)
2. Abramovitz, E., Deutch, D., Gilad, A.: Interactive inference of SPARQL queries using provenance. In: ICDE (2018)
3. Amsterdamer, Y., Callen, Y.: SPARQLIt: Interactive SPARQL query refinement. In: ICDE (2021)
4. Apache Jena. https://jena.apache.org/
5. Arenas, M., Diaz, G.I., Kostylev, E.V.: Reverse engineering SPARQL queries. In: WWW (2016)
6. Arenas, M., Grau, B.C., Kharlamov, E., Marciuska, S., Zheleznyakov, D., Jiménez-Ruiz, E.: SemFacet: semantic faceted search over YAGO. In: WWW (2014)
7. Baik, C., Jagadish, H.V., Li, Y.: Bridging the semantic gap with SQL query logs in natural language interfaces to databases. In: ICDE (2019)
8. Blazor. https://dotnet.microsoft.com/apps/aspnet/web-apps/blazor/
9. Bonifati, A., Ciucanu, R., Staworko, S.: Interactive inference of join queries. In: EDBT (2014)
10. Code and query repository for SPARQLIt. https://github.com/ycallen/dexa22
11. DBpedia. https://wiki.dbpedia.org/
12. Diaz, G.I., Arenas, M., Benedikt, M.: SPARQLByE: querying RDF data by example. In: PVLDB, vol. 9, no. 13 (2016)
13. Diefenbach, D., Singh, K.D., Maret, P.: WDAqua-core1: a question answering service for RDF knowledge bases. In: WWW Comp (2018)
14. Dimitriadou, K., Papaemmanouil, O., Diao, Y.: Explore-by-example: an automatic query steering framework for interactive data exploration. In: SIGMOD (2014)
15. Dubey, M., Dasgupta, S., Sharma, A., Höffner, K., Lehmann, J.: AskNow: a framework for natural language query formalization in SPARQL. In: ESWC (2016)
16. El-Roby, A., Ammar, K., Aboulnaga, A., Lin, J.: Sapphire: querying RDF data made simple. In: PVLDB, vol. 9, no. 13 (2016)
17. Elasticsearch. https://www.elastic.co/elasticsearch/
18. Elbassuoni, S., Ramanath, M., Weikum, G.: Query relaxation for entity-relationship search. In: ESWC (2011)
19. Ferré, S.: Expressive and scalable query-based faceted search over SPARQL endpoints. In: ISWC (2014)
20. Golenberg, K., Sagiv, Y.: A practically efficient algorithm for generating answers to keyword search over data graphs. In: ICDT (2016)
21. Haag, F., Lohmann, S., Siek, S., Ertl, T.: QueryVOWL: visual composition of SPARQL queries. In: ESWC (2015)
22. Hu, S., Zou, L., Yu, J.X., Wang, H., Zhao, D.: Answering natural language questions by subgraph matching over knowledge graphs. IEEE Trans. Knowl. Data Eng. 30(5) (2018)
23. Iyer, S., Konstas, I., Cheung, A., Krishnamurthy, J., Zettlemoyer, L.: Learning a neural semantic parser from user feedback. In: ACL (2017)
24. Jayaram, N., Khan, A., Li, C., Yan, X., Elmasri, R.: Querying knowledge graphs by example entity tuples. In: ICDE (2016)
25. Kacholia, V., Pandit, S., Chakrabarti, S., Sudarshan, S., Desai, R., Karambelkar, H.: Bidirectional expansion for keyword search on graph databases. In: PVLDB (2005)

26. Kim, H., So, B., Han, W., Lee, H.: Natural language to SQL: where are we today? In: PVLDB, vol. 13, no. 10 (2020)

27. Li, F., Jagadish, H. V.: NaLIR: an interactive natural language interface for querying relational databases. In: SIGMOD (2014)

28. Mottin, D., Lissandrini, M., Velegrakis, Y., Palpanas, T.: Exemplar queries: give me an example of what you need. In: PVLDB, vol. 7, no. 5 (2014)

29. Rafes, K., Abiteboul, S., Boulakia, S.C., Rance, B.: Designing scientific SPARQL queries using autocompletion by snippets. In: eScience (2018)

30. Schenkel, R., Theobald, A., Weikum, G.: Semantic similarity search on semistructured data with the XXL search engine. Inf. Retr. **8**(4) (2005)

31. Sen, J., et al.: ATHENA++: natural language querying for complex nested SQL queries. In: PVLDB, vol. 13, no. 11 (2020)

32. SPARQL Query Language for RDF. https://www.w3.org/TR/rdf-sparql-query/

33. Suchanek, F.M., Kasneci, G., Weikum, G.: YAGO: a core of semantic knowledge. In: WWW (2007)

34. Usbeck, R., Gusmita, R.H., Ngomo, A.N., Saleem, M.: 9th challenge on question answering over linked data (QALD-9). In: ISWC (2018)

35. Vargas, H., Buil-Aranda, C., Hogan, A., López, C.: RDF explorer: a visual SPARQL query builder. In: Ghidini, C., et al. (eds.) ISWC 2019. LNCS, vol. 11778, pp. 647–663. Springer, Cham (2019). https://doi.org/10.1007/978-3-030-30793-6_37

36. Vollmers, D., Jalota, R., Moussallem, D., Topiwala, H., Ngomo, A.N., Usbeck, R.: Knowledge graph question answering using graph-pattern isomorphism. arXiv preprint arXiv:2103.06752 (2021)

37. Wang, Y., Khan, A., Wu, T., Jin, J., Yan, H.: Semantic guided and response times bounded top-k similarity search over knowledge graphs. In: ICDE (2020)

38. Weiss, Y.Y., Cohen, S.: Reverse engineering SPJ-queries from examples. In: PODS (2017)

39. Wikidata. https://www.wikidata.org/wiki

40. Yin, X., Gromann, D., Rudolph, S.: Neural machine translating from natural language to SPARQL. Future Gen. Comput. Syst. **117** (2021)

41. Zheng, W., Zou, L., Peng, W., Yan, X., Song, S., Zhao, D.: Semantic SPARQL similarity search over RDF knowledge graphs. In: PVLDB, vol. 9, no. 11 (2016)

Anonymisation of Heterogeneous Graphs with Multiple Edge Types

Guillermo Alamán Requena, Rudolf Mayer$^{(\boxtimes)}$ (iD), and Andreas Ekelhart (iD)

SBA Research, Vienna, Austria
{galamanrequena,rmayer,aekelhart}@sba-research.org

Abstract. Anonymisation is a strategy often employed when sharing and exchanging data that contains personal and sensitive information, to avoid possible record identification or inference. Besides the actual attributes contained within a dataset, also certain other aspects might reveal information on the data subjects. One example of this is the structure within a graph, i.e. the connection between nodes. These might allow to re-identify a specific person, e.g. by knowledge of the number of connections for some individuals within the dataset.

Thus, anonymisation of the structure is an important aspect of achieving privacy. In this paper, we therefore present an algorithm that extends upon the current state of the art by considering multiple types of connections (relations) between nodes.

Keywords: Graph structure anonymisation · Multiple relational types

1 Introduction

The amount of data collected is ever increasing, and data represented as graphs, e.g. social networks or processes e.g. in the knowledge work domain [5], are no exception. Several interesting data analysis tasks utilise such graphs, which represent connections between individuals, organisations, and other entities. As this data is highly personal, data protection becomes an important aspect. Historically, tabular data was among the first types to be addressed, with methods such as k-anonymity or differential privacy being developed.

In graphs, besides the values within nodes, which could be treated in a similar manner as tabular data, also the structural information encoded in the connections (edges) is of concern. Depending on the background knowledge of an attacker, it might be possible to re-identify individuals based on this structure alone, e.g. especially those individuals that have unusual patterns of connections [6]. Thus, recent years have also shown an increase in works

SBA Research (SBA-K1) is a COMET Centre within the framework of COMET - Competence Centers for Excellent Technologies Programme and funded by BMK, BMDW, and the federal state of Vienna; COMET is managed by FFG. This work is supported by FFG under Grant No. 871299 (project KnoP-2D).

© The Author(s), under exclusive license to Springer Nature Switzerland AG 2022
C. Strauss et al. (Eds.): DEXA 2022, LNCS 13426, pp. 130–135, 2022.
https://doi.org/10.1007/978-3-031-12423-5_10

addressing structural anonymisation, such as adaptations of the concept of k-anonymity [2,7], and combinations of node and structure anonymisation, in various types of graphs [1,8,9].

In this paper, we specifically expand on previous approaches for achieving structural anonymity for graphs with multiple types of edge connections. While most existing works consider homogeneous graphs, i.e. with only one type (e.g. *foaf:knows*), in a heterogeneous graph, nodes can be linked by varying types of connections. Heterogeneity complicates structure anonymisation, as attacks may take advantage of this addition information. Based on the ideas from [3], our goal is to develop a method to anonymise heterogeneous Resource Description Framework (RDF)[1] graphs. The idea developed in [3] is that the one-hop neighbourhood of any resource to be anonymised should be indistinguishable from the one-hop neighbourhood of at least k-1 other resources. For that purpose, they developed a greedy heterogeneous graph modification algorithm for a simplified RDF graph which includes only 4 types of semantic connections. However, although the general guidelines of the algorithm are stated in pseudo code, no implementation is publicly available. Based on their approach, our contributions in this work consist of (i) an extension to the approach of [3] to increase flexibility and usability of the anonymisation method, and (ii) an open-source implementation in Python[2].

2   K-RDF-Neighbourhood Anonymisation with Multiple Edge Types

In this section, we describe our method and extensions of the anonymisation method [3]. Since the focus of our method is to anonymise heterogeneous RDF graphs, we first present the formal definition of a heterogeneous graph by [4]:

Definition 1. *A heterogeneous graph is defined as a directed graph $G = (V, E, A, \Delta)$ where each node $\nu \in V$ and each edge $\epsilon \in E$ are associated with their type mapping functions $\theta(\nu) : V \to A$ and $\omega(\epsilon) : E \to \Delta$, respectively.*

In the following, we describe our method on an example heterogeneous RDF graph utilising FOAF[3], with vertices of type *foaf:Person*, representing people, edges of type *foaf:knows*, representing relations between individuals, and edges of type *foaf:CurrentProject*, indicating projects an individual is working on. Other edges primarily serve to describe properties, such as *foaf:Age*, or *foaf:Name*, while *custom:has_disease* is an example of a custom property outside the FOAF specification. Following the procedure of [10] and [3], we will demonstrate the anonymisation on the one-hop neighbourhood of *foaf:Person* resources[4].

---

[1] https://www.w3.org/RDF/.

[2] https://github.com/sbaresearch/graph-anonymisation.

[3] http://www.foaf-project.org/.

[4] Note that apart from reducing computational complexity, it is logical to target individuals, since it is the most common setting when facing ananonymisation task.

Our method initially gets a list of connection types to consider for structure anonymisation, all other attributes will be removed[5]. Edge connections in the one-hop-neighbourhood of any node ν of a target graph can be classified into three different categories, depending on the type of information they describe:

- **Attribute connections:** edges connecting a node (e.g. *foaf:Person*) to a descriptive characteristic of this individual which is stored as a Literal.
- **Unidirectional connections:** directed edges connecting a node to other entities (e.g. *foaf:Person* to a project via *foaf:CurrentProject*).
- **Bidirectional connections:** edges symmetrically connecting nodes (e.g. *foaf:Person* via *foaf:knows*).

Definition 2. *A heterogeneous RDF graph is said to be k-anonymous if there are at least k identical one-hop-neighbourhoods in the target graph for each node $\nu \in N$. We consider that two attributes of the one-hop-neighbourhood of a pair of nodes x and y, are identical if they are generalised to the same level. We consider two unidirectional connections of the one-hop-neighbourhood of a pair of nodes x and y to be identical if they point exactly to the same resources. We consider the bidirectional connections of the one-hop neighbourhood of a pair of nodes x and y to be identical if their one-hop-neighbourhoods are isomorphic.*

In order to fulfil the anonymisation criteria defined above, we rely on three different algorithms (similar to [3]):

The Neighbourhood Code Extraction Algorithm compares one-hop-neighbourhoods of target nodes (e.g. *foaf:Person*) across the target graph. For this purpose, we encode the node neighbourhood information into a more efficient data structure than the raw RDF graph. We chose a hashtable due to its low indexing complexity $(O(n))$. Depending on the type of edge connection, the information contained in the one-hop-neighbourhood of a node ν is stored in a different way:

- **Attribute connections**: the attributes of each individual are stored as *key value* pairs (i.e., *foaf:Age* "40").
- **Unidirectional connections**: the resources to which each unidirectional connection of an individual points to are stored in a list. The type of connection is the *key* (i.e., *foaf:CurrentProject*) and the list of resources is the *value* associated with it (i.e., ["Project1", "Project3", "Project7"]).
- **Bidirectional connections**: Multiple isomorphic tests have to be conducted for each bidirectional connection. At this time, no polynomial time algorithm for the general isomorphic problem [10] is known. In our approach, we utilise the same string representation of the edges as proposed in [3] and based on [10]. The main idea is to encode the information of each sub-graph G_{bidi_i} created by considering only one type of bidirectional connection across the one-hop-neighbourhood of a node ν, so that the one-hop-neighbourhood of

---

[5] Node value anonymisation, if necessary, is a pre-requisite step and not covered by our structure anonymisation method.

two *foaf:Person* nodes can be considered isomorphic in terms of that type of bidirectional connections if the generated codes are identical in structure. The way this encoding is constructed consists of finding the *minimum* depth-first search (DFS) tree of each component and concatenating it in a list where all these minimum trees are stored. We simplify the search of the minimum DFS tree by dynamically discarding candidate paths. In the worst case scenario in which all the DFS trees in the subgraph fulfil the criteria, one of the paths is taken randomly and the encoding algorithm becomes $O(n!)$ which is the same complexity as the original algorithm proposed by [3]. Following the guidelines of [3], we call the dictionary encoding the one-hop-neighbourhood of a node ν the Full Neighbourhood Code of ν ($FNHC_\nu$).

Dissimilarity Computation Algorithm. To compute the dissimilarity between each of the nodes, we use the information stored in the Full Neighbourhood Code hashtable. The dissimilarity between the one-hop-neighbourhood of two nodes x and y is the weighted sum of the dissimilarity of each connection in that neighbourhood:

$$sim(FNHC_x, FNHC_y) = \sum_{i=0}^{N} \alpha_i * sim_i(FNHC_{x_i}, FNHC_{y_i}) \qquad (1)$$

where N is the set of connection types present in the one-hop-neighbourhood, α_i is the weight of the dissimilarity of attribute i (sim_i) to the total dissimilarity between the nodes x and y. There are three types of dissimilarity functions, one for each type of edge connection described above:

- **Dissimilarity of attribute connections** is the normalised distance of two attributes x_i and y_i given a defined hierarchy tree. It ranges between 0 (identical) and 1 (reached highest level of hierarchy).
- **Dissimilarity of unidirectional connections** between two nodes x and y, given a set of Literals to which they point, is defined as the number of connections of that type to be deleted so that two nodes x and y are connected to exactly the same Literals or resources.
- **Dissimilarity of bidirectional connections** between a node x and another node y, given the one-hop-neighbourhood, is determined by the amount of edges one needs to delete so that one-hop-neighbourhoods of both nodes are identical (isomorphic).

To compute the complete dissimilarity between two nodes x and y, one needs to calculate the similarity of each of the connections using the corresponding methods explained above and apply the weighted sum provided in Eq. (1).

The Graph Modification Algorithm is also based on the ideas presented by [3] with some modifications to improve scalability. The main goal of this algorithm is to transform the one-hop-neighbourhood of a group of k given nodes, so that the anonymisation criteria is fulfilled for all of them. We refer to this group of nodes as *anonymised neighbourhoods* or *equivalent classes*.

- For the **generalisation of attribute connections**, the attributes of each of the k nodes are generalised to the lowest level's possible common value in the hierarchy tree provided.
- When **generalising unidirectional connections**, one should remove the necessary edges, so that each of the k nodes are connected exactly to the same Literals and resources via those unidirectional connections. We follow the idea of only deleting edges to avoid introducing false information (added edges) in the graph. In addition, as pointed out by [3], the approach of deleting edges fits with the open world assumption which suggests that missing statements can also be true.
- The **generalisation of bidirectional connections** is the most complex one. As for unidirectional connections, it relies on the same type of calculations used when computing dissimilarity. That means, for each node in the neighbourhood of size k, one should delete all the necessary edges so that the one-hop-neighbourhood of each of them is isomorphic in terms of each of the bidirectional connections. With our method, it is enough to take one of the nodes as reference and perform a pairwise comparisons to every other node twice (double-pass). At every comparison, the one-hop-neighbourhood of the reference node and the other node under comparison are updated via edge deletion so that they are isomorphic. The idea is that after the first pass, the reference one-hop-neighbourhood takes the minimum isomorphic representation and in the second round, this structure is acquired by all the other nodes.

We would like to point out two of the major challenges that arise when anonymising bidirectional connections. First, when deleting edges during the described double pass, the edges of other one-hop-neighbourhoods may be affected as well, and this can lead to more edge deletions than necessary and hence, additional information loss. To avoid this issue, we only store which edges to delete during the double pass, but they are only deleted when the algorithm has finished. Edge deletion may still cause some additional edges to be deleted in the neighbourhood, and therefore, they might not be isomorphic anymore. However, since the calculation of which edges to delete ensures that they are actually isomorphic in the first place, deleting additional edges of the structure of each of the one-hop-neighbourhoods does not reveal any additional information, and we can still consider them isomorphic in terms of the anonymisation goal.

Secondly, deleting edges may affect the one-hop-neighbourhood of other nodes that are not in the same neighbourhood as the k target nodes: (i) The one-hop-neighbourhood of non-anonymised nodes is affected – then, one needs to simply update the one-hop-neighbourhood or (ii) this affects the one-hop-neighbourhood of anonymised nodes, then this is the exact same situation as in 1).

Due to these improvements, our method is able to deal with larger graphs than the earlier approach by [3]. Furthermore, the consideration of the outlined challenges leads to reduced information loss.

3   Conclusions

Anonymisation of graph data differs from relational data as also the structure of graphs can be utilize by an attacker to perform e.g. a re-identification attack.

In this paper, we have thus presented an algorithm for anonymising the structure of graphs. We extended previous work by allowing on the one hand heterogeneous graph structures with multiple types of edges, and on the other hand also scaled up the algorithm.

Future work will focus on evaluating our approach in diverse settings against benchmark datasets, and measure the effect of the anonymisation on utility.

References

1. Campan, A., Truta, T.M.: Data and structural k-anonymity in social networks. In: Bonchi, F., Ferrari, E., Jiang, W., Malin, B. (eds.) PInKDD 2008. LNCS, vol. 5456, pp. 33–54. Springer, Heidelberg (2009). https://doi.org/10.1007/978-3-642-01718-6_4
2. Feder, T., Nabar, S.U., Terzi, E.: Anonymizing graphs, October 2008. arXiv:0810.5578
3. Heitmann, B., Hermsen, F., Decker, S.: k - RDF-Neighbourhood anonymity: combining structural and attribute-based anonymisation for linked data. In: Workshop on Society, Privacy and the Semantic Web - Policy and Technology (PrivOn), Vienna, Austria (2017). http://ceur-ws.org/Vol-1951/PrivOn2017_paper_3.pdf
4. Hu, Z., Dong, Y., Wang, K., Sun, Y.: Heterogeneous graph transformer. In: The Web Conference 2020, pp. 2704–2710, WWW. ACM, Taipei, Taiwan, April 2020. https://doi.org/10.1145/3366423.3380027
5. Hübscher, G., et al.: Graph-based managing and mining of processes and data in the domain of intellectual property. Inf. Syst. **106**, 101844 (2022). https://doi.org/10.1016/j.is.2021.101844
6. Ji, S., Mittal, P., Beyah, R.: Graph data anonymization, de-anonymization attacks, and de-anonymizability quantification: a survey. IEEE Commun. Surv. Tutorials **19**(2), 1305–1326 (2017). https://doi.org/10.1109/COMST.2016.2633620
7. Liu, K., Terzi, E.: Towards identity anonymization on graphs. In: ACM SIGMOD International Conference on Management of Data, p. 93, SIGMOD. ACM Press, Vancouver, Canada (2008). https://doi.org/10.1145/1376616.1376629
8. Mohapatra, D., Patra, M.R.: Anonymization of attributed social graph using anatomy based clustering. Multimedia Tools Appl. **78**(18), 25455–25486 (2019). https://doi.org/10.1007/s11042-019-07745-4
9. Zheleva, E., Getoor, L.: Preserving the privacy of sensitive relationships in graph data. In: Bonchi, F., Ferrari, E., Malin, B., Saygin, Y. (eds.) PInKDD 2007. LNCS, vol. 4890, pp. 153–171. Springer, Heidelberg (2008). https://doi.org/10.1007/978-3-540-78478-4_9
10. Zhou, B., Pei, J.: Preserving privacy in social networks against neighborhood attacks. In: International Conference on Data Engineering, pp. 506–515, ICDE. IEEE, Cancun, Mexico, April 2008. https://doi.org/10.1109/ICDE.2008.4497459

Deep Learning

A Divergent Index Advisor Using Deep Reinforcement Learning

Zahra Sadri[(✉)] and Le Gruenwald

University of Oklahoma, Norman, OK, USA
{zahra.sadri,ggruenwald}@ou.edu

Abstract. A divergent design index is a tuning method that employs replication to specialize the index configuration of each replica in a replicated database for a subset of a workload to minimize the total processing cost of the workload. Studies show that this tuning method improves the workload performance in comparison with the case that all replicas have the same index configuration. Current divergent design algorithms do not have any mechanism to learn about the effectiveness of the recommended index sets. Moreover, they solely rely on the query optimizer's cost estimation, which can be inaccurate.

To tackle these problems, we introduce a new divergent index advisor, DINA, that learns the goodness of the workload partitioning among replicas and the efficiency of their index configurations by employing a Deep Reinforcement Learning (DRL) algorithm. The DRL agent explores various possible workload partitions and learns the benefit of their index configurations via performance observation. We conduct experiments using the TPC-H and TPC-DS database benchmarks to evaluate the performance of DINA. The experiments show that DINA yields better query execution time than the existing algorithms.

Keywords: Learned divergent index advisor · Deep reinforcement learning · Replicated database

1 Introduction

Divergent design index tuning extends the index selection problem (ISP) for a single node [2, 9, 15, 16, 21, 23, 25] to the index selection for a replicated database where the database is replicated on multiple nodes, each of which is called a replica [8, 24]. Thus, similar to the ISP, the goal is to minimize the processing cost of the workload. Unlike ISP for a single node, it utilizes the replication feature to create a set of index configurations such that each index configuration is specialized for a subset of the workload. To achieve this goal, it divides a workload among the replicas and recommends an index configuration for each subset on each replica in a way that minimizes the processing cost of the entire workload. The specialization helps deriving the indexes that are specifically needed for the queries in each subset of the workload; hence the query processing costs are reduced compared to the case where one index configuration is recommended for the whole workload.

© The Author(s), under exclusive license to Springer Nature Switzerland AG 2022
C. Strauss et al. (Eds.): DEXA 2022, LNCS 13426, pp. 139–152, 2022.
https://doi.org/10.1007/978-3-031-12423-5_11

Similar to ISP [7], divergent index tuning is an NP-hard method [8]. Therefore, heuristic and optimization approaches [8, 24] have been proposed to solve it. Current divergent index advisors suffer from two major problems. One problem is that they have no mechanism to learn about the goodness of the workload partitioning and their index configurations. Suppose the divergent advisor partitions the workload such that the derived indexes impact their benefit negatively, or the recommended indexes cause a performance regression. Because these advisors do not have any way to receive feedback on their decision, they might choose the same workload partitioning or the same set of indexes on a subsequent cycle. The second problem is that these techniques solely rely on the query optimizer, which is not flawless due to statistical errors [10]. To address the former problem, we proposed DRLindex [30, 31] which employed a Deep Reinforcement Learning (DRL) algorithm to select index configurations for a cluster database. In these works, we formulate index selection for multiple nodes as a DRL problem and define the core components such as the agent and the environment. The critical part of DRL is an appropriate reward function. We propose a reward function that considers the estimated processing cost of the workload and the load-skew factor. DRLindex [31] shows that DRL is a promising solution for divergent index selection, but it solely relies on the query optimizer. To address this gap, we extend our work and present DRL Divergent Index Advisor (DINA), a learned divergent index advisor. Similar to [8, 24], DINA exploits the data replication when recommending index configurations for a replicated database, and it is capable to refine its decision as proposed in [30, 31]. The contributions of this work include the following:

- To avoid the possible query cost estimation errors due to using a query optimizer, DINA employs two training phases. It learns the efficiency of various possible workload partitions and their index configurations by creating the indexes and observing the real execution times.
- DINA is able to exploit multi-column indexes and unlike DRLindex is not limited to single-column indexes.
- We present an experimental performance study using the TPC-H and TPC-DS benchmarks to compare our proposed algorithm with the existing divergent index advisors [8, 24]. The results demonstrate the efficiency of our algorithm.

The remainder of the paper is organized as follows: Sect. 2 introduces the related work; Sect. 3 describes the architecture and the components of the proposed index advisor; Sect. 4 presents the experimental results evaluating the performance of the advisor; and Sect. 5 concludes the paper and discusses future research directions.

2   Related Work

Index tuning algorithms have been studied since the '70s and different index advisors have been proposed based on the workload characteristics and databases. These index advisors either apply heuristic, optimization, or machine learning methods. An index advisor for a single node [2, 9, 15, 16, 21, 23, 25, 26] recommends one index configuration for the workload to minimize the processing cost of the workload. Divergent design index advisors [8, 24] introduce the idea of utilizing the replication to create several index configurations such that each index configuration is recommended for a subset of a workload. DivDesign [8] is a heuristic algorithm; it divides the workload among nodes, recommends an index configuration for each subset of the workload, and computes the estimated processing cost of each query on each replica using the what-if tool [5]. Then, it redistributes queries among replicas based on the estimated processing cost. The algorithm stops when the subset of the workload stays unchanged. RITA [24] formulates the divergent design problem as a Binary Integer Program (BIP) and uses existing linear-optimization software to solve the BIP. RITA uses INUM, a fast what-if tool [22] to estimate the cost of the workload. Unlike a What-if tool [5], INUM is not implemented in all databases which makes the application of RITA limited. The main drawbacks of heuristic and optimization methods are that they do not have mechanisms to learn from their mistakes, and they rely solely on the query optimizer, which might lead to a performance regression [10]. To overcome these issues, learned index advisors have emerged. In [3], the authors proposed a cost model that can learn using reinforcement learning and evaluated the learned cost model in index selection problem. Lift [25] proposed a learned index advisor for a document database and investigated the impact of using the demonstration data in reducing the training time of the model. Other learned index advisors such as SMARTIX [21], DBA bandits [23], OpenGauss [16], and DRL-based index advisor [15] apply different reinforcement algorithms in the process of learning an index set. These techniques are proposed for a single node and do not consider replication and how to distribute queries among nodes. DINA employs replication and decides how to divide queries of the workload among the nodes and selects an index configuration for each node. It can choose index sets with two main goals: minimizing the processing cost of the workload and avoiding skewed load among nodes.

3   DRL Divergent Index Selection Framework

We design a divergent index advisor called DRL Divergent Index Advisor (DINA) that divides queries among nodes and selects a set of index configurations for the replicas of a replicated database. Finding an optimal combination among all possible partitions of the query workload and combinations of candidate indexes is the critical factor of a divergent index advisor. We employ a DRL algorithm to equip our advisor to search this large search space efficiently. DRL has been successful in searching large search spaces in database areas such as knob tuning [17, 32], query optimization [14, 19], partitioning [11, 12], and Indexing [15, 16, 21, 23, 25]. DRL explores the search area efficiently; that is, it does not just greedily grab the first best-founded solution. Instead, it explores other choices to find a possible better solution. We assume that the database is fully replicated in all the nodes. Hence, each query can be processed by any replica and there is no need to move data among nodes. Throughout this paper, node and replica are used interchangeably.

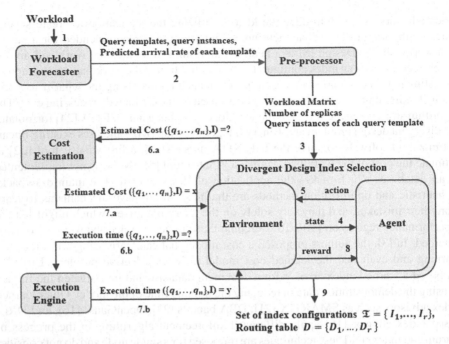

Fig. 1. The architecture of DINA

Figure 1 depicts the components of DINA and the steps of the workflow in numbers. First, DINA invokes the Workload Forecaster [18] to predict the coming workload. This module receives the workload and provides a set of query templates, the queries of those templates, and the total number of queries per template for the coming workload. A query template is a representative of a group of query instances that have the same format but may have different parameter values. For example, *Select * from T_1 where $T_1.att1 > \#$* is a query template in which the value of # can change in different queries. Throughout this paper, a template means a query template unless otherwise mentioned. The detail of the forecaster can be found in [18]. Second, the Pre-processing module receives the workload information. It determines a set of candidate indexes (CIS) for the templates in the workload. After identifying the candidate indexes, it keeps track of which candidate index(es) appear in the plan of each template. Also, it stores the benefit of the candidate indexes for related query templates either positive or negative in a *Workload Matrix* (see Sect. 3.1). The Divergent Index Selection module formulates the index selection problem for the replicated database as a DRL problem (see Sect. 3.2). It distributes the query templates among the replicas and recommends a set of index configurations. This module, first, interacts with the cost estimation module and then the execution engine. The Cost Estimation module returns the estimated processing cost of the query (see Sect. 3.3). The Execution Engine runs a query and returns its execution time. In the following, we explain each of these modules in detail.

3.1 Pre-processing Module

This module receives query templates, queries, and the number of queries per query template. Then, it extracts the characteristics of the workload to generate a workload matrix. In the following, we describe the candidate index selection method, the frequency vector, and the workload matrix, respectively.

The candidate index selection method chooses a subset of all possible indexes (single or multi-columns) that reduces the execution time of the queries of at least one query template. As queries in each query template can benefit from similar indexes [26], the method processes the query templates instead of each query. To speed up, we call one of the existing index advisors to get all possible indexes for each query template without any space budget limitation and make sure that those indexes are used in the execution plan of at least one template. Here, we keep track of which candidate index appears in the plan of which template. Then, we process the impact of these candidate indexes. To calculate the impact of each candidate index, first, we execute all the queries in the workload without indexes. Then, we call the agent to start exploring the action space which is the set of candidate indexes. This exploration is to extract the required information for the learning step. The agent creates indexes and executes those queries that have the common columns with the created candidate index. Next, it computes the impact of the candidate index for the corresponding queries by dividing the execution time of the query before creating the index by the execution time of the query after the index creation. We end up with the impact of each candidate index for the corresponding query templates. Finally, we put all this information together and create a workload matrix, which shows the average estimated impact of each candidate index on the queries of a template. The workload matrix is a matrix of size $qt \times m$, where qt is the number of templates in the workload and m is the number of candidate indexes.

3.2 DRL Divergent Index Selection Module

This module explains how DINA selects index configurations using DRL. First, we explain DRL and how it works briefly. Then, we describe each DRL component in our design in detail.

Deep Reinforcement Learning

Reinforcement Learning (RL) has an agent and an environment that interacts together [27]. The agent is the learner part of RL which makes a sequence of decisions to optimize an objective such as maximizing a score in a game. The environment is defined by its component including a set of actions $\mathcal{A}$, a set of states $\mathcal{S}$, and a reward function. At each timestep, the agent selects an action a_i for a state s_i, and receives a reward value rw_i that evaluates the taken action. The selected action a_i is applied to the environment and its state changes from a current state s_i to a next state s_{i+1}. A policy π guides an agent on which action to take at each state. The quality of a policy is quantified by a value function that associates an expected cumulative discounted reward to each state. Here, the discounted reward is the value of the reward multiplied by a parameter called discounted rate $0 \le \gamma \le 1$. The value of the discount rate implies the impact of the future rewards, which are estimated. The objective of the agent is to find the optimal policy that governs the agent to maximize the expected cumulative discounted reward known as a return in the long term.

Q-learning [27] is an algorithm to solve RL. At each time step t, the Q-function approximates the value of each action using the following equation:

$$Q(s_t, a_t) = Q(s_t, a_t) + \alpha[rw_{t+1} + \gamma max_{a_{t+1}}Q(s_{t+1}, a_{t+1}) - Q(s_t, a_t)] \qquad (1)$$

where, $Q(s_t, a_t)$ is the Q-value for state s_t and action a_t, α is a learning rate, rw is the reward value, γ is a discount rate, and $max_{a_{t+1}}Q(s_{t+1}, a_{t+1})$ is the maximum value of estimated future returns for the next state.

Deep Q-learning (DQN) [20] is a Q-learning approach that utilizes a neural network $Q_\theta(s, a)$ with weights θ in the learning process of its agent. The neural network helps the DRL-agent to predict the values of actions. Weights in the neural networks are updated by Gradient Decent or ADAM on the loss.

Environment. This component defines the state and action space of the problem and returns a reward value for each action. First, it depicts the states of the replicas according to their index configurations. Second, it identifies possible actions for each state that the agent can choose, i.e., the query templates for distribution and their candidate indexes that can be indexed on each replica. Third, it interacts with the databases in the replicated database to apply the taken actions by the agent -either simulating the index or creating it depending on the training phase-. Finally, it returns the reward value to the agent. The reward value is computed by the reward function, which has two objectives: minimizing both the processing cost of the workload and the load-skew. The reward value is a weighted sum of these two objectives. To compute this reward based on the training phase, the environment interacts either with the Cost Estimation (see Sect. 3.3) or the Execution Engine which executes queries and returns the execution time of the queries.

State Representation. The states of the environment should reflect the impact of the selected actions by the agent. Therefore, the state is the current index configurations of the replicas. We represent the current index configurations of the replicas by a matrix of size $r \times m$, where r is the number of replicas and m is the number of candidate indexes, and we call it *the state matrix*. An entry [i, j] in the state matrix is set to 1 if a candidate index $j \in [1, m]$ has been chosen to be an index for a replica $i \in [1, r]$ and otherwise, it is set to 0.

Set of Actions. The actions include selecting a query template to be executed on a specific replica and creating a proper index for that query.

Reward Function. In our design, we pursue two main objectives: (a) minimizing the processing cost of the workload (in the pre-training phase) and the execution time of the workload (in the re-training phase), and (b) reducing the load skew among the replicas. Thus, we want the agent to learn two points. First, the index configurations should be selected to minimize the estimated processing cost or execution time of the workload. Second, the index configurations should be selected such that the workload distributes among nodes as evenly as possible to prevent the overload/underload conditions on replicas. To achieve these two goals, we define a reward function that consists of the following two parts:

- Workload processing cost: a scalar reward is given with respect to the improvement in the estimated processing cost or execution time of the workload in the presence of the recommended index configurations.
- Workload Skew: a reward value is assigned based on the total workload skew among replicas. The less the workload skew, the higher the reward.

In the following, we describe how to measure the reward of each objective.

Workload Processing Cost. The agent should select a set of index configurations for replicas such that the total estimated processing cost of the workload is minimized. To obtain the minimum processing cost, each query must be processed by a replica that has the best index configuration, that is, a replica which minimizes the processing cost of the query. The routing table identifies that replica.

First, we compute the estimated processing cost of the workload using the following equation:

$$TotalCost(W, M \cup \mathfrak{T}) = \sum_{q \in QT} \sum_{R \in [1,r]} cost(q, I_R) \qquad (2)$$

where q is a query of a query template in the workload, and I_R is the index configuration of the replica to which the query is routed. The reward is computed using the following Eq. (3):

$$reward(\mathfrak{T}) = \frac{TotalCost(W, M - \{\mathfrak{T}\}) - TotalCost(W, M \cup \mathfrak{T})}{TotalCost(W, M - \{\mathfrak{T}\})} \qquad (3)$$

where $(W, M - \{\mathfrak{T}\})$ denotes the estimated processing cost of the workload when $\mathfrak{T}$ is not materialized.

Workload-Skew. We want to minimize the workload skew among the replicas when recommending index configurations. We define workload skew as a considerable difference between the amounts of the workload that each replica processes. For instance, suppose in a database cluster with three nodes, the first node is completely idle, the second node processes a small part of the workload, while the third node is almost overloaded by processing the large portion of the workload. We want to avoid these unbalanced situations. The ideal case is to have the workload distributed evenly among all replicas; thus, the workload skew is zero. It is almost impossible to achieve zero skew; but we want to decrease it as much as possible. The goal is to create index configurations for the replicas in a way that not only minimizes the workload processing cost but also balances the workloads among the replicas as much as possible.

To compute the workload-skew, first, we compute the amount of workload that each replica should process in the best case, i.e., when the workload is evenly distributed among the replicas, as follows:

$$workload(R)_{bestcase} = \frac{TotalCost(W, \mathfrak{T})}{r} \qquad (4)$$

Second, we calculate the amount of the workload that each replica R with the index configuration I_R should process:

$$workload(R)_{real} = \sum_{q \wedge R} cost(q, I_R) \tag{5}$$

where $q \wedge R$ means the queries that are processed on replica R. Third, we compute the workload skew on each replica as follows:

$$workload_{skew}(R) = \frac{|Workload(R)_{real} - Workload(R)_{bestcase}|}{Workload(R)_{bestcase}} \tag{6}$$

In Eq. (6), the value of the workload-skew (R) either greater or less than zero implies that the replica suffers from over skew or under skew, respectively. We want to reduce these cases. The value of zero shows there is no skew on a replica, which is the desirable case. After finding the amount of the workload skew on each replica, we calculate the sum of the workload-skew values of the replicas. Finally, the reward of the workload skew is computed in Eq. (7):

$$reward(S) = \frac{1}{\sum_{R \in r} workload\_skew(R)} \tag{7}$$

Final Reward. Eventually, the reward is a weighted sum of $reward(\mathfrak{T})$ and $reward(S)$ as follows:

$$reward = \alpha \times reward(\mathfrak{T}) + \beta \times reward(S) \tag{8}$$

In Eq. (8), α and β are obtained by trial and error and their values define a trade-off between cost reduction and load-balancing.

DRL-Agent. In our case, because of a large search space, the agent uses a neural network to predict the value of each action at each state. In the following, we explain the training process of the agent, which has two phases: pre-training and re-training.

Pre-training Phase. In this phase, we train the agent as described in Algorithm 1 (Fig. 2). The learning process is episodic (line 3) [27]. Each episode consists of a specific number of time steps which in our case defines the maximum time that the agent can spend on selecting index configurations. An episode starts with an initial state s_0 where there is no index configuration on any replica (line 4) and ends when all queries are divided among replicas, or space budget limit reaches on all replicas, or after the time steps end.

At each time step, the agent receives the state vector s_t and applies the ε-greedy policy [27] to choose an action a_t (line 7–8). Based on this policy, a random action is chosen with the probability $0 < \varepsilon < 1$. An action selects a template to be processed on a specific replica and chooses an index(es) for that template depending on the available space budget on a replica. Then, it simulates the virtual index(es) on that replica and updates the state vector to show the new state s_{t+1} of the existing indexes on the replicas (line 9). Next, the reward rw_t of the selected index(es) is computed as explained in Reward Function Section.

Algorithm 1 DINA Pre-training

Input: memory D, number of episodes N, number of time steps T, epsilon decay epsilon ε, a set of actions A, space budget B, workload W, initial state s_0
Output: a trained model
1: Randomly initialize Q-network Q_θ
2: Randomly initialize target network $Q_{\theta'}$
3: **for** ep = 1 to N **do**
4:      Reset to state s_0 // No index configuration on any node
5:      end-of-episode = False
6:      **for** t = 1 to T
7:          with probability ε , select a random action a_t from A
8:              Otherwise a_t= argmax$_a$ $Q_\theta(s_t, A)$
9:          s_{t+1} ◄ simulate the action and update the state matrix
                    to show the new index configuration
10:          rw_t ◄ compute the reward as explained in Section 3.5.1
11:          **If** $W = \phi$ or B is full in all replicas:
12:              end-of-episode = True
13:          **end if**
14:          Store transition (s_t, a_t, r_t, s_{t+1}) in D|
15:          $s_t = s_{t+1}$
16:          Sample-minibatch (s_i, a_i, r_i, s_{i+1}) from D
17:          Train Q-network with ADAM and loss
18:      **end for**
19: Decrease ε by epsilon decay
20: Update weights of target model
21: **end for**

Fig. 2. DINA pre-training algorithm

The pre-training phase uses the query optimizer to estimate the estimated processing cost of the workload. A set flag shows the end of the episode (line 11–13). The agent's experiences $(s_t, a_t, rw_t, s_{t+1})$ are stored in the memory (line 14), and a mini batch of experiences is randomly sampled from the memory (line 16) to train the agent (line 17). Over time that agent learns about actions, the value of ε is gradually decreased by multiplying ε to a factor called epsilon decay (line 19). At the end of this training phase, the agent is trained to find the best query distribution and index configurations for replicas for the forecasted workload.

Re-training Phase. In the previous phase, we avoid creating actual indexes and rely on the query optimizer for cost estimation. As we know, the estimates of the query optimizer are not completely reliable [10]. Therefore, in this phase, we re-train the model by creating the indexes and measuring the execution time of the queries. Consequently, the reward is computed using the execution time, which means that the query estimated cost is replaced by the actual query execution time in Eqs. (2) and (3).

3.3   The Cost Estimation Module

This module receives the queries and the recommended index configuration for a replica as inputs. Then, it estimates the processing cost of the queries using a "What-if" tool [5]. Since each node has a query optimizer, the cost estimation can be done in parallel on all replicas.

4 Experimental Results

We conduct experiments to evaluate the effectiveness of DINA in comparison with the two existing automatic approaches. In this section, we describe the experimental setup and discuss the results of the experiments.

4.1 Experimental Setup

Benchmarks. We use the TPC-H [28] and TPC-DS [29] database benchmarks. The former consists of 8 tables and 22 templates, and the latter includes 24 tables and 99 templates. TPC-DS is more complicated and diverse in comparison with TPC-H. For both the benchmarks, we load 10 GB databases and generate the related query workloads using the provided query generator.

Experimental Setups. DINA is written in Python and employs Keras [13] and Tensorflow [1] to create a neural network. The neural network is the function approximator with three hidden layers, each with 64 neurons. It uses ReLU [4] as the activation function in hidden layers and a Linear function in the output layer. Training is performed using ADAM. The hyperparameter values are 0.001 for learning rate, 0.99 discount rate, 100000 replay buffer size, and 32 minibatch size. We use the free version of IBM DB2 Express-C V11.1 for a DBMS in our experiments.

We create clusters consisting of two to five nodes in the Clemson cluster of Cloudlab [6]. Each node is equipped with two Intel E5-2660 v2 10-core, 128 GB of DDR4 and a 10 Gbps interconnect.

Baseline. We compare our work with two existing index advisors DivDesign [8] and RITA [24] which are proposed for replicated databases and described in Sect. 2 "Related Work". Similar to [8], we run DivDesign five times and report the best result. We report the performance of each algorithm when each query is processed by the replica that has the best index configuration for that query.

Experimental Parameters. The dynamic parameters in our experiments are the space index budget B and number of replicas N. B is tested with the values of 2 GB, 4 GB, 6 GB, 8 GB, and 10 GB, while N is tested with the values of 1, 2, 3, 4, and 5 nodes.

Performance Metrics. We compare the performance of DINA with the baselines using the averaged total execution time of all queries for three runs for the set of index configurations recommended by DINA and the baselines.

Results. First, we compare the performance of DINA-cost trained by using the estimated processing costs provided by the query optimizer and DINA-exe trained using the execution time of the queries in the workload with the baselines for both benchmarks.

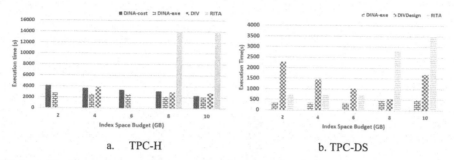

Fig. 3. (a) Performance of DINA vs Baselines for TPC-H benchmark; (b) Performance of DINA vs baselines for TPC-DS for varying index space budget and N = 3.

Figure 3(a) depicts the comparison of DINA and baselines when the number of nodes is 3 and we vary the space budget. RITA and DIVDesign (DIV) did not select a proper index for queries of a specific query template in the workload which results in a very high execution time. Therefore, we report the results of these algorithms whenever the execution time was under 15,000 s. For those cases that are not depicted, the averaged execution time was above 88,000 s. Overall, for this experiment, RITA did not perform well as it did not recommend good index configurations for the workload. In all cases, DINA-exe shows the best results among all the algorithms. For instance, when the space budget is 4 GB, DINA-exe can utilize the aggregate index space budget on nodes 28% and 32% better than DINA-cost and DIVDesign, respectively. In addition, DINA-cost can utilize the aggregate index space budget either as good as DIVDesign or better. As the index space budget increases, the performance of DIVDesign and DINA-cost get closer to that of DINA-exe, thus it might make sense to train the agent using query execution time when the index space budgets are low and use other algorithm for when the index space budgets are high.

Figure 3(b) shows the same experiments for TPC-DS. Here, we just report the result of DINA-exe because it has better performance than DINA-cost.

DINA-exe has the best performance in comparison with the baselines, which means it can find better partitions for the query workloads and recommend better index configurations. For lower index space budgets (2 GB–6 GB), on average DINA-exe improves the performance 50% and 75% compared to RITA and DIVDesign, respectively. RITA and DIVDesign have lower performance for higher index space budgets. The main reason is that, as the index space budget increases, they recommend more indexes for the workload.

In Fig. 4, we evaluated the performance of the studied algorithms when we varied the number of nodes and keep the index space budget fixed at 2 GB. DINA yields the best query execution time in all cases except for the case when the number of nodes is 5. Overall, on average, DINA performs 66% and 32% better than DivDesign and RITA, respectively.

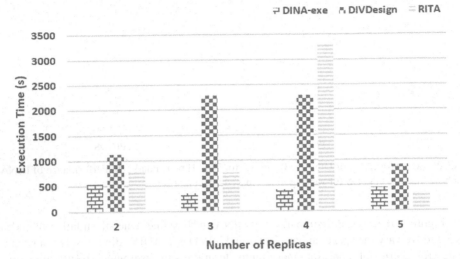

Fig. 4. Performance of DINA vs Baselines for TPC-DS benchmark for varying number of nodes and space budget = 2 GB

5 Conclusion and Future Research

In this paper, we introduced a new divergent index advisor called DINA that derives index configurations for replicas in a replicated database using Deep Reinforcement Learning (DRL). DINA is able to use feedback of its previous execution during the learning process to make better future decisions. A version of DINA called DNA-Exe does not rely on query estimation cost, which often is not accurate, provided by the query optimizer. The experiments using the TPC-H and TPC-DS benchmarks show that DINA performs better than the existing divergent index advisors especially when the index space budget is low.

In the future, we will extend this work for dynamic workloads, and evaluate the solution for more workloads that include updates. Moreover, we want to extend the work to cloud databases and consider features such as multi-tenancy and service-level agreements (SLAs). Also, we look for a more efficient reinforcement learning algorithm, which can converge faster.

References

1. Abadi, M., et al.: TensorFlow: a system for large-scale machine learning. In: 12th USENIX Symposium on Operating Systems Design and Implementation (OSDI 16), pp. 265–283 (2016)
2. Agrawal, S., Chaudhuri, S., Kollar, L., Marathe, A., Narasayya, V., Syamala, M.: Database tuning advisor for Microsoft SQL server 2005. In: Proceedings of the 2005 ACM SIGMOD International Conference on Management of Data (SIGMOD), pp. 930–932 (2005)
3. Basu, D., et al.: Cost-model oblivious database tuning with reinforcement learning. In: Chen, Q., Hameurlain, A., Toumani, F., Wagner, R., Decker, H. (eds.) DEXA 2015. LNCS, vol. 9261, pp. 253–268. Springer, Cham (2015). https://doi.org/10.1007/978-3-319-22849-5_18

4. Agarap, A.F.: Deep learning using rectified linear units (ReLU). arXiv preprint arXiv:1803. 08375 (2018)
5. Chaudhuri, S., Narasayya, V.: AutoAdmin "what-if" index analysis utility. ACM SIGMOD Rec. **27**(2), 367–378 (1998)
6. CloudLab. https://www.cloudlab.us/
7. Comer, D.: The difficulty of optimum index selection. ACM Trans. Database Syst. (TODS) **3**(4), 440–445 (1978)
8. Consens, M.P., Ioannidou, K., LeFevre, J., Polyzotis, N.: Divergent physical design tuning for replicated databases. In: Proceedings of the 2012 ACM SIGMOD International Conference on Management of Data, pp. 49–60 (2012)
9. Dash, D., Polyzotis, N., Ailamaki, A.: CoPhy: a scalable, portable, and interactive index advisor for large workloads. PVLDB **4**, 362–372 (2011)
10. Ding, B., Das, S., Marcus, R., Wu, W., Chaudhuri, S., Narasayya, V.R.: AI meets AI: leveraging query executions to improve index recommendations. In: Proceedings of the 2019 International Conference on Management of Data, pp. 1241–1258 (2019)
11. Durand, G.C., et al.: GridFormation: towards self-driven online data partitioning using reinforcement learning. In: Proceedings of the First International Workshop on Exploiting Artificial Intelligence Techniques for Data Management, pp. 1–7 (2018)
12. Hilprecht, B., Binnig, C., Röhm, U.: Learning a partitioning advisor for cloud databases. In: Proceedings of the 2020 ACM SIGMOD International Conference on Management of Data (SIGMOD), pp. 143–157 (2020)
13. Keras. https://keras.io/
14. Krishnan, S., Yang, Z., Goldberg, K., Hellerstein, J.M., Stoica, I.: Learning to optimize join queries with deep reinforcement learning. CoRR arXiv preprint arXiv:1808.03196 (2018)
15. Lan, H., Bao, Z., Peng, Y.: An index advisor using deep reinforcement learning. In: Proceedings of the 29th ACM International Conference on Information & Knowledge Management, pp. 2105–2108 (2020)
16. Li, G., et al.: openGauss: an autonomous database system. Proc. VLDB Endow. **14**(12), 3028–3042 (2021)
17. Li, G., Zhou, X., Li, S., Gao, B.: QTune: a query-aware database tuning system with deep reinforcement learning. Proc. VLDB Endow. **12**(12), 2118–2130 (2019)
18. Ma, L., Van Aken, D., Hefny, A., Mezerhane, G., Pavlo, A., Gordon, G.J.: Query-based workload forecasting for self-driving database management systems. In: Proceedings of the 2018 International Conference on Management of Data, pp. 631–645 (2018)
19. Marcus, R., Papaemmanouil, O.: Deep reinforcement learning for join order enumeration. In: Proceedings of the First International Workshop on Exploiting Artificial Intelligence Techniques for Data Management, pp. 1–4 (2018)
20. Mnih, V., et al.: Human-level control through deep reinforcement learning. Nature **518**(7540), 529–533 (2015)
21. Licks, G.P., et al.: SMARTIX: a database indexing agent based on reinforcement learning. Appl. Intell. **50**(8), 2575–2588 (2020). https://doi.org/10.1007/s10489-020-01674-8
22. Papadomanolakis, S., Dash, D., Ailamaki, A.: Efficient use of the query optimizer for automated physical design. In: Proceedings of the 33rd International Conference on Very Large Data Bases, pp. 1093–1104 (2007)
23. Perera, R.M., Oetomo, B., Rubinstein, B.I., Borovica-Gajic, R.: DBA bandits: self-driving index tuning under ad-hoc, analytical workloads with safety guarantees. In: IEEE 37th International Conference on Data Engineering (ICDE), pp. 600–611 (2021)
24. Tran, Q.T., Jimenez, I., Wang, R., Polyzotis, N., Ailamaki, A.: RITA: an index-tuning advisor for replicated databases. In: Proceedings of the 27th International Conference on Scientific and Statistical Database Management, pp. 1–12 (2015)

25. Schaarschmidt, M., Kuhnle, A., Ellis, B., Fricke, K., Gessert, F., Yoneki, E.: LIFT: reinforcement learning in computer systems by learning from demonstrations. arXiv preprint arXiv: 1808.07903 (2018)
26. Schnaitter, K., Abiteboul, S., Milo, T., Polyzotis, N.: On-line index selection for shifting workloads. In: IEEE 23rd International Conference on Data Engineering Workshop, pp. 459–468 (2007)
27. Sutton, R.S., Barto, A.G.: Reinforcement Learning: An Introduction. MIT Press, Cambridge (2018)
28. TPC: TPC-H benchmark. http://www.tpc.org/tpch/
29. TPC: TPC-DS benchmark. http://www.tpc.org/tpcds/
30. Sadri, Z., Gruenwald, L., Leal, E.: Online index selection using deep reinforcement learning for a cluster database. In: 2020 IEEE 36th International Conference on Data Engineering Workshops (ICDEW), pp. 158–161 (2020)
31. Sadri, Z., Gruenwald, L., Lead, E.: DRLindex: deep reinforcement learning index advisor for a cluster database. In: Proceedings of the 24th Symposium on International Database Engineering & Applications, pp. 1–8 (2020)
32. Zhang, J., et al.: An end-to-end automatic cloud database tuning system using deep reinforcement learning. In: Proceedings of the 2019 International Conference on Management of Data, pp. 415–432 (2019)

Deep Active Learning Framework for Crowdsourcing-Enhanced Image Classification and Segmentation

Zhiyao Li, Xiaofeng Gao$^{(\boxtimes)}$, and Guihai Chen

MoE Key Lab of Artificial Intelligence, Department of Computer Science
and Engineering, Shanghai Jiao Tong University, Shanghai, China
gao-xf@cs.sjtu.edu.cn

Abstract. Crowdsourcing is a distributed problem solving model that encompasses many types of tasks, and from a machine learning perspective, the development of crowdsourcing provides a new way to obtain manually labeled data with the advantages of lower annotation costs and faster annotation speed very recently, especially in the field of computer vision for image classification and segmentation. Therefore, it is necessary to investigate how to combine machine learning algorithms with crowdsourcing effectively and cost-effectively. In this paper, we propose a deep active learning (AL) framework by combining active learning strategies, CNN models and real datasets, to test the effectiveness of the active learning strategies through multiple scenario comparisons. Experiment results demonstrate the effectiveness of our framework in reducing the data annotation burden. Moreover, Our findings suggest that the strength is often observed in the case of relatively large data scale.

Keywords: Crowdsourcing · Active learning · Image classification · Image segmentation · Deep learning

1 Introduction

Crowdsourcing is the process of completing particular tasks by recruiting crowd workers, breaking down big tasks into smaller and simpler subtasks, allocating subtasks to workers operating in parallel, and recalling the results [5]. Supervised machine learning often requires large amounts of manually labeled data to train and evaluate models, such as image annotation in the field of computer vision, pattern recognition, and linguistic annotations for natural language processing [4,20,30]. In recent years, crowdsourcing access to manually labeled data has gained widespread attention in the machine learning community and become a popular paradigm to generate a substantial quantity of superior quality data for training and evaluating models because of its advantages of lower cost and faster speed. Therefore, crowdsourcing has become increasingly significant in machine

© The Author(s), under exclusive license to Springer Nature Switzerland AG 2022
C. Strauss et al. (Eds.): DEXA 2022, LNCS 13426, pp. 153–166, 2022.
https://doi.org/10.1007/978-3-031-12423-5_12

learning domain. However, crowdsourcing greatly depends on human workers for manual data labeling, which is often time-consuming and labor-intensive [34,35]. In this context, it is crucial to study how to combine machine learning algorithms with crowdsourcing effectively and cost-effectively while maintaining model performance.

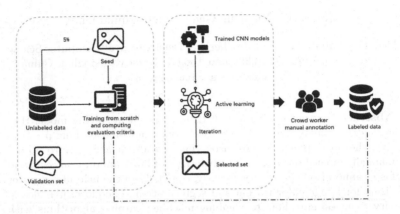

Fig. 1. Deep active learning framework by combining active learning strategie and CNN models for crowdsourcing-enhanced image classification and segmentation.

Selecting the most informative samples relevant to the research question is a common approach for reducing the burden of data annotation in crowdsourcing services. Accordingly, since the objective of adopting active learning algorithms in the supervised machine learning community intends to obtain greater accuracy with less manually labeled data, it would be suitable to incorporate various active learning strategies in crowdsourcing services [11,34]. For instance, Zhao et al. (2020) [35] integrated several active learning strategies (e.g., Entropy, Least Confidence, Kullback-Leibler Divergence, and Vote Entropy) into the crowd-sourcing environment. Wang et al. (2017) [31] confirmed that active learning facilitated a significant reduction of human annotation in deep image classification based on Convolutional Neural Networks (CNNs) by comparing different informative sample selection criteria. Yet, there is a possibility for active learning strategies to perform differently according to the models and datasets utilized, which is rarely considered in the improvement of crowdsourcing-enhanced data annotation via active learning.

Correspondingly, the focus of this study is to combine crowdsourced data annotation with active learning in an effective and cost-efficient manner. The primary objective is to investigate the efficacy of active learning strategies in the context of improving crowdsourcing services used in image classification and segmentation and attempts to propose a general deep active learning framework for crowdsourcing-enhanced image classification and segmentation tasks (Fig. 1). To this end, this study defines multiple scenarios by combining different active learning strategies (i.e., Bayesian Active Learning by Disagreement, Core-set,

Maximum Entropy, Least Confidence, Minimum Margin and random strategy), CNN models (i.e., Resnet and DeepLabV3+) and real datasets (i.e., CIFAR-10, CIFAR-100, WHU satellite and aerial imagery dataset). The main contributions of this paper can be summarized as follows:

- In the context of CNN-based image classification and segmentation, active learning strategies reduce the burden of data annotation compared to random sampling, which could be incorporated in the process of upgrading crowd-sourcing services;
- It also demonstrates that the efficacy of active learning strategies varies in different scenarios, with the performance of active learning approaching that of random sampling when the size of the initial annotation set is small.

The rest of the paper is organized as follows: Sect. 2 illustrates the state-of-the-art usage of active learning in the reduction of data annotation burden and the combination of active learning and crowdsourcing. Section 3 presents active learning strategies, CNN models, and datasets utilized for implementing multi-scenario comparisons. Section 4 shows the details of the experiments and the corresponding results. Section 5 presents conclusions and perspectives.

2   Related Work

Active learning aims to optimize the performance of the model by labeling the smallest number of samples, while minimizing the cost of labeling [7,17]. Conventional machine learning requires relatively few labeled samples, thus scholars have paid little attention to active learning. Recently, with the expansion of the Internet, huge amounts of unlabeled data have appeared, particularly in information extraction, medical images, speech recognition, etc., and obtaining a considerable quantity of high-quality labeled datasets require significant human and financial resources. Therefore, active learning has steadily received due attention. In terms of application scenarios, active learning can be classified into membership query synthesis, stream-based selective sampling, and pool-based active learning [22]. When it comes to selecting the most informative sample, there are seven primary active learning query strategies [22,26]: uncertainty sampling [10,23,26,27], query-by-committee [3,27], expected model change [27], expected error reduction [12,18], variance reduction [9,24,27], density-weighted methods [1,6,25,27,32] and hybrid query strategies [1,28,33,36]. Some important methods that deserve our attention, such as least confident, margin sampling, entropy, vote entropy, average Kullback-Leibler divergence, deep Bayesian active learning, etc., are summarized in Table 1.

Table 1. Active learning query strategies.

Related work	AL query strategies						
	Uncertainty sampling	Query-by-committee	Expected model change	Expected error reduction	Variance reduction	Density-weighted methods	Hybrid query strategies
	- Least confident - Margin sampling - Entropy	- Vote entropy - Average kullback-leibler divergence	- Decision-theoretic approach - Expected gradient length	- Decision-theoretic approach	- Inverse matrix	- Core-set - Batch query size	- Deep bayesian - RankCGAN
Culotta et al. 2005 [10]	✓						
Scheffer et al. 2001 [23]	✓						
Settles et al. 2008 [27]	✓	✓	✓		✓	✓	
Burbidge et al. 2007 [3]		✓					
Moskovitch et al. 2007 [18]				✓			
Guo et al. 2008 [12]				✓			
Cover et al. 2006 [9]					✓		
Schein et al. 2007 [24]					✓		
Xu et al. 2007 [32]						✓	
Jordan et al. 2019 [1]						✓	✓
Sener et al. 2018 [25]							✓
Chitta et al. 2019 [6]							✓
Shui et al. 2019 [28]							✓
Yin et al. 2017 [33]						✓	
Zhdanov et al. 2019 [36]						✓	

Some research has combined crowdsourcing and active learning, focusing primarily on the issue of crowdsourced data labelling. Due to the high cost of hiring expert workers, researchers wish to replace expert workers with crowdsourced labour for data annotation. The authors of [16] proposed a novel method based on the active cross-query learning strategy, which allows every worker, rather than domain experts, to label a portion of the selected query data. [2] presented an innovative cooperative scheme based on active learning and crowdsourcing, designed to offer a solution for the cold start issue, i.e. initialising the classification of a set of unlabeled large-scale data sets. Calpric, an automated classification tool developed by [21] using active learning and crowdsourcing, can perform annotation with the same level of accuracy as skilled human annotators while minimising labelling costs. To improve the accuracy of integrated labels, [29] proposed a novel active learning framework that takes workers' confidence information into account. [8] proposed two classification methods based on crowdsourcing and active learning and evaluated their suitability. However, the existing research does not consider the impact of different data sample sizes and models on the outcomes, nor does it compare the differences between active learning strategies. This provides research inspiration.

3   Proposed Method

3.1   Notation

We list the symbols used throughout the paper in Table 2.

Table 2. List of symbols.

Symbol	Definition
r	The serial number of rounds
j	The serial number of random initial labeled sets
i	The serial number of samples
k	The number of selected samples in every round
N_{Lr}	The number of labeled data in the round r
N_{Ur}	The number of unlabeled data in the round r
$L_r^j = \{(x_i, y_i)\}_{i=1}^{N_{Lr}}$	The labeled set in the group j and round r
$U_r^j = \{(x_i, y_i)\}_{i=1}^{N_{Lr}}$	The unlabeled set in the group j and round r
Φ_r	The trained model in the round r
$\Psi_\theta(\cdot, \cdot, \cdot)$	The selection function with parameter θ, such as Min-Margin
$P_\theta(y_b \mid x_a)$	The probability that sample x_a belongs to label y_b

3.2 Neural Networks for Image Tasks

ConvNets are increasingly identified as commodities in the computer vision sector. Numerous efforts have been directed at enhancing the original architecture to realize improved precision. This study uses ResNet and DeepLabV3+.

ResNets introduces a residual connection which can change the output $f(x)$,

$$h(x) = f(x) + x \tag{1}$$

where x is input.

Then it can address the degradation issue. Rather than expecting that few stacked layers straightforwardly map to a chosen underlying mapping, the approach in this study explicitly allow the stacked layers map to a residual mapping. Since the level of difficulty to optimize the original is higher than optimizing the residual mapping, in the event of an optimal identity mapping, it would be easier to push the residual to zero than fitting an identity mapping through origin layers. Feedforward neural networks comprising "residual connections" can conduct identity mapping, adding the outputs to the stacked layers' outputs, while do not add computational complexity or additional parameters.

DeepLabV3+ is one of the top performing networks in image segmentation. It encodes multi-view information by DeepLabV3, and makes use of the corresponding low-level features when decoding. DeepLabV3 benefits from the combination of atrous spatial pyramid pooling (ASPP) with different rates of dilation convolution, i.e.

$$y[i] = \sum_{k=1}^{K} x[i + r \cdot k] w[k] \tag{2}$$

where x is input data, $w = (w_1, \cdots, w_k, \cdots w_K) \in \mathbb{R}^K$ is the kernel and r is the flexible rate, respectively . By controlling the rate parameter r, one can arbitrarily control the receptive fields of the convolution layer.

Therefore, it both increases the resolution of final feature map and deals with the multi-scale objects. We choose pretrained MobileNetV2 as the backbone of DeepLabV3. MobileNetV2 is a lightweight Deep Convolutional Neural Networks (DCNNs). It replaces the standard convolution with Depthwise separable convolution to accelerate the computations. Residual blocks are also applied in MobileNetV2.

3.3   Active Learning

We first have an initial labeled data set $L_0^0 = \{(x_i, y_i)\}_{i=1}^{N_{L0}}$, a large candidate set of unlabeled data $U_0^0 = \{x_i\}_{i=1}^{N_{U0}}$, a training active learning model Φ_0 that is based on pool. Then we evaluate $x_i \in U_0$ and sample k (size of budget) instances, which are labeled by a master using sampling function $\Psi_\theta \left(L_0^0, U_0^0, \Phi_0 \right)$. We extend L_0^0 to L_0^1 labeled set by adding the selected samples which have oracle-annotated labels. The generated L_0^1 set is utilized to retrain model Φ. Repeat the above sample annotation training loop until the sampling budget is exhausted or the training has converged.

In this paper, we evaluate multiple different sampling functions: Least Confidence, Min-Margin, Max-Entropy, Bayesian Active Learning by Disagreement, Core-set, etc.

Least Confidence. The examples are sorted in descending order in this method by the probability of not predicting the most confident sequence from the model:

$$x^* = \arg\max_i (1 - P_\theta(y_i \mid x)) \tag{3}$$

Min-Margin. Multiple models are trained on stratified bootstrap samples of the labeled training set. Then candidate examples are selected, which have the smallest margin among all the bootstrapped models [14]:

$$margin(h, x) = h\left(x; \hat{y}_1(x)\right) - h\left(x; \hat{y}_2(x)\right) \tag{4}$$

where $\hat{y}_1(x)$ and $\hat{y}_2(x)$ are the first two highest scoring classes under the estimator h: $\hat{y}_1(x) = \arg\max_g h(x; g)$ and $\hat{y}_2(x) = \arg\max_{g, g \neq \hat{y}_1(x)} h(x; g)$.

This method includes two hyper-parameters: the number of bootstrapped predictor κ and the fraction β of bootstrap sample size, which are consistent in our experiments.

Max-Entropy. Entropy is a measurable physical property that is associated with disorder, randomness, or uncertainty. Greater entropy indicates that the uncertainty of the system is very large. Thus, we select the samples with large entropy as the annotation data [13].

$$x^* = argmax_x - \sum_i P_\theta \left(y_i \mid x\right) \cdot \ln P_\theta \left(y_i \mid x\right) \tag{5}$$

Bayesian Active Learning by Disagreement (BALD). In BALD, we use dropout layers and Monte Carlo dropout (MC dropout) to train the model Φ and approximate the sampling from posterior, respectively.

In this paper, pool points are expected to maximize information acquired about the parameters θ of model, in other words, maximize mutual information between estimations and model posterior.

$$\mathcal{I}\left[y, \theta \mid x\right] = \mathcal{H}\left[y \mid x\right] - E_{p(\theta)}[\mathcal{H}[y \mid \mathbf{x}, \theta]] \tag{6}$$

where $\mathcal{H}[y \mid x, \theta]$ is the entropy of predicted y given parameters θ. However, some parameters produce disagreeing estimations with high certainty. We should point to points that have high variance in the softmax layer of input. Correspondingly, the highest probability of each random forward pass of the model is assigned to a different category.

Core-Set. Core-set method exploits data points' geometry and selects samples which can contain all data points. Essentially, this algorithm would like to seek some of points, i.e., cover points, which minimizes the distance between each data point and its corresponding nearest cover points.

$$\min_{s^1:|s^1| \leq b|} \max_i \min_{j \in s^1 \cup s^0} \Delta \left(x_i, x_j\right) \tag{7}$$

where x_i is point in the dataset and x_j is its nearest cover point.

4   Performance Evaluation

In this section, the paper validates effective crowdsourcing image processing via active learning from two perspectives: image classification and image segmentation. The training procedure of active learning method is summarized in Algorithm 1. All methods are implemented in PyTorch 1.8 with Python 3.8 based on Pycls library[1] [19], which are trained on single GPU NVIDIA 2070S with 8G memory.

4.1   Crowdsourcing Image Classification via Active Learning

Datasets Description. We use two public datasets, including CIFAR10 and CIFAR100, which are labeled subsets of the 80 million tiny images dataset and were collected by Alex Krizhevsky, Vinod Nair, and Geoffrey Hinton [15] (Table 3).

---

[1] https://github.com/acl21/deep-active-learning-pytorch.

Algorithm 1: Fine-tuned active learning method.

Input:
Unlabeled set $U_0^0 = \{x_i\}_{i=1}^{N_{U_0}}$;
Labeled set $L_0^0 = \{(x_i, y_i)\}_{i=1}^{N_{L_0}}$;
Pre-trained CNN model Φ_0;
Selection function Ψ_θ;
Selection ratio α;
Output:
Labeled candidates L;
Fine-tuned CNN model Φ_r at round r;

1 **repeat**
2   **for** $r = 1, 2, \ldots$ **do**
3          $C_r \leftarrow$ selected k samples from U by selection function Ψ_θ;
4          $p \leftarrow \Phi_r(C_r)$;
5          $a \leftarrow mean(p) \left\{ a = \frac{1}{k} \sum_{i=1}^{k} p_i \right\}$;
6   **if** $a > 0.5$ **then**
7             $S_r \leftarrow$ top α percent of the samples of C_r;
8          $L \leftarrow L \cup S_r$;
9          $U \leftarrow U \setminus C_r$;
10   Retrain the model Φ_r using the updated L set;
11 **until** *model performance is satisfactory*;

CIFAR10. CIFAR10 dataset is a part of the 80 million small images datasets. The CIFAR10 dataset includes 60000 32×32 colored images, which are labeled into 10 completely mutually exclusive classes, with 6000 images per class. These 10 classes mainly describe the main objects in the image, such as automobile, bird, airplane, cat, dog, frog, deer, horse, truck, or ship. Especially, there are no overlaps between the automobile class and the truck class. The CIFAR10 dataset has 50000 training samples and 10000 test samples.

CIFAR100. CIFAR100 dataset is quite similar to CIFAR-10, except it has 100 classes containing 600 images each. There are 500 training images and 100 testing images per class. These 100 classes in CIFAR100 dataset are classified into 20 superclasses. Each image has two label information at the same time: the "fine" label is accurate classification information, and the "coarse" label is rough superclasses classification information. For example, superclass fish contains aquarium fish class, flatfish class, ray class, shark class and trout class.

Experimental Setup. We use ResNets in our experiments and the optimizer we use is SGD with exponential decay, in which learning rate starts at 0.025 and decay rate is set to 0.0003. The momentum and gamma factor are set to 0.9 and 0.1 respectively. We pre-process input by random horizontal flip and

Table 3. Performance comparisons of different AL strategies on CIFAR10 and CIFAR100. The bold value marks the best one in each model.

Dataset	Model	Active learning strategies	Accuracy (%)
CIFAR10	Resnet18	Max Entropy	91.55
		Least Confidence	**91.74**
		Min Margin	91.46
		BALD	91.08
		Coreset	91.13
		Random	89.67
	Resnet50	Max Entropy	91.65
		Least Confidence	91.71
		Min Margin	91.20
		BALD	**91.88**
		Coreset	91.68
		Random	83.65
CIFAR100	Resnet18	Max Entropy	**55.35**
		Least Confidence	54.20
		Min Margin	54.68
		BALD	54.78
		Coreset	54.76
		Random	54.27
	Resnet50	Max Entropy	54.85
		Least Confidence	**54.90**
		Min Margin	54.56
		BALD	54.48
		Coreset	54.10
		Random	53.75

normalization, where $p = 0.5$ and the input is divided by 255. In training process, the batch size is set to be 96. We train 200 epochs of the base classifier on labeled set. For iterations of active learning, the best model is fine-tuned (selected by the accuracy of validation set), where the maximum number of iterations is set to 5. Note that we evaluate the model on validation set every 2 period epochs.

Performance Comparison. We first make comparison of different active learning methods against random method. As shown in Fig. 2, our experiments indicate that compared to random method, AL methods achieve remarkable classification accuracy in CIFAR10 dataset. For example, it achieves best accuracy of 91.74% in Resnet18 when integrated with Least Confidence method and 91.88% in Resnet50 when integrated with Bayesian Active Learning by Disagreement (BALD), which has improved by 2.07% and 8.23% compared to random method

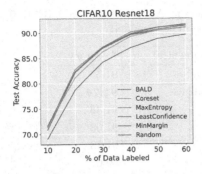
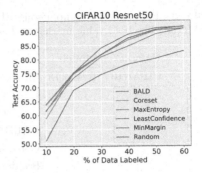

Fig. 2. Performance comparisons of different AL strategies on CIFAR10 dataset. There is no discernible difference between different AL strategies.

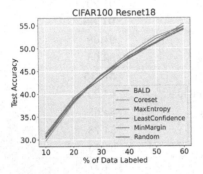
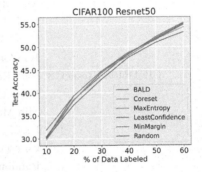

Fig. 3. Performance comparisons of different AL strategies on CIFAR100 dataset. AL strategies have little effect in the case of small data scale.

respectively. Notably, relatively poor accuracy is achieved in CIFAR100 dataset (as shown in Fig. 3), it is because that there is more samples of each class in CIFAR10, which promotes the model training.

4.2   Crowdsourcing Image Segmentation via Active Learning

Datasets Description. Regarding image segmentation, we use two public satellite datasets, the WHU satellite dataset and aerial imagery dataset, from the New Zealand Land Information Services website[2].

WHU Satellite Dataset. The WHU satellite dataset (East Asia) consists of 17388 images from cities in East Asia and cover 29085 buildings with 512×512 cells with a spatial resolution of 2.7 m per pixel. Among them 21556 buildings are separated for training and the rest 7529 buildings are used for testing.

WHU Aerial Imagery Dataset. The WHU aerial imageries consist of 8189 images from New Zealand cities and cover over 187000 buildings with 512×512 cells with a spatial resolution of 0.3 m per pixel. The ready-to-use samples are

---

[2] https://www.linz.govt.nz/.

divided into three parts: a training set (130500 buildings), a validation set (14500 buildings) and a test set (42000 buildings).

Experimental Setup. We use DeepLabV3+ for image segmentation experiment. About hyper-parameter setup, similar to the previous experiments, the optimizer is SGD with exponential decay. In training process, the batch size, epoch and learning rate are set to be 64, 100 and 0.1 respectively. For iterations of active learning, the maximum number of iterations is set to be 19. Meanwhile, both the active start size and active selection size are set to be 5%. We computed the mean intersection over union (mIoU) to evaluate the performance of models in each iteration of each experiment. They were computed as follows:

$$mIoU = \frac{1}{k+1} \sum_{i=0}^{k} \frac{p_{ii}}{\sum_{j=0}^{k} p_{ij} + \sum_{j=0}^{k}(p_{ji} - p_{ii})} \tag{8}$$

where k is the total number of classes ($k = 2$ in this study). p_{ii} and p_{ij} represent the total numbers of pixels belonging to true pixel class i that are predicted to belong to i and j, respectively.

Performance Comparison. As shown in Fig. 4, the experimental results of segmentation on two datasets also indicate that AL strategies is better than random method, while the differences between different AL strategies are not obvious. These results also suggest that there should be enough samples when using AL strategies for segmentation problem (Table 4).

Table 4. Performance comparisons of different active learning strategies on WHU satellite and WHU aerial imagery dataset against random method. Convergence is reached after 18 iterations. The bold value marks the best one in each model.

Dataset	Model	Active learning strategies	MIoU (%)
WHU satellite dataset	DeepLabV3+	Max Entropy	83.30
		Min Margin	83.16
		MCdropout	**83.45**
		Random	82.65
WHU aerial imagery dataset	DeepLabV3+	Max Entropy	93.85
		Min Margin	**93.90**
		MCdropout	93.78
		Random	93.05

4.3 Analysis and Discussion

Variance in Different Labeled Data. Since the labeled data is incomplete in real-world, we explore the change of variance in classification and segmentation

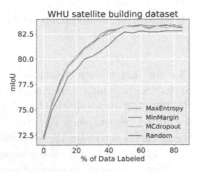
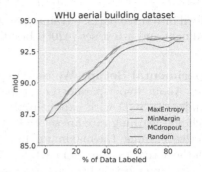

Fig. 4. Performance comparisons of different AL strategies on WHU satellite and WHU aerial imagery dataset. There is no discernible difference between different AL strategies.

accuracy with different initial labeled data. As shown in Fig. 2, Fig. 3, and Fig. 4, we can summarize the following points:

- To sum up, the test accuracy increases with the increase of initial labeled data, proving that AL strategies performs better than random method, while there is no discernible difference between different AL strategies;
- AL strategies have little effect in the case of small data scale and do not even beat the random strategy well, as can be derived from the experiments on CIFAR100.

5   Conclusions and Perspectives

Creating training data has increasingly been a critical bottleneck in machine learning. By selecting the most informative instances for labelling, active learning allows an efficient approach to creating training data. Meanwhile, crowdsourcing, as a distributed problem-solving method, is economically feasible for data labelling. In this paper, active learning has been applied to improve the data labelling task of the crowdsourcing system via the use of different strategies such as Max Entropy, Least Confidence, Min Margin, etc. to study the effectiveness of active learning algorithms in reducing the burdens of data labelling. Finally, we conducted comparative experiments on image classification and segmentation on four real datasets, demonstrating the effectiveness of the proposed framework, and emphasising that it works better when the data size is relatively large.

Acknowledgements. This work was supported by the National Key R&D Program of China [2019YFB2102200], the National Natural Science Foundation of China [61872238, 61972254], Shanghai Municipal Science and Technology Major Project [2021SHZDZX0102] and the ByteDance Research Project [CT20211123001686].

References

1. Ash, J.T., Zhang, C., Krishnamurthy, A., Langford, J., Agarwal, A.: Deep batch active learning by diverse, uncertain gradient lower bounds. arXiv preprint arXiv:1906.03671 (2019)
2. Brangbour, E., Bruneau, P., Tamisier, T., Marchand-Maillet, S.: Active learning with crowdsourcing for the cold start of imbalanced classifiers. In: Luo, Y. (ed.) CDVE 2020. LNCS, vol. 12341, pp. 192–201. Springer, Cham (2020). https://doi.org/10.1007/978-3-030-60816-3_22
3. Burbidge, R., Rowland, J.J., King, R.D.: Active learning for regression based on query by committee. In: Yin, H., Tino, P., Corchado, E., Byrne, W., Yao, X. (eds.) IDEAL 2007. LNCS, vol. 4881, pp. 209–218. Springer, Heidelberg (2007). https://doi.org/10.1007/978-3-540-77226-2_22
4. Callison-Burch, C., Dredze, M.: Creating speech and language data with Amazon's mechanical turk. In: Proceedings of the NAACL HLT 2010 Workshop on Creating Speech and Language Data with Amazon's Mechanical Turk, pp. 1–12 (2010)
5. Chang, J.C., Amershi, S., Kamar, E.: Revolt: collaborative crowdsourcing for labeling machine learning datasets. In: Proceedings of the 2017 CHI Conference on Human Factors in Computing Systems, pp. 2334–2346 (2017)
6. Chitta, K., Alvarez, J.M., Haussmann, E., Farabet, C.: Training data distribution search with ensemble active learning. arXiv preprint arXiv:1905.12737 (2019)
7. Costa, J., Silva, C., Antunes, M., Ribeiro, B.: On using crowdsourcing and active learning to improve classification performance. In: 2011 11th International Conference on Intelligent Systems Design and Applications, pp. 469–474. IEEE (2011)
8. Cósta, J., Silva, C., Antunes, M., Ribeiro, B.: On using crowdsourcing and active learning to improve classification performance. In: 11th International Conference on Intelligent Systems Design and Applications, ISDA 2011, Córdoba, Spain, 22–24 November 2011, pp. 469–474. IEEE (2011)
9. Cover, T.M., Thomas, J.A.: Elements of Information Theory, 2nd edn. Wiley-Interscience (2006)
10. Culotta, A., McCallum, A.: Reducing labeling effort for structured prediction tasks. In: AAAI, vol. 5, pp. 746–751 (2005)
11. Gilyazev, R., Turdakov, D.Y.: Active learning and crowdsourcing: a survey of optimization methods for data labeling. Program. Comput. Softw. **44**(6), 476–491 (2018)
12. Guo, Y., Schuurmans, D.: Discriminative batch mode active learning. In: NIPS, vol. 20, pp. 593–600. Citeseer, MIT Press (2008)
13. Holub, A., Perona, P., Burl, M.C.: Entropy-based active learning for object recognition. In: IEEE Conference on Computer Vision and Pattern Recognition, CVPR Workshops 2008, Anchorage, AK, USA, 23–28 June 2008, pp. 1–8 (2008)
14. Jiang, H., Gupta, M.R.: Minimum-margin active learning. arXiv arXiv:1906.00025 (2019)
15. Krizhevsky, A., Hinton, G., et al.: Learning multiple layers of features from tiny images (2009)
16. Li, B., Zhang, A., Chen, W., Yin, H., Cai, K.: Active cross-query learning: a reliable labeling mechanism via crowdsourcing for smart surveillance. Comput. Commun. **152**, 149–154 (2020)
17. McCallumzy, A.K., Nigamy, K.: Employing EM and pool-based active learning for text classification. In: Proceedings of the International Conference on Machine Learning (ICML), pp. 359–367. Citeseer (1998)

18. Moskovitch, R., Nissim, N., Stopel, D., Feher, C., Englert, R., Elovici, Y.: Improving the detection of unknown computer worms activity using active learning. In: Hertzberg, J., Beetz, M., Englert, R. (eds.) KI 2007. LNCS (LNAI), vol. 4667, pp. 489–493. Springer, Heidelberg (2007). https://doi.org/10.1007/978-3-540-74565-5_47
19. Munjal, P., Hayat, N., Hayat, M., Sourati, J., Khan, S.: Towards robust and reproducible active learning using neural networks. arXiv arXiv:2002.09564 (2020)
20. Patterson, G., Hays, J.: Sun attribute database: discovering, annotating, and recognizing scene attributes. In: 2012 IEEE Conference on Computer Vision and Pattern Recognition, pp. 2751–2758. IEEE (2012)
21. Qiu, W., Lie, D.: Deep active learning with crowdsourcing data for privacy policy classification. CoRR abs/2008.02954 (2020)
22. Ren, P., et al.: A survey of deep active learning. arXiv preprint arXiv:2009.00236 (2020)
23. Scheffer, T., Decomain, C., Wrobel, S.: Active hidden Markov models for information extraction. In: Hoffmann, F., Hand, D.J., Adams, N., Fisher, D., Guimaraes, G. (eds.) IDA 2001. LNCS, vol. 2189, pp. 309–318. Springer, Heidelberg (2001). https://doi.org/10.1007/3-540-44816-0_31
24. Schein, A.I., Ungar, L.H.: Active learning for logistic regression: an evaluation. Mach. Learn. **68**(3), 235–265 (2007)
25. Sener, O., Savarese, S.: Active learning for convolutional neural networks: a core-set approach. In: International Conference on Learning Representations (2018)
26. Settles, B.: Active learning literature survey. Technical report, pp. 55–66 (2010)
27. Settles, B., Craven, M., Ray, S.: Multiple-instance active learning. Adv. Neural. Inf. Process. Syst. **20**, 1289–1296 (2007)
28. Shui, C., Zhou, F., Gagné, C., Wang, B.: Deep active learning: unified and principled method for query and training. arXiv preprint arXiv:1911.09162 (2019)
29. Song, J., Wang, H., Gao, Y., An, B.: Active learning with confidence-based answers for crowdsourcing labeling tasks. Knowl. Based Syst. **159**, 244–258 (2018)
30. Vaughan, J.W.: Making better use of the crowd: how crowdsourcing can advance machine learning research. J. Mach. Learn. Res. **18**(1), 7026–7071 (2017)
31. Wang, K., Zhang, D., Li, Y., Zhang, R., Lin, L.: Cost-effective active learning for deep image classification. IEEE Trans. Circ. Syst. Video Technol. **27**(12), 2591–2600 (2016)
32. Xu, Z., Akella, R., Zhang, Y.: Incorporating diversity and density in active learning for relevance feedback. In: Amati, G., Carpineto, C., Romano, G. (eds.) ECIR 2007. LNCS, vol. 4425, pp. 246–257. Springer, Heidelberg (2007). https://doi.org/10.1007/978-3-540-71496-5_24
33. Yin, C., et al.: Deep similarity-based batch mode active learning with exploration-exploitation. In: 2017 IEEE International Conference on Data Mining (ICDM), pp. 575–584. IEEE (2017)
34. Zhang, J., Wu, X., Sheng, V.S.: Learning from crowdsourced labeled data: a survey. Artif. Intell. Rev. **46**(4), 543–576 (2016). https://doi.org/10.1007/s10462-016-9491-9
35. Zhao, Y., Prosperi, M., Lyu, T., Guo, Y., Zhou, L., Bian, J.: Integrating crowdsourcing and active learning for classification of work-life events from tweets. In: Fujita, H., Fournier-Viger, P., Ali, M., Sasaki, J. (eds.) IEA/AIE 2020. LNCS (LNAI), vol. 12144, pp. 333–344. Springer, Cham (2020). https://doi.org/10.1007/978-3-030-55789-8_30
36. Zhdanov, F.: Diverse mini-batch active learning. arXiv preprint arXiv:1901.05954 (2019)

Sentiment and Knowledge Based Algorithmic Trading with Deep Reinforcement Learning

Abhishek Nan[1,2], Anandh Perumal[1,2], and Osmar R. Zaiane[1,2]([⊠]) [iD]

[1] University of Alberta, Edmonton, Canada
{abhisheknan,anandhpe,zaiane}@ualberta.ca
[2] Alberta Machine Intelligence Institute, Edmonton, Canada

Abstract. Algorithmic trading, due to its inherent nature, is a difficult problem to tackle; there are too many variables involved in the real-world which makes it almost impossible to have reliable algorithms for automated stock trading. The lack of reliable labelled data that considers physical and physiological factors that dictate the ups and downs of the market, has hindered the supervised learning attempts for dependable predictions. To learn a good policy for trading, we formulate an approach using reinforcement learning which uses traditional time series stock price data and combines it with news headline sentiments, while leveraging knowledge graphs for exploiting news about implicit relationships.

Keywords: Reinforcement learning · Trading · Stock price prediction · Sentiment analysis · Knowledge graph · Natural Language Processing

1 Introduction

Machine learning is mainly about building predictive models from data. When the data are time series, models can also forecast sequences or outcomes. Predicting how the stock market will perform is an application where people have naturally attempted machine learning but it turned out to be very difficult because involved in the prediction are many factors, some rational and some appearing irrational. Machine learning has been used in the financial market since the 1980s [3], trying to predict future returns of financial assets using supervised learning such as artificial neural networks [2], support vector machines [14] or even decision trees [21]; but so far, there has been only limited success. There are multiple causes for this. For instance, in supervised machine learning, we usually have labelled datasets with balanced class distributions. When it comes to the stock market, there is no such labelled data for when someone should have bought/sold their holdings. This leads credence to the problem being fit for the reinforcement learning framework [26], a behavioral-based learning paradigm relying on trial and error and supplemented with a reward mechanism. Reinforcement learning has the ability to generate this missing *labelling* once we

© The Author(s), under exclusive license to Springer Nature Switzerland AG 2022
C. Strauss et al. (Eds.): DEXA 2022, LNCS 13426, pp. 167–180, 2022.
https://doi.org/10.1007/978-3-031-12423-5_13

define a proper reward signal. But there are still other issues in this context which are specific to stock markets. They are prone to very frequent changes and often these changes cannot be inferred from the historical trend alone. They are affected by real-world factors such as political, social and even environmental factors. For instance, an earthquake destroying a data-center could result in stock prices dropping for a company; a new legislation about trade can positively impact the value of a company. Noise to signal ratio is very high in such conditions and it becomes difficult to learn anything meaningful under such circumstances. Such environments can be modelled as Partially Observable Markov Decision Processes (POMDPs) [31], where the agent only has limited visibility of all environmental conditions. A POMDP models an agent decision process in which it is assumed that the system dynamics are determined by a discrete time stochastic control process, but the agent cannot directly observe the underlying state [13]. Our contribution is the use of sentiment analysis done on news related to a traded company and its services in conjunction with a reinforcement learning algorithm to learn an appropriate policy to trade stocks of the given company. To find the relevant news title on-which to apply sentiment analysis, we use a traversal of a knowledge graph.

After highlighting the related work in Sect. 2, we present our approach in Sect. 3 combining news headlines and their sentiment after finding their relation with the relevant stock hinging on a knowledge graph, and finally learning a good policy for buying and selling using Reinforcement Learning. As a proof of concept, we present an empirical evaluation using the stocks of Microsoft, Amazon and Tesla between 2014 and 2018 in Sect. 4. Section 5 highlights the analysis of the observed results. Finally, we present perspective on future work.

2   Related Work

There have been many approaches in the past which try to model traditional time series approaches for stock price prediction [12,23,28]. The main idea in these approaches is to predict the stock price at the next time step given the past trend. This prediction is then fed to a classifier which tries to predict the final buy/sell/hold action. Most modern deep learning techniques try to use some form of recurrent networks to model the sequential trend in the data. The authors of [4] used LSTMs with great success to make predictions in the Chinese stock market. Approaches integrating some form of event data has been explored as well to some extent. For instance, the authors of [16] used manually extracted features from news headlines to integrate event information and spliced them with several other economic indicators according to prior knowledge and combined them together as the input to neural networks.

An alternative approach is to use Reinforcement Learning. Fischer shows in a comprehensive survey on the use of RL in financial markets that there are many attempted approaches but the problem is far from being solved [7]. From a reinforcement learning (RL) perspective, [27] proposed an Adaptive Network Fuzzy Inference System (ANFIS) supplemented by the use of RL as a non-arbitrage algorithmic trading system. The authors in [5] use a deep learning component which automatically senses the changing market dynamics for feature learning and these features are used as input to an RL module which learns to make trading decisions. [10] explored the use of actor-critic methods for stock trading and serves as one of the primary motivations behind our research.

Furthermore, public opinion can often provide valuable indication as to how a company might be posed to perform in the market. Attempts have been made previously to directly classify each comment on a stock trading forum as a indicator of a buy/sell/hold decision [25]. Rather than use text data in its entirety as a variable to make decisions, the general sentiment of the text can be extracted as a score [22] and combined with other related data.

Our approach in a way tries to take the best of these methods and extend them into a single dynamic system paired with knowledge graphs. We extract sentiments from event information and use knowledge graphs to detect implicit relationships between event information and a given traded company. We then combine this information with the time series stock data, and allow our agent to learn an optimal policy using deep reinforcement learning. We also take advantage of more recent RL techniques such as the DQN (Deep Q-Learning) introduced by [20]. This approach of combining knowledge graph driven sentiment data with deep RL is our novel proposal and has not been explored in any literature we surveyed.

3   Proposed Approach

Our approach combines concepts from a few different domains; hence, we give a short overview of each of them and connect how they are used in our approach.

3.1   Reinforcement Learning

The typical reinforcement learning setting involves an agent and an environment loop (Fig. 1a), where the agent interacts with the environment via some action and then it receives back some observation from the environment which tells it how the environment has/has not changed as an effect of that action. Such a sequence of action-observations is known as an episode; which terminates when a failure condition is met or the goal is achieved. In most cases, this observation contains a human designed $reward(r_t)$, which gives the agent some indication as to how good or bad that $action(a_t)$ might have been. Reward designing is an active area of research itself, but for simple cases we can just assign the reward to be 1 in case of successful completion of the task, and 0 for all other interactions. Also the information from the observation can be used to maintain some sense of a $state(s_t)$, which is akin to the agents perception of the world. This continuous back and forth interaction between the agent and the environment with the sole purpose of trying to maximise the $return$ (sum of rewards over an entire episode) makes the agent learn an optimal behaviour in the given environment for a particular task.

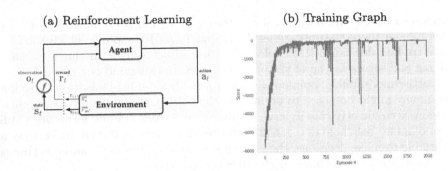

(a) Reinforcement Learning   (b) Training Graph

Fig. 1. (a) Partially Observable Markov Decision Process [6]; (b) Training score vs the number of episodes for the agent with sentiment information.

In some cases, the agent only has limited visibility of the environment. Such scenarios are modelled as a Partially Observable Markov Decision Process (POMDP) [17]. In these cases, the agent maintains its perception of the world via observations which are mappings from the underlying true environmental state. For instance, in case of a stock trading agent, the environment is the stock market and the actions are buying, selling, or holding. Since the agent does not have a perfect idea of everything that is going on in the world, the POMDP of the real world is represented by features of the state such as stock prices, sentiments, historical trend, etc.

Q-learning. is a model free reinforcement learning algorithm. Given an environment, the agent tries to learn a policy which maximises the total reward it gets from the environment at the end of an episode (a sequence of interactions). For instance, in our problem setting, an episode during training would be the sequence of interactions the agent makes with the stock market starting from January 1, 2014 and ending on December 31, 2017. The agent would try to learn a behaviour which maximises the value of its portfolio at the end date.

The intuition behind Q-learning is that the agent tries to learn the utility of being in a certain state and taking a particular action in that state and then following the behavioural policy learnt so far till the end of the episode (called the action value of that state). So, Q-learning tries to learn the action value of every state and action. It does this by exploring and exploiting at the same time. For instance, a trading agent starts on Day 1 and it has two options: *Buy* and *Sell*. It takes the *Buy* option (say arbitrarily) the first time it experiences Day 1 and receives a reward of 10 units. For optimal performance, the agent will *usually* follow the best possible option available to it. *Usually*; because if it always followed the best option that it *thinks* is available to it, it will not learn the value of taking the other options available to it in that state. For instance, in the above example, *Sell* could have led to a reward of 20, but it would have never known this if it always took the *Buy* option after it first experienced Day 1 with a *Buy* action. This dilemma is known as the exploration and exploitation trade-off. A naive, yet effective way of solving this is to always take the "greedy" option, except also act randomly a small percentage of the time, say with a probability of 0.1. This is known as the ϵ-greedy approach with $\epsilon = 0.1$. This finally brings us to the Q-learning equation, which updates the action values of each state and action pair.

$$Q(S_t, a_t) \leftarrow Q(S_t, a_t) + \alpha[R_{t+1} + \gamma \max_a Q(S_{t+1}, a) - Q(S_t, a_t)]$$

Function Approximation. A shortcoming of the above mentioned Q-learning methodology is the obvious fact that it relies on the idea of a distinct state. What this means is that the Q-learning update can only be applied to an environment where each state(s_t) can be distinctly labelled. This would mean we would have to maintain a huge table of every possible state and action combination that can be encountered and their action values. This does not generalize very well and is not tractable for real world problems. For instance, given today's state of the world to a stock trading agent, it might make some decision and learn from it, but it is very unlikely that the exact same conditions will ever be presented to it again. The solution to this is to use a function approximator, which given the current environmental observation and the chosen action maps them to an action value. The parameters of the approximator can then be updated similar to supervised learning once we have observed the actual reward. In our experiments, we use an artificial neural network for function approximation.

For large state spaces since optimizing artificial neural networks via just back-propagation becomes unstable so we adapt the modifications to a Deep Q-Network (DQN) as presented by [20]. These modifications include gradient clipping, experience replay and using a Q-network which periodically updates an independent target network.

3.2   Sentiment Analysis

Sentiment analysis is an automated process to annotate text predicted to be expressing a positive or negative opinion. Also known as opinion mining, sentiment analysis categorizes text into typically two classes positive vs. negative, and often a third class: neutral. Discovering the polarity of a text is often used to analyze product or service reviews, like restaurants, movies, electronics, etc. but also other written text like blog posts, memos, etc. There are two main types of sentiment analysis approaches, namely lexicon-based using a dictionary of words with their polarities; and machine learning based which build a predictive model using a labelled train dataset [30].

Each sentence or sequence of sentences in a language in general, has a positive or a negative connotation associated with it; sometimes neutral. A news headline, the full news article itself, or even this paper, typically express an opinion to some degree. Natural Language Processing techniques are used to extract such connotations in an automated manner [9]. Once extracted, it can serve as a vital data point for applications such as in marketing to understand customers' opinion, as mentioned above. In our case we would like to use sentiment analysis to assess whether a news headline is favorable or admonitory to the company for which we are trading stocks.

Consequently, in our case, each news headline is posited to be either positive, negative or neutral from the perspective of the company we are considering trading stocks for. Positive sentiments can predict a general upturn in stock prices for a company, and similarly negative sentiments can possibly indicate a downturn [8,24]. While sentiments can be directly extracted from any text corpus (news headlines in this case), a lot more implicit information can be obtained by pairing knowledge graphs with this approach.

3.3   Knowledge Graphs

Lexical thesauri and ontologies are databases of terms interconnected with semantic relationships. Some examples include WordNet[1] for English terms and the Unified Medical Language System (UMLS)[2] for terms in the medical domain. They are often represented in a graph with entities and relationships. A knowledge base or a knowledge graph, are more complex graphs where the entities are not simple terms but a composite of knowledge. Some examples include DBpedia[3] or Google Knowledge Graph, which we use in this work.

The Google Knowledge Graph was specifically created to enhance the results of a Google search. Traditionally a Web search used to be limited to string matching keywords in an entire corpora to a given query. However, since entities in the real world are linked to each other and this link can be expressed in different ways, simple string matching is not adequate for an intelligent search. This interconnection is characterized in the knowledge graph which represents a graph-like data structure where each node is an entity and the edges between the nodes indicate the relationships between them. For instance, a naive search for "Bill Gates" using simple string matching would not bring up Microsoft. However, with a knowledge graph, since "Bill Gates", being the principal founder of Microsoft, he is a very relevant node close to the "Microsoft" node in a knowledge graph and hence, "Microsoft" would be brought up as a relevant search result. This way entities which are related to a company, but not explicitly mentioned in the news headline, can be identified as potential factors impacting stock prices. In our case, headlines covering Excel, Windows, Azure, Steve Ballmer, or Satya Nadella, or other entities connected to Microsoft in the knowledge graph, would be passed to the sentiment analysis and their polarity exploited in the learning algorithm.

---

[1] https://wordnet.princeton.edu/.
[2] https://www.nlm.nih.gov/research/umls/.
[3] https://wiki.dbpedia.org/.

4 Empirical Evaluation

4.1 Data

Stock Data: We used stock data from the Yahoo Finance API[4] dated from January 1, 2014 to December 31, 2017 for our training environment. The data for the test period is from January 1, 2018 to December 31, 2018. In our experiment, for both training and testing cases we used Microsoft Corporation's (MSFT), Amazon.com Inc. (AMZN) and Tesla Inc. (TSLA) stock data. - i.e., we trained our agent to trade theses stocks.

Sentiment Data: For news information, we scraped historical news headlines from the Reuters Twitter account[5] using a python scraper [29]. The time period of the news headlines corresponds exactly to the stock data, i.e., training data from January 1, 2014 to December 31, 2017 and testing data from January 1, 2018 to December 31, 2018.

Next, for each news headline we remove stopwords and tokenize it. Each token is then checked for the existence of an *Organization* node relationship with the specific company of interest (Microsoft Corporation in our example case) in a knowledge graph within a pre-specified distance. In our experiment we chose a distance measure of 5. Selecting a walk-length longer than this resulted in too much noise, and any shorter meant there would be very few implicit relationships found. This value was tuned empirically on the basis of some manual experiments we performed. For our experiments, we used the Google Knowledge Graph[6]. Once we find that any token in a headline is within this pre-specified distance of our organization (example: Microsoft), by extension we deem the entire headline as relevant to the organization in consideration. This is a naive approach, but allows us to make better use of news data that might not be directly linked to Microsoft, but might have indirect consequences. For instance, a news headline talking about Azure, which is Microsoft's cloud service offering, would not get identified as a news affecting MSFT stock prices, but by using a knowledge graph, we can uncover this implicit relationship.

Once we have headlines relevant to Microsoft, we use an ensemble sentiment analyser for sentiment classification. Since some headlines proved to be tricky to classify correctly by any single available sentiment classifier, we tried this approach of using an ensemble comprising of IBM Watson [11], TextBlob [1] and NLTK [18]. We classify each news headline as positive and negative news and use the classification from the classifiers above, choosing whichever one has the highest confidence. If there are multiple headlines on the same day, we use the majority of the sentiment score from all headlines for that day leading to a net +1 if majority is positive sentiment and −1 if majority is negative sentiment. An example positive headline dated 2016-07-14 can read: "Microsoft wins landmark appeal over seizure of foreign emails.", while an example of a headline expressing

---

[4] https://finance.yahoo.com/quote/MSFT/history/.
[5] https://twitter.com/reuters.
[6] https://developers.google.com/knowledge-graph/.

a negative sentiment dated 2015-12-31 is "Former employees say Microsoft didn't tell victims about hacking."

4.2   MDP Formulation

Episode: A single episode consists of the agent interacting with the stock trading environment once per day starting from January 1, 2014 and lasts until December 31, 2017 (for the training period). The agent explores different policies and improves its existing policy as more and more episodes elapse.

State: Our current environment describes each state using 6 variables:

1. Current amount of money the agent has;
2. Current number of stocks the agent has;
3. Opening stock price on today's date;
4. Difference between today's opening price and average opening price of last 5 d' window;
5. Difference between today's opening price and average opening price of last 50 d' window;
6. Average sentiment towards the company for today's date.

While (1) (2) (3) are values necessary for maintaining the state of the agent (4) and (5) were added to give it some indication of the trend in the stock prices. (4) provides the trend over a short time window (5 days), while (5) provides the trend information over a longer time window (50 days). (6) provides the sentiment information calculated as described in the previous section. In short, relevance of headlines are assessed with a knowledge graph. The sentiment expressed in the relevant pieces are used in (6). Figure 2 shows the entire workflow for the construction of the agent's state before it goes into the DQN.

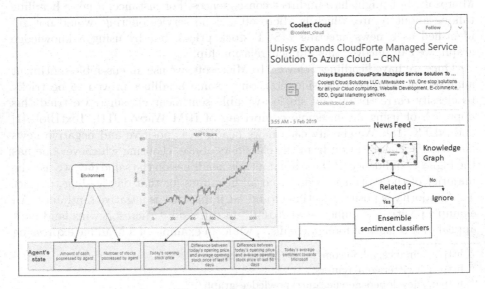

Fig. 2. Construction of agent's state

Action Space: The agent, our stock trading bot, interacts with this environment on a per day basis. It has the option to take three actions:
(1). Buy a stock; (2). Sell a stock; (3). Do nothing/hold.

Rewards: The intuition behind rewards is to provide a feedback signal to the agent to allow it to learn which actions are good/bad based on when they were taken. So, in our case, a net increase in portfolio at the end of the trading period should lead to a positive reward, while a net loss would lead to a negative reward. So, our initial attempts focused on this reward scheme where the agent's reward was the net profit/loss after 3 years (2014–2017). But, this reward signal proved too sparse to train on, since the agent got just one single reward after 3 years of activity and it is difficult for it to know which action taken when (over 3 years) contributed to the final reward. The agent just learnt to "Do nothing", since as the result of a general increasing trend in the MSFT stock price, it led to a small net increase in the portfolio and this was a local optima for the agent which it could not move out of due to the sparse reward signals.

Finally, after plenty of experimentation with the reward scheme, we arrived at one where it was rewarded for not just making a profit, but also for buying/selling on a day to day basis. If on any given day, it decided to Buy or Sell, it was given a reward of $+1$ for making a profit and -1 for making a loss. It was given a small negative reward of -0.1 for "Doing nothing" to discourage it from being passive for extended periods. A reward of -10 was given if it ran out of money, but still had stocks. A reward of -100 was given in case it went completely bankrupt with 0 stocks in hand and no money to buy a single stock.

Deep Q-Network (DQN) Architecture: The DQN used two identical neural networks (Q-network and target network) each with 3 hidden layers for function approximation. Each hidden layer had a size of 64 units and used ReLU activation. The input layer had 6 input nodes corresponding to each state feature. The output layer had 3 nodes corresponding to the action space. The experience replay buffer size was restricted to a size of 1000.

Training: The DQN was trained with mini-batch gradient descent using Adam [15] on the Huber loss. During training the agent started off with $1000 USD and 10 MSFT, AMZN or TSLA shares on January 1, 2014 and interacted with the environment till December 31, 2017. The agent in the form of a DQN is trained over 2000 epochs.

Figure 1 presents how the return (sum of rewards for all actions taken as specified by the reward scheme). The agent initially starts with a large negative value (representing a high loss portfolio) and then gradually converges towards a better policy (which possibly yields profits) as more training episodes elapse.

5   Results and Analysis

The primary hypothesis of this work was that providing sentiment information to the agent on a daily basis would add to its performance ceiling and it would be able to make more profit via trading. Therefore, we compare both approaches, i.e., an agent with sentiment data provided and another agent without any sentiment data provided. We evaluate both on our test data set, which spans from January 1, 2018 to December 31, 2018. If sentiments do add any additional value to the environment, it should be able to make more profit.

Table 1. Sharpe Ratios for different approaches

Agent	Sharpe ratio MSFT	Sharpe ratio AMZN	Sharpe ratio TSLA
Random policy	−2.249	−1.894	−2.113
Without sentiment	−1.357	1.487	0.926
With sentiment	**2.432**	**2.212**	**1.874**

5.1   Training Data Analysis

Before looking at the performance on the test data, we also analyse the performance of both models on the training data as well. Figure 3 shows this comparative analysis. The *Baseline Portfolio Value* is the starting portfolio value of the agent (i.e., the net value of 1000$ and 10 stocks on starting day). The *Random Policy* is an agent which takes random actions (Buy, Sell, Hold) on each interaction. As expected, a random policy agent goes broke soon enough and makes no profit. The agent with no sentiment input, does learn a policy good enough to make profit, but nowhere near good enough as compared to the agent which had sentiment input.

Figure 3 shows the same trend with the same training done on data about Microsoft, Amazon and Tesla stocks,. We can distinctly see that the learned policy using sentiment from news headlines outperforms the policy that only considers stock data.

5.2   Test Data Analysis

For Microsoft, during the test period, the stock prices at the beginning start quite a bit higher (approx. $85 in January 2018) as compared to the training period start (approx. $40 in January 2014). Despite this, the MSFT agent (Fig. 3) learned a policy good enough to generate profit, both with and without sentiment data. However, in general, it was not able to generalize to the test data as well as we saw during training, and its profits dropped. But still, the agent with sentiment information ends up making more profit than the agent without sentiment data. Similar trends are present for both the other stocks as well.

5.3   Sharpe Ratio

The Sharpe ratio is another measure that is often used in trading as a means of evaluating the risk adjusted return on investment. It can be used as a metric to evaluate the performance of different trading strategies. It is calculated as the expected return of a portfolio minus the risk-free rate of return, divided by the standard deviation of the portfolio investment. In modern portfolio theory [19], a Sharpe ratio of 1 is considered decent. About 2 or higher is very good and 3 is considered excellent. Table 1 presents this data for our agents' policies.

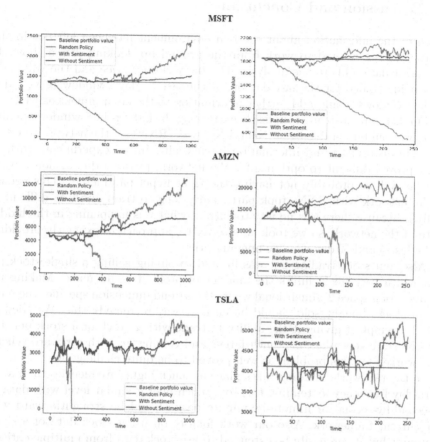

Fig. 3. Performance (Total portfolio value) of different agents for different stocks: MSFT, AMZN and TSLA (Left column: Train data - period (January 1, 2014 to December 31, 2017; Right column: Test data - period (January 1, 2018 to December 31, 2018.)

The result for the random policy is as expected. It learns a terrible policy and its Sharpe ratio is the least good among all three approaches. Surprisingly, the agent without sentiment data learns a pretty poor policy as well (albeit still better than the random policy), despite making profits. On closer analysis, it turns out that the MSFT stock had a general upward trend already and due to this reason a not-so-good policy could also produce profits, despite making sub-optimal decisions as indicated by the Sharpe ratio. Finally, we come to the agent which learnt a trading policy along with the sentiment data. Not only did it procure the highest profits as stated earlier, but also its decision making was very good as evidenced by its Sharpe ratio of 2.4 for MSFT, 2.2 for AMZN, and close to 2 for TSLA.

6   Discussion and Conclusion

Much of the information about the real environment has been left out in this effort since we wanted to work from the ground up, looking just at how the sentiment data adds to the analysis. The daily closing price and the volume of data being traded for the last day or for the last "x-day" window (e.g., 5-day, 50-day windows) could add further information to the environment as well.

Furthermore, instead of explicitly extracting the last "x-day" window opening price, we could use a Recurrent Neural Network (RNN) for the network to retain on some historical trend information intrinsically. Initial experiments with an RNN proved difficult to optimize for the network, possibly due to noise in the data as well as probably not having the right hyper-parameters. This version of the network with RNNs took particularly long to train and was difficult to analyse because there was no way to extract what was happening in the hidden state of the network, so we took an alternative approach of explicitly providing it the last 5 and 50 day average opening price.

Also, our stock trading bot was limited to buying/selling a single stock per day, which very likely limits the amount of profit it could make. Making the agents action space 2-dimensional where the second dimension specifies the number of stocks bought/sold should be an easy way to remedy this. We tried an initial attempt at giving it the ability to trade with 1 stock or 5 stock per day, but the state space became much larger and coming up with a reward scheme that worked for this problem as well proved to be quite challenging.

In the real world, trading takes place at much higher frequencies than at an intra-day frequency; extending this to a much finer granular level with data on a second-by-second or minute-by-minute basis should be straightforward with our current framework. Also our work focuses on using stock data of a single company, but it can easily be extended to use stock data from multiple entities.

Also, in the knowledge graph we kept the relationship distance threshold quite limited so as to restrain the noise added to the data in terms of news headlines. Provided with a knowledge graph which has weighted nodes, which tell if there is a positive or negative relationship between the entity in question and the company stocks are being traded for, we can potentially exploit much longer distance relationships and in a much more accurate manner.

We present an approach of extracting implicit relationships between entities from news headlines via knowledge graphs and exploiting sentiment analysis, positive or negative, on these headlines, and then using this information, train a reinforcement learning agent. The trained reinforcement learning agent can perform better in terms of profits incurred as compared to an agent which does not have this additional information on headline sentiments. The whole pipeline as such is a novel approach and the empirical study demonstrates its validity.

Acknowledgments. The work is supported in part by the Natural Sciences and Engineering Research Council of Canada (NSERC). Osmar Zaïane is supported by the Amii Fellow Program and the Canada CIFAR AI Chair Program.

References

1. Textblob. https://github.com/sloria/textblob
2. Atsalakis, G.S., Valavanis, K.P.: Surveying stock market forecasting techniques - part ii: Soft computing methods. Expert Syst. Appl. **36**(3), 5932–5941 (2009)
3. Braun, H., Chandler, J.S.: Predicting stock market behavior through rule induction: an application of the learning-from-example approach*. Decis. Sci. **18**(3), 415–429 (1987)
4. Chen, K., Zhou, Y., Dai, F.: A lstm-based method for stock returns prediction: a case study of china stock market. In: 2015 IEEE International Conference on Big Data (Big Data), pp. 2823–2824 (2015)
5. Deng, Y., Bao, F., Kong, Y., Ren, Z., Dai, Q.: Deep direct reinforcement learning for financial signal representation and trading. IEEE Trans. Neural Netw. Learn. Syst. **28**(3), 653–664 (2017)
6. Filos, A.: Reinforcement learning for portfolio management. GitHub repository (2018)
7. Fischer, T.G.: Reinforcement learning in financial markets- a survey working paper, FAU Discussion Papers in Economics, No. 12/2018, Institute for Economics, Erlangen (2018)
8. Fisher, K.L., Statman, M.: Investor sentiment and stock returns. Financ. Anal. J. **56**(2), 16–23 (2000)
9. Godbole, N., Srinivasaiah, M., Skiena, S.: Large-scale sentiment analysis for news and blogs. Icwsm **7**(21), 219–222 (2007)
10. Grek, T.: A blundering guide to making a deep actor-critic bot for stock trading. GitHub repository (2018)
11. IBM: Ibm watson. https://www.ibm.com/watson/services/tone-analyzer/
12. Jain, S., Kain, M.: Rediction for stock marketing using machine learning. Int. J. Recent Innov. Trends Comput. Commun. **6**(4), 131–135 (2018)
13. Kaelbling, L., Littman, M., Cassandra, A.: Planning and acting in partially observable stochastic domains. Artif. Intell. **101**(1–2), 99–134 (1998)
14. Karazmodeh, M., Nasiri, S., Hashemi, S.M.: Stock price forecasting using support vector machines and improved particle swarm optimization. J. Autom. Control Eng. 1(2) (2013)
15. Kingma, D.P., Ba, J.: Adam: A method for stochastic optimization (2014). arXiv preprint arXiv:1412.6980
16. Kohara, K., Ishikawa, T., Fukuhara, Y., Nakamura, Y.: Stock price prediction using prior knowledge and neural networks. Intell. Syst. Accounting Finance Manage. **6**(1), 11–22 (1997)
17. Littman, M.L.: A tutorial on partially observable Markov decision processes. J. Math. Psychol. **53**(3), 119–125 (2009)
18. Loper, E., Bird, S.: NLTK: The Natural Language Toolkit (2002). http://citeseerx.ist.psu.edu/viewdoc/summary?doi=10.1.1.4.585
19. Maverick, J.: Investopedia. https://www.investopedia.com/ask/answers/010815/what-good-sharpe-ratio.asp
20. Mnih, V., et al.: Human-level control through deep reinforcement learning. Nature **518**(7540), 529 (2015)

21. Panigrahi, S., Mantri, J.: A text based decision tree model for stock market forecasting. In: 2015 International Conference on Green Computing and Internet of Things. IEEE (2015)
22. Sabherwal, S., Sarkar, S.K., Zhang, Y.: Do internet stock message boards influence trading? evidence from heavily discussed stocks with no fundamental news. J. Bus. Finance Accoun. **38**(9–10), 1209–1237 (2011)
23. Sharma, M.: Survey on stock market prediction and performance analysis. Int. J. Adv. Res. Comput. Eng. Technol. **3**(1), 131–135 (2014)
24. Si, J., Mukherjee, A., Liu, B., Li, Q., Li, H., Deng, X.: Exploiting topic based twitter sentiment for stock prediction. In: Proceedings of the 51st Annual Meeting of the Association for Computational Linguistics, vol. 2, pp. 24–29 (2013)
25. Sprenger, T.O., Tumasjan, A., Sandner, P.G., Welpe, I.M.: Tweets and trades: the information content of stock microblogs. Eur. Financ. Manag. **20**(5), 926–957 (2014)
26. Sutton, R.S., Barto, A.G., et al.: Introduction to reinforcement learning, vol. 135. MIT press Cambridge (1998)
27. Tan, Z., Quek, C., Cheng, P.Y.: Stock trading with cycles: a financial application of anfis and reinforcement learning. Expert Syst. Appl. **38**(5), 4741–4755 (2011)
28. Taran, C., Roy, D., Srinivasan, N.: Stock price prediction - a novel survey. Int. J. Appl. Eng. Res. **10**(4), 11375–11383 (2015)
29. Taspinar, A.: Twitterscraper. https://github.com/taspinar/twitterscraper
30. Yaddolahi, A., Shahraki, A.G., Zaiane, O.R.: Current state of text sentiment analysis from opinion to emotion mining. ACM Comput. Surv. **50**(2), 1–33 (2017)
31. Åström, K.: Optimal control of Markov processes with incomplete state information. J. Math. Anal. Appl. **10**(1), 174–205 (1965)

DeepCore: A Comprehensive Library for Coreset Selection in Deep Learning

Chengcheng Guo[1], Bo Zhao[2], and Yanbing Bai[1(✉)]

[1] Center for Applied Statistics, School of Statistics,
Renmin University of China, Beijing, China
{chengchengguo,ybbai}@ruc.edu.cn
[2] School of Informatics, The University of Edinburgh, Edinburgh, Scotland
bo.zhao@ed.ac.uk

Abstract. Coreset selection, which aims to select a subset of the most informative training samples, is a long-standing learning problem that can benefit many downstream tasks such as data-efficient learning, continual learning, neural architecture search, active learning, etc. However, many existing coreset selection methods are not designed for deep learning, which may have high complexity and poor generalization performance. In addition, the recently proposed methods are evaluated on models, datasets, and settings of different complexities. To advance the research of coreset selection in deep learning, we contribute a comprehensive code library (The code is available in https://github. com/PatrickZH/DeepCore.), namely *DeepCore*, and provide an empirical study on popular coreset selection methods on CIFAR10 and ImageNet datasets. Extensive experiments on CIFAR10 and ImageNet datasets verify that, although various methods have advantages in certain experiment settings, random selection is still a strong baseline.

Keywords: Coreset selection · Data-efficient learning · Deep learning

1 Introduction

Deep learning has shown unprecedented success in many research areas such as computer vision, etc. As it evolves, not only neural networks but also the training datasets are becoming increasingly larger, which requires massive memory and computation to achieve the state-of-the-art. One promising technique to reduce the computational cost is coreset selection [20,21,32,38] that aims to select a small subset of the most informative training samples S from a given large training dataset T. The models trained on the coreset are supposed to have close generalization performance to those trained on the original training set.

Coreset selection has been widely studied since the era of traditional machine learning, whose research generally focuses on how to approximate the distribution of the whole dataset with a subset, for example, they assume that data are

---

C. Guo and B. Zhao—Equal contribution.

© The Author(s), under exclusive license to Springer Nature Switzerland AG 2022
C. Strauss et al. (Eds.): DEXA 2022, LNCS 13426, pp. 181–195, 2022.
https://doi.org/10.1007/978-3-031-12423-5_14

from a mixture of Gaussians in a given metric space [3,4,7,13,52]. However, for those classic coreset selection methods proposed for traditional machine learning tasks, their effectiveness in deep learning is doubtful, due to the high computational complexity and fixed data representations. Recently, the research of coreset selection for deep learning tasks emerges [21,38,49]. The newly developed coreset selection methods are evaluated in different settings in terms of models, datasets, tasks, and selection fractions, resulting in their performances hardly being compared fairly.

We focus our studies on image classification tasks. To address the above problems, in this paper, we provide an exhaustive empirical study on popular coreset selection methods in the same settings. We contribute a comprehensive code library, namely *DeepCore*, for advancing the research of coreset selection in deep learning. Specifically, we re-implement 12 popular coreset selection methods in a unified framework based on PyTorch [37]. These methods are compared in settings of various selection fractions from 0.1% to 90% on CIFAR10 [25] and ImageNet-1K [39] datasets. Besides the reported results in the paper, our library supports popular deep neural architectures, image classification datasets and coreset selection settings.

2   Review of Coreset Selection Methods

In this section, we first formulate the problem of coreset selection. Then, brief surveys of methods and applications of coreset selection are provided respectively.

2.1   Problem Statement

In a learning task, we are given a large training set $T = \{(x_i, y_i)\}_{i=1}^{|T|}$, where $x_i \in \mathcal{X}$ is the input, $y_i \in \mathcal{Y}$ is the ground-truth label of x_i, where $\mathcal{X}$ and $\mathcal{Y}$ denote the input and output spaces, respectively. Coreset selection aims to find the most informative subset $S \subset T$ with the constraint $|S| < |T|$, so that the model θ^S trained on S has close generalization performance to the model θ^T trained on the whole training set T.

2.2   Survey: Methodologies

Geometry Based Methods. It is assumed that data points close to each other in the feature space tend to have similar properties. Therefore, geometry based methods [1,7,41,46] try to remove those data points providing redundant information then the left data points form a coreset S where $|S| \ll |T|$.

HERDING. The HERDING method selects data points based on the distance between the coreset center and original dataset center in the feature space. The algorithm incrementally and greedily adds one sample each time into the coreset that can minimize distance between two centers [7,52].

K-CENTER GREEDY. This method tries to solves the *minimax facility location* problem [12], i.e. selecting k samples as S from the full dataset T such that the largest distance between a data point in $\mathcal{T}\backslash\mathcal{S}$ and its closest data point in $\mathcal{S}$ is minimized:

$$\min_{\mathcal{S}\subset\mathcal{T}} \max_{x_i\in\mathcal{T}\backslash\mathcal{S}} \min_{x_j\in\mathcal{S}} \mathcal{D}(x_i, x_j), \tag{1}$$

where $\mathcal{D}(\cdot, \cdot)$ is the distance function. The problem is NP-hard, and a greedy approximation known as K-CENTER GREEDY has been proposed in [41]. K-CENTER GREEDY has been successfully extended to a wide range of applications, for instance, active learning [1,41] and efficient GAN training [46].

Uncertainty Based Methods. Samples with lower confidence may have a greater impact on model optimization than those with higher confidence, and should therefore be included in the coreset. The following are commonly used metrics of sample uncertainty given a certain classifier and training epoch, namely LEAST CONFIDENCE, ENTROPY and MARGIN [9], where C is the number of classes. We select samples in descending order of the scores:

$$s_{least\ confidence}(x) = 1 - \max_{i=1,...,C} P(\hat{y} = i|x)$$

$$s_{entropy}(x) = -\sum_{i=1}^{C} P(\hat{y} = i|x) \log P(\hat{y} = i|x) \tag{2}$$

$$s_{margin}(x) = 1 - \min_{y\neq\hat{y}}(P(\hat{y}|x) - P(y|x)).$$

Error/Loss Based Methods. In a dataset, training samples are more important if they contribute more to the error or loss when training neural networks. Importance can be measured by the loss or gradient of each sample or its influence on other samples' prediction during model training. Those samples with the largest importance are selected as the coreset.

FORGETTING EVENTS. Toneva et al. [49] count how many times the *forgetting* happens during the training, i.e. the misclassification of a sample in the current epoch after having been correctly classified in the previous epoch, formally $acc_i^t > acc_i^{t+1}$, where acc_i^t indicates the correctness (True or False) of the prediction of sample i at epoch t. The number of forgetting reveals intrinsic properties of the training data, allowing for the removal of unforgettable examples with minimal performance drop.

GRAND and EL2N *Scores.* The GRAND score [38] of sample (x, y) at epoch t is defined as

$$\chi_t(x, y) \triangleq \mathbb{E}_{\theta_t}||\nabla_{\theta_t}\ell(x, y; \theta_t)||_2. \tag{3}$$

It measures the average contribution from each sample to the decline of the training loss at early epoch t across several different independent runs. The score calculated at early training stages, e.g. after a few epochs, works well, thus this method requires less computational cost. An approximation of the GRAND score is also provided, named EL2N score, which measures the norm of error vector:

$$\chi_t^*(x, y) \triangleq \mathbb{E}_{\theta_t}||p(\theta_t, x) - y||_2. \tag{4}$$

IMPORTANCE SAMPLING. In importance sampling (or adaptive sampling), we define $s(\boldsymbol{x}, y)$ is the upper-bounded (worst-case) contribution to the total loss function from the data point $(\boldsymbol{x}, y)$, aka sensitivity score. It can be formulated as:

$$s(\boldsymbol{x}, y) = \max_{\theta \in \boldsymbol{\theta}} \frac{\ell(\boldsymbol{x}, y; \boldsymbol{\theta})}{\sum_{(\boldsymbol{x}', y') \in \mathcal{T}} \ell(\boldsymbol{x}', y'; \boldsymbol{\theta})}, \tag{5}$$

where $\ell(\boldsymbol{x}, y)$ is a non-negative cost function with parameter $\boldsymbol{\theta} \in \boldsymbol{\theta}$. For each data point in $\mathcal{T}$, the probability of being selected is set as $p(\boldsymbol{x}, y) = \frac{s(\boldsymbol{x}, y)}{\sum_{(\boldsymbol{x}, y) \in \mathcal{T}} s(\boldsymbol{x}, y)}$. The coreset $\mathcal{S}$ is constructed based on the probabilities [3,34]. Similar ideas are proposed in *Black box learners* [10] and JTT [30], where wrongly classified samples will be upweighted or their sampling probability will be increased.

Decision Boundary Based Methods. Since data points distributed near the decision boundary are hard to separate, those data points closest to the decision boundary can also be used as the coreset.

ADVERSARIAL DEEPFOOL. While exact distance to the decision boundary is inaccessible, Ducoffe and Precioso [11] seek the approximation of these distances in the input space $\mathcal{X}$. By giving perturbations to samples until the predictive labels of samples are changed, those data points require the smallest adversarial perturbation are closest to the decision boundary.

CONTRASTIVE ACTIVE LEARNING. To find data points near the decision boundary, Contrastive Active Learning (CAL) [31] selects samples whose predictive likelihood diverges the most from their neighbors to construct the coreset.

Gradient Matching Based Methods. Deep models are usually trained using (stochastic) gradient descent algorithm. Therefore, we expect that the gradients produced by the full training dataset $\sum_{(\boldsymbol{x}, y) \in \mathcal{T}} \nabla_\theta \ell(\boldsymbol{x}, y; \boldsymbol{\theta})$ can be replaced by the (weighted) gradients produced by a subset $\sum_{(\boldsymbol{x}, y) \in \mathcal{S}} w_{\boldsymbol{x}} \nabla_\theta \ell(\boldsymbol{x}, y; \boldsymbol{\theta})$ with minimal difference:

$$\min_{\mathbf{w}, \mathcal{S}} \mathcal{D}(\frac{1}{|\mathcal{T}|} \sum_{(\boldsymbol{x}, y) \in \mathcal{T}} \nabla_\theta \ell(\boldsymbol{x}, y; \boldsymbol{\theta}), \frac{1}{|\mathbf{w}|_1} \sum_{(\boldsymbol{x}, y) \in \mathcal{S}} w_{\boldsymbol{x}} \nabla_\theta \ell(\boldsymbol{x}, y; \boldsymbol{\theta})) \tag{6}$$

$$s.t. \quad \mathcal{S} \subset \mathcal{T}, \ w_{\boldsymbol{x}} \geq 0,$$

where $\mathbf{w}$ is the subset weight vector, $|\mathbf{w}|_1$ is the sum of the absolute values and $\mathcal{D}(\cdot, \cdot)$ measures the distance between two gradients.

CRAIG. Mirzasoleiman et al. [32] try to find an optimal coreset that approximates the full dataset gradients under a maximum error ε by converting gradient matching problem to the maximization of a monotone submodular function F and then use greedy approach to optimize F.

GRADMATCH. Compared to CRAIG, the GRADMATCH [20] method is able to achieve the same error ε of the gradient matching but with a smaller subset. GRADMATCH introduces a squared l2 regularization term over the weight vector

$\mathbf{w}$ with coefficient λ to discourage assigning large weights to individual samples. To solve the optimization problem, it presents a greedy algorithm – *Orthogonal Matching Pursuit*, which can guarantee $1 - exp(\frac{-\lambda}{\lambda+k\nabla^2_{max}})$ error with the constraint $|S| \leq k$, k is a preset constant.

Bilevel Optimization Based Methods. Coreset selection can be posed as a bilevel optimization problem. Existing studies usually consider the selection of subset (optimization of samples S or selection weights $\mathbf{w}$) as the outer objective and the optimization of model parameters θ on S as the inner objective. Representative methods include cardinality-constrained bilevel optimization [5] for continual learning, RETRIEVE for semi-supervised learning (SSL) [22], and GLISTER [21] for supervised learning and active learning.

RETRIEVE. The RETRIEVE method [22] discusses the scenario of SSL under bilevel optimization, where we have both a labeled set $\mathcal{T}$ and an unlabled set $\mathcal{P}$. The bilevel optimization problem in RETRIEVE is formulated as

$$\mathbf{w}^* = \arg\min_{\mathbf{w}} \sum_{(\boldsymbol{x},y)\in\mathcal{T}} \ell_s(\boldsymbol{x},y;\arg\min_{\theta}(\sum_{(\boldsymbol{x},y)\in\mathcal{T}} \ell_s(\boldsymbol{x},y;\theta) + \lambda\sum_{x\in\mathcal{P}} w_x\ell_u(\boldsymbol{x};\theta))),$$
(7)

where ℓ_s is the labeled-data loss, e.g. cross-entropy and ℓ_u is the unlabeled-data loss for SSL, e.g. consistency-regularization loss. λ is the regularization coefficient.

GLISTER. To guarantee the robustness, GLISTER [21] introduces a validation set $\mathcal{V}$ on the outer optimization and the log-likelihood $\ell\ell$ in the bilevel optimization:

$$S^* = \arg\max_{S\subset\mathcal{T}} \sum_{(\boldsymbol{x},y)\in\mathcal{V}} \ell\ell(\boldsymbol{x},y;\arg\max_{\theta} \sum_{(\boldsymbol{x},y)\in S} \ell\ell(\boldsymbol{x},y;\theta)).$$
(8)

Submodularity Based Methods. Submodular functions [17] are set functions $f : 2^\mathcal{V} \to \mathbb{R}$, which return a real value for any $\mathcal{U} \subset \mathcal{V}$. f is a submodular function, if for $\mathcal{A} \subset \mathcal{B} \subset \mathcal{V}$ and $\forall x \in \mathcal{V}\backslash\mathcal{B}$:

$$f(\mathcal{A} \cup \{x\}) - f(\mathcal{A}) \geq f(\mathcal{B} \cup \{x\}) - f(\mathcal{B}).$$
(9)

Submodular functions naturally measure the diversity and information, thus can be a powerful tool for coreset selection by maximizing them. Many functions obey the above definition, e.g.Graph Cut (GC), Facility Location (FL), Log Determinant [16], etc. For maximizing submodular functions under cardinality constraint, greedy algorithms have been proved to have a bounded approximation factor of $1 - \frac{1}{e}$ [35].

FASS. Wei et al. [51] discuss the connection between likelihood functions and submodularity, proving that under a cardinality constraint, maximizing likelihood function is equivalent to maximization of submodular functions for Naïve Bayes or Nearest Neighbor classifier, naturally providing a powerful tool for coreset selection. By introducing submodularity into Naive Bayes and Nearest

Neighbor, they propose a novel framework for active learning namely FILTERED ACTIVE SUBMODULAR SELECTION (FASS).

PRISM. Kaushal et al. [19] develop PRISM, a submodular method for *targeted subset selection*, which is a learning scenario similar to active learning. In targeted subset selection, a subset S will be selected to be labeled from a large unlabeled set $\mathcal{P}$, with additional requirement that S has to be aligned with the targeted set $\mathcal{T}$ of specific user intent.

SIMILAR. Kothawade et al. [24] introduce SIMILAR, a unified framework of submodular methods that successfully extends submodularity to broader settings which may involve rare classes, redundancy, out-of-distribution data, etc.

Proxy Based Methods. Many coreset selection methods require to train models on the whole dataset for calculating features or some metrics for one or many times. To reduce this training cost, SELECTION VIA PROXY methods [9,40] are proposed, which train a lighter or shallower version of the target models as proxy models. Specifically, they create proxy models by reducing hidden layers, narrowing dimensions, or cutting down training epochs. Then, coresets are selected more efficiently on these proxy models.

2.3   Survey: Applications

Data-efficient Learning. The basic application of coreset selection is to enable efficient machine learning [20,32,38,49]. Training models on coresets can reduce the training cost while preserving testing performance. Especially, in Neural Architecture Search (NAS) [44], thousands to millions deep models have to be trained and then evaluated on the same dataset. Coreset can be used as a proxy dataset to efficiently train and evaluate candidates [9,40], which significantly reduces computational cost.

Continual Learning. Coreset selection is also a key technique to construct memory for continual learning or incremental learning [2,5,55], in order to relieve the catastrophic forgetting problem. In the popular continual learning setting, a memory buffer is maintained to store informative training samples from previous tasks for rehearsal in future tasks. It is proven that continual learning performance heavily relies on the quality of memory, i.e. coreset [23].

Active Learning. Active learning [42,43] aims to achieve better performance with the minimal query cost by selecting informative samples from the unlabeled pool $\mathcal{P}$ to label. Thus, it can be posed as a coreset selection problem [11,24,31,41,51].

Besides the above, coreset selection is studied and successfully applied in many other machine learning problems, such as robust learning against noise [22,24,33], clustering [3,4,47], semi-supervised learning [6,22], unsupervised learning [18], efficient GAN training [46], regression tasks [8,34] etc.

3   DeepCore Library

In the literature, coreset selection methods have been proposed and tested in different experiment settings in terms of dataset, model architecture, coreset size, augmentation, training strategy, etc. This may lead to unfair comparisons between different methods and unconvincing conclusions. For instance, some methods may have only been evaluated on MNIST with shallow models, while others are tested on the challenging ImageNet dataset with deep neural networks. Even though tested on the same dataset, different works are likely to use different training strategies and data augmentations which significantly affect the performance. Furthermore, it causes future researchers inconvenience in identifying and improving the state-of-the-art.

Therefore, we develop *DeepCore*, an extensive and extendable code library, for coreset selection in deep learning, reproducing dozens of popular and advanced coreset selection methods and enabling a fair comparison of different methods in the same experimental settings. DeepCore is highly modular, allowing to add new architectures, datasets, methods and learning scenarios easily. We build DeepCore on PyTorch [37].

Coreset Methods. We list the methods that have been re-implemented in DeepCore according to the categories in 2.2, they are 1) geometry based methods CONTEXTUAL DIVERSITY (CD) [1], HERDING [52] and K-CENTER GREEDY [41]; 2) uncertainty based methods LEAST CONFIDENCE, ENTROPY and MARGIN [9]; 3) error/loss based methods FORGETTING [49] and GRAND [38]; 4) decision boundary based methods CAL [31] and DEEPFOOL [11]; 5) gradient matching based methods CRAIG [32] and GRADMATCH [20]; 6) bilevel optimization methods GLISTER [21]; and 7) submodularity based methods with GRAPH CUT (GC) and FACILITY LOCATION (FL) functions [16]. We also have RANDOM selection as the baseline.

Datasets. We provide the experiment results on CIFAR10 [25] and ImageNet-1K [39] in this paper. Besides, our DeepCore has provided the interface for other popular computer vision datasets, namely MNIST [29], QMNIST [54], FashionMNIST [53], SVHN [36], CIFAR100 [25] and TinyImageNet [27].

Network Architectures. We provide the code of popular architectures, namely MLP, LeNet [28], AlexNet [26], VGG [45], Inception-v3 [48], ResNet [14], WideResNet [56] and MobileNet-v3 [15].

4   Experiment Results

In this section, we use our DeepCore to evaluate different coreset selection methods in multiple learning settings on CIFAR10 and ImageNet-1K datasets. ResNet-18 is used as the default architecture in all experiments.

Table 1. Coreset selection performances on CIFAR10. We train randomly initialized ResNet-18 on the coresets of CIFAR10 produced by different methods and then test on the real testing set.

Fraction	0.1%	0.5%	1%	5%	10%	20%	30%	40%	50%	60%	90%	100%
Random	21.0 ± 0.3	30.8 ± 0.6	36.7 ± 1.7	64.5 ± 1.1	75.7 ± 2.0	**87.1 ± 0.5**	90.2 ± 0.3	92.1 ± 0.1	93.3 ± 0.2	94.0 ± 0.2	95.2 ± 0.1	95.6 ± 0.1
CD [1]	15.8 ± 1.2	20.5 ± 0.7	23.6 ± 1.9	38.1 ± 2.2	58.8 ± 2.0	81.3 ± 2.5	90.8 ± 0.5	93.3 ± 0.4	94.3 ± 0.2	94.6 ± 0.6	95.4 ± 0.1	95.6 ± 0.1
Herding [52]	20.2 ± 2.3	27.3 ± 1.5	34.8 ± 3.3	51.0 ± 3.1	63.5 ± 3.4	74.1 ± 2.5	80.1 ± 2.2	85.2 ± 0.9	88.0 ± 1.1	89.8 ± 0.9	94.6 ± 0.4	95.6 ± 0.1
k-Center greedy [41]	18.5 ± 0.3	26.8 ± 1.2	31.1 ± 1.2	51.4 ± 2.1	75.8 ± 2.4	87.0 ± 0.3	90.9 ± 0.4	92.8 ± 0.1	93.9 ± 0.2	94.1 ± 0.1	95.4 ± 0.1	95.6 ± 0.1
Least confidence [9]	14.2 ± 0.9	17.2 ± 1.8	19.8 ± 2.2	36.2 ± 1.9	57.6 ± 3.1	81.9 ± 2.2	90.3 ± 0.4	93.1 ± 0.5	94.5 ± 0.1	94.7 ± 0.1	95.5 ± 0.1	95.6 ± 0.1
Entropy [9]	14.6 ± 2.2	17.5 ± 1.3	21.1 ± 1.3	35.3 ± 3.0	57.6 ± 2.8	81.9 ± 0.4	89.8 ± 1.6	93.2 ± 0.2	94.4 ± 0.3	**95.0 ± 0.1**	95.4 ± 0.1	95.6 ± 0.1
Margin [9]	17.2 ± 1.1	21.7 ± 1.6	28.2 ± 1.0	43.4 ± 3.3	59.9 ± 2.9	81.7 ± 3.2	90.9 ± 0.4	93.0 ± 0.2	94.3 ± 0.3	94.8 ± 0.3	95.5 ± 0.1	95.6 ± 0.1
Forgetting [49]	21.4 ± 0.5	29.8 ± 1.0	35.2 ± 1.6	52.1 ± 2.2	67.0 ± 1.5	86.6 ± 0.6	**91.7 ± 0.3**	93.5 ± 0.2	94.1 ± 0.1	94.6 ± 0.2	95.3 ± 0.1	95.6 ± 0.1
GraND [38]	17.7 ± 1.0	24.0 ± 1.1	26.7 ± 1.3	39.8 ± 2.3	52.7 ± 1.9	78.2 ± 2.9	91.2 ± 0.7	**93.7 ± 0.3**	**94.6 ± 0.1**	**95.0 ± 0.2**	95.5 ± 0.2	95.6 ± 0.1
Cal [31]	22.7 ± 2.7	33.1 ± 2.3	37.8 ± 2.0	60.0 ± 1.4	71.8 ± 1.0	80.9 ± 1.1	86.0 ± 1.9	87.5 ± 0.8	89.4 ± 0.6	91.6 ± 0.9	94.7 ± 0.3	95.6 ± 0.1
DeepFool [11]	17.6 ± 0.4	22.4 ± 0.8	27.6 ± 2.2	42.6 ± 3.5	60.8 ± 2.5	83.0 ± 2.3	90.0 ± 0.7	93.1 ± 0.2	94.1 ± 0.1	94.8 ± 0.2	95.5 ± 0.1	95.6 ± 0.1
Craig [32]	22.5 ± 1.2	27.0 ± 0.7	31.7 ± 1.1	45.2 ± 2.9	60.2 ± 4.4	79.6 ± 3.1	88.4 ± 0.5	90.8 ± 1.4	93.3 ± 0.6	94.2 ± 0.2	95.5 ± 0.1	95.6 ± 0.1
GradMatch [20]	17.4 ± 1.7	25.6 ± 2.6	30.8 ± 1.0	47.2 ± 0.7	61.5 ± 2.4	79.9 ± 2.6	87.4 ± 2.0	90.4 ± 1.5	92.9 ± 0.6	93.2 ± 1.0	93.7 ± 0.5	95.6 ± 0.1
Glister [21]	19.5 ± 2.1	27.5 ± 1.4	32.9 ± 2.4	50.7 ± 1.5	66.3 ± 3.5	84.8 ± 0.9	90.9 ± 0.3	93.0 ± 0.2	94.0 ± 0.3	94.8 ± 0.2	**95.6 ± 0.2**	95.6 ± 0.1
FL [16]	22.3 ± 2.0	31.6 ± 0.6	38.9 ± 1.4	60.8 ± 2.5	74.7 ± 1.3	85.6 ± 1.9	91.4 ± 0.4	93.2 ± 0.3	93.9 ± 0.2	94.5 ± 0.3	95.5 ± 0.2	95.6 ± 0.1
GC [16]	**24.3 ± 1.5**	**34.9 ± 2.3**	**42.8 ± 1.3**	**65.7 ± 1.2**	**76.6 ± 1.5**	84.0 ± 0.5	87.8 ± 0.4	90.6 ± 0.3	93.2 ± 0.3	94.4 ± 0.3	95.4 ± 0.1	95.6 ± 0.1

4.1 CIFAR10 Results

For CIFAR10 experiments, we use SGD as the optimizer with batch size 128, initial learning rate 0.1, Cosine decay scheduler, momentum 0.9, weight decay 5×10^{-4} and 200 training epochs. We select subsets with fractions of 0.1%, 0.5%, 1%, 5%, 10%, 20%, 30%, 40%, 50%, 60%, 90% of the whole training set respectively. The training on the whole dataset can be considered as the upper-bound. For data augmentation, we apply random crop with 4-pixel padding and random flipping on the 32×32 training images.

For some methods, the gradient, prediction probability, or feature vector of each sample is required to implement sample selection. For a fair comparison, we use the ResNet-18 models trained on the whole dataset for 10 epochs to extract above-mentioned metrics. When gradient vector $\nabla_\theta \ell(\boldsymbol{x}, y; \theta)$ is required, we use the gradients of the parameters in the final fully-connected layer as suggested in many previous studies [20,21,32]. This allows gradient vectors to be easily obtained without back-propagation throughout the whole network. While Deep-Core supports both balanced and imbalance sample selection, experiments in this paper all adopt balanced selection, namely, the same number of samples are selected for every class.

Table 1 shows the detailed results of different methods on CIFAR10, and Fig. 1 depicts the performance curves. The mean and standard deviation is calculated with 5 random seeds. Good experimental results come from the submodular function based methods, in both small and large learning setting. Especially in small fractions of 0.1%–1%, the advantage of submodular function based methods is obvious. Graph Cut (GC) is more prominent among them, and achieves the best results when selecting 0.1% to 10% of the training data. In particular, Graph Cut outperforms the other methods by more than 5% in the testing accuracy when 50 samples are selected per class, i.e. 1% of the whole training set. CAL also shows superiority in small fractions between 0.1%–5%, with performance comparable to Facility Location (FL). However, its superiority dis-

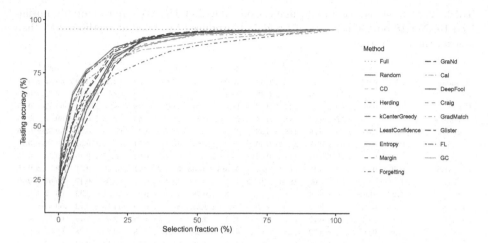

Fig. 1. Coreset selection performances in curves on CIFAR10. We train randomly initialized ResNet-18 on the coresets of CIFAR10 produced by different methods and then test on the real testing set. Detailed numbers are provided in Table 1.

appears when the coreset size increases, especially when selecting more than 30% training data. Except the above methods, all other methods fail to outperform the random sampling baseline in small settings between 0.1% and 1%. Forgetting method outperforms others in 30%-fraction setting. Between 40% and 60%, GRAND and uncertainty score based methods stand out. In all fraction settings, GRADMATCH and HERDING barely beat the random sampling. For GRADMATCH, the experiment setting in the original paper is adaptive sampling, where subsets iteratively updated along with network training. Here, for a fair comparison, coresets are selected and then fixed for all training epochs. HERDING is originally designed for fixed representations from a mixture of Gaussians, thus its performance heavily depends on the embedding function. Note that the above findings are based on one hyper-parameter setting, the findings may change if hyper-parameters change. For example, HERDING may have better performances if the model for feature extraction is fully trained. We study the influence of some hyper-parameters later.

4.2   ImageNet Results

For ImageNet, we train ResNet-18 models on coresets with batch size 256 for 200 epochs. The training images are randomly cropped and then resized to 224×224. The left-right flipping with the probability of 0.5 is also implemented. Other experimental settings and hyper-parameters are consistent with CIFAR10 experiments. Due to the long running time of DEEPFOOL on ImageNet, its results are not provided. For K-CENTER GREEDY and CONTEXTUAL DIVERSITY, here we do not provide the results when only 1 sample is selected from each class (i.e. fraction of 0.1%), because their first sample is drawn randomly from each class

Table 2. Coreset selection performances on ImageNet-1K. We train randomly initialized ResNet-18 on the coresets of ImageNet produced by different methods and then test on the real testing set.

	0.1%	0.5%	1%	5%	10%	30%	100%
Random	0.76 ± 0.01	3.78 ± 0.14	8.85 ± 0.46	40.09 ± 0.21	52.10 ± 0.22	**64.11 ± 0.05**	69.52 ± 0.45
CD	–	1.18 ± 0.06	2.16 ± 0.18	25.82 ± 2.02	43.84 ± 0.12	62.13 ± 0.45	69.52 ± 0.45
Herding	0.34 ± 0.01	1.70 ± 0.13	4.17 ± 0.26	17.41 ± 0.34	28.06 ± 0.05	48.58 ± 0.49	69.52 ± 0.45
k-center greedy	–	1.57 ± 0.09	2.96 ± 0.24	27.36 ± 0.08	44.84 ± 1.03	62.12 ± 0.46	69.52 ± 0.45
Least confidence	0.29 ± 0.04	1.03 ± 0.25	2.05 ± 0.38	27.05 ± 3.25	44.47 ± 1.42	61.80 ± 0.33	69.52 ± 0.45
Entropy	0.31 ± 0.02	1.01 ± 0.17	2.26 ± 0.30	28.21 ± 2.83	44.68 ± 1.54	61.82 ± 0.31	69.52 ± 0.45
Margin	0.47 ± 0.02	1.99 ± 0.29	4.73 ± 0.64	35.99 ± 1.67	50.29 ± 0.92	63.62 ± 0.15	69.52 ± 0.45
Forgetting	0.76 ± 0.01	4.69 ± 0.17	14.02 ± 0.13	**47.64 ± 0.03**	**55.12 ± 0.13**	62.49 ± 0.11	69.52 ± 0.45
GraNd	1.04 ± 0.04	7.02 ± 0.05	**18.10 ± 0.22**	43.53 ± 0.19	49.92 ± 0.21	57.98 ± 0.17	69.52 ± 0.45
Cal	**1.29 ± 0.09**	7.50 ± 0.26	15.94 ± 1.30	38.32 ± 0.78	38.77 ± 0.56	44.89 ± 3.72	69.52 ± 0.45
Craig	1.13 ± 0.08	5.44 ± 0.52	9.40 ± 1.69	32.30 ± 1.24	38.77 ± 0.56	44.89 ± 3.72	69.52 ± 0.45
GradMatch	0.93 ± 0.04	5.20 ± 0.22	12.28 ± 0.49	40.16 ± 2.28	45.91 ± 1.73	52.69 ± 2.16	69.52 ± 0.45
Glister	0.98 ± 0.06	5.91 ± 0.42	14.87 ± 0.14	44.95 ± 0.28	52.04 ± 1.18	60.26 ± 0.28	69.52 ± 0.45
FL	1.23 ± 0.03	5.78 ± 0.08	12.72 ± 0.21	40.85 ± 1.25	51.05 ± 0.59	63.14 ± 0.03	69.52 ± 0.45
GC	1.21 ± 0.09	**7.66 ± 0.43**	16.43 ± 0.53	42.23 ± 0.60	50.53 ± 0.42	63.22 ± 0.26	69.52 ± 0.45

as initialization. Hence, they are identical to RANDOM baseline for fraction 0.1% on ImageNet. We run all experiments for 3 times with random seeds.

Experiment results are given in Table 2. The results show that error based methods, FORGETTING and GRAND, generally have better performance on ImageNet. Especially, FORGETTING overwhelms RANDOM when fewer than 10% data are selected as the coreset. However, none of methods will outperform RANDOM when the coreset size is large, i.e. 30% data. RANDOM is still a strong and stable baseline. The same to that on CIFAR10, these findings on ImageNet may vary for different hyper-parameters.

4.3   Cross-architecture Generalization

We conduct cross-architecture experiments to examine whether methods with good performance are model-agnostic, i.e., whether coresets perform well when being selected on one architecture and then tested on other architectures. We do experiments on four representative methods (FORGETTING, GLISTER, GRAND and GRAPH CUT) with four representative architectures (VGG-16 [45], Inception-v3 [48], ResNet-18 [14] and WideResNet-16-8 [56]) under two selection fractions (1% and 10%). All other unspecified settings are the same to those in Sec. 4.1. In Tab. 3, the rows represent models used to obtain coresets, and the columns indicate models on which coresets are evaluated. We can see submodular selection with Graph Cut provides stably good testing results, regardless of which model architecture is used to perform the selection. However, GRAND shows preference of the model on which gradient norms are computed. Coresets obtained on Inception-v3 generally have the best performance, while those obtained on ResNet-18 are the worst. The possible reason is that the ranking of gradient norm is sensitive to the architecture. The architecture used to implement selection also has obvious influence on GLISTER and FORGETTING methods.

4.4 Sensitiveness to Pre-trained Models

As previously mentioned, some coreset selection methods rely on a pre-trained model to obtain metrics, e.g. feature, gradient and loss, for selecting samples. This experiment explores the influence of the pre-trained models, which are pre-trained for different epochs, on the final coreset performance. Similar to Sect. 4.3, four representative methods (FORGETTING, GLISTER, GRAND and GRAPH CUT) and two selection fractions (1% and 10%) are tested in this experiment. Except for different pre-training epochs, all other settings and hyperparameters are consistent with those in Sect. 4.1. We report our results in Table 4. For FORGETTING, good results can be achieved with models pre-trained for only 2 epochs, i.e. selecting samples based on whether the first forgetting event occurs on each sample. Spending more epochs in calculating forgetting events does not lead to improvements. The forgetting events can only be counted for more than 2 training epochs, thus no results are provided for FORGETTING in epoch 0 and 1. GRAND also performs best with models pre-trained for 2 epochs. The results indicate that it is not necessary to pre-train a model for too many epochs to obtain the metrics.

Table 3. Cross-architecture generalization performance (%) of four representative methods (FORGETTING, GRAND, GLISTER and GRAPH CUT). The coreset is selected based on one (row) architecture and then evaluated on another (column) architecture.

C\T	VGG-16	Inception-v3	ResNet-18	WRN-16-8	VGG-16	Inception-v3	ResNet-18	WRN-16-8
Random	1%				10%			
Random Selection	15.36 ± 2.03	32.98 ± 1.20	36.74 ± 1.69	45.77 ± 1.17	78.03 ± 0.92	76.01 ± 0.82	75.72 ± 2.02	82.72 ± 0.54
Forgetting	1%				10%			
VGG-16	17.56 ± 3.42	31.37 ± 0.63	35.07 ± 1.38	40.30 ± 1.94	72.71 ± 2.26	70.68 ± 1.85	$\mathbf{71.53 \pm 0.42}$	80.71 ± 1.11
Inception-v3	21.81 ± 3.04	$\mathbf{33.27 \pm 1.70}$	$\mathbf{36.94 \pm 1.28}$	$\mathbf{41.52 \pm 1.71}$	72.94 ± 0.63	71.15 ± 2.84	70.40 ± 2.09	$\mathbf{81.51 \pm 0.95}$
ResNet18	$\mathbf{22.81 \pm 3.46}$	32.64 ± 1.33	35.20 ± 1.59	39.45 ± 0.62	70.87 ± 1.27	66.87 ± 1.82	66.99 ± 1.48	79.19 ± 0.38
WRN-16-8	20.53 ± 3.49	28.46 ± 1.48	31.79 ± 1.11	35.92 ± 1.97	67.68 ± 1.37	64.38 ± 1.82	65.59 ± 2.03	75.59 ± 1.09
GraNd	1%				10%			
VGG-16	$\mathbf{18.61 \pm 3.84}$	29.78 ± 0.90	33.77 ± 0.87	38.07 ± 1.75	69.74 ± 1.48	65.90 ± 1.88	65.45 ± 1.33	76.63 ± 0.74
Inception-v3	15.94 ± 2.50	$\mathbf{31.46 \pm 0.98}$	$\mathbf{34.73 \pm 1.04}$	$\mathbf{40.16 \pm 1.83}$	$\mathbf{73.51 \pm 0.75}$	70.52 ± 3.15	70.07 ± 2.91	$\mathbf{79.62 \pm 1.27}$
ResNet18	14.42 ± 3.10	25.91 ± 1.59	26.69 ± 1.30	30.40 ± 0.75	61.05 ± 1.91	58.48 ± 3.95	52.73 ± 1.86	70.96 ± 1.14
WRN-16-8	14.59 ± 4.03	28.68 ± 1.43	32.30 ± 1.87	35.88 ± 3.18	61.49 ± 1.81	57.19 ± 2.42	57.82 ± 2.27	69.19 ± 1.92
Glister	1%				10%			
VGG-16	14.5 ± 3.86	31.08 ± 2.30	34.10 ± 1.71	39.45 ± 2.55	71.71 ± 1.83	70.23 ± 1.78	69.31 ± 2.19	77.74 ± 0.68
Inception-v3	$\mathbf{19.74 \pm 4.01}$	$\mathbf{32.05 \pm 1.12}$	$\mathbf{35.52 \pm 2.09}$	41.24 ± 1.39	$\mathbf{73.15 \pm 1.94}$	$\mathbf{71.32 \pm 1.77}$	71.03 ± 1.39	$\mathbf{78.57 \pm 1.45}$
ResNet18	15.16 ± 4.47	30.41 ± 2.08	32.93 ± 2.36	37.64 ± 1.83	67.37 ± 2.48	66.34 ± 2.18	66.26 ± 3.47	75.36 ± 1.52
WRN-16-8	14.16 ± 4.15	28.39 ± 2.50	32.83 ± 0.98	37.05 ± 2.72	70.70 ± 2.40	64.25 ± 2.53	66.88 ± 2.97	75.07 ± 2.96
Graph Cut	1%				10%			
VGG-16	27.47 ± 4.00	37.38 ± 2.09	$\mathbf{43.02 \pm 1.30}$	51.80 ± 0.82	$\mathbf{77.91 \pm 0.71}$	76.64 ± 1.25	$\mathbf{78.66 \pm 0.55}$	$\mathbf{81.06 \pm 0.78}$
Inception-v3	25.00 ± 3.91	37.26 ± 1.23	42.06 ± 0.69	51.67 ± 1.20	75.15 ± 1.09	73.69 ± 1.42	75.49 ± 0.91	78.33 ± 0.40
ResNet-18	$\mathbf{29.01 \pm 3.63}$	37.54 ± 0.62	42.78 ± 1.30	51.50 ± 1.37	75.29 ± 1.05	73.94 ± 1.11	76.65 ± 1.48	79.13 ± 0.75
WRN-16-8	22.64 ± 3.82	$\mathbf{37.71 \pm 1.73}$	40.78 ± 1.79	$\mathbf{53.02 \pm 1.80}$	76.64 ± 0.92	75.84 ± 0.84	77.19 ± 1.14	80.77 ± 0.30

5 Extended Related Work

An alternative way to reduce training set size is dataset condensation (or distillation) [50,57,58]. Instead of selecting subsets, it learns to synthesize informative training samples that can be more informative than real samples in the original

Table 4. Sensitiveness to pre-trained models. Performance (%) of different methods using pre-trained models with varying pre-training epochs.

Pre-train Epochs	0	1	2	5	10	15	20	50	100	150	200
1%											
Forgetting	-	-	36.06 ± 0.65	36.81 ± 1.82	35.20 ± 1.59	32.96 ± 1.20	32.22 ± 1.01	24.23 ± 0.64	20.41 ± 0.91	19.84 ± 0.56	19.47 ± 0.30
GraNd	28.17 ± 0.20	31.05 ± 1.36	31.24 ± 2.36	29.70 ± 1.02	26.69 ± 1.30	26.11 ± 1.46	26.39 ± 0.89	26.81 ± 1.97	26.52 ± 1.10	26.08 ± 0.65	27.17 ± 1.84
Glister	27.63 ± 0.85	33.97 ± 2.68	33.31 ± 1.08	32.93 ± 1.51	32.93 ± 2.36	32.28 ± 2.09	31.15 ± 2.24	31.46 ± 1.56	32.89 ± 1.24	33.37 ± 1.91	34.06 ± 2.17
Graph Cut	33.61 ± 1.40	43.15 ± 1.31	43.00 ± 0.76	44.33 ± 1.55	42.78 ± 1.30	41.33 ± 2.01	41.30 ± 2.80	42.23 ± 1.72	40.46 ± 0.93	41.74 ± 1.46	40.53 ± 2.27
10%											
Forgetting	-	-	72.62 ± 2.79	72.72 ± 1.44	66.99 ± 1.48	60.87 ± 1.92	54.62 ± 2.48	44.10 ± 1.21	42.29 ± 1.01	41.97 ± 0.70	41.99 ± 1.02
GraNd	62.54 ± 2.15	63.15 ± 1.99	71.34 ± 1.82	67.97 ± 1.86	52.73 ± 1.86	64.76 ± 1.83	65.20 ± 1.21	66.33 ± 2.29	57.21 ± 1.75	58.36 ± 1.49	65.34 ± 0.55
Glister	59.35 ± 2.31	60.83 ± 3.18	68.79 ± 1.15	68.81 ± 2.75	66.26 ± 3.47	61.99 ± 3.05	68.03 ± 1.72	65.05 ± 1.66	66.26 ± 2.92	68.16 ± 2.78	68.16 ± 3.03
Graph Cut	63.39 ± 1.54	62.52 ± 1.02	68.26 ± 1.11	72.91 ± 1.13	76.65 ± 1.48	77.06 ± 1.09	68.73 ± 0.87	77.48 ± 0.51	76.66 ± 1.64	76.16 ± 2.14	76.33 ± 1.52

training set. Although remarkable progress has been achieved in this research area, it is still challenging to apply dataset condensation on large-scale and high-resolution datasets, e.g. ImageNet-1K, due to the expensive and difficult optimization.

6   Conclusion

In this work, we contribute a comprehensive code library – *DeepCore* for coreset selection in deep learning, where we re-implement dozens of state-of-the-art coreset selection methods on popular datasets and network architectures. Our code library enables a convenient and fair comparison of methods in various learning settings. Extensive experiments on CIFAR10 and ImageNet datasets verify that, although various methods have advantages in certain experiment settings, random selection is still a strong baseline.

Acknowledgment. This research was supported by Public Health & Disease Control and Prevention, Major Innovation & Planning Interdisciplinary Platform for the "Double-First Class" Initiative, Renmin University of China (No. 2022PDPC), fund for building world-class universities (disciplines) of Renmin University of China. Project No. KYGJA2022001. This research was supported by Public Computing Cloud, Renmin University of China.

References

1. Agarwal, S., Arora, H., Anand, S., Arora, C.: Contextual diversity for active learning. In: Vedaldi, A., Bischof, H., Brox, T., Frahm, J.-M. (eds.) ECCV 2020. LNCS, vol. 12361, pp. 137–153. Springer, Cham (2020). https://doi.org/10.1007/978-3-030-58517-4_9
2. Aljundi, R., Lin, M., Goujaud, B., Bengio, Y.: Gradient based sample selection for online continual learning. Adv. Neural. Inf. Process. Syst. **32**, 11816–11825 (2019)
3. Bachem, O., Lucic, M., Krause, A.: Coresets for nonparametric estimation-the case of dp-means. In: ICML, PMLR, pp. 209–217 (2015)
4. Bateni, M., Bhaskara, A., Lattanzi, S., Mirrokni, V.S.: Distributed balanced clustering via mapping coresets. In: NIPS, pp. 2591–2599 (2014)

5. Borsos, Z., Mutny, M., Krause, A.: Coresets via bilevel optimization for continual learning and streaming. In: Advances in Neural Information Processing Systems, vol. 33 (2020)
6. Borsos, Z., Tagliasacchi, M., Krause, A.: Semi-supervised batch active learning via bilevel optimization. In: ICASSP 2021, pp. 3495–3499. IEEE (2021)
7. Chen, Y., Welling, M., Smola, A.: Super-samples from kernel herding. In: The Twenty-Sixth Conference Annual Conference on Uncertainty in Artificial Intelligence (2010)
8. Chhaya, R., Dasgupta, A., Shit, S.: On coresets for regularized regression. In: International Conference on Machine Learning, PMLR, pp. 1866–1876 (2020)
9. Coleman, C., et al.: Selection via proxy: efficient data selection for deep learning. In: ICLR (2019)
10. Dasgupta, S., Hsu, D., Poulis, S., Zhu, X.: Teaching a black-box learner. In: ICML, PMLR (2019)
11. Ducoffe, M., Precioso, F.: Adversarial active learning for deep networks: a margin based approach (2018). arXiv preprint arXiv:1802.09841
12. Farahani, R.Z., Hekmatfar, M.: Facility location: concepts, models, algorithms and case studies (2009)
13. Feldman, D., Faulkner, M., Krause, A.: Scalable training of mixture models via coresets. In: NIPS, Citeseer, pp. 2142–2150 (2011)
14. He, K., Zhang, X., Ren, S., Sun, J.: Deep residual learning for image recognition. In: Proceedings of the IEEE Conference on Computer Vision and Pattern Recognition, pp. 770–778 (2016)
15. Howard, A., et al.: Searching for mobilenetv3 (2019). http://arxiv.org/abs/1905.02244
16. Iyer, R., Khargoankar, N., Bilmes, J., Asanani, H.: Submodular combinatorial information measures with applications in machine learning. In: Algorithmic Learning Theory, pp. 722–754. PMLR (2021)
17. Iyer, R.K., Bilmes, J.A.: Submodular optimization with submodular cover and submodular knapsack constraints. In: Advances in Neural Information Processing Systems, vol. 26 (2013)
18. Ju, J., Jung, H., Oh, Y., Kim, J.: Extending contrastive learning to unsupervised coreset selection (2021). arXiv preprint arXiv:2103.03574
19. Kaushal, V., Kothawade, S., Ramakrishnan, G., Bilmes, J., Iyer, R.: Prism: A unified framework of parameterized submodular information measures for targeted data subset selection and summarization (2021). arXiv preprint arXiv:2103.00128
20. Killamsetty, K., Durga, S., Ramakrishnan, G., De, A., Iyer, R.: Grad-match: gradient matching based data subset selection for efficient deep model training. In: ICML, pp. 5464–5474 (2021)
21. Killamsetty, K., Sivasubramanian, D., Ramakrishnan, G., Iyer, R.: Glister: generalization based data subset selection for efficient and robust learning. In: Proceedings of the AAAI Conference on Artificial Intelligence (2021)
22. Killamsetty, K., Zhao, X., Chen, F., Iyer, R.: Retrieve: Coreset selection for efficient and robust semi-supervised learning (2021). arXiv preprint arXiv:2106.07760
23. Knoblauch, J., Husain, H., Diethe, T.: Optimal continual learning has perfect memory and is np-hard. In: International Conference on Machine Learning, PMLR, pp. 5327–5337 (2020)
24. Kothawade, S., Beck, N., Killamsetty, K., Iyer, R.: Similar: Submodular information measures based active learning in realistic scenarios (2021). arXiv preprint arXiv:2107.00717

25. Krizhevsky, A., Hinton, G., et al.: Learning multiple layers of features from tiny images (2009)
26. Krizhevsky, A., Sutskever, I., Hinton, G.E.: Imagenet classification with deep convolutional neural networks. In: Pereira, F., Burges, C.J.C., Bottou, L., Weinberger, K.Q. (eds.) Advances in Neural Information Processing Systems, vol. 25. Curran Associates, Inc. (2012)
27. Le, Y., Yang, X.: Tiny imagenet visual recognition challenge. CS 231N **7**(7), 3 (2015)
28. LeCun, Y., Boser, B., Denker, J.S., Henderson, D., Howard, R.E., Hubbard, W., Jackel, L.D.: Backpropagation applied to handwritten zip code recognition. Neural Comput. **1**(4), 541–551 (1989)
29. LeCun, Y., Bottou, L., Bengio, Y., Haffner, P., et al.: Gradient-based learning applied to document recognition. Proc. IEEE **86**(11), 2278–2324 (1998)
30. Liu, E.Z., et al.: Just train twice: Improving group robustness without training group information. In: ICML, pp. 6781–6792 (2021)
31. Margatina, K., Vernikos, G., Barrault, L., Aletras, N.: Active learning by acquiring contrastive examples (2021). arXiv preprint arXiv:2109.03764
32. Mirzasoleiman, B., Bilmes, J., Leskovec, J.: Coresets for data-efficient training of machine learning models. In: ICML, PMLR (2020)
33. Mirzasoleiman, B., Cao, K., Leskovec, J.: Coresets for robust training of deep neural networks against noisy labels (2020)
34. Munteanu, A., Schwiegelshohn, C., Sohler, C., Woodruff, D.P.: On coresets for logistic regression. In: NeurIPS (2018)
35. Nemhauser, G.L., Wolsey, L.A., Fisher, M.L.: An analysis of approximations for maximizing submodular set functions-i. Math. Program. **14**(1), 265–294 (1978)
36. Netzer, Y., Wang, T., Coates, A., Bissacco, A., Wu, B., Ng, A.Y.: Reading digits in natural images with unsupervised feature learning (2011)
37. Paszke, A., et al.: Pytorch: An imperative style, high-performance deep learning library. In: Advances in Neural Information Processing Systems, vol. 32 (2019)
38. Paul, M., Ganguli, S., Dziugaite, G.K.: Deep learning on a data diet: finding important examples early in training (2021). arXiv preprint arXiv:2107.07075
39. Russakovsky, O., et al.: ImageNet Large Scale Visual Recognition Challenge. In: IJCV (2015)
40. Sachdeva, N., Wu, C.J., McAuley, J.: Svp-cf: selection via proxy for collaborative filtering data (2021). arXiv preprint arXiv:2107.04984 (2021)
41. Sener, O., Savarese, S.: Active learning for convolutional neural networks: a core-set approach. In: ICLR (2018)
42. Settles, B.: Active learning literature survey (2009)
43. Settles, B.: From theories to queries: Active learning in practice. In: Active Learning and Experimental Design Workshop in Conjunction with AISTATS 2010, JMLR Workshop and Conference Proceedings, pp. 1–18 (2011)
44. Shim, J.h., Kong, K., Kang, S.J.: Core-set sampling for efficient neural architecture search (2021). arXiv preprint arXiv:2107.06869
45. Simonyan, K., Zisserman, A.: Very deep convolutional networks for large-scale image recognition (2014). arXiv preprint arXiv:1409.1556
46. Sinha, S., Zhang, H., Goyal, A., Bengio, Y., Larochelle, H., Odena, A.: Small-gan: Speeding up gan training using core-sets. In: ICML, PMLR (2020)
47. Sohler, C., Woodruff, D.P.: Strong coresets for k-median and subspace approximation: goodbye dimension. In: 2018 IEEE 59th Annual Symposium on Foundations of Computer Science (FOCS), pp. 802–813. IEEE (2018)

48. Szegedy, C., Vanhoucke, V., Ioffe, S., Shlens, J., Wojna, Z.: Rethinking the inception architecture for computer vision. In: Proceedings of the IEEE Conference on Computer Vision and Pattern Recognition, pp. 2818–2826 (2016)
49. Toneva, M., Sordoni, A., des Combes, R.T., Trischler, A., Bengio, Y., Gordon, G.J.: An empirical study of example forgetting during deep neural network learning. In: ICLR (2018)
50. Wang, T., Zhu, J.Y., Torralba, A., Efros, A.A.: Dataset distillation (2018). arXiv preprint arXiv:1811.10959
51. Wei, K., Iyer, R., Bilmes, J.: Submodularity in data subset selection and active learning. In: International Conference on Machine Learning, PMLR (2015)
52. Welling, M.: Herding dynamical weights to learn. In: Proceedings of the 26th Annual International Conference on Machine Learning, pp. 1121–1128 (2009)
53. Xiao, H., Rasul, K., Vollgraf, R.: Fashion-mnist: a novel image dataset for benchmarking machine learning algorithms (2017). arXiv preprint arXiv:1708.07747
54. Yadav, C., Bottou, L.: Cold case: The lost mnist digits. In: Advances in Neural Information Processing Systems, vol. 32 (2019)
55. Yoon, J., Madaan, D., Yang, E., Hwang, S.J.: Online coreset selection for rehearsal-based continual learning (2021). arXiv preprint arXiv:2106.01085
56. Zagoruyko, S., Komodakis, N.: Wide residual networks (2016). arXiv preprint arXiv:1605.07146
57. Zhao, B., Bilen, H.: Dataset condensation with differentiable siamese augmentation. In: International Conference on Machine Learning (2021)
58. Zhao, B., Mopuri, K.R., Bilen, H.: Dataset condensation with gradient matching. In: International Conference on Learning Representations (2021). https://openreview.net/forum?id=mSAKhLYLSsl

Context Iterative Learning
for Aspect-Level Sentiment Classification

Wenting Yu[(✉)], Xiaoye Wang, Peng Yang, Yingyuan Xiao, and Jinsong Wang

Tianjin University of Technology, Tianjin, China
ywthelium@163.com, jiaoliu456@163.com, 29139475@qq.com,
{yyxiao,jswang}@tjut.edu.cn

Abstract. Aspect-based sentiment analysis is to predict the sentiment polarity of different aspects of a sentence. Many irrelevant words are mistaken for opinion words in long sentences. According to extensive research, irrelevant words are far removed from the central words. This paper proposes a solution: First, we design the Context Iterative Learning network (CILN). Context attention module (CAM) is proposed, which employs Context Features Dynamic Mask (CDM) to cover words far from the center word and Context Features Dynamic Weighted (CDW) to reduce the weight of words far away. The calculation of CAM is done alternately to reduce the influence of distant irrelevant words. Finally, the obtained feature sequences are linked with the global sentence sequence. The Accuracy and Macro-F1 indicators obtained from the experiments based on benchmark datasets demonstrate the efficacy of the proposed method.

Keywords: Aspect-based sentiment analysis · Feature extraction · Distribution reduction

1 Introduction

Aspect-based sentiment analysis (ABSA) is a text classification task, which divides the sentiment polarity of content into positive, neutral and negative [2].

The attention mechanism is now a crucial model in solving sentiment analysis tasks [9]. However, the attention mechanism does not always accurately predict aspect polarity [1]. Attention mechanisms can neither capture position information between words nor learn the relationship between sequence information and words in a sentence. Existing ABSA models mainly enhance aspect representation learning, such as MetNet [6]. MetNet may learn disturbing information together. This paper proposes a CAM module built on the CDM/CDW block and multi-head attention. A more accurate aspect-context feature representation is extracted through multiple iterations of CAM by reducing the influence of irrelevant words on sentiment prediction.

We propose the Context Iterative Learning Network (CILN). It is inspired by MemNet [8], AEN-Bert [7] and LCF-Bert [11]. First, we enter the context

© The Author(s), under exclusive license to Springer Nature Switzerland AG 2022
C. Strauss et al. (Eds.): DEXA 2022, LNCS 13426, pp. 196–202, 2022.
https://doi.org/10.1007/978-3-031-12423-5_15

sequences and aspect terms sequences so that both sequences can traverse multiple CAMs at the same time. The CDM and CDW modules are used alternately in each CAM. The CDM module is used in the first CAM module. The CDW module is used in the following CAM module. After the multi-layer CAM, the new aspect-context sequence merge with the global sequence.

The main contributions are as follows: We design the CILN, which extracts contextual features by iteration and enhances the attention of aspect terms. The CAM module is designed, and the CDM and CDW modules are used in combination with the multi-head attention mechanism to reduce the influence of irrelevant words on aspect prediction.

2   Related Work

Deep learning is primarily used for sentiment analysis now. AOA [5] is an attention-over-attention neural network for aspect-oriented sentiment classification. LCF-BERT [11] is an aspect-based sentiment classification mechanism based on Multi-head Self-Attention (MHSA)-local context focus (LCF). Zhang [12] uses a graph convolutional network to extract sentence features, and uses graph convolution to investigate the influence of the dependency tree. MET-Net [6] designs a hierarchical structure that iteratively enhances the representation of aspects and contexts.

3   Context Attention Modules (CAM)

CAM is illustrated in Fig. 1. The input sequence is divided into two data streams. In the first data stream, context sequences are passed through Intra-multi-headed attention mechanism (Intra-MHA), position-wise feed-forward networks (PFFN) and CDM in turn. In the second data stream, context sequences and aspect terms sequences are passed through Inter-multi-headed attention mechanism (Inter-MHA) and PFFN in turn. And $\oplus$ denotes the multi-headed attention mechanism(MHA) that connects two data streams information. CAM is performed iteratively, and the CDM/CDW in each CAM is performed alternately. The purpose of alternate execution is to avoid extracting a single context feature.

3.1   Intra-Multi-Headed Attention Mechanism (Intra-MHA) and Inter-Multi-Headed Attention Mechanism (Inter-MHA)

Inter-MHA [7] is a multi-headed attention calculation that takes into account context and aspect terms. The context sequences and the aspect sequences are learned together to solve the long dependency problem. The formula for Inter-MHA is as follows: $h_j = \text{Multihead-Attention}\left(v_{c_i}, v_{a_j}\right)$ [7]. Where v_{a_j} is the aspect sequence vector, v_{c_i} is the context sequence vector.

Intra-MHA [7] learns important features from different heads and can selectively emphasize the sentence's relatively important features. Intra-MHA is expressed as: $h_i = \text{Multihead-Attention}\left(v_{c_i}, v_{c_i}\right)$ [7].

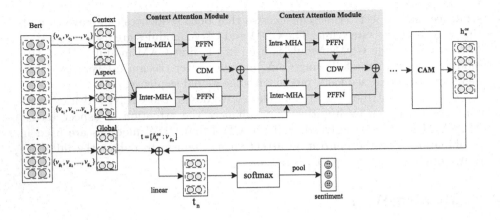

Fig. 1. Structure of Context Iterative Learning Network (CILN)

3.2   Context Features Dynamic Mask/Context Features Dynamic Weighted (CDM/CDW)

In CAM, the semantic relative distance(SRD) determines the CDM's conceal-ment range and the range of dynamic weight reduction. The SRD is the word distance between the context word tokens and the specific aspect terms.

CDM [11] uses the specific aspect terms as the center and the SRD as the radius to calculate the next attention mechanism for words that are within the SRD distance, and irrelevant words that are masked. The input local context matrix is V^l. CDM based on certain SRD threshold α is expressed as: $V_i^m = \begin{cases} O, SRD > \alpha \\ E, SRD \leq \alpha \end{cases}$. Where O represents zero vector, and E represents one vector. m represents CDM. The mask matrix is multiplied with the local context matrix output in the last step: $V^M = [V_1^m, V_2^m, \ldots, V_i^m] \cdot V^l$.

CDW [11] takes the aspect terms as the center and SRD as the radius to reduce the weight of words outside the SRD distance. The input con-text matrix is V^l. CDW based on certain SRD threshold α is expressed as: $V_i^w = \begin{cases} E - \frac{SRD_i - \alpha}{N} \cdot E, SRD > \alpha \\ E, SRD \leq \alpha \end{cases}$, $V^W = [V_1^w, V_2^w, \ldots, V_i^w] \cdot V^l$. Where SRD_i the i-th SRD distance, N is the length of the sentence. w represents CDM.

3.3   Position-Wise Feed-Forward Networks (PFFN) and Aspect-Context Representation Output

PFFN transforms the information from the previous step and provides rich fea-ture representations. PFFN is made up of two layers of Feed Forward Neural Net-works (FFNNs). The input of PFFN is expressed as s_c. PFFN can be expressed as: $PFFN_c = \text{Relu}(W_{c1} \times s_c + b_{c1}) W_{c2} + b_{c2}$. Where W_{c1} and W_{c2} are trainable weights of two FFNNs. b_{c1} and b_{c2} are learnable biases of two FFNNs.

$V_i^{W/M}$ is the context vector processed by the CDM/CDW, and P_c is the aspect terms vector after the PFFN. The specific aspect-context is expressed as: $h_i^{ca} = \text{Multihead-Attention}\left(V_i^{W/M}, P_c\right)$. The MHA here also has its own independent parameters. We input the obtained h_i^{ca} into the next CAM.

4   Context Iterative Learning Network (CILN)

We send the comment sentence into Bert to convert the words into vectors (The context sequence is $V_c = \{v_{c_1}, v_{c_2}, \ldots, v_{c_t}\}$. The aspect term sequence is $V_a = \{v_{a_1}, v_{a_2}, \ldots, v_{c_m}\}$. The global sequence is $V_g = \{v_{g_1}, v_{g_2}, v_{g_3}, \ldots, v_{g_n}\}$.) in Fig. 1. Then the converted context sequences and aspect terms sequences are fed into the CAM. After several iterations the aspect-context sequence will be obtained. The representation is expressed jointly with the global sequences ($\oplus$ indicates a connection operation), and finally the resulting final representation is classified into sentiment polarities. 3 represents three kinds of sentiment polarities.

4.1   Pooling Layer and Training

We connect CAM's output with the global sequences as $t = [h_n^{ca}, V_g]$. Where h_n^{ca} is the aspect-context representation after several CAM Iterations. Finally, we input the final representation into the softmax layer for sentiment classification. The softmax classification can be expressed as: $Y = \text{softmax}(t) = \frac{\exp f(t)}{\sum_{x=1}^{3} \exp f(t)}$, $f(t) = W_s \times t + b_s$. Where W_s and b_s are learnable weights and biases.

   The objective optimization function of this paper is the cross-entropy loss with L_2 regularization, and the function is defined as: $L(\theta) = -\sum_{i=1}^{3} \widehat{y_x} \log y_x + \lambda \sum_{\theta \in \Theta} \theta^2$. Where y_x is the one-hot vector. λ is the parameter of L_2 regularization, and θ is the parameter set of the model in this paper.

5   Experiment

5.1   Datasets and Experimental Settings

To better evaluate the model in this paper. We use three benchmark datasets: SemEval2014 Task 4 (14Rest and 14Lap) and ACL Twitter dataset (Twitter) [4]. The datasets have been adopted by the models proposed by the majority of researchers and are the most frequently used datasets in ABSA.

   Most of the hyperparameters follow the common hyperparameter settings for sentiment analysis tasks. The learning rate is set to 2×10^{-5}, and the hidden dimensions and the embedding dimensions are set to 768. The dropout rate is set to 0.1, the L_2 regularization is set to 1×10^{-5}, and the batch size is set to 16. A total of 12 epochs were trained. The performance of the model is evaluated by using accuracy and macro F1 indicators.

Table 1. Experimental results (%). This article uses "–" to indicate unrecorded experimental results. All experimental results are the results of rerunning on our equipment.

Model	Laptop		Restaurant		Twitter	
	ACC	F1	ACC	F1	ACC	F1
MemNet	67.08	59.12	78.04	65.63	70.24	67.78
RAM	66.73	57.43	75.18	57.48	67.34	63.76
AOA	63.17	49.43	73.12	53.17	65.61	61.47
Aen-Bert	78.06	74.93	80.45	69.35	72.54	71.05
LCF-Bert	79.00	74.60	83.93	74.68	73.55	72.65
MCRF-SA	75.43	71.78	80.71	70.28	–	–
MetNet	76.18	71.83	79.11	67.84	66.76	63.52
BiGCN	74.92	71.76	79.37	68.56	73.55	71.79
Our	**79.78**	**76.44**	**84.91**	**78.87**	**75.43**	**74.14**
−w/o CAM	78.68	73.82	83.48	74.49	72.69	71.48
+1 CAM	79.78	75.01	84.29	77.36	72.11	70.01
+2 CAM	**79.78**	**76.44**	**84.91**	**78.87**	**75.43**	**74.14**
+3 CAM	78.68	74.94	84.20	77.55	74.13	73.34
+4 CAM	78.53	75.31	83.93	76.44	72.83	72.17
+5 CAM	78.53	74.50	84.02	75.84	72.98	72.10

5.2   Baseline and Result

To comprehensively evaluate our method, this paper compares the proposed method with the model baselines: MemNet (2016) [8], RAM (2017) [3], AOA (2018) [5], Aen-Bert (2019) [7], LCF-BERT (2019) [11], MCRF-SA (2020) [10], MetNet (2020) [6], BiGCN (2020) [13].

The results that our model with 2 layers of CAM outperforms all baselines in Table 1. Twitter's performance is not as good as that of other datasets. Because Twitter has irregular grammatical expressions and many misspellings, which leads to poor performance on Twitter compared to the other two datasets. The accuracy of our model is 12.70% higher than MemNet on the laptop dataset, 6.87% on the Restaurant, and 5.17% on the Twitter. LCF-BERT is the second best performing. The accuracy of our model on the Laptop, Restaurant, and Twitter increased by 0.78%, 0.98%, and 1.88%. We believe that our model outperforms LCF because LCF only uses CDW/CDM once. Whereas we iteratively use CAM and alternate CDM/CDW for each CAM, enriching context feature extraction and resulting in higher ACC and F1 scores.

To explore the application effect of CAM in this model, ablation experiments are carried out on the basis of the best CAM superimposing two layers, including CAM resection. "−w/o" stands for delete a module. The experimental results clearly show that CAM ablation will affect the performance, which shows CAM is helpful to improve the ABSA.

As shown in Table 1, different numbers (from +1 to +5) of CAM layers are tried. The results of 2 layers are the best. First of all, the effect of CAM increases as the number of layers increases. When the number of layers is increased after the model effect has been brought to the best number of layers, the effect gradually decreases and unstable results appear. The model proposed in this paper only models the context feature layer directly related to the specific word in each CAM. Thereby increasing the number of CAM layers can improve ABSA performance. Adding more layers, model is overfitting and the result decreases.

6   Conclusion

We propose the CILN to improve the impact of irrelevant words on ABSA. To obtain a better representation, we employ a hierarchical structure CAM to iteratively learn aspects and contexts. The results demonstrate that CILN is useful for ABSA.

Acknowledgements. This work is supported by "Tianjin Project + Team" Key Training Project under Grant No. XC202022.

References

1. Bahdanau, D., Cho, K.H., Bengio, Y.: Neural machine translation by jointly learning to align and translate. In: Proceedings of ICLR (2015)
2. Barbieri, F., Camacho-Collados, J., Anke, L.E., Neves, L.: TweetEval: unified benchmark and comparative evaluation for tweet classification. In: Proceedings of EMNLP Findings (2020)
3. Chen, P., Sun, Z., Bing, L., Yang, W.: Recurrent attention network on memory for aspect sentiment analysis. In: Proceedings of EMNLP, pp. 452–461 (2017)
4. Dong, L., Wei, F., Tan, C., Tang, D., Zhou, M., Xu, K.: Adaptive recursive neural network for target-dependent Twitter sentiment classification. In: Proceedings of ACL, pp. 49–54 (2014)
5. Huang, B., Ou, Y., Carley, K.M.: Aspect level sentiment classification with attention-over-attention neural networks. In: Thomson, R., Dancy, C., Hyder, A., Bisgin, H. (eds.) SBP-BRiMS 2018. LNCS, vol. 10899, pp. 197–206. Springer, Cham (2018). https://doi.org/10.1007/978-3-319-93372-6_22
6. Jiang, B., Hou, J., Zhou, W., Yang, C., Wang, S., Pang, L.: METNet: a mutual enhanced transformation network for aspect-based sentiment analysis. In: Proceedings of COLING, pp. 162–172 (2020)
7. Song, Y., Wang, J., Jiang, T., Liu, Z., Rao, Y.: Targeted sentiment classification with attentional encoder network. In: Tetko, I.V., Kůrková, V., Karpov, P., Theis, F. (eds.) ICANN 2019. LNCS, vol. 11730, pp. 93–103. Springer, Cham (2019). https://doi.org/10.1007/978-3-030-30490-4_9
8. Tang, D., Qin, B., Liu, T.: Aspect level sentiment classification with deep memory network. In: Proceedings of EMNLP, pp. 214–224 (2016)
9. Vaswani, A., et al.: Attention is all you need. In: Proceedings of NeurIPS (2017)
10. Xu, L., Bing, L., Lu, W., Huang, F.: Aspect sentiment classification with aspect-specific opinion spans. In: Proceedings of EMNLP, pp. 3561–3567 (2020)

11. Zeng, B., Yang, H., Xu, R., Zhou, W., Han, X.: LCF: a local context focus mechanism for aspect-based sentiment classification. Appl. Sci. **9**, 3389 (2019)
12. Zhang, C., Li, Q., Song, D.: Aspect-based sentiment classification with aspect-specific graph convolutional networks. In: EMNLP/IJCNLP, no. 1 (2019)
13. Zhang, M., Qian, T.: Convolution over hierarchical syntactic and lexical graphs for aspect level sentiment analysis. In: Proceedings of EMNLP, pp. 3540–3549 (2020)

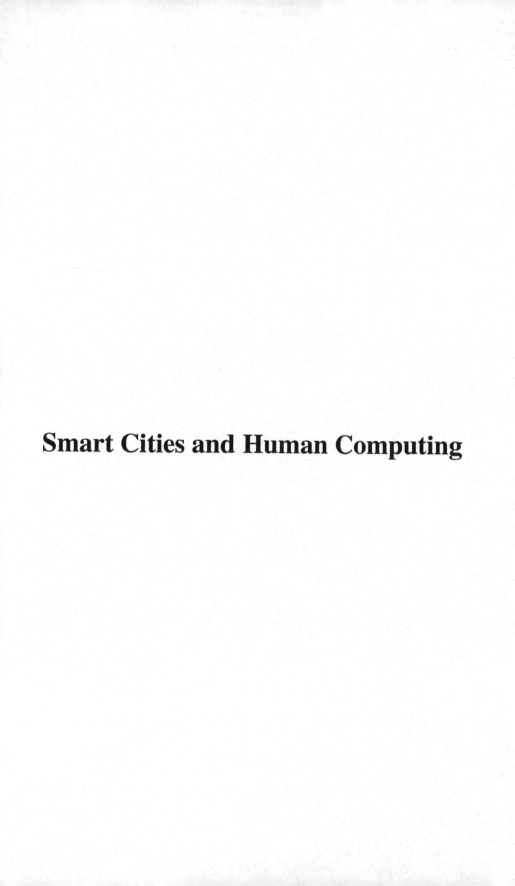

Smart Cities and Human Computing

EcoLight: Eco-friendly Traffic Signal Control Driven by Urban Noise Prediction

Chahinez Ounoughi[1,2]([✉])[iD], Ghofrane Touibi[2], and Sadok Ben Yahia[1][iD]

[1] Department of Software Science, Tallinn University of Technology, Tallinn, Estonia
{chahinez.ounoughi,sadok.ben}@taltech.ee
[2] Université de Tunis El Manar, Faculté des Sciences de Tunis,
LR11ES14, 2092 Tunis, Tunisia
ghofrane.touaibi@etudiant-fst.utm.tn

Abstract. Traffic congestion is of utmost importance for modern societies due to population and economic growth. Thus, it contributes to environmental problems like increasing greenhouse gas emissions and noise pollution. Traffic signal control plays a vital role in improving traffic flow in urban networks. Hence, optimizing cycle timing at many intersections is paramount to reducing congestion and increasing sustainability. In this paper, we introduce an alternative to conventional traffic signal control, namely *EcoLight*, that provides significant improvements in noise levels, CO_2 emissions, and fuel consumption, resulting from the incorporation of future noise predictions. A *Sequence to Sequence Long Short Term Memory (SeqtoSeq-LSTM)* prediction model, combined with a deep reinforcement learning algorithm, allows the system to achieve higher efficiency than its competitors based on real-world data from Helsinki, Finland.

Keywords: CO_2 emissions · Congestion · Fuel consumption · Reinforcement learning · SUMO Simulation · Traffic signal control · Urban noise

1 Introduction

Traffic congestion levels have been rising precipitously in the last few years due to an imbalance between the rise in travel demand and the availability of transportation services. According to [18], the cost of congestion in cities such as Stuttgart and Paris is around 2% of their GDP. The general rule is that cities should develop strategies based on their visions and goals to reduce congestion. Implementation of new infrastructure is often slow and costly. Therefore, urban planners and policymakers are interested in making existing infrastructure more efficient [16]. One of the proposed hypotheses is that *"An improved traffic light system will lead to better traffic management and, therefore, more peaceful urban areas"* [1]. Hence, optimizing cycle timing at intersections can potentially contribute significantly to reducing congestion and improving environmental quality at the same time. Real-time control of traffic signals plays a vital role in

© The Author(s), under exclusive license to Springer Nature Switzerland AG 2022
C. Strauss et al. (Eds.): DEXA 2022, LNCS 13426, pp. 205–219, 2022.
https://doi.org/10.1007/978-3-031-12423-5_16

reducing congestion by responding in real-time to several factors, including constantly changing traffic network dynamics. Moreover, the rapid increase in transport requirements has brought challenges to the sustainable development of our society concerning emissions and energy consumption induced by traffic. The European Environment Agency (EEA) reports that road traffic noise continues to be the primary contributor to noise pollution. Around 100 million people are exposed to road traffic noise above 55 decibels (dB) in the 33 member countries of the EEA. Among them, 32 million (about one-third) are subjected to extremely high levels of noise exceeding 65 dB [8]. Furthermore, according to the World Health Organization (WHO), exposure to loud noise causes high blood pressure, hearing loss, heart disease, sleep disturbances, and stress. Hence, measuring road traffic noise is a good indicator of traffic congestion intensity.

Numerous traffic signal control solutions have been used and proposed to overcome the traffic congestion issue. Worthy of mentioning, integrated Arduino in cameras with machine learning (e.g., object detection deep learning algorithms), and genetic algorithms for traffic signal timing optimization to help experts manage congestion. Recently, researchers have begun investigating reinforcement learning (RL) techniques for controlling traffic signals. These techniques appear to be more effective than traditional transportation methods. Its main advantage is that it learns how to take real-time action by observing the environment's reaction to previous actions.

One major issue of most RL-based traffic signal control approaches is that the setting considers, in each phase, only *mobility* and *current* traffic conditions when designing the next control strategy. We elaborate on these two characteristics by integrating two novel aspects into the RL techniques: *(i) Sustainability*: is achieved by incorporating noise as an environmental input feature; and *(ii) Proactivity*: is achieved by predicting future levels of noise so that the model is better prepared to make decisions based on current observations as well as future noise predictions. Therefore, in this paper, we propose a new eco-friendly RL-based traffic signal control model driven by urban noise traffic prediction, namely *EcoLight*. Our proposed approach reduces traffic congestion by reducing noise levels, CO2 emissions, and fuel consumption. By and large, the main contributions of *EcoLight* are as follows:

- At the noise prediction stage, we take advantage of the sequence to sequence architecture and propose splitting the time-series noise traffic data into fixed-sized sequences, where the size is determined based on an analysis of road network traffic behavior. Our method includes building a stacked layers architecture based on LSTM to extract temporal dependencies from noise data. Then, by using the past noise sequences as input, we would return a future traffic noise sequence.
- At the traffic signal control stage, we heavily rely on a deep reinforcement learning control model that takes as an input traffic-related information, i.e., the queue length, average waiting time, the phase, number of vehicles, and the vehicles' position at an intersection, besides the traffic noise estimation to predict the upcoming traffic signal action.

– We run our simulation experiments on a publicly available dataset of a road intersection collected in Helsinki, Finland. The harvested evaluation criteria (noise levels, CO2 emissions, and fuel consumption) outperform those obtained by the pioneering ones in the literature.

The rest of this paper proceeds as follows. In Sect. 2, we scrutinize the related work that paid attention to both traffic noise prediction and traffic signal control approaches. As an introduction to traffic signal control, Sect. 3 introduces key notions that will simplify the understanding of our research goal. Section 4 thoroughly describes the proposed *EcoLight* approach. In the penultimate section, we present the experimental evaluation and discuss the proposed model's performance against its competitors. The final section includes a conclusion and recommendations for future research.

2   Related Work

Modern societies nowadays are characterized by a great deal of noise. In addition to being a nuisance, it can also negatively impact the environment and human health. While evidence of noise's harmful effects is increasing, spatial understanding of its distribution is limited. This section introduces, first, brief overview noise prediction methods for traffic congestion enhancement, followed by methods for traffic signal control.

2.1   Noise Prediction

Noise pollution from road traffic is the most prevalent source of outdoor ambient noise in Europe. Different prediction models may produce different noise levels depending on traffic noise's location and emission sources. At present, very little research focuses on developing models that help determine the effects of traffic noise on society. Worth mentioning, Staab et al. [20] used a land-use regression (LUR) model and context-aware feature engineering to construct a geostatistical model mapping approach to represent the arrangement of sources and the surrounding environment. In this article, the authors deal with small communities that have not been adequately mapped in Europe. To improve traffic noise modeling, another solution was proposed by Ahmed et al. [2] that developed a deep neural network-based optimization approach that integrated the wrapper for the feature-subset selection (WFS) method. Using this method, weekday noise maps are created for different times of the day, such as mornings, afternoons, evenings, and nights. Khan et al. [10] conducted a comparison study between three different noise estimation models used throughout Europe. In this study, the main focus was to explore potential patterns in the performance of the models for specific configuration types. Based on vehicular traffic volume, percentage of heavy vehicles, and vehicles' average speed, a neuro-fuzzy inference system that identifies at what noise level the traffic (Leq dBA) will be detected has been proposed by Singh et al. [19]. Comparing it with conventional soft-computing techniques validates its suitability for planning mitigation measures for both new and existing roads. Finally, Zhang et al. [29] examined the accuracy of different machine learning recurrent architectures for predicting traffic noise using real-life traffic data with multiple variables. According to the study, using a multivariate bidirectional GRU model (Gated Recurrent Unit) with a many-to-many architecture achieved the best computation efficiency and accuracy.

The noise generated by traffic is a complex phenomenon. In modeling traffic noise, large and high-dimensional data are gathered. In this case, deep recurrent learning architectures are the best tools for analyzing large datasets and discovering nonlinear relationships.

2.2   Traffic Signal Control

Traffic signal control is an integral part of an intelligent transportation system that improves traffic efficiency. However, some challenges accompany these systems, such as protecting against high roadside cameras, keeping malicious vehicles from getting in, and preventing single points of failure. Literature has examined several traffic signal control systems to cope with those challenges. Two different approaches have been developed: a fixed-time (rule-based) strategy and a traffic-responsive strategy [13].

As part of a fixed-time strategy, several signal plans (e.g., from 8:00 to 10:00 am) are predetermined based on historical traffic flow data. Thus, a traffic signal is periodically changed per the predetermined signal plans. Worth mentioning, Le et al. [12] proposed a decentralized traffic signal control using a Back-pressure scheme for urban roads networks, which has received widespread recognition as a method for achieving an optimal throughput control policy in data networks. They concluded that the proposed scheme of fixed cycle times and cyclic phases stabilizes the traffic for any possible transportation demand. However, since such traditional transportation systems do not work in real-time, they can only be used when the demand is relatively stable within each time interval.

By using current traffic information, the traffic-responsive strategy overcomes the above limitation. In this strategy, the major challenge is forecasting incoming vehicles or traffic status. Bravo et al. [5] proposed a city-wide traffic control management program that assists traffic managers in making decisions, namely HITUL. Utilizing meta-heuristic algorithms and nature-inspired techniques, the HITUL system uses different technologies to gather data and optimize traffic signal priorities using existing traffic information. Various reinforcement-learning methods have recently been proposed to improve the traffic signal control and achieved better results than traditional transportation methods. Worth mentioning, *IntelliLight* [24], an RL-based method with an extended phase-sensitive gate that provides an overall measure of traffic signal control performance based on factors such as the waiting time and the number of vehicles at intersections. *Presslight* [22] is another RL-based method that uses the current phase, the number of vehicles on outgoing lanes, and the number of vehicles on incoming lanes as the state, and uses the Max-pressure (MP) as the reward for achieving coordination between neighbors. *Colight* [23] utilizes graph attentional networks to facilitate communication. In this case, it uses the attention mechanism to represent neighboring information to achieve the goal of cooperative traffic signal control.

Table 1. Representative traffic signal control methods.

Citation	Method	Simulator	Road net. (# inters.)	Evaluation
[12]	Back-pressure scheme	SUMO	Real (2)	Avg. travel time
[5]	Meta-heuristic algorithm	SUMO	Real (961)	Emissions, Waiting time
[24]	RL with extended phase-sensitive gate	SUMO	Synthetic (1), Real (24)	Reward, Queue Length, Delay, Duration
[22]	RL with MP-based reward	CityFlow	Sythetic (1), Real (3, 5, 16)	Avg. travel time
[23]	RL with graph attentional networks	CityFlow	Real (196)	Avg. travel time
[25]	RL trained with Demonstrations	CityFlow	Real (1)	Travel time
[15]	RL with object detection	Pygame	Synthetic (1)	Avg. waiting time
[3]	Queue-length responsive	Real env	Real (1)	Avg. waiting time
[7]	RL-FRAP with MP coordination	CityFlow	Real (2510)	Avg. travel time, Throughput
[26]	RL-FRAP with MAML	CityFlow	Real (1)	Travel time
[28]	MUMOMAML with clustering for Parameter initialization	CityFlow	Real (1, 5, 16)	Avg. travel time

DemoLight [25] learns a stochastic policy (demonstrations) that maps states to an action probability distribution based on a generated analogy between agents and humans. *FRAP* [30] is a reinforcement learning-based method designed to learn the inherent logic of the traffic signal control problem, called phase competition. The advantage of this method is that it combines similar transactions irrespective of the intersection structure or local traffic conditions.

ThousandLight [7] is one of the most recent works that has been tested on the real-road network with 2510 traffic signals. By leveraging the 'pressure' concept, they developed RL-FRAP-based agents capable of signal coordination at a regional level. Furthermore, the authors demonstrated that individual agents can achieve implicit coordination through reward design, thereby decreasing dimensionality. Another RL-FRAP with model-agnostic meta-learning (MAML) is proposed in [26]. This model is able to transfer knowledge between different intersections by focusing on action spaces and state spaces instead of traffic flow, for example, training an agent at a four-way intersection and testing it at a five-way intersection. To improve the generalization ability of traffic signal control models, [28] proposed a meta-RL framework called *GeneraLight*. *GeneraLight* enhances generalization performance by combining flow clustering parameters initialization with multi-modal MAML (MUMOMAML). Table 1 summarizes the comparison of factors that influence the evaluation of traffic signal control strategies: method, simulation environment, road network, and evaluation metrics. Recent studies have shown promising results when using reinforcement learning techniques for traffic signal control. However, the use of these techniques relies only on the *current* traffic conditions. Therefore, through our approach, we contribute several novel *sustainable* and *proactive* aspects to this line of research.

3   Formalization of the Problem

This section introduces the fundamental notions used to formalize the traffic signal control problem.

A road network consists of several junctions indexed by J. Each junction $j \in J$ consists of a number of in-roads, R_j. Note that the R_j are mutually disjoint, and denote $R = \cup_{j \in J} R_j$. Multi-lane roads with different turns, such as left- or right-turn-only lanes, are represented by multiple in-roads. Therefore, in-roads may model one or more lanes of traffic flow. A junction may serve different combinations of in-roads at the same time. It refers to service *phases* when several in-roads are maintained simultaneously. For a junction j, a service phase can be represented as a vector $\sigma = (\sigma_r, r \in j)$, where σ_r is the rate at which cars at j can be serviced by the in-road r. Specifically, $\sigma_r > 0$ if the in-road r is green during phase σ, or $\sigma_r = 0$ otherwise. Accordingly, at each time step t, the system has to determine how much time it will spend serving each phase in S_j over the next interval, with the constraint that each phase must last for some non-zero length of time. Where S_j denotes the set of phases at junction j.

4   EcoLight Approach

Deep reinforcement learning has proven to be a promising method for controlling traffic signal. By extending the previously proposed reinforcement learning solutions, we improve the robustness of the traffic signal control system by using future traffic noise predictions. Our proposed traffic signal control driven by noise prediction, namely *EcoLight*, takes advantage of all traffic features along with the predicted amount of future generated noise. Integrating these *sustainable* and *proactive* aspects into our deep RL Q-network will enhance its decision-making capabilities and raise the green awareness of the city's stakeholders. Figure 1 illustrates the final approach framework.

4.1   Traffic Noise Prediction

A time series is an ordered sequence of numerical observations collected and stored at regular intervals over time. It characterizes by its *"Frequency"* (the time separating two consecutive data points). Time-series data must be defined clearly and with equal frequency. The time intervals we most often deal with for traffic-related data are 1, 5, 10 to 60 min. According to the sequence-to-sequence

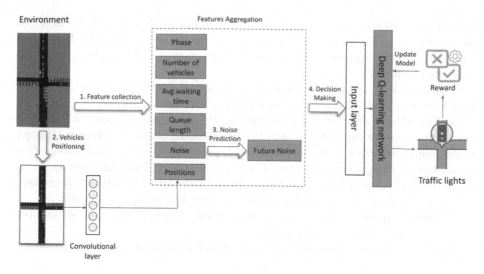

Fig. 1. EcoLight general framework.

architecture that we adopt in our algorithm, predicting hourly traffic noise would grant the input and output data as follows:

$$\text{Past hour} \xrightarrow{topredict} \text{Future hour}$$

$$[x_{(1,1)}, x_{(1,2)}, \ldots, x_{(1,s)}] \xrightarrow{topredict} [y_{(2,1)}, y_{(2,2)}, \ldots, y_{(2,s)}]$$

$$[x_{(2,1)}, x_{(2,2)}, \ldots, x_{(2,s)}] \xrightarrow{topredict} [y_{(3,1)}, y_{(3,2)}, \ldots, y_{(3,s)}]$$

$$\ldots \xrightarrow{topredict} \ldots$$

where $x_{(hour,observation)}$ and $y_{(hour,observation)}$ denote the past and future noise, respectively. And s represents the number of noise observations in one hour. Our approach embraces the Sequence to Sequence architecture to pre-process the time-series noise data. After splitting the time-series traffic data into fixed-sized sequences, we leverage an LSTM-based architecture to predict traffic noise of a future specific period (e.g., hourly, daily, etc.). Effectively it pinpoints long-term temporal dependencies accurately. We train and update the model using the back-propagation algorithm as an optimizer and a loss function to minimize the prediction error. Finally, we evaluate the model's predicted sequences, comparing them with the actual traffic noise ones using the prevalent evaluation metrics.

4.2 Traffic Signal Control

A reinforcement learning model consists of online and offline stages. A traffic state can be defined as a combination of five features: queue length, waiting time, number of vehicles, the vehicles' positions, and the phase. As soon as the prediction algorithm has been executed, the noise prediction will be explored as a state input to the model. Then, we use the reward to describe how much that action a has improved the traffic. In summary, the *EcoLight* approach is described as follows:

1. **Offline stage:** the traffic was allowed to flow through the system according to a fixed timetable to train the model and collect data samples.

2. **Online stage:** at every time interval Δ_t, the traffic signal agent will observe the state s from the environment and take action a according to ϵ-greedy strategy combining exploration (random action with probability ϵ) and exploitation (the estimation of the potential reward of doing this action given the state s).

3. **Memorization:** the agent will observe the environment and get the reward r from it. Then, the tuple (state, action, reward) will be stored in memory.

4. **Network update:** after several timestamps, the network will be updated according to the logs in the memory.

Algorithm 1 summarizes the steps of the reinforcement learning approach.

Algorithm 1. EcoLight: Traffic signal control

Require: predicted_roads_noise: predictions output; Simulation.
Ensure: CO2, Noise, Fuel_consumption
 1: Initialize action-value function Q
 2: Initialize updated Q'
 3: *Prnoise* extracted from predicted_roads_noise
 4: Initialize experience memory M
 5: Initialize the Agent to interact with the environment
 6: $\epsilon \leftarrow$ setting new Epsilon
 7: **for** (i=0; i < N; i++) **do**
 8:     **while** simulation not terminated **do**
 9:         Observe state s
10:         $s \leftarrow (Q\_leng, W\_time, N\_Veh, Pos\_veh, Prnoise)$
11:         With probability ϵ select action a_t
12:         Choose $QValues(M)$, action a
13:         Observe reward r, next state s_+
14:         Store transition(s,a,r,s_+) in M
15:     **end while**
16:     **if** UpdateTime **then**
17:         Update(network)
18:         Reset $Q' \leftarrow Q$
19:     **end if**
20: **end for**
21: Noise, CO2, Fuel_consumption $\leftarrow$ Evaluation(Simulation)
22: **return** Noise, CO2, Fuel_consumption

---

5   Experimental Evaluation

This section describes our experimental setup and evaluation process for comparing our *EcoLight* approach to pioneering baselines using real-world data.

5.1 Dataset

Experiments on real-world data are needed to determine *EcoLight*'s efficiency against the pioneering baselines. The Helsinki Region InfoShare [9] provided us with a complete database of urban traffic noise in Helsinki. The provided dataset is composed of several shapefiles [14], which present a storage format for geographic data between November 2011 and January 2012. These files can contain lines, points, polylines, and polygons representing different map features. Therefore, we performed a data transformation process to extract the complete traffic information, such as road names and noise values. The applied process can be resumed in these four steps: *(i)* convert the Helsinki OpenStreet map to shapefile (Fig. 2); *(ii)* project the noise file on the shapefile; *(iii)* using *QGIS3*, run the intersection tool to extract the full dataset noise and roads details; and finally *(iv)* export the intersection results to *.csv* file to be used for the noise prediction model.

(a) OpenStreetMap   (b) Shapefile

Fig. 2. Conversion of the Helsinki OpenStreet map to shapefile.

5.2 Experimental Setups

Our experiments carried out under the configuration of *Ubuntu 18.04.3 LTS* (CPU: *Intel Xeon Processor (Skylake)* × 8, RAM: 16Go), in which *Python* (3.7) and *Keras* (2.3.1) with the simulator *SUMO* [21] have been installed.

Prediction Settings. We adopt the use of a fully connected network of an *LSTM Tanh* activation layer with the size of 40 units and output layer *Sigmoid* activation layer for the prediction task. The *Adam* optimizer [11], as well as *mean squared error (MSE)* as the loss function, are used to fine tune the training model within 100 epochs for the three considered dataset splits according to the period of the day (Morning, Evening, and Night).

Simulation Settings. *"Lonnrotinkatu"* is the intersection in Helsinki that is chosen to create a network in *SUMO*. First, the simulation presents the environment, including the state. Then the **EcoLight** model, according to that state, will predict the action of the lights then get its reward (as depicted in Fig. 3). Table 2 presents the parameters setting of the model and reward coefficient hence the simulation. We found out that the action time interval Δ_t has minimal influence on the performance of our model as long as Δ_t is between 5 to 25 s.

5.3   Baseline Methods for Comparison

To accurately validate the performance of our proposed **EcoLight** approach, we led a comparison with the existing traffic signal control baseline methods; the Deep RL-based **IntelliLight** [24], a Max-green-based algorithm Priority-driven Enhanced Traffic Signal Scheduling Algorithm **PETSSA** [17], and the defaults fixed-time-based traffic signal control model in the *SUMO* simulator with no intervention **BASIC**. For the sake of a fair comparison, we tested all the baseline methods using the same datasets.

5.4   Evaluation

Noise Prediction: The prediction performance of our model compared to a time-series forecasting baseline are evaluated using the *mean squared error (MSE)* and the *mean absolute error (MAE)* defined respectively by (1) and (2).

Table 2. Simulation settings.

Parameter	Value
Model update interval	300 s
Action time interval Δ_t	5 s
γ for future reward	0.80
ϵ for exploration	0.05
Sample size	300
Memory length	1000

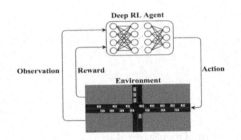

Fig. 3. Simulation process.

$$MSE = \frac{1}{J} \sum_{j=1}^{J} (n_j - \hat{n}_j)^2 \quad (1) \qquad MAE = \frac{1}{J} \sum_{j=1}^{J} |n_j - \hat{n}_j| \quad (2)$$

where J is the size of the tested junctions, n_j is the ground-truth junction's noise, and $\hat{n}_j$ is the predicted noise level yield by the model of the j-th junction.

Traffic Signal Control: Traffic poses a significant burden on society through its environmental impact, including air and noise pollution and the consumption of nonrenewable materials. With the use of *SUMO*, we can measure the generated pollution and the fuel consumption by using different models and interfaces. Among the information that can be obtained are: *(i) Trip information:* sum of pollutants emitted/fuel consumed by a single vehicle; *(ii) Lane emissions*: pollutants emitted and fuel consumed at a lane, aggregated over time; and *(iii) Lane noise*: noise generated along a lane, accumulated over a period of time.

Therefore, the traffic signal control performance evaluation of our approach against the pioneering ones is based on the emitted *noise, CO2 emissions*, and *fuel consumption* of each model on the considered dataset.

5.5   Results and Discussion

Table 3 glances the noise prediction performance of our **SeqtoSeq-LSTM** app-roach versus the **AutoRegressive Integrated Moving Average (ARIMA)** [4] non-parametric model using both mentioned evaluation metrics for each period of the day. This baseline combines the advantages of both autoregressive and moving average models in stationary random sequence analysis. In prac-tice, most time-series aren't stationary. **ARIMA** overcomes this limitation by introducing a differencing process [27]. A good look at our results underscores that our model sharply outperforms **ARIMA** in predicting future noise with high improvement percentages for both morning and night periods of the day. Notwithstanding, the **ARIMA** model gives a slightly similar performance to our proposed model for the evening period of the day. In the sequel, we evaluate the effectiveness of our **EcoLight** traffic signal control in response to several environmental and economic factors.

Table 3. Noise prediction performance.

Model	MAE			MSE		
Evaluation	Morning	Evening	Night	Morning	Evening	Night
ARIMA	65.89	2.31	72.94	4439.42	11.24	5537.93
SeqtoSeq-LSTM	**1.15**	**1.07**	**1.62**	**6.94**	**6.39**	**10.27**

Effectiveness over Traffic Noise. From the achieved results (Table 4), the **BASIC** shows the worst performance on the considered intersection as it is based on a fixed-timing strategy that does not adapt according to current and poten-tial future situation of the traffic. The results underscore that the **PETSSA** model reduces better the noise level for both lanes of the fourth in-road of the intersection. Figure 4(a) depicts the improvement percentages of **IntelliLight**, **PETSSA**, and **EcoLight** models compared to the **BASIC** logic strategy. Over-all, our proposed approach outperforms all the baselines for the produced noise at the considered intersection.

Table 4. Produced noise performance.

Model	Lane11	Lane12	Lane21	Lane22	Lane31	Lane32	Lane41	Lane42
Basic	70.38	69.38	72.86	69.54	68.91	71.57	70.25	70.55
PETSSA	70.20	67.99	72.80	68.28	67.88	70.38	**67.24**	**68.50**
IntelliLight	70.09	67.95	72.94	67.92	68.53	70.37	68.02	69.00
EcoLight	**68.77**	**67.90**	**72.62**	**67.08**	**67.52**	**68.09**	69.82	68.92

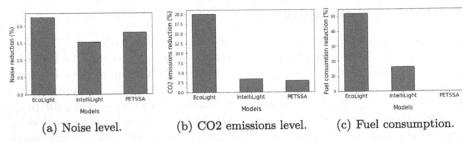

(a) Noise level.   (b) CO2 emissions level.   (c) Fuel consumption.

Fig. 4. Reduction vs. BASIC.

Effectiveness over CO2 Emission. According to our approach, significant reductions in CO2 are recorded for the majority of lanes compared to the other baselines (Fig. 4(b)). Although **EcoLight** isn't the best for some lanes, its performance is barely worse than the best achieved by the **IntelliLight** (Table 5). As depicted in Fig. 4(b), the improve rates of **IntelliLight**, **PETSSA**, and **EcoLight** models are comparable to those of the **BASIC**.

Table 5. Produced CO2 emission performance.

Model	Lane11	Lane12	Lane21	Lane22	Lane31	Lane32	Lane41	Lane42
BASIC	74,579,545.10	19,333,168.50	145,881,252.44	16,540,482.62	49,821,628.05	78,431,824.14	18,681,756.18	26,546,180.90
PETSSA	73,854,895.81	**18,137,521.04**	143,853,266.32	15,721,922.60	48,154,074.55	75,249,215.03	17,950,045.97	24,612,349.73
IntelliLight	73,954,895.81	**18,137,521.04**	142,853,266.32	**15,521,922.60**	49,254,074.55	75,249,215.03	**17,850,045.97**	**22,712,349.73**
EcoLight	**62,053,611.60**	18,692,031.10	**106,341,550.52**	15,628,314.04	**41,154,824.82**	**58,907,941.13**	18,167,102.88	22,801,416.51

Effectiveness over Fuel Consumption. A comparison of the improvement percentages of fuel consumption by **IntelliLight**, **PETSSA**, and **EcoLight** models to that of **BASIC** logic is shown in Fig. 4(c). **PETSSA** performs the same as **BASIC** with no improvement in terms of fuel consumption. We notice that the **IntelliLight** model gives a significant power reduction in two different lanes on the considered intersection (as shown in Table 6). While operating **EcoLight**, vehicular fuel consumption can be reduced by more than 50%.

Table 6. Produced fuel consumption performance.

Model	Lane11	Lane12	Lane21	Lane22	Lane31	Lane32	Lane41	Lane42
BASIC	32,925.02	9,759.01	105,496.35	3,357.92	20,427.88	55,544.55	5,723.92	74,821.11
PETSSA	32,925.02	9,759.01	105,496.35	3,357.92	20,427.88	55,544.55	5,723.92	74,821.11
IntelliLight	32,157.82	10,171.84	90,043.45	**3,303.29**	20,325.73	38,920.39	**5,498.10**	58,134.54
EcoLight	**26,675.58**	**8,034.87**	**45,713.41**	6,717.90	**17,691.55**	**25,323.00**	7,809.22	**9,801.32**

6   Conclusion

In this paper, we introduced an eco-friendly traffic signal control driven by urban noise prediction, namely *EcoLight*. We address the traffic signal control problem using a well-designed deep reinforcement learning approach that integrates future noise predictions. We conduct our experiments on Helsinki's geographical data. The yielded results provide evidence for the reliability and sustainability of the use of future noise predictions. Indeed, carried out experiments underscore the incapacity of the baselines to perform better in terms of noise, CO2 emissions, and fuel consumption compared to our *EcoLight* approach.

We point out a critical future direction to make *EcoLight* more relevant to the real world. The *EcoLight* is designed and tested to consider a simplified case of one intersection in Helsinki, whereas real-world network design is significantly more complex. Multiple intersections have been addressed by combining several reinforcement learning agents at a limited number of intersections. Meanwhile, sales of electric cars jumped 43% to more than 3.2 million of 370 different car models in 2020 [6]. This type of vehicles tend to be environmentally friendly and provide less noise. Future work will seek to improve the reduction by proposing a hybrid approach that enhances our *EcoLight* with traffic-related features prediction other than noise, combined with the *PETSSA* method to benefit from the Max-green strategy to reduce delay times, thereby limiting congestion levels.

Acknowledgment. This work was supported by grants to TalTech - TalTech Industrial (H2020, grant No 952410) and Estonian Research Council (PRG1573).

References

1. Ahmad Rafidi, M.A., Abdul Hamid, A.H.: Synchronization of traffic light systems for maximum efficiency along jalan bukit gambier, penang, malaysia. SHS Web Conf. **11**, 01006 (2014). https://doi.org/10.1051/shsconf/20141101016
2. Ahmed, A.A., Pradhan, B., Chakraborty, S., Alamri, A., Lee, C.W.: An optimized deep neural network approach for vehicular traffic noise Trend modeling. IEEE Access **9**(1995), 107375–107386 (2021). https://doi.org/10.1109/ACCESS. 2021.3100855
3. Alaidi, A.H., Aljazaery, I., Alrikabi, H., Mahmood, I., Abed, F.: Design and implementation of a smart traffic light management system controlled wirelessly by arduino. Int. J. Inter. Mobile Technol. (iJIM) **14**(07), 32–40 (2020)
4. Box, G.E.P., Pierce, D.A.: Distribution of residual autocorrelations in autoregressive-integrated moving average time series models. J. Am. Stat. Assoc. **65**(332), 1509–1526 (1970). https://doi.org/10.1080/01621459.1970.10481180
5. Bravo, Y., Ferrer, J., Luque, G., Alba, E.: Smart mobility by optimizing the traffic lights: a new tool for traffic control centers. In: Alba, E., Chicano, F., Luque, G. (eds.) Smart-CT 2016. LNCS, vol. 9704, pp. 147–156. Springer, Cham (2016). https://doi.org/10.1007/978-3-319-39595-1_15
6. CALSTART: Drive to zero's zero-emission technology inventory (zeti) (2020). https://globaldrivetozero.org/tools/zero-emission-technology-inventory/

7. Chen, C., et al.: Toward a thousand lights: Decentralized deep reinforcement learning for large-scale traffic signal control. In: Proceedings of the AAAI Conference on Artificial Intelligence, vol. 34, pp. 3414–3421 (2020)
8. EEA: Road traffic remains biggest source of noise pollution in europe (2017). https://www.eea.europa.eu/highlights/road-traffic-remains-biggest-source
9. Helsinki, E.O.: Helsinki region infoshare (May 2022). https://hri.fi/
10. Khan, J., Ketzel, M., Jensen, S.S., Gulliver, J., Thysell, E., Hertel, O.: Comparison of Road Traffic Noise prediction models: CNOSSOS-EU, Nord 2000 and TRANEX. Environ. Pollut. **270**, 116240 (2021). https://doi.org/10.1016/j.envpol.2020.116240
11. Kingma, D.P., Ba, J.: Adam: A method for stochastic optimization. In: Bengio, Y., LeCun, Y. (eds.) 3rd International Conference on Learning Representations, ICLR 2015, San Diego, CA, USA, 7–9 May 2015, Conference Track Proceedings (2015)
12. Le, T., Kovács, P., Walton, N., Vu, H.L., Andrew, L.L., Hoogendoorn, S.S.: Decentralized signal control for urban road networks. Trans. Res. Part C: Emer. Technol. **58**, 431–450 (2015). https://doi.org/10.1016/j.trc.2014.11.009
13. Liu, Q., Cai, Y., Jiang, H., Lu, J., Chen, L.: Traffic state prediction using ISOMAP manifold learning. Phys. A **506**, 532–541 (2018). https://doi.org/10.1016/j.physa.2018.04.031
14. Lonnrotinkatu: Helsinki metropolitan traffic noise dataset, January 2012. https://hri.fi/
15. Ng, S.C., Kwok, C.P.: An intelligent traffic light system using object detection and evolutionary algorithm for alleviating traffic congestion in hong kong. Int. J. Comput. Intell. Syst. **13**(1), 802–809 (2020). https://doi.org/10.2991/ijcis.d.200522.001
16. Ounoughi, C., Yeferny, T., Ben Yahia, S.: Zed-tte: zone embedding and deep neural network based travel time estimation approach. In: 2021 International Joint Conference on Neural Networks (IJCNN), pp. 1–10 (2021). https://doi.org/10.1109/IJCNN52387.2021.9533456
17. Salin, S.: Petssa: Priority-driven enhanced traffic signal scheduling algorithm, May 2022. https://github.com/habe33/tammsaare-sopruse
18. Sanvicente, E., Kielmanowicz, D., Rodenbach, J., Chicco, A., Ramos, E.: Key technology and social innovation drivers for car sharing. deliverable 2.2 of the stars h2020 project. Tech. rep. (2020)
19. Singh, D., Upadhyay, R., Pannu, H.S., Leray, D.: Development of an adaptive neuro fuzzy inference system based vehicular traffic noise prediction model. J. Ambient. Intell. Humaniz. Comput. **12**(2), 2685–2701 (2021). https://doi.org/10.1007/s12652-020-02431-y
20. Staab, J., Schady, A., Weigand, M., Lakes, T., Taubenböck, H.: Predicting traffic noise using land-use regression-a scalable approach. J. Ex. Sci. Environ. Epidemiol. **32**, 1–12 (2021). https://doi.org/10.1038/s41370-021-00355-z
21. SUMO: Simulation of urban mobility, May 2022. https://sumo.dlr.de/docs/index.html
22. Wei, H., Chen, C., Zheng, G., Wu, K., Gayah, V., Xu, K., Li, Z.: Presslight: learning max pressure control to coordinate traffic signals in arterial network. In: Proceedings of the 25th ACM SIGKDD International Conference on Knowledge Discovery & Data Mining, KDD 2019. pp. 1290–1298. Association for Computing Machinery, New York (2019). https://doi.org/10.1145/3292500.3330949, https://doi.org/10.1145/3292500.3330949

23. Wei, H., et al.: Colight: learning network-level cooperation for traffic signal control. In: Proceedings of the 28th ACM International Conference on Information and Knowledge Management, CIKM 2019, pp. 1913–1922. Association for Computing Machinery, New York (2019). https://doi.org/10.1145/3357384.3357902, https://doi.org/10.1145/3357384.3357902

24. Wei, H., Zheng, G., Yao, H., Li, Z.: Intellilight: a reinforcement learning approach for intelligent traffic light control. In: Proceedings of the 24th ACM SIGKDD International Conference on Knowledge Discovery & Data Mining, KDD 2018, pp. 2496–2505. Association for Computing Machinery, New York (2018). https://doi.org/10.1145/3219819.3220096

25. Xiong, Y., Zheng, G., Xu, K., Li, Z.: Learning traffic signal control from demonstrations. In: Proceedings of the 28th ACM International Conference on Information and Knowledge Management, CIKM 2019, pp. 2289–2292. Association for Computing Machinery, New York (2019). https://doi.org/10.1145/3357384.3358079, https://doi.org/10.1145/3357384.3358079

26. Zang, X., Yao, H., Zheng, G., Xu, N., Xu, K., Li, Z.: Metalight: value-based meta-reinforcement learning for traffic signal control. In: Proceedings of the AAAI Conference on Artificial Intelligence, vol. 34, pp. 1153–1160 (2020)

27. Zhang, B., Zhao, C.: Dynamic turning force prediction and feature parameters extraction of machine tool based on arma and hht. Proc. Institu. Mech. Eng. Part C: J. Mech. Eng. Sci. **234**(5), 1044–1056 (2020)

28. Zhang, H., Liu, C., Zhang, W., Zheng, G., Yu, Y.: Generalight: improving environment generalization of traffic signal control via meta reinforcement learning. In: Proceedings of the 29th ACM International Conference on Information & Knowledge Management, pp. 1783–1792 (2020)

29. Zhang, X., Kuehnelt, H., De Roeck, W.: Traffic noise prediction applying multivariate bi-directional recurrent neural network. Appli. Sci. (Switzerland) **11**(6) (2021). https://doi.org/10.3390/app11062714

30. Zheng, G., et al.: Learning phase competition for traffic signal control. In: Proceedings of the 28th ACM International Conference on Information and Knowledge Management, CIKM 2019 pp. 1963–1972. Association for Computing Machinery, New York, (2019). https://doi.org/10.1145/3357384.3357900, https://doi.org/10.1145/3357384.3357900

Mining Fluctuation Propagation Graph Among Time Series with Active Learning

Mingjie Li[1], Minghua Ma[2], Xiaohui Nie[3], Kanglin Yin[3], Li Cao[3],
Xidao Wen[1]([✉]), Zhiyun Yuan[4], Duogang Wu[4], Guoying Li[4], Wei Liu[4],
Xin Yang[4], and Dan Pei[1]

[1] Tsinghua University, Beijing, China
wenxidao@mail.tsinghua.edu.cn
[2] Microsoft Research Asia, Beijing, China
[3] BizSeer, Beijing, China
[4] China Construction Bank, Beijing, China

Abstract. Faults are inevitable in a complex online service system. Compared with the textual incident records, the knowledge graph provides an abstract and formal representation for the empirical knowledge of how fluctuations, especially faults, propagate. Recent works utilize causality discovery tools to construct the graph for automatic troubleshooting but neglect its correctness.

In this work, we focus on structure discovery of the fluctuation propagation graph among time series. We conduct an empirical study and find that the existing methods either miss a large proportion of relations or discover almost a complete graph. Thus, we propose a relation recommendation framework named *FPG-Miner* based on active learning. The experiment shows that operators' feedback can make a mining method to recommend the correct relations earlier, accelerating the trustworthy application of intelligent algorithms like automatic troubleshooting. Moreover, we propose a novel classification-based approach named *CAR* to speed up relation discovery. For example, when discovering 20% correct relations, our approach shortens 2.3–42.2% of the verification quota compared with the baseline approaches.

Keywords: Fluctuation propagation graph · Causal discovery · Active learning · Online service systems

1 Introduction

Faults are inevitable in complex online service systems. Currently, operators summarize how they locate the root cause in the form of text for each concrete fault, *e.g.*, the *troubleshooting guide* [12]. However, it can be hard to utilize the text for automated troubleshooting. In contrast, a *fluctuation propagation graph* (FPG) is an abstract and formal representation of the empirical knowledge towards automatic troubleshooting. An FPG describes how fluctuations,

---

Most work was done when Minghua Ma and Xiaohui Nie were at Tsinghua University.

ⓒ The Author(s), under exclusive license to Springer Nature Switzerland AG 2022
C. Strauss et al. (Eds.): DEXA 2022, LNCS 13426, pp. 220–233, 2022.
https://doi.org/10.1007/978-3-031-12423-5_17

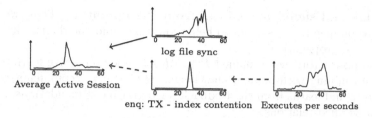

Fig. 1. Part of the FPG with four monitoring metrics for an Oracle database. Above each metric name presents the time series at the same period. Time (horizontal axis) is shown in minutes.

including faults, propagate among monitoring variables. In literature, the concept of FPG has already been used in many works, *e.g.*, locating root causes automatically [1,17,19,30,32], discovering alert correlation [29], and handling alert storms [37].

A *time series* is a chronological sequence of values for the same metric. Time series can be easily understood for operators and is the most widely available data for operation work. Meanwhile, many previous works convert logs and alerts into time series for analysis and visualization [9,18]. Thus, this work focuses on structure discovery of the FPG among time series.

Figure 1 shows a real scenario of the FPG for the Oracle database, collected from our collaboration with the database administrators (DBAs). The Oracle database exposes plenty of metrics, measuring resource usage, counting events, timing duration of each task, and recording any other status of a database instance [23]. The performance of a database degrades significantly when the Average Active Session (AAS) is too high. Some events may contribute to the high AAS, such as 1) "log file sync", *i.e.*, the database writer process waits for the log file to synchronize with the database, and 2) "enq: TX - index contention", *i.e.*, a transaction waits for an index used by another transaction. "enq: TX - index contention" can be the consequence of high workload, indicated by the number of executes per second (EPS) of SQL commands. Thus, certain performance degradation may result from propagation from high EPS to high AAS, shown as the dashed path in Fig. 1.

There are mainly two ways to construct an FPG in the literature. Some works construct the graph manually [30,33]. Expert operators reach a consensus on the graph based on their domain knowledge. Many recent works have attempted to learn the graph from monitoring data [1,4,6,17,19,31,32], neglecting its correctness. For example, the PC algorithm [13] is widely used [1,4,17,31].

FPG construction faces two main challenges. **The first challenge comes from the lack of effective tools for unsupervised mining**. Our empirical study (Sect. 3) shows that using existing mining methods for FPG is unsatisfactory. Meanwhile, a graph-based algorithm may fail to achieve its goal, *e.g.*, it fails to locate the root cause. In such an out-of-the-loop situation [8], a trustworthy FPG can still provide basic situation awareness for operators. **The second challenge is that relation verification requires extensive domain**

knowledge and significant efforts from the operators. Thus, we need a relation recommendation system to help operators build the domain knowledge of fluctuation propagation.

We propose a framework named *FPG-Miner* based on active learning, combining data mining with domain knowledge. *FPG-Miner* recommends relations to operators and learns from the feedback for better recommendations, accumulating the verified relations.

Moreover, we propose a novel approach, *CAR*, to implement *FPG-Miner*. *CAR* partitions the time series into small windows and capture the correlation between every two metrics in each window. The temporary correlation provides the basis for statistical features. Further, *CAR* takes XGBoost [5] as a supervised classifier to recommend unverified relations, utilizing accumulated feedback.

We alter several methods in our empirical study for *FPG-Miner* as baseline approaches. In the experiment, we simulate operators' feedback to compare different approaches based on two real-world datasets. The result validates that *FPG-Miner* can enhance mining performance. Moreover, *CAR* outperforms baseline approaches.

We conclude our contributions as follows.

1. We conduct an empirical study to evaluate the gap, neglected in the literature, between a mined FPG and the ground truth. The existing methods either miss a large proportion of relations or discover almost a complete graph on two real-world datasets.
2. Due to the gap mentioned above, we design an FPG construction framework named *FPG-Miner* to accelerate relation discovery by active learning.
3. We propose *CAR*, a novel implementation for *FPG-Miner* based on XGBoost. The experiment shows that *CAR* speeds up relation discovery.

2   Related Work

Causal Discovery. We consider FPG construction as a causal discovery problem. Many causal discovery methods have been proposed [7,13,14,20,35,38]. Besides synthetic datasets, some works also use real-world datasets from other fields for evaluation, such as biology [38] and geography [25]. Readers can find thorough discussion in the recent survey [10]. To obtain a more rational causal graph for online service system operations, CauseInfer [4] enforces TCP latency as the common descendant of other metrics in the same service.

Active Learning. The intuition behind active learning is that the learner can perform better with less labeled data if it can choose what to learn [27]. We borrow the idea from active learning to discover correct relations as early as possible. A basic active learning strategy is to learn from the most relevant data points [26]. However, this strategy suffers from learning those that an active learning model already knows. A natural solution is to learn from the most uncertain data points, named *uncertainty sampling* [15]. Readers can find more information on active learning from the survey [27] and the recent tutorial [3].

Graph for Troubleshooting. There are similar concepts to the FPG in litera-ture. The *diagnosis graph* [33] is named by its functionality for troubleshooting. In contrast, the *attributed graph* [32] emphasizes the origin of service dependency and deployment location. Some works use the *causal(ity) graph* [1,4,19,21] or the *impact graph* [31] according to the property of fluctuation propagation. The service dependency graph is also an FPG [16] but more coarse than the graph among metrics discussed in this work.

3 Empirical Study of Mining Methods

In this section, we compare different mining methods empirically. Following is the research question.

RQ1 How do existing mining methods perform among monitoring metrics?

3.1 Experimental Setup

Dataset. We adopt two datasets in this work. The metrics in each dataset make up a directed graph with relations as the edges. A *positive* sample refers to a relation in the ground truth graph. In both datasets, the reverse relation of a positive sample is not in the graph, *i.e.*, it is a *negative* sample.

The **Oracle database dataset** ($\mathcal{D}_{OD}$) comes from a top global commercial banking system with many services. Each service utilizes two exclusive Oracle database instances for data management. We choose one database instance with a real workload for the empirical study before digging into the data.

$\mathcal{D}_{OD}$ includes 51 kinds of metrics. Each time series contains 1040 data points with an interval of 6 min. We invited DBAs of the target system to label the relations according to their expert knowledge. They labeled 490 relations that are part of the ground truth. Among those labeled relations, 210 are positive, such as the relation between "log file sync" and "AAS" in Sect. 1. On the other hand, both directions of the rest 280 labeled relations are negative.

The **telecommunication network dataset** ($\mathcal{D}_{TN}$) is publicly available, collected from real telecommunication networks [11]. $\mathcal{D}_{TN}$ contains the time series for 55 kinds of anonymous variables, which count the numbers of different alarms in 10 min. The underlying causal relations are provided according to expert experience, among which 563 are positive. The original dataset covers more than five months. We filter in four weeks in our experiment as the whole dataset takes too long for some mining methods to finish. Each time series in the final dataset contains 4032 data points.

Mining Methods. We adopt four representative groups of methods to explore the mining performance to obtain the FPG, as shown in Table 1. An intuitive group of methods for constructing the FPG among metrics is correlation analysis. Causality considers confounders to rule out spurious relations [22], which suits the FPG better than correlation. As a result, we compare three groups of causal discovery methods as suggested by a recent survey [10]: constraint-based, score-based, and FCM (Functional Causal Model) based.

Table 1. Comparison among existing mining methods

Group	Methods
Correlation	Pearson correlation (*Pearson-r* and *Pearson-p*), Cross-Correlation (*CC*) [29], CoFlux [29]
Constraint-based	PC (*PC-gauss* [13] and *PC-RCIT* [28]), PCTS [19] (*PCTS-PCMCI* [25] and *PCTS-PCMCI+* [24])
Score-based	GES [7]
FCM-based	NOTEARS [38], NRI [14], TCDF [20]

Evaluation Metrics. We adopt the classical *Precision, Recall*, and *F1-score* metrics to evaluate the performance of each method. In terms of efficiency, we record the execution time, denoted as *Time Cost*. Denote the ground truth graph as $\mathcal{G}_G = <V, E_G>$ and the mined graph as $\mathcal{G}_M = <V, E_M>$, where V is the set of variables and E_G (E_M) is a set of directed edges among V. The output of each method contains four parts: True Positives ($TP = |E_G \cap E_M|$), True Negatives, False Positives ($FP = |E_M \setminus E_G|$), and False Negatives ($FN = |E_G \setminus E_M|$). Precision, Recall, and F1-score are further calculated by Eq. (1). As for the $\mathcal{D}_{OD}$, we cast TP, FP, and FN on the labeled edges E_L in evaluation, *i.e.*, $TP' = |E_L \cap E_G \cap E_M|$, $FP' = |E_L \cap E_M \setminus E_G|$, and $FN' = |E_L \cap E_G \setminus E_M|$.

$$Precision = TP/(TP + FP) \tag{1a}$$

$$Recall = TP/(TP + FN) \tag{1b}$$

$$F1\text{-}score = 2 \times Precison \times Recall/(Precision + Recall) \tag{1c}$$

3.2 Results

The experiment is conducted on an Ubuntu server with 22 cores, 57 GB memory, x86-64 architecture. Only the implementation of NRI and TCDF is compatible with GPU. Thus, we conduct the whole experiment with the CPU only for a fair comparison of execution time.

Each method in the experiment *suffers from either a low precision or low discovery ability on both datasets*, as shown in Table 2. Existing methods fail to achieve a precision higher than 0.5 on both datasets. Meanwhile, the methods with the highest precision have intolerably low discovery ability. On the other hand, a longer execution time cannot guarantee better performance.

One reason for the bad performance is the lack of domain knowledge during the mining process. For example, the relation between "enq: TX - index contention" and EPS in Fig. 1 is not linear, *i.e.*, there are no wait events until the workload achieves a certain high volume. As a result, methods with linear models such as NOTEARS [38] cannot handle the relations well. As for the deep learning models like NRI [14], it is hard to localize their "bugs" [34,36].

Table 2. Comparison among existing mining methods

Method	$\mathcal{D}_{OD}$				$\mathcal{D}_{TN}$			
	Precision	Recall	F1-score	Time Cost	Precision	Recall	F1-score	Time Cost
Pearson-r	0.206	0.348	0.259	<1 s	**0.500**	0.007	0.014	<1 s
Pearson-p	0.214	0.890	0.345	<1 s	0.236	0.416	0.301	<1 s
CC [29]	0.225	0.638	0.333	<1 s	0.417	0.009	0.017	5 s
CoFlux [29]	0.142	0.095	0.114	0:01:35	0.184	**0.535**	0.274	0:05:30
PC-gauss [13]	0.203	0.062	0.095	~1 s	0.262	0.066	0.105	15 s
PC-RCIT [13,28]	0.133	0.019	0.033	0:32:41	0.300	0.027	0.049	1:12:43
PCTS-PCMCI [19,25]	0.217	**0.952**	**0.353**	0:03:18	0.228	0.496	**0.312**	0:12:32
PCTS-PCMCI+ [19,24]	0.235	0.243	0.239	0:23:40	0.229	0.410	0.294	3:16:24
GES [7]	0.248	0.257	0.252	~1 s	0.213	0.105	0.140	~1 s
NOTEARS [38]	0.127	0.090	0.106	1:39:22	0.309	0.030	0.055	0:06:25
NRI [14]	0.213	0.252	0.231	>1 day	0.277	0.346	0.308	>4 d
TCDF [20]	**0.333**	0.010	0.019	0:03:14	0.357	0.027	0.050	0:04:41

Algorithm 1. Mine the FPG with active learning

1: **procedure** MINE($data, n$)   ▷ n is the number of recommendations per iteration
2:     $relations \leftarrow \emptyset$
3:     $miner \leftarrow$ TRAIN($data$)
4:     **repeat**
5:         $candidates \leftarrow miner$.RECOMMEND($n$)
6:         **for all** $relation \in candidates$ **do**
7:             **if** Operators confirm $relation$ **then**
8:                 $relations \leftarrow relations \cup \{relation\}$
9:         $miner$.LEARN($data, relations$)
10:    **until** Stopping criteria is satisfied
11:    **return** $relations$

In contrast, experienced operators can tell a relation from a spurious one based on their rich domain knowledge. In the discussion on the labels of $\mathcal{D}_{OD}$, DBAs refer to historical troubleshooting cases, advice from Oracle customer support, and other information in memory as proof. Thus, we propose to bring operators' feedback (missing knowledge in the data) into the mining procedure.

4   *FPG-Miner*: Mine with Active Learning

We propose a framework named *FPG-Miner* to mine the FPG among time series with domain knowledge, as described in Algorithm 1. The framework integrates three core steps—training, recommendation, and learning—into a whole process called a *miner*. For each recommendation $A \rightarrow B$ (A and B stand for metrics), a miner expects one of the following three feedback from the operators: 1) $A \rightarrow B$ is correct (and $B \rightarrow A$ is a negative sample), 2) $A \rightarrow B$ is reversed, *i.e.*, $B \rightarrow A$ is positive, and 3) both $A \rightarrow B$ and $B \rightarrow A$ are negative. The miner will learn from the feedback and recommend new relations. Verification can cost a lot of time. Hence, the recommendation procedure contains multiple iterations to achieve an incremental application of verified relations.

In the rest of this section, we will first explain the rationale behind this general active learning framework (Sect. 4.1). After that, we provide a novel implementation for the training step (Sect. 4.2).

4.1   Recommendation Framework

Ideally, a miner should recommend correct relations (including reversed ones) in preference to incorrect ones. The process can stop after operators confront the first incorrect recommendation. It is hard to achieve such an ideal recommendation process. Each practical miner may mix correct and incorrect relations. As a result, an incorrect recommendation is insufficient to tell whether we have discovered all the positive relations. Thus, the process has to continue after the first incorrect recommendation arises (Line 10 in Algorithm 1).

The natural criterion is that operators have verified all of the relations. Given the number of metrics N, the number of relations is bounded by $N(N-1)/2$. Thus, the process will terminate after $\lceil N(N-1)/(2n) \rceil$ iterations, where n is the number of relations to recommend per iteration.

A miner shall learn from mistakes to avoid new incorrect recommendations, shortening the overall verification times to discover each positive relation (Line 9 in Algorithm 1). Inspired by the uncertainty sampling in active learning research [15, 27], recommending uncertain relations may bring more information to the miner for long-term benefit. A miner is supposed to provide confidence between 0 and 100% for each relation, encoding the labels of verified ones. Based on the confidence, we consider the following strategies.

Confidence-First. A miner first recommends the relation with the highest confidence, aiming at filtering out the unimportant or spurious relations.

Uncertainty-First. A miner first recommends the most uncertain relation to improve itself. A straightforward uncertainty measurement is the distance between confidence and 50%.

Mixed. The mixed strategy combines the two strategies above. Specifically, every $n = 3$ relations that a miner recommends for verification contain two with the highest confidence and one with the highest uncertainty.

Random. The random strategy is the baseline strategy. It is also applied when more than one relations share the same highest confidence or uncertainty.

4.2   Continuous Association Rule Classifier

Mining methods used in Sect. 3 are designed for the unsupervised task. We can alter those methods as miners for *FPG-Miner*. Moreover, we propose the Continuous Association Rule (*CAR*) classifier, as shown in Fig. 2.

Inspired by association rule mining [2], *CAR* calculates statistical features, such as supports, for each ordered pair of metrics, *i.e.*, directed relations. As *CAR* does not filter out any relations, *e.g.*, based on some thresholds like the minimum support, all positive relations remain in our consideration. Meanwhile, we design

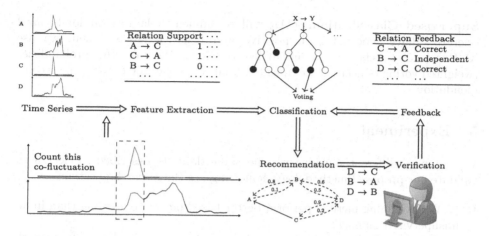

Fig. 2. The overview of *CAR* under the framework of *FPG-Miner*

Table 3. Features of a directed relation

Feature $(A \to B)$	Definition	Feature $(A \to B)$	Definition
Coverage	$P(A)$	Support	$P(AB)$
Consequence coverage	$P(B)$	Lift	$P(AB)/[P(A)P(B)]$
Confidence	$P(B\|A)$	IR	$P(A)/P(B)$
Reversed confidence	$P(A\|B)$	KULC	$[P(B\|A) + P(A\|B)]/2$

a novel approach to calculate features from time series directly, different from counting transactions in the original association rule mining.

Feature Extraction. The natural fluctuation of time series can be so large that it conceals causal relations. For example, there are spikes in both the index-contention event number and the EPS at the 30th minute in Fig. 1. Meanwhile, the EPS alone has another one at the 42nd minute. The second spike increases the outlier number of the EPS, having no contribution to the correlation of the two metrics, but is misleading. The intuition behind *CAR* is to capture the co-fluctuation of the causal metric and its effect when the causal one changes large enough, *e.g.*, the spikes at the 30th minute.

We partition each time series into sliding windows with the size of L_E, *e.g.*, $L_E = 10$. L_E implies how long we assume that the pattern of any time series is static. In each sliding window, we calculate the Pearson p-value pair-wisely. Two metrics are taken as correlated in this window if the p-value is less than α, *e.g.*, $\alpha = 0.05$. The *Support* value of every two metrics is the ratio of correlated windows. We count the ratio of windows correlated with any other metrics as the *Coverage* of a given metric. The features in Table 3 are calculated based on the *Support* and *Coverage*. For example, *Confidence* is defined as the *Support* divided by the *Coverage*, *i.e.*, $P(B|A) = P(AB)/P(A)$.

Supervised Classification. CAR will recommend relations randomly until operators have reported both positive and negative labels. Then, we use XGBoost [5] to classify the unlabeled relations. We take the probabilities (weighted voting of decision trees) as the final confidence for CAR in the recommendation.

5 Experiment

We compare different miners with the same datasets described in Sect. 3 to validate our proposed methodology. Following are the research questions.

RQ2 Will a mining method perform better based on active learning than in an unsupervised manner?

RQ3 How does CAR perform compared with other miners under the framework of *FPG-Miner*?

RQ4 Are there some relations more important than other ones?

5.1 Experimental Setup

Miners. We alter PC-gauss, GES, and NRI as miners for comparison, which represent three kinds of causal discovery methods [10], respectively. We classify these miners and their variants into three groups.

Static miners recommend relations in the predefined order without learning. In Algorithm 1, operators may confirm each relation once. Thus, mining with active learning will find more relations than in an unsupervised manner. We wrap PC-gauss, GES, and NRI as static miners to compare the two manners fairly. Moreover, we adopt a random miner with equal probability for related or not, denoted as *Random*.

PC and GES provide only binary output, *i.e.*, the existence of a relation. A *binary miner* utilizing such a mining method provides confidence of one or zero. Hence, a binary miner recommends randomly from relations it considers positive, *i.e.*, the confidence-first strategy.

Probabilistic miners calculate probabilities as confidence, supporting various recommendation strategies. NRI can provide voting from time windows as confidence for each relation. Meanwhile, the XGBoost model of CAR provides weighted voting from decision trees. The NRI miner tunes its neural network for two epochs based on existing parameters in each learning step. In contrast, CAR trains a new classifier with all the available labels. We choose the recommendation strategy for the best performance.

Evaluation Metrics. We simulate Algorithm 1, and miners interact directly with the ground truth. In each iteration, $n = 3$ relations are presented to the mock operators for labeling. The simulation stops when a miner has recommended all the labeled positive relations.

Table 4. Comparison among miners without/with active learning

Miner	Learning	$\mathcal{D}_{OD}$						$\mathcal{D}_{TN}$					
		AUC	T@k					AUC	T@k				
			10%	20%	30%	50%	100%		10%	20%	30%	50%	100%
PC-gauss	Without	0.589	47	**99**	161	291	490	**0.639**	106	248	407	703	1483
	With	**0.648**	41	102	**145**	**237**	489	0.619	112	259	428	746	1485
GES	Without	**0.690**	28	**76**	125	214	490	**0.651**	90	223	370	684	1479
	With	0.639	46	87	142	244	**488**	0.636	128	227	401	720	1483
NRI	Without	0.589	74	118	175	273	488	0.658	138	291	407	633	1485
	With	**0.741**	53	**83**	**113**	**192**	478	**0.731**	85	**177**	**285**	**575**	1482

Let $C(i)$ be the number of correct undirected relations among the first i recommendations. Denote the total number of correct undirected relations with labels as N_C. The ideal series of $C(i)$ is $C^*(i)$, as shown in Eq. (2). *Area Under Curve* (AUC) compares $C(i)$ against $C^*(i)$, as shown in Eq. (3). A high AUC indicates that a miner can learn FPG quickly. *T@k* is the number of times it takes a miner to recommend k correct relations, *i.e.*, $C(T@k) = k$. The lower T@k indicates that the miner can discover correct relations faster at the beginning.

$$C^*(i) = \begin{cases} i & \text{if } 1 \le i \le N_C \\ N_C & \text{if } i > N_C \end{cases} \tag{2}$$

$$AUC = \frac{\sum_{i=1}^{N_C} C(i)}{\sum_{i=1}^{N_C} C^*(i)} \tag{3}$$

The evaluation metrics in this section are different from those in Sect. 3. We argue that the operators have to verify each relation in the FPG. As a result, the verified ones will have a precision of 100%, which becomes trivial in comparison. Meanwhile, operators have limited time to verify relations. The recall metric measures the number of relations a miner discovers (k) given a certain verification quota (q) during the journey to obtain the whole ground truth, *i.e.*, $Recall@q = k/N_C$. Slightly different from the recall, we measure the number of verification that a miner uses to discover certain relations, *i.e.*, $T@k = q$, to address the restriction on the verification quota. T@k also implies the precision of recommendations, *i.e.*, $Precision@q = k/T@k$.

5.2   Results

Improvement with Active Learning. Table 4 compares active learning and the corresponding unsupervised manner to answer RQ2. We find that *active learning can enhance some but not all relation mining methods.* The NRI miner is improved by operators' feedback significantly. In contrast, the GES miner performs better without operators' feedback. GES utilizes a score function to estimate data likelihood given a causal graph. The score function performs differently from operators. For example, adding an extra relation ($A_1 \rightarrow A_{32}$) into

Table 5. Comparison among miners under the framework of *FPG-Miner*

Miner	$\mathcal{D}_{OD}$							$\mathcal{D}_{TN}$						
	AUC	T@k					Time Cost	AUC	T@k					Time Cost
		10%	20%	30%	50%	100%	/iteration		10%	20%	30%	50%	100%	/iteration
Random	0.617	45	89	148	269	490	<1 s	0.617	148	276	443	756	1480	<1 s
PC-gauss	0.648	41	102	145	237	489	3 s	0.639	106	248	407	703	1483	<1 s
GES	0.690	28	76	125	214	490	<1 s	0.651	90	223	370	684	1479	<1 s
NRI	0.741	53	83	113	192	478	0:05:02	0.731	**85**	177	285	575	1482	0:24:11
CAR	**0.774**	**26**	**59**	**104**	**187**	**477**	<1 s	**0.792**	86	**173**	**269**	**455**	**1464**	<1 s

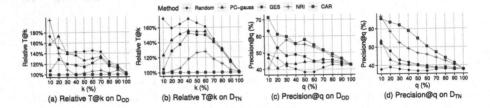

Fig. 3. Relative T@k and Precision@q for each miner on both datasets. For the relative T@k, we hold *CAR*'s T@k as one

the ground truth graph of $\mathcal{D}_{TN}$ can also increase the score. Thus, feedback may break the intrinsic mechanism of GES. In this way, we will discuss GES as a static miner in the rest of this section. The performance of the PC-gauss miner depends on the dataset. We will discuss PC-gauss with operators' feedback in $\mathcal{D}_{OD}$ while taking PC-gauss as a static miner in $\mathcal{D}_{TN}$.

Overall Results. Table 5 summarizes the miners' performance to answer RQ3. Figure 3(a) and Fig. 3(b) show the relative T@k for each miner with *CAR*'s T@k as one. For the sake of clarity, we also present $Precision@q = k/T@k$ in Fig. 3(c) and Fig. 3(d), where q is the verification quota. *CAR* discovers positive relations faster than baseline miners on both datasets, enhanced by the feedback from operators. GES has a low T@10%. However, it falls behind as *CAR* and NRI receives much feedback.

Contribution of Feature Extraction. We replace the feature extraction of *CAR* to demonstrate its effect, denoting the degraded miner as Association Rule (*AR*). Specifically, AR takes data points that are 1.5× of interquartile range far from the median in the sliding window as outliers. It further calculates the features in Table 3 based on those outliers. Table 6 shows the comparison between *CAR* and *AR*. The proposed feature extraction shortens T@20% by 26% and 25% on $\mathcal{D}_{OD}$ and $\mathcal{D}_{TN}$, respectively.

5.3 Case Study: Root Cause Analysis

We utilize the root cause analysis (RCA) task as a downstream application of the mined graph to explore RQ4. We adopt MicroCause [19] to localize root

Table 6. Comparison between CAR and its variant AR

Miner	$\mathcal{D}_{OD}$						$\mathcal{D}_{TN}$					
	AUC	T@k					AUC	T@k				
		10%	20%	30%	50%	100%		10%	20%	30%	50%	100%
CAR	**0.774**	**26**	**59**	104	**187**	477	**0.792**	86	**173**	269	455	1464
AR	0.738	39	80	**103**	198	489	0.678	123	232	356	573	1484

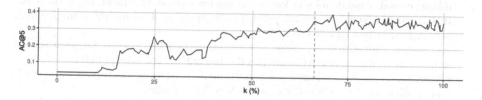

Fig. 4. As the number of correct relations (k) recommended by CAR increases, the graph quality changes, indicated by RCA performance (AC@5).

cause metrics. $AC@5$ refers to the probability that the top 5 results given by MicroCause include the root cause metrics [19]. We take AC@5 as the quality indicator of the mined graph. AC@5 is further measured on 99 high AAS faults.

Figure 4 shows that at least 33.8% of the relations seem neither helpful nor harmful to the RCA task in this case study. After CAR finds 66.2% of the relations (the dashed line in Fig. 4), the increasing trend of AC@5 stops. We would conclude that the answer to RQ4 is positive. However, an advanced algorithm in the future may still need the whole graph to take effect.

6   Conclusion

A fluctuation propagation graph (FPG) is a formal representation of the empirical knowledge towards automatic troubleshooting. This work focuses on structure discovery of the verified FPG among monitoring metrics. Our first empirical study shows that the existing methods have poor precision and recall on two real-world datasets. Thus, we propose a framework named $FPG\text{-}Miner$, combining operators' feedback to enhance the discovery ability. As shown in the case study, some relations are more important than others, strengthening our motivation. Under the framework of $FPG\text{-}Miner$, we propose a novel classification-based approach named CAR to speed up relation discovery. The experiment result confirms that active learning can enhance mining performance. Meanwhile, CAR recommends correct relations earlier compared with the baseline approaches. We believe that our methodology can be applied to other domains. However, the generalizability of our findings shall be examined in future work.

Acknowledgment. We thank Ruming Tang for proofreading this paper. This work is supported by the National Key R&D Program of China under Grant 2019YFB1802504, and the State Key Program of National Natural Science of China under Grant 62072264.

References

1. Aggarwal, P., et al.: Localization of operational faults in cloud applications by mining causal dependencies in logs using golden signals. In: Hacid, H., et al. (eds.) ICSOC 2020. LNCS, vol. 12632, pp. 137–149. Springer, Cham (2021). https://doi.org/10.1007/978-3-030-76352-7_17
2. Agrawal, R., Imieliński, T., Swami, A.: Mining association rules between sets of items in large databases. ACM SIGMOD Rec. **22**(2), 207–216 (1993)
3. Chakraborty, S.: Active learning for multimedia computing: survey, recent trends and applications, pp. 4785–4786. ACM, New York (2020)
4. Chen, P., Qi, Y., Zheng, P., Hou, D.: CauseInfer: automatic and distributed performance diagnosis with hierarchical causality graph in large distributed systems. In: INFOCOM, pp. 1887–1895 (2014)
5. Chen, T., Guestrin, C.: XGBoost: a scalable tree boosting system. In: KDD, pp. 785–794 (2016)
6. Cheng, W., Zhang, K., Chen, H., Jiang, G., Chen, Z., Wang, W.: Ranking causal anomalies via temporal and dynamical analysis on vanishing correlations. In: KDD, pp. 805–814 (2016)
7. Chickering, D.M.: Optimal structure identification with greedy search. J. Mach. Learn. Res. **3**, 507–554 (2003)
8. Endsley, M.R.: From here to autonomy: lessons learned from human-automation research. Hum. Factors **59**(1), 5–27 (2017)
9. Farshchi, M., Schneider, J.G., Weber, I., Grundy, J.: Experience report: anomaly detection of cloud application operations using log and cloud metric correlation analysis. In: ISSRE, pp. 24–34 (2015)
10. Guo, R., Cheng, L., Li, J., Hahn, P.R., Liu, H.: A survey of learning causality with data: problems and methods. ACM Comput. Surv. **53**(4), 1–37 (2021)
11. Huawei Technologies Noah's Ark Lab: Datasets for causal structure learning. https://github.com/huawei-noah/trustworthyAI/tree/master/Causal_Structure_Learning/Datasets. Accessed Feb 2022
12. Jiang, J., et al.: How to mitigate the incident? An effective troubleshooting guide recommendation technique for online service systems. In: ESEC/FSE, pp. 1410–1420 (2020)
13. Kalisch, M., Bühlmann, P.: Estimating high-dimensional directed acyclic graphs with the PC-algorithm. J. Mach. Learn. Res. **8**, 613–636 (2007)
14. Kipf, T., Fetaya, E., Wang, K.C., Welling, M., Zemel, R.: Neural relational inference for interacting systems. In: ICML, vol. 80, pp. 2688–2697, 10–15 July 2018
15. Lewis, D.D., Gale, W.A.: A sequential algorithm for training text classifiers. In: SIGIR. pp. 3–12 (1994)
16. Liu, D., et al.: MicroHECL: high-efficient root cause localization in large-scale microservice systems. In: ICSE-SEIP (2021)
17. Ma, M., Xu, J., Wang, Y., Chen, P., Zhang, Z., Wang, P.: Automap: diagnose your microservice-based web applications automatically. In: WWW, pp. 246–258 (2020)
18. Mahimkar, A., et al.: Troubleshooting chronic conditions in large IP networks. In: CONEXT (2008)

19. Meng, Y., et al.: Localizing failure root causes in a microservice through causality inference. In: IWQoS, pp. 1–10 (2020)
20. Nauta, M., Bucur, D., Seifert, C.: Causal discovery with attention-based convolutional neural networks. Mach. Learn. Knowl. Extr. 1(1), 312–340 (2019)
21. Nie, X., Zhao, Y., Sui, K., Pei, D., Chen, Y., Qu, X.: Mining causality graph for automatic web-based service diagnosis. In: IPCCC, pp. 1–8 (2016)
22. Pearl, J.: Causality: Models, Reasoning, and Inference, 2nd edn. Cambridge University Press (2009)
23. Roeser, M.B., McDermid, D., Surampudi, S.: Oracle database database reference (2021). https://docs.oracle.com/en/database/oracle/oracle-database/21/refrn/index.html
24. Runge, J.: Discovering contemporaneous and lagged causal relations in autocorrelated nonlinear time series datasets. In: UAI, vol. 124, pp. 1388–1397 (August 2020)
25. Runge, J., Nowack, P., Kretschmer, M., Flaxman, S., Sejdinovic, D.: Detecting and quantifying causal associations in large nonlinear time series datasets. Sci. Adv. 5(11), eaau4996 (2019)
26. Salton, G., Buckley, C.: Improving retrieval performance by relevance feedback. J. Am. Soc. Inf. Sci. 41(4), 288–297 (1990)
27. Settles, B.: Active Learning. Synthesis Lectures on Artificial Intelligence and Machine Learning. Morgan & Claypool Publishers LLC. (2012)
28. Strobl, E.V., Zhang, K., Visweswaran, S.: Approximate kernel-based conditional independence tests for fast non-parametric causal discovery. J. Causal Infer. 7(1), 1–24 (2019)
29. Su, Y., et al.: CoFlux: robustly correlating KPIs by fluctuations for service troubleshooting. In: IWQoS (2019)
30. Wang, H., et al.: Groot: an event-graph-based approach for root cause analysis in industrial settings. In: ASE, pp. 419–429 (2021)
31. Wang, P., et al.: CloudRanger: root cause identification for cloud native systems. In: CCGRID, pp. 492–502 (2018)
32. Wu, L., Tordsson, J., Elmroth, E., Kao, O.: MicroRCA: root cause localization of performance issues in microservices. In: NOMS, pp. 1–9 (2020)
33. Yan, H., Breslau, L., Ge, Z., Massey, D., Pei, D., Yates, J.: G-RCA: a generic root cause analysis platform for service quality management in large IP networks. IEEE/ACM Trans. Netw. 20(6), 1734–1747 (2012)
34. Zhang, J.M., Harman, M., Ma, L., Liu, Y.: Machine learning testing: survey, landscapes and horizons. IEEE Trans. Softw. Eng. 48, 1–36 (2020)
35. Zhang, J.: On the completeness of orientation rules for causal discovery in the presence of latent confounders and selection bias. Artif. Intell. 172(16), 1873–1896 (2008)
36. Zhang, Y., Ren, L., Chen, L., Xiong, Y., Cheung, S.C., Xie, T.: Detecting numerical bugs in neural network architectures. In: ESEC/FSE (November 2020)
37. Zhao, N., et al.: Understanding and handling alert storm for online service systems. In: ICSE-SEIP, pp. 162–171 (2020)
38. Zheng, X., Aragam, B., Ravikumar, P.K., Xing, E.P.: DAGs with no tears: continuous optimization for structure learning. In: NIPS, vol. 31, pp. 9472–9483 (2018)

Towards Efficient Human Action Retrieval Based on Triplet-Loss Metric Learning

Iris Kico[✉] [iD], Jan Sedmidubsky[iD], and Pavel Zezula

Masaryk University, Botanicka 68a, 602 00 Brno, Czechia
iriskico@mail.muni.cz, zezula@fi.muni.cz

Abstract. Recent pose-estimation methods enable digitization of human motion by extracting 3D skeleton sequences from ordinary video recordings. Such spatio-temporal skeleton representation offers attractive possibilities for a wide range of applications but, at the same time, requires effective and efficient content-based access to make the extracted data reusable. In this paper, we focus on content-based retrieval of pre-segmented skeleton sequences of human actions to identify the most similar ones to a query action. We mainly deal with the extraction of content-preserving action features, which are learned using the triplet-loss approach in an unsupervised way. Such features are (1) effective as they achieve a similar retrieval quality as the features learned in a supervised way, and (2) of a fixed size which enables the application of indexing structures for efficient retrieval.

Keywords: Human motion data · Skeleton sequences · Action similarity · Action retrieval · Triplet-loss learning · LSTM

1 Introduction

Human motion can be digitized into a discrete sequence of simplified skeleton poses, where each pose keeps 2D or 3D space coordinates of important body joints in a specific time moment. Until recently, such spatio-temporal data were captured by specialized hardware technologies, so the amount of digitized motion data was fairly limited. However, recent pose-estimation software methods [4,23] enable extracting skeleton data from ordinary video recordings, which opens unprecedented application potential in many domains. For example, in sports to automatically assess a figure-skating performance without emotions of human referees; in healthcare to remotely evaluate the progress in rehabilitation exercising; in smart-cities to detect potential threats like a running group of people, or in computer animation to find previously-captured animations relevant for building a new movie scene [3]. All these potential applications require content-based

Supported by ERDF "CyberSecurity, CyberCrime and Critical Information Infrastructures Center of Excellence" (No. CZ.02.1.01/0.0/0.0/16_019/0000822).

© The Author(s), under exclusive license to Springer Nature Switzerland AG 2022
C. Strauss et al. (Eds.): DEXA 2022, LNCS 13426, pp. 234–247, 2022.
https://doi.org/10.1007/978-3-031-12423-5_18

processing techniques that perform effectively and efficiently on large datasets of skeleton sequences.

Current research mainly focuses on content-based processing of pre-segmented skeleton sequences, called *actions*, that are perceived as semantically-indivisible motions with respect to the context of a given application. The most popular tasks are: recognizing classes of actions [7,9,25], detecting actions in a stream [6,19,24], or searching for query-relevant subsequences within a long skeleton sequence [1,16,17]. These tasks often employ *query-by-example retrieval* as the underlying operation: given a query action, the objective is to search the dataset of actions that are the most similar to the query one. The retrieval operation has to solve two important issues: effectiveness (i.e., the quality of the retrieved results) and efficiency (i.e., the query response time). In this paper, we focus on the action-retrieval operation from both points of view.

Related Work

A fundamental prerequisite for skeleton-data retrieval is an ability to determine the *similarity* between two actions. The similarity can be numerically calculated using time-warping functions, like the Dynamic Time Warping (DTW), on the level of action poses represented by raw joint coordinates [2]. To better reflect the similarity semantics, content-preserving *features* are extracted on the level of whole actions. The features may be manually designed by a domain expert in a handcrafted way [15]. However, the handcrafted features can hardly represent more complex dependencies in movement patterns and, therefore, have been practically abandoned and replaced by *deep features* that can be automatically learned using well-trained neural-network models [22]. The network models, like convolutional (CNN) [9,26], graph-convolutional [12], or long short-term memory (LSTM) [19] neural networks, are often trained for the classification of actions into a predefined set of classes. The learned parameters of hidden network layers can then be utilized to extract the action feature. Such features are typically represented as fixed-size high-dimensional vectors (e.g., 4,096D features in [17]) and efficiently compared by the Euclidean [17] or Hamming [9] distance functions to determine the similarity of action pairs.

The deep features are almost exclusively learned in a supervised way, which requires the set of labeled training actions to be defined in advance. Categorization of such training actions determines the semantics of similarity perception. However, in scenarios where no semantic labeling is available, unsupervised approaches are the only possibility. In such cases, auto-encoders [20] or transformers [5] are trained to learn an action embedding (i.e., fixed-size action feature) by reducing the dimensionality of original action data into the fixed-size feature and reconstructing the original action from such feature. An alternative way is to employ the Siamese or triplet-loss learning strategy to learn the action feature using examples of similar and dissimilar action pairs. To find suitable action pairs for training, it is necessary to use additional domain-expert knowledge or some simple metric that can at least roughly estimate the low-level action similarity. Based on the recent survey [15], there is a very limited number of skeleton-data approaches [1] that extract the action feature using the triplet-loss strategy in an unsupervised way (sometimes also referred to as self-supervised way).

The most relevant approach to our work is the application of the triplet-loss learning strategy in [1]. This approach firstly partitions an original action into fixed-size segments, extracts the feature for each segment, and determines the similarity of actions by comparing their corresponding sequences of segment features using the expensive Earth mover's distance. To learn the segment feature, the positive examples are selected as temporally-close segments belonging to the same action and negative ones as randomly chosen segments belonging to different actions. In this paper, we introduce advanced strategies for selecting positive and negative examples by employing low-level similarity functions. We extract the feature on the level of the whole action, which enables orders-of-magnitude more efficient comparison of actions using the Euclidean distance function, in comparison with [1]. In addition, we employ a long short-term memory network as the internal model for triplet-loss learning, which has demonstrated higher effectiveness for skeleton data [18] than convolutional networks used in [1].

Paper Contributions

In this paper, we focus on action retrieval by proposing a new approach for the extraction of deep action features in an unsupervised way. The proposed unsupervised approach has much higher applicability than most existing purposely-trained classifiers that require labeled actions to be defined in advance. We experimentally evaluate the quality of extracted features by achieving high retrieval effectiveness, which is competitive to the quality of supervised approaches, as well as high efficiency, which can further be improved by straightforward application of indexing schemes. The specific paper contributions include: (i) introduction of the Uniform Time Warping function to efficiently determine low-level similarity of raw-action data (ii) definition of new strategies for selection of positive and negative actions for triplet-loss feature learning, and (iii) incorporating the LSTM network into the triplet-loss learning process (in contrast to CNN-based approaches in [1]).

2   Action Retrieval

In this section, we formally define the problem of action retrieval using k-nearest neighbor queries. Then, we present the baseline retrieval approach employing time-warping distance functions for determining the low-level similarity of actions. Such functions will serve as underlying similarity concepts for the needs of deep-feature extraction in an unsupervised way (Sect. 3).

2.1   Problem Definition

We represent skeleton data of a single *action* A as a sequence $A = (P_1, \ldots, P_n)$ of n consecutive 3D *poses* P_i, where the i-th pose $P_i \in \mathbb{R}^{j \cdot 3}$ is captured at time moment i $(1 \leq i \leq n)$ and consists of xyz-coordinates of j tracked *joints*. In this paper, we use the body model with $j = 31$ joints. The retrieval problem is then defined based on k-nearest neighbor search (kNN) as follows. Having a dataset $\mathcal{A} = \{A_1, \ldots, A_l\}$ of l *actions* $\{A_1, \ldots, A_l\}$ and a *query* action A_Q, the objective is to search the dataset and finds its k actions ($k \ll l$) that are the most similar to the query A_Q:

$$\{\mathcal{A}' \in \mathcal{A} | k = |\mathcal{A}'|, \forall A' \in \mathcal{A}', \forall A_i \in \mathcal{A} \backslash \mathcal{A}' : dist(A_Q, A') \leq dist(A_Q, A_i)\}, \quad (1)$$

where the similarity between the query A_Q and any dataset action A_i $(1 \leq i \leq l)$ is quantified by a distance function $dist(A_Q, A_i)$.

2.2   Retrieval Process

For simplicity, we implement the retrieval process using the sequential scan approach by comparing the query action against all the dataset actions and selecting k most similar as the nearest neighbors. As future work, we outline how the retrieval process can be simply speed-up by adopting an indexing structure.

To apply the sequential-scan approach, there is a need to define the distance function $dist()$. We firstly use the standard Dynamic Time Warping (DTW) as applied in [14]. However, as later demonstrated in the experiments, this function does not perfectly align semantically-related skeleton sequences as it prefers shorter actions with respect to a query. Since this paper works with well-segmented actions, it is also meaningful to apply a warping function that uniformly aligns the action poses in the temporal dimension. For this purpose, we define the *Uniform Time Warping* (UTW) function which determines the distance between actions $A = (P_1, \ldots, P_n)$ and $A' = (P'_1, \ldots, P'_{n'})$ as:

$$UTW(A, A') = \begin{cases} \frac{1}{n} \cdot \sum_{i=1}^{n} poseDist(P_i, P'_{i \cdot \lfloor n'\backslash n \rfloor}) & n \geq n' \\ \frac{1}{n'} \cdot \sum_{i=1}^{n'} poseDist(P'_i, P_{i \cdot \lfloor n\backslash n' \rfloor}) & n < n', \end{cases} \quad (2)$$

where the distance function $poseDist()$ quantifies the similarity between two poses based on the sum of the Euclidean distances between their corresponding joint coordinates. Simply, UTW maps all the poses of a longer action to the temporally-corresponding poses of a shorter one and computes the average pose distance of such mappings. In contrast to DTW, such average distance guarantees that shorter nor longer actions are preferred. The other advantage is linear time complexity compared to the quadratic time complexity of DTW.

The DTW and UTW functions determine the similarity of actions numerically and do not take any semantics into account. Moreover, both functions are hardly indexable since they do not satisfy the triangle-inequality postulate of a metric space. For these reasons, we learn and extract semantic-preserving action features in the form of fixed-size vectors that can be efficiently compared by the Euclidean distance function and thus potentially indexed using any vector- or metric-space index structure [11]. In the following, we introduce the feature extraction approach that learns action semantics in an unsupervised way with the help of both DTW and UTW distance functions.

3   Learning Action Features

The main paper's objective is to extract fixed-size features from unlabeled actions of a variable length. The features should preserve the information contained in the original raw skeleton actions while reducing data dimensionality. For this purpose, we employ the Long Short-Term Memory (LSTM) neural network, which has already proven to be a successful and lightweight solution for action recognition in skeleton data in a supervised way [15]. To train the LSTM network in an unsupervised way, we adopt the triplet-loss learning approach that learns the data semantics by minimizing the distance between similar examples while maximizing the distance between dissimilar ones. In supervised approaches [13], the similar and dissimilar examples can be simply determined based on labels. However, in our scenario, we need to determine such examples based on the low-level similarity of skeleton data. In the rest of this section, we describe the principles of triplet-loss learning in combination with LSTM and introduce several strategies for generating training data from unlabeled actions.

3.1   Triplet-Loss Learning

To train the adopted LSTM network, we use the triplet-loss function to calculate the model error. To compute the loss function value, we extract an action *embedding* – a fixed-size high-dimensional vector that reflects the semantics of the given action in a given time moment, and it will correspond to the desired action feature after the network is fully trained. The embedding $f(A)$ is extracted for each action A independently by feeding the action through the LSTM network and taking the content of the last hidden layer as the output. The network is gradually trained using the provided *triplets* consisting of an *anchor* action A, *positive* action example A^P, and *negative* action example A^N. The positive action should be similar to the anchor (i.e., $dist(A, A^P)$ is "low") while the negative action should be dissimilar to the anchor (i.e., $dist(A, A^N)$ is "high enough"). The loss-function value is denoted as *loss* and formally defined as:

$$loss(A, A^P, A^N) = max\left(0, Eucl\left(f(A), f(A^P)\right) - Eucl\left(f(A), f(A^N)\right) + m\right),$$

where $Eucl\left(f(A), f(A^P)\right) = \|f(A) - f(A^P)\|_2$ represents the Euclidean distance between the action embeddings $f(A), f(A^P)$ and m is a margin that should correspond to a requested distance between the embeddings of the positive and negative examples. Since the network is learned in a fixed number of iterations (i.e., epochs), the loss value for each iteration is calculated as the mean value of losses computed over all the provided triplets. In this paper, we follow the offline triplet generation approach, which means that the triplets are generated before the training process and thus remain the same for all the iterations.

Training and Validation Details. The network model is trained using the provided triplets that are automatically generated from the dataset $\mathcal{A}$ of unlabeled actions according to a given strategy (individual triplet-generation

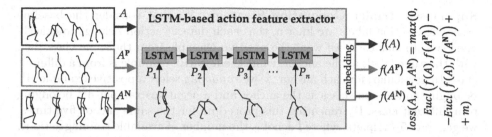

Fig. 1. Triplet-loss learning approach with an LSTM-based neural network as the internal model. The network is used to extract the embeddings $f(A), f(A^P)$ and $f(A^N)$ for the positive A^P and negative A^N example with respect to the given anchor A.

strategies are presented in Sect. 3.2). For each triplet, the anchor, positive, and negative actions are independently processed by the LSTM network to extract their embeddings, as graphically illustrated in Fig. 1. Each such action is processed by feeding its individual poses $(P_1, \ldots, P_n)$ – in the form of joint-coordinate vector P_i – into individual LSTM cells. The output of a given cell is forwarded to the input of the next cell, and the output of the last n-th cell constitutes the action embedding.

To validate the accuracy of the model trained using the provided triplets, we follow how the loss value decreases over individual iterations. The loss can be computed in the same way also for test actions which can also be provided in the form of triplets. As the loss value is not further improving over individual iterations, the training can be stopped. In this paper, we also use the hard stop condition on the number of iterations (set to 100 in all the experiments). We additionally validate the model accuracy using the 1NN approach with the knowledge of action labels. In this scenario, the feature embeddings are extracted independently for each training and test action after each iteration. Then, the Euclidean distance between a test and each training action is calculated on the level of their embeddings, and the nearest-neighbor's label is assigned to the test action. The accuracy is expressed as the ratio between the number of correctly-assigned labels and the number of all test actions.

3.2 Triplet Generation Strategies

We create the triplets using different strategies that are mostly based on identifying positive and negative examples using the low-level similarity function $dist()$. In all the cases, the objective is to have the positive example closer to the anchor than the negative example: $dist(A, A^P) < dist(A, A^N)$. There are so many triplets that can be generated using this rule, so we need to select the most contributing ones, as the network training is sensitive to the triplet selection. To employ the whole dataset, we generate a fixed number of 5 triplets for each training action considered as the anchor. To get rough intuition about possible achievable accuracy, we also consider the supervised triplet-generation strategy in which the labels of training actions are known.

Supervised Triplet Generation Strategy. We first consider the baseline case where action labels are known, thus each dataset action belongs to exactly one class. In this case, we want the positive example from the same class (i.e., with the same label) as the anchor, while the negative example is from a different class. In particular, for each anchor A, we *randomly* select a positive example $A^{\mathbf{P}}$ belonging to the same class as the anchor and a negative example $A^{\mathbf{N}}$ belonging to a different class. By repeating this procedure for the same anchor five times, we generate $5 \cdot l$ triplets, where l denotes the number of available training actions.

Unsupervised Random Triplet Generation Strategy. If the labels of actions are not known, the most straightforward way is to generate the triplets randomly. In particular, for each training action selected as the anchor, one randomly chosen dataset action is considered as the positive example and one randomly chosen action as the negative example. We again generate the $5 \cdot l$ triplets by repeating this procedure five times for each anchor.

Unsupervised kNN Triplet Generation Strategies. The important idea for further triplet-generation strategies is to determine the positive and negative examples based on a low-level similarity function $dist()$, such as DTW or UTW. Specifically, for each training action considered as the anchor A, we calculate the distance $dist(A, A')$ to each other training action A' to determine the list of k-nearest neighbors (kNN), where $k \leq l$ can even correspond to the number of training actions (see Eq. 1). Then, the positive example should be selected from the beginning of the list, while the negative ones from the tail of the list. At the same time, the following two important issues should be taken into account during the selection process.

- Some triplet-loss studies in different domains (e.g., in face recognition [13]) suggest the negatives to be more distant from the anchor but not that distant from the positive. Such "hard" negatives should help the network learn more complex dependencies.
- The construction of nearest-neighbor lists using low-level similarity functions, such as DTW or UTW, does not need to perfectly reflect the semantics of a target application.

Both issues motivate us to take the positive example from the beginning of the list and the negative one not too far but, at the same time, not too close to the anchor. For example, assume the dataset of 1,000 training actions, so we set $k = 40$ to obtain the list of only 40 nearest neighbors and select the positive from the first 5 neighbors while the negative is within the range of 20–40NN.

By generating 5 triplets for each training anchor, we simply consider the first five positives from the nearest-neighbor list. However, we have many possibilities for the selection of negatives. We propose the following four cases for the given anchor A and the selected close positive example $A^{\mathbf{P}}$.

1. $rnd(range)$ – the negative is selected randomly as a random action from the given nearest-neighbor range.

2. $maxP(range)$ – the negative $A^{\mathbf{N}}$ is selected as the action from the given nearest-neighbor range having the maximum distance to the positive $A^{\mathbf{P}}$, i.e., $\{\forall A' \in range : dist(A^{\mathbf{P}}, A^{\mathbf{N}}) \geq dist(A^{\mathbf{P}}, A')\}$. The idea is to have the negative more distant from the positive.

3. $minP(range)$ – orthogonal case to the previous one by selecting the negative $A^{\mathbf{N}}$ from the given range with the minimum distance to the positive $A^{\mathbf{P}}$, i.e., $\{\forall A' \in range : dist(A^{\mathbf{P}}, A^{\mathbf{N}}) \leq dist(A^{\mathbf{P}}, A')\}$. This constitutes a "harder" triplet for learning.

4. $maxPminA(range)$ – the negative $A^{\mathbf{N}}$ is selected from the given range with the maximum distance to the positive $A^{\mathbf{P}}$ but, at the same time, with the minimum distance to the anchor. This constitutes a different form of "harder" triplet compared to the previous case.

Action-Length Limitation Filtering. One of the problems of low-level similarity functions is that they need to deal with the comparison of actions that may significantly vary in length (e.g., from tens of poses to hundreds of poses). Although time-warping mechanisms of DTW and UTW partly solve this problem, some shorter actions can be quantified as more similar to a longer action than two longer semantically-relevant actions. To avoid selecting actions that differ much in length, we propose to apply length-limitation filtering. In particular, when constructing the k-nearest neighbor list, we consider only the actions having a similar length as the anchor $A = (P_1, \ldots, P_n)$. We define the thresholds on minimum t_{min} and maximum t_{max} deviation from the anchor, so the action $A' = (P'_1, \ldots, P'_{n'})$ is considered to be added into the nearest-neighbor list only if it satisfies the length restriction: $t_{min} \cdot n \leq n' \leq t_{max} \cdot n$. Although the values of the thresholds depend on a target application, we suggest this setting: $t_{min} \sim 0.7$ and $t_{max} \sim 1.5$ (i.e., the action can be maximally about roughly 50 % shorter or longer with respect to the anchor).

4   Experimental Evaluation

We evaluate the proposed triplet-generation strategies' influence on the quality of extracted action features. The quality is quantified by evaluating kNN queries and compared with existing supervised and unsupervised approaches.

4.1   Dataset

We adopt the popular HDM05 dataset [10] that provides the HDM05-122 ground truth [18] with 2,328 actions divided into 122 classes. Each action is captured with the 120 frame-per-second rate and consists of 3D positions of 31 joints estimated in each pose. The actions correspond to daily/exercising activities and significantly differ in length – 13 frames (0.1 s) and 900 frames (7.5 s) for the shortest and longest action. We have chosen this challenging dataset as it provides the highest number of classes to be recognized. As suggested in [8, 18], we also pre-process the dataset by downsampling the actions to 12 Hz and unifying the skeleton position, orientation, and size in each action pose.

4.2   Methodology

We evaluate the effectiveness of the proposed feature-learning approach on a content-based kNN retrieval scenario. We first split the dataset actions into training and test sets in a balanced way, so each set consists of 1,164 actions. The unlabeled training actions are used for generating triplets using different strategies, as described in Sect. 3.2. In particular, we generate 5 different triplets for each training action considered as the anchor, which results in 5,820 triplets in total. Each trained model is then used to extract the features (i.e., embeddings) of all training and test actions as the output of the last hidden LSTM-network layer. The features are finally used to evaluate kNN retrieval by calculating the Euclidean distance between the features of a given test action (i.e., *query*) and each training action. The accuracy of a single query is computed as the ratio of the number of nearest neighbors that belong to the same class as the query and value k (action labels are used only for evaluation purposes). As there are about 9 actions available for each class on average, we limit k to 10. Overall accuracy is calculated as a mean value over accuracies of all 1,164 queries.

Training Details. All the experiments were run on a six-core PC with Intel(R) Core(TM) i7-8700K CPU at 3.70 GHz, 16 GB RAM, with NVIDIA GeForce GTX 1060 6 GB GPU. The proposed approach was implemented using Python 3.7 and the PyTorch 1.5.1 framework. Each network was trained in 100 iterations using the Adam optimizer, and the learning rate was set to 0.00001. The size of hidden LSTM-network layers was fixed to 1,024 dimensions in all the experiments. The training time took up to 10 h.

4.3   Experimental Results

As a baseline, we evaluate retrieval accuracy of the *supervised* triplet generation strategy. For each training anchor action, 5 positives are randomly chosen from the same class as the anchor, and one negative is randomly selected from a different class for each anchor-positive pair. This guarantees 100 % *data accuracy* of generated triplets as the positives and negatives are selected based on the action labels. After training and evaluating the retrieval scenario, retrieval accuracy achieves **81.96 %** for $k = 1$.

Before evaluating any unsupervised approach, we determine retrieval accuracy by extracting the action features from the untrained LSTM network (i.e., the network with randomly initialized parameters). Achieved 1NN retrieval accuracy of $\sim$47 % is surprisingly high, but this is caused by the fact that the network works only as a "dimensionality-reduction method" which transforms the original skeleton-data space into a fixed-size feature space. So the objective of unsupervised approaches is to increase retrieval accuracy from 47 % up to 82 %, achieved by the supervised approach.

We start to evaluate the unsupervised approaches using the basic *random* triplet generation strategy "5 · $rnd()$ $rnd()$", where 5 positives are randomly selected for each anchor (and one random negative for each anchor-positive pair).

Table 1. 1NN retrieval accuracy based on the unsupervised triplet-loss feature-learning approach using different triplet-generation strategies. All the strategies generate the same number of 5,820 triplets, so 1NN retrieval accuracy can be directly compared. Data accuracy of positive (negative) samples represent the ratio of triplets in which the positive (negative) sample belongs (does not belong) to the same class as the anchor. Triplet data accuracy represents the ratio of triplets where both the positive and negative samples are correctly selected.

	Strategy for selection of		Data accuracy (%)			1NN retrieval accuracy (%)
	Positive	Negative	Positive	Negative	Triplet	
–	$5 \cdot rnd()$	$rnd()$	0.85	98.86	0.84	67.87
DTW	1–5NN	$rnd(20$–40NN$)$	58.14	94.09	54.28	73.80
UTW	1–5NN	$rnd(20$–40NN$)$	60.96	93.23	56.58	76.29
DTW+LL	1NN	$5 \cdot rnd(20$–40NN$)$	76.28	94.34	71.56	82.82
	1–5NN	$rnd(20$–40NN$)$	57.38	93.61	53.06	82.82
	20–24NN	$rnd(80$–100NN$)$	16.75	98.97	16.65	82.82
	1–5NN	$maxPminA(20$–40NN$)$	57.38	86.33	52.82	81.44
	1–5NN	$minP(20$–40NN$)$	57.38	82.75	41.41	83.08
	1–5NN	$maxP(20$–40NN$)$	57.38	88.57	54.87	81.87
UTW+LL	1NN	$5 \cdot rnd(20$–40NN$)$	75.00	93.31	70.05	82.99
	1–5NN	$rnd(20$–40NN$)$	55.81	93.73	51.68	83.76
	20–24NN	$rnd(80$–100NN$)$	15.84	98.95	15.74	82.30
	1–5NN	$maxPminA(20$–40NN$)$	55.81	87.26	50.91	81.87
	1–5NN	$minP(20$–40NN$)$	55.81	82.06	39.74	82.73
	1–5NN	$maxP(20$–40NN$)$	55.81	90.03	53.99	82.39

Although this random approach seems to be an extreme baseline, it can surprisingly achieve "reasonable" retrieval accuracy of 67.87 %, as depicted in the first line of Table 1. It is caused by the fact that there is roughly 50 % probability that the positive example is semantically more similar to the anchor than the negative one, which implies that half of the generated triplets will contribute to the learning process. We further focus on the generation of triplets using the *unsupervised* kNN-based strategies, where DTW or UTW are used as low-level similarity functions for constructing the nearest-neighbor list for each anchor. Constructing such lists for all the anchors takes tens of hours for DTW, while only minutes for UTW. We select positives from the 5 nearest neighbors and "hard" negatives that are not too far from the anchor – so we have decided to select the negatives from the interval between the 20-th and 40-th nearest neighbor. When the negative sample is selected randomly from this interval, we achieve 1NN retrieval accuracy of 73.80 % and 76.29 % (lines 2–3 in Table 1) for the DTW- and UTW-based nearest-neighbor list, respectively. Such accuracies are quite high but still not comparable to the supervised baseline.

We further show how important it is to select the positives with a comparable length as the anchor length. By applying the action length limitation (LL) filtering to select positives, we achieve high 1NN retrieval accuracy of 82.82 %

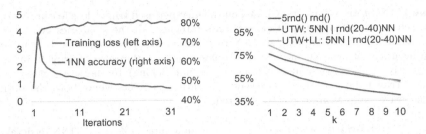

Fig. 2. Evolution of 1NN retrieval accuracy in comparison with the decreasing training-loss value in the first 30 iterations (left). Accuracy of kNN retrieval (for $k \in [1, 10]$) for the three selected triplet-generation strategies (right).

and 82.99 % for DTW and UTW, respectively (first lines of the DTW+LL and UTW+LL settings). The length-limitation filtering helps time-warping functions find a better alignment between poses of actions, which even leads to the accuracy comparable to the supervised approach. Up to now, the negatives are selected randomly from the interval of 20–40NN. Since we want to avoid random selection (e.g., due to repeatability of experiments), we evaluate the proposed strategies for selecting negatives. In particular, for each action from the range of 20–40NN, we calculate its distance to the positive using DTW/UTW and select such action as the negative based on the following three strategies: minimum distance to the positive "$minP$", maximum distance to the positive "$maxP$", or maximum distance to the positive and at the same time minimum distance to the anchor "$maxPminA$". As depicted in Table 1, the $minP$ strategy achieves high accuracy of 83.08 % and 82.73 % for DTW+LL and UTW+LL variants.

In general, we can say that retrieval accuracy depends on the accuracy of the triplets that contribute to feature learning. In the case of harder triplets, where negatives are closer to the positives and, at the same time, close to the anchor, we achieve the highest accuracy. The results also indicate that it is sufficient to select the positive as the action which is more similar to the anchor than the negative sample, regardless of their real labels (lines starting with 20–24NN for both variants). For an increasing value of k, accuracy is generally decreasing, as illustrated for the three selected approaches in Fig. 2 (right). In Fig. 2 (left), we can see how the training-loss value decreases over individual iterations in comparison with increasing 1NN retrieval accuracy.

4.4   State-of-the-Art Comparison

As already discussed, we evaluate the quality of action features in two directions: the ability to preserve the semantic content with respect to a given application (effectiveness), and the time needed to locate the most similar dataset actions with respect to the query (efficiency). Therefore, we compare relevant action feature extractors and action retrieval approaches in both these directions.

The feature extractors need to be separated into the two orthogonal approaches of supervised and unsupervised learning; the supervised methods

Table 2. Comparison with the state-of-the-art approaches on the HDM05 dataset.

Learning	Approach	Acc. (%)	Efficiency
Supervised	LSTM classifier [18]	89.86	High
	PB-GCN classifier [21]	88.17	
	1NN on LSTM classifier features + Eucl. [18]	86.42	
	F-DMT-Net classifier [26]	85.30	
	CNN classifier [8]	83.33	
Unsuper.	1NN on raw skeleton data + DTW [14]	77.70	*Low*
	1NN on motion words + DTW [14]	77.61	*Medium*
	1NN on triplet-loss LSTM features + Eucl.	**83.08**	**High**

generally achieve higher precision, but require labeled training data that limit their applicability. In Table 2, we compare the accuracy results of both supervised and unsupervised approaches on the challenging HDM05 dataset. The best result of 89.86 % is achieved using the purposely-trained LSTM classifier [18] which does not directly support the retrieval functionality. By employing this trained classifier to extract the embeddings of actions and evaluating the 1NN retrieval scenario on such embeddings, accuracy decreases to 86.42 %. Although our approach (last line in bold) is unsupervised, its accuracy of 83.08 % approaches accuracy of supervised classifiers. Among the unsupervised methods, our approach is the clear winner.

Besides high effectiveness, our approach enables straightforward application of vector- or metric-based indexing methods as the extracted action features are represented as fixed-size vectors. Even without any indexing, the Euclidean-based comparison of action features is orders of magnitude more efficient than DTW-based alignment methods applied in [14]. In addition, the internal LSTM model used in the training process allows our approach to simply adjust the size of the hidden network layer, which enables extracting a requested size of action features based on the needs of a target application.

5   Conclusions

We have proposed an unsupervised approach for extraction of fixed-size and content-preserving features from skeleton-data actions. By combining an LSTM network, triplet-loss learning, and proposed triplet-generation strategies, we can even achieve similar accuracy as the approaches trained in a supervised way. In particular, applying time-warping functions with action-length filtering contributes to a more suitable selection of positive and negative samples, where the negative sample should not be too distant from both the anchor and positive. The results also indicate that it is sufficient to select the positive sample simply as the action which is closer to the anchor than the negative, disregarding their real labels. In addition, the fixed-size nature of extracted features together with the Euclidean-based comparison open great possibilities for indexing action features in the future.

References

1. Aristidou, A., Cohen-Or, D., Hodgins, J.K., Chrysanthou, Y., Shamir, A.: Deep motifs and motion signatures. ACM Trans. Graph. **37**(6), 187:1–187:13 (2018). https://doi.org/10.1145/3272127.3275038
2. Barnachon, M., Bouakaz, S., Boufama, B., Guillou, E.: Ongoing human action recognition with motion capture. Pattern Recogn. **47**(1), 238–247 (2014)
3. Budikova, P., Sedmidubsky, J., Zezula, P.: Efficient indexing of 3d human motions. In: International Conference on Multimedia Retrieval (ICMR), pp. 10–18. ACM (2021)
4. Chang, S., et al.: Towards accurate human pose estimation in videos of crowded scenes. In: 28th ACM International Conference on Multimedia (MM), pp. 4630–4634. ACM (2020). https://doi.org/10.1145/3394171.3416299
5. Cheng, Y.B., Chen, X., Chen, J., Wei, P., Zhang, D., Lin, L.: Hierarchical transformer: Unsupervised representation learning for skeleton-based human action recognition. In: IEEE International Conference on Multimedia and Expo (ICME), pp. 1–6 (2021). https://doi.org/10.1109/ICME51207.2021.9428459
6. Häring, S., Memmesheimer, R., Paulus, D.: Action segmentation on representations of skeleton sequences using transformer networks. In: IEEE International Conference on Image Processing (ICIP), pp. 3053–3057 (2021). https://doi.org/10.1109/ICIP42928.2021.9506687
7. Khaire, P., Kumar, P.: Deep learning and rgb-d based human action, human–human and human–object interaction recognition: a survey. J. Visual Commun. Image Repr.**86**, 1–25 (2022). https://doi.org/10.1016/j.jvcir.2022.103531, https://www.sciencedirect.com/science/article/pii/S1047320322000724
8. Laraba, S., Brahimi, M., Tilmanne, J., Dutoit, T.: 3d skeleton-based action recognition by representing motion capture sequences as 2d-rgb images. Comput. Anim. Virtual Worlds **28**(3–4), e1782 (2017)
9. Lv, N., Wang, Y., Feng, Z., Peng, J.: Deep hashing for motion capture data retrieval. In: IEEE International Conference on Acoustics, Speech and Signal Processing (ICASSP), pp. 2215–2219. IEEE (2021). https://doi.org/10.1109/ICASSP39728.2021.9413505
10. Müller, M., Röder, T., Clausen, M., Eberhardt, B., Krüger, B., Weber, A.: Documentation Mocap Database HDM05. Tech. Rep. CG-2007-2, Universität Bonn (2007)
11. Novak, D., Zezula, P.: Rank aggregation of candidate sets for efficient similarity search. In: Decker, H., Lhotská, L., Link, S., Spies, M., Wagner, R.R. (eds.) DEXA 2014. LNCS, vol. 8645, pp. 42–58. Springer, Cham (2014). https://doi.org/10.1007/978-3-319-10085-2_4
12. Peng, W., Hong, X., Zhao, G.: Tripool: graph triplet pooling for 3d skeleton-based action recognition. Pattern Recogn. **115**, 107921 (2021). https://doi.org/10.1016/j.patcog.2021.107921
13. Schroff, F., Kalenichenko, D., Philbin, J.: Facenet: a unified embedding for face recognition and clustering. In: IEEE Conference on Computer Vision and Pattern Recognition (CVPR), pp. 815–823 (2015). https://doi.org/10.1109/CVPR.2015.7298682
14. Sedmidubsky, J., Budikova, P., Dohnal, V., Zezula, P.: Motion words: a text-like representation of 3D skeleton sequences. In: ECIR 2020. LNCS, vol. 12035, pp. 527–541. Springer, Cham (2020). https://doi.org/10.1007/978-3-030-45439-5_35

15. Sedmidubsky, J., Elias, P., Budikova, P., Zezula, P.: Content-based management of human motion data: Survey and challenges. IEEE Access **9**, 64241–64255 (2021). https://doi.org/10.1109/ACCESS.2021.3075766, https://doi.org/10.1109/ACCESS.2021.3075766

16. Sedmidubsky, J., Elias, P., Zezula, P.: Similarity searching in long sequences of motion capture data. In: Amsaleg, L., Houle, M.E., Schubert, E. (eds.) SISAP 2016. LNCS, vol. 9939, pp. 271–285. Springer, Cham (2016). https://doi.org/10.1007/978-3-319-46759-7_21

17. Sedmidubsky, J., Elias, P., Zezula, P.: Searching for variable-speed motions in long sequences of motion capture data. Inf. Syst. **80**, 148–158 (2019). https://doi.org/10.1016/j.is.2018.04.002

18. Sedmidubsky, J., Zezula, P.: Augmenting Spatio-Temporal Human Motion Data for Effective 3D Action Recognition. In: 21st IEEE International Symposium on Multimedia (ISM), pp. 204–207. IEEE Computer Society (2019). https://doi.org/10.1109/ISM.2019.00044

19. Song, S., Lan, C., Xing, J., Zeng, W., Liu, J.: Spatio-temporal attention-based LSTM networks for 3d action recognition and detection. IEEE Trans. Image Process. **27**(7), 3459–3471 (2018). https://doi.org/10.1109/TIP.2018.2818328, https://doi.org/10.1109/TIP.2018.2818328

20. Tanfous, A.B., Zerroug, A., Linsley, D., Serre, T.: How and what to learn: taxonomizing self-supervised learning for 3d action recognition. In: IEEE/CVF Winter Conference on Applications of Computer Vision (WACV), pp. 2888–2897 (2022). https://doi.org/10.1109/WACV51458.2022.00294

21. Thakkar, K.C., Narayanan, P.J.: Part-based graph convolutional network for action recognition. In: British Machine Vision Conference (BMVC), pp. 1–13. BMVA Press (2018). http://bmvc2018.org/contents/papers/1003.pdf

22. Wang, J., Chen, Y., Hao, S., Peng, X., Hu, L.: Deep learning for sensor-based activity recognition: a survey. Pattern Recogn. Lett. **119**, 3–11 (2019)

23. Wang, J., Jin, S., Liu, W., Liu, W., Qian, C., Luo, P.: When human pose estimation meets robustness: Adversarial algorithms and benchmarks. In: IEEE/CVF Conference on Computer Vision and Pattern Recognition (CVPR), pp. 11855–11864 (2021)

24. Wang, W., Chang, F., Liu, C., Li, G., Wang, B.: Ga-net: a guidance aware network for skeleton-based early activity recognition. IEEE Trans. Multimedia, 1–13 (2021). https://doi.org/10.1109/TMM.2021.3137745

25. Wen, Y.H., Gao, L., Fu, H., Zhang, F.L., Xia, S., Liu, Y.J.: Motif-gcns with local and non-local temporal blocks for skeleton-based action recognition. IEEE Trans. Pattern Anal. Mach. Intell. 1–15 (2022). https://doi.org/10.1109/TPAMI.2022.3170511

26. Zhang, T., et al.: Deep manifold-to-manifold transforming network for skeleton-based action recognition. IEEE Trans. Multi. **22**(11), 2926–2937 (2020). https://doi.org/10.1109/TMM.2020.2966878

KAPP: Knowledge-Aware Hierarchical Attention Network for Popularity Prediction

Shuodian Yu[1], Jianxiong Guo[2], Xiaofeng Gao[1]([✉]), and Guihai Chen[1]

[1] Shanghai Key Laboratory of Scalable Computing and Systems,
Department of Computer Science and Engineering, Shanghai Jiao Tong University,
Shanghai, China
timplex233@sjtu.edu.cn, {gao-xf,gchen}@cs.sjtu.edu.cn
[2] Advanced Institute of Natural Sciences, Beijing Normal University,
Zhuhai 519087, China
jianxiongguo@bnu.edu.cn

Abstract. With the social networks becoming a major source of information in recent years, predicting the popularity of information in social networks has appeared intriguing to researchers in both academia and industry. However, existing methods still lack the utilization of external knowledge features, and are hard to extract the internal knowledge correlation of social information.

In this paper, we propose a knowledge-aware hierarchical attention network for popularity prediction (KAPP), which integrates the representation of the knowledge graph into popularity prediction. We aim to learn information representation based on temporal point process and knowledge graphs simultaneously from social content. In information cascading, we design a hierarchical attention mechanism to simulate the attention of human beings and the influences of users in social networks, and naturally establish the model structure from knowledge characteristics to popularity prediction.

In All, through attention mechanism for knowledge graph expression and analogy learning of temporal point process, our work makes an efficient prediction for information in social networks with the deep learning method. On the real-world data set of Weibo, we evaluate our model with intensive experiments and metrics, which outperforms previous methods including traditional approaches and deep learning methods.

1 Introduction

Nowadays, as social networks become a major source of information, online social platforms, e.g., Twitter, Facebook and so on remarkably facilitate the production and delivery of information. There are millions of information produced on these platforms every day, which makes predicting the popularity of pieces of information valuable for us to detect popular information in advance. Following most of the previous research, the popularity of an item is generally measured by the number of times it has been reposted in this network.

© The Author(s), under exclusive license to Springer Nature Switzerland AG 2022
C. Strauss et al. (Eds.): DEXA 2022, LNCS 13426, pp. 248–255, 2022.
https://doi.org/10.1007/978-3-031-12423-5_19

In the past few years, there have been a series of efforts devoted to the popularity prediction problem. Borrowing ideas from financial modeling and epidemiology, generative approaches [1,7] based on temporal point process were proposed to characterize and model the process that information diffusing. The end-to-end deep representation learning methods [4] further improve the performance while fully utilizing the temporal and the content information. Deep-Hawkes [3] learns the parameters of temporal point process automatically and achieved high performance. Nonetheless, it only predicts based on the diffusing paths while ignoring the information content itself. The difficulties of popularity prediction nowadays still fall into the usage and representation of derived features with reasonability and interpretability.

Knowledge graph (KG) is a kind of knowledge base that stores all kinds of knowledge with relationships among different entities, which bridges the gap between human commonsense knowledge and structured network. We aim to utilize knowledge graph to improve the representation learning of social information. At present, most of knowledge graphs tend to use a factual triple in the form of (head, relation, tail) to express the relation between two entities, which enables us to learn low-dimensional embeddings of entities without loss of graph structure.

In this paper, we propose a knowledge-aware hierarchical attention network for popularity prediction (KAPP). We introduce the knowledge graph to express the user characteristics with their social information content. Each user is represented as a sequence of entities that are extracted from their historical tweet or retweet contents, which makes the model aware of the user preferences and interests. A hierarchical attention mechanism is introduced to model the user historical records and the retweet relationship between users individually. This hierarchical architecture naturally models the preferences and the influence of different users in information cascades. The KAPP model concurrently inherits the predictive power of deep learning methods and the high interpretability of temporal point process and knowledge graph, connecting the traditional feature-based methods and temporal-based methods. We verify the predictive power of our model by using it to predict the popularity in the real-world Weibo dataset. Our contribution of this thesis can be summarized as follows:

(1) **Application of knowledge graph:** The proposed model applies knowledge graph to characterize the representation of both users and social message contents. It bridges the gap between common sense knowledge and time-series based popularity prediction so that the model is aware of diverse and interpretable human knowledge.

(2) **Attention mechanism for information modeling:** We construct a hierarchical attention network to model the attention distributed on different knowledge graph entities and different participators in the retweet cascade. It naturally makes an analogy of temporal point process and expresses the popular factors of different entities and user influence in the social network that both will affect the future popularity of messages.

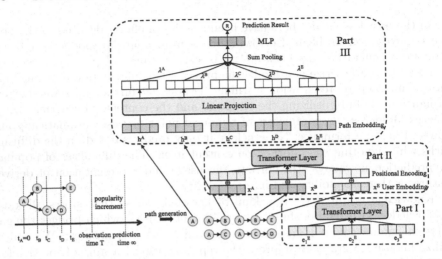

Fig. 1. The Architectural Overview of the KAPP Model. Each user is represented by his historical content entities with a transformer layer in Part I. The cascade of a piece of information is sampled into several retweet paths in Part II, which are transformed with another attention network and contribute to the final prediction in Part III.

2   Methodology

In this section we will introduce the proposed model for popularity prediction in detail as illustrated in Fig. 1.

2.1   Knowledge-Aware User Embedding

People are topic-sensitive in social networks as they tend to be interested in specific topic categories. Based on messages that a user retweeted before an observation time, it is feasible to distill the user's interested topics through the contents of historical social information. First, for a piece of message m, we link the words with entities in knowledge graph. TransE [2] is applied to learn the representation of knowledge entities. Then the message m can be denoted as $content(m) = \{e_1, e_2, \dots\}$. To fully model the characters or interests of users with contents, we derive the entity representations from the historical records of users. Formally, we use $\mathcal{M}_u$ as the set of retweeted messages of a user u, and collect all the entities mentioned in past information to indicate the user as the input of the model:

$$h'_u = \bigcup_{m \in \mathcal{M}_u} \{e | e \in content(m)\}. \tag{1}$$

Nevertheless, the number of exposed entities of different users may have large divergence. Therefore we have to unify the number of entities for all users through term frequency-inverse document frequency (TF-IDF) [6] to select the most important entities for the users $h_u = \{e_1, e_2, \dots, e_n\}$

The importance of different entities is changing dynamically with variant users. Here self-attention mechanism is applied to learn the importance of each entity to model the interests of users. We use a Transformer block introduced in [9] to aggregate the users' interested entity collections, which results in the knowledge-aware user representation as:

$$u^i = \sum_{j=1}^{n} \text{Trans}_{user}(e_j^i),\qquad(2)$$

where the parameters of this function is trained and shared among all the users.

2.2   Attention-Based Retweet Path Encoding

Borrowing the idea from DeepCas [4], representing a cascade graph as a set of cascade paths is a proper way to model the diffusion of information. We sample the retweet paths to represent the graph as illustrated in Fig. 1. For an inputted cascade $\mathcal{C}^i$, we encode the entire retweet path p_j^i for each user u_j^i where $0 \leq j \leq R_T^i$ and R_T^i denotes the cascade size before the observation time T. The path p_j^i is a sequence that describes the entire retweet relationship from the original user to u_j^i.

We still adopt another Transformer layer $\text{Trans}_{path}(p_j^i)$ to model the cascade paths. External position information of the items is injected to make use of the user order in retweet paths since the attention mechanism does not care about the relative order as [9]. For each retweet path p_j^i, we use the last states of $\text{Trans}_{path}(p_j^i)$ as the representation of the diffusion path denoted. The representation c_i for the whole cascade $\mathcal{C}^i$ is obtained by a sum pooling for all retweet paths:

$$c^i = \sum_{j=1}^{R_T^i} \text{Trans}_{path}(p_j^i).\qquad(3)$$

2.3   Prediction with Point Process

Since the observation time can be very long in the real data, we manually split the time sequence of length T into several disjoint intervals to make the time information discrete and learnable. The time range $[0, T)$ is divided into a collection of L equaling subranges $\{[t_0, t_1), [t_0, t_1), \ldots, [t_0, t_L)\}$ through a mapping function f:

$$f\left(T - t_j^i\right) = l, \quad \text{if} \quad t_{l-1} \leq T - t_j^i < t_l,\qquad(4)$$

where t_j^i is the time gap from the origin to the j-th retweet event of m^i.

Following the DeepHawkes [3], a non-parametric way is used to learn the time decay effect in Hawkes process directly without handcrafted functions. For the cascade $\mathcal{C}^i$, the arrival rate of future retweet events can be assembled by summing up all the observed point process. For each retweet path p_j^i while the corresponding time is t_j, we use the learned path encoding $\text{Trans}(p_j^i)$ multiplied by the time decay effect $\lambda_{f(T-t_j^i)}$ to represent its embedding after the observation time to get the representation for the cascade $\mathcal{C}^i$. A MLP layer is used to obtain the one-dimensional output as:

$$\Delta \hat{R}_T^i = \text{MLP} \left(\sum_{j=1}^{R_T^i} \lambda_{f(T-t_j^i)} \text{Trans}(p_j^i) \right). \tag{5}$$

We take log-transformation to ΔR_T to reduce the impacts of outliers and the objective function to be minimized is:

$$\mathcal{L} = \frac{1}{M} \sum_{i=1}^{M} \left(\log \Delta \hat{R}_T^i - \log \Delta R_T^i \right)^2. \tag{6}$$

3   Experiments

3.1   Experiment Setup

To thoroughly evaluate the KAPP model, we simulate it by applying to a real-world data set, Weibo [13], of popularity prediction. The task is to predict the future size of the cascades for the messages in Weibo. We use CN-DBpedia [12] as our knowledge graph, which contains 1575402 entities and 2115915 triple relations after pre-processing. Following the practice of previous works, we adopt four metrics to measure the prediction precision: Mean Square Log-Transformed Error (MSLE) and Median Square Log-Transformed Error (mSLE) used in [3], Mean Absolute Percentage Error (MAPE) [11], and Wrong Percentage Error (WroPerc) [8]. When we train the representation of knowledge graph, we choose TransE [2]. Stochastic gradient descent (SGD) with L2-norm is used to train both the knowledge graph embeddings and the prediction model.

3.2   Numerical Results

Comparison Among Different Models. The results of comparisons of different models are shown in Tables 1 and our model outperforms all the baselines. We compared with Feature-Based method described in [3], SEISMIC [14], and Deep-Hawkes [3]. Note that for SEISMIC, due to it lacks the optimization about future popularity, it does not perform well for cascade prediction and we only use mSLE as the evaluation metric. The results compared to both feature-based method and time-based method show that our KAPP model has a good integration of two aspects. It outperforms end-to-end deep learning method DeepHawkes, which indicates that the knowledge features we introduce indeed contribute to precision of social network popularity prediction problem.

Table 1. Overall performance comparison (the smaller the better)

Method	MSLE	mSLE	MAPE	WroPerc
Feature-based	3.475	1.224	1.013	35.7%
SEISMIC	–	0.703	–	–
DeepHawkes	1.314	**0.563**	0.292	26.9%
KAPP	**1.246**	0.564	**0.288**	25.9%

Table 2. Comparison among variants

Method	MSLE	mSLE	MAPE	WroPerc
Original KAPP with 2 heads, 1 block	**1.246**	**0.564**	0.288	25.9%
KAPP-GRU	1.251	0.571	**0.251**	26.1%
KAPP without KG	1.306	0.628	0.291	26.1%
KAPP + TransH	1.285	0.607	0.301	27.1%
KAPP + TransR	1.253	0.571	0.293	**25.4%**

Analysis of Components. To verify that the attention mechanism contributes to the prediction results, we replace the Transformer layer in the part of retweet path encoding with GRU to compare with original model as KAPP-GRU in Table 2. It evidences that the attention mechanism has ability to model the short sequence and capture the relatedness between items as well.

To prove that the knowledge graph works, we deploy a ablation model KAPP without which directly use a randomly initialized learnable vector to represent users. We also adopts TransH [10] and TransR [5] to learn the representation of knowledge graph entities. As shown in Table 2, our original model outperforms all kinds of variants. It proves that the model indeed makes use of the knowledge graph information since it have a lower error than the model without knowledge graph. As for the comparison between TransE, TransH, TransR, although the later two models learned more robust representation of knowledge graph, the TransE method still achieves slightly better results. One possible reason is that, the entities embeddings are fed into the prediction model and propagate with a deep network. Subtle representation is not suitable or needed in such situation. Hence TransE is chosen as our final version of knowledge graph representation learning.

4   Conclusion

This paper proposes a knowledge-aware hierarchical attention network for popularity prediction to predict the incremental popularity of messages in social networks. It introduces the knowledge graph to learn the representations of users based on their historical information at semantic-level, which provides us with a way to understand the information cascades that which entities or topics promote the diffusion process. The hierarchical self-attention mechanism naturally expresses the importance of different entities and users in social network.

Acknowledgements. This work was supported by the National Key R&D Program of China [2019YFB2102200], the National Natural Science Foundation of China [61872238, 61972254], Shanghai Municipal Science and Technology Major Project [2021SHZDZX0102] and the ByteDance Research Project [CT20211123001686].

References

1. Bao, P., Shen, H.W., Jin, X., Cheng, X.Q.: Modeling and predicting popularity dynamics of microblogs using Self-excited Hawkes Processes. In: International Conference on World Wide Web (WWW), pp. 9–10 (2015)
2. Bordes, A., Usunier, N., Garcia-Duran, A., Weston, J., Yakhnenko, O.: Translating embeddings for modeling multi-relational data. In: Advances in Neural Information Processing Systems (NIPS), pp. 2787–2795 (2013)
3. Cao, Q., Shen, H., Cen, K., Ouyang, W., Cheng, X.: DeepHawkes: bridging the gap between prediction and understanding of information cascades. In: ACM on Conference on Information and Knowledge Management (CIKM), pp. 1149–1158 (2017)
4. Li, C., Ma, J., Guo, X., Mei, Q.: DeepCas: an end-to-end predictor of information cascades. In: International Conference on World Wide Web (WWW), pp. 577–586 (2017)
5. Lin, Y., Liu, Z., Sun, M., Liu, Y., Zhu, X.: Learning entity and relation embeddings for knowledge graph completion. In: AAAI Conference on Artificial Intelligence (AAAI) (2015)
6. Rajaraman, A., Ullman, J.D.: Mining of Massive Datasets. Cambridge University Press, Cambridge (2011)
7. Shen, H., Wang, D., Song, C., Barabási, A.L.: Modeling and predicting popularity dynamics via reinforced Poisson processes. In: AAAI Conference on Artificial Intelligence (AAAI) (2014)
8. Tatar, A., de Amorim, M.D., Fdida, S., Antoniadis, P.: A survey on predicting the popularity of web content. J. Internet Serv. Appl. 5(1), 1–20 (2014). https://doi.org/10.1186/s13174-014-0008-y
9. Vaswani, A., et al.: Attention is all you need. In: Advances in Neural Information Processing Systems (NIPS), pp. 5998–6008 (2017)
10. Wang, Z., Zhang, J., Feng, J., Chen, Z.: Knowledge graph embedding by translating on hyperplanes. In: AAAI Conference on Artificial Intelligence (AAAI) (2014)
11. Wu, Q., Yang, C., Zhang, H., Gao, X., Weng, P., Chen, G.: Adversarial training model unifying feature driven and point process perspectives for event popularity prediction. In: International Conference on Information and Knowledge Management (CIKM), pp. 517–526 (2018)
12. Xu, B., et al.: CN-DBpedia: a never-ending Chinese knowledge extraction system. In: Benferhat, S., Tabia, K., Ali, M. (eds.) IEA/AIE 2017. LNCS (LNAI), vol. 10351, pp. 428–438. Springer, Cham (2017). https://doi.org/10.1007/978-3-319-60045-1_44

13. Zhang, J., Liu, B., Tang, J., Chen, T., Li, J.: Social influence locality for modeling retweeting behaviors. In: International Joint Conference on Artificial Intelligence (IJCAI) (2013)
14. Zhao, Q., Erdogdu, M.A., He, H.Y., Rajaraman, A., Leskovec, J.: SEISMIC: a self-exciting point process model for predicting tweet popularity. In: ACM SIGKDD International Conference on Knowledge Discovery and Data Mining (KDD), pp. 1513–1522 (2015)

Advanced Machine Learning

A Heterogeneous Network Representation Learning Approach for Academic Behavior Prediction

Li Huang and Yan Zhu$^{(\boxtimes)}$

Southwest Jiaotong University, Chengdu 611756, China
yzhu@swjtu.edu.cn

Abstract. Predicting authors' academic behavior (e.g. co-authorship, citation) based on heterogeneous academic network can help scholars to grasp interesting research directions and participate in various co-operations. Most of the existing network representation methods use the structural and content features of nodes, but have not fully exploited the edges (relationships) between nodes (entities) and investigated the semantic compatibility of different edge types yet. To solve the above problems, a heterogeneous network representation learning method (HNE-ABP) is proposed to improve feature extraction and academic behavior prediction performance. HNEABP has three strengths: 1) capture rich neighbor information via balanced sampling and Skip-Gram, 2) apply knowledge graph embedding (KGE) technique to learn pairwise node information and to weight the importance of first-order neighbors, 3) solve the semantic incompatibility of edges based on KGE. Validation experiments on three academic network datasets show that HNE-ABP outperforms the popular network representation methods, which gives the credit to HNEABP for learning richer feature information effectively, so as to improve the performance of academic behavior prediction.

Keywords: Heterogeneous network representation learning · Link prediction · Knowledge graph · Balanced sampling

1 Introduction

In today's academic society, scientific research activities tend to be diversified, multi-cooperation, and interdisciplinary. Academic social network has become an important data resource containing massive academic information. Academic network is classified as homogeneous and heterogeneous network. A homogeneous network contains only one type of nodes and one type of edges between nodes. A heterogeneous network contains rich structural and semantic information, such as multiple types of nodes (entities) and edges (relationships). Figure 1(a) depicts a homogeneous network, while Fig. 1(b) depicts a heterogeneous one, where three types of nodes (author, paper, venue) and four types of edges (e.g. publication, collaboration) are shown.

Mining heterogeneous academic networks can discover research trend and scholars' behavior pattern (e.g. collaboration, citation), or recommend interdisciplinary cooperator, and so on. In this paper, we devote ourselves to extract and represent rich and

© The Author(s), under exclusive license to Springer Nature Switzerland AG 2022
C. Strauss et al. (Eds.): DEXA 2022, LNCS 13426, pp. 259–272, 2022.
https://doi.org/10.1007/978-3-031-12423-5_20

discriminative features from heterogeneous academic networks for efficient academic behavior prediction. Academic behavior prediction is also called edge/link prediction, which predicts the author's future academic behaviors. For example, we predict there may be an edge between A2 and A3 as new cooperation (demonstrated with a red dotted line), when A1/A2 and A1/A3 have cooperative relationships (see Fig. 1(b)).

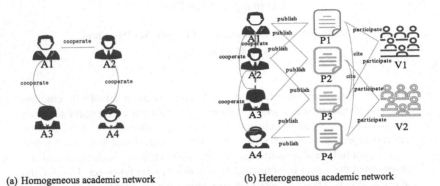

(a) Homogeneous academic network   (b) Heterogeneous academic network

Fig. 1. Academic network

One of the conventional link prediction methods is to determine a possible link between nodes based on similarity index, such as common neighbor (CN), Adamic Adar (AA). The shortages are high real-time and imprecise. Another method uses matrix operations and probability models, which are computationally expensive and time-consuming.

In some present researches, low-dimensional vector is used effectively to capture node features in heterogeneous networks and improves the link prediction accuracy. However, many methods still have not exploited the hidden information of nodes and edges. Methods like Deepwalk [1] and Node2vec [2] sample neighbors with random walk, which focuses on the nodes connected by a large number of edges of a certain type and result in the edges of some minor types are not sampled or less sampled. In JUST [3] and BHin2vec [4], the unbalanced sampling is solved by a balanced walk, all node features are learnt in the same representation space, but different edges between the nodes represent different semantics. In their work, not only the semantic incompatibility of edges occurs, but also the reinforcement of learning of first-order neighbors is not considered.

To deal with the above issues simultaneously, we propose a heterogeneous network representation learning method (HNEABP) that combines balanced sampling and knowledge graph embedding technology for academic behavior prediction. The contributions are as follows.

(a) Two balanced walk algorithms are proposed to sample neighbors of different types in a balanced manner and extract neighbor information comprehensively.
(b) Node pair information is preserved by using knowledge graph embedding (KGE) technology. In this way, the learning of first-order neighbor is strengthened.

(c)   Nodes are mapped to different representation spaces with KGE method TransR [5], where we can effectively learn the semantic information of nodes and edges and solve the problem of semantic incompatibility.
(d)   Validation experiments on three datasets show that HNEABP can capture the multi-order neighbor information and learn richer features than the baseline methods.

The structure of this paper is as follows. The related work is introduced in Sect. 2. The working mechanism of HNEABP is addressed in detail in Sect. 3. Section 4 discusses the verification experiments and results. Finally, we conclude the proposed methods and give future research directions in Sect. 5.

2   Related Work

Recently studies on node representation learning methods in heterogeneous networks are very active. This paper investigates and analyzes related work on neighbor sampling, neighbor weighting, and semantic incompatibility.

Neighbor sampling. The number of nodes or edges of different types is not equal in heterogeneous network. In random walk sampling methods, the larger the amount of edges of one type is, the greater the sampling probability of their connecting nodes, and vice versa. Such a sampling technique results in some nodes are learnt repeatedly and some nodes information is rarely or even never sampled. Unbalanced sampling cannot extract node feature comprehensively. Therefore, balanced sampling methods are introduced, for example, JUST adds jump and stay strategies in random walks to overcome the bias rooting in high visible nodes sampling. BHin2vec uses inverse training ratio to update the walk probability for balanced sampling. However, the different importance of heterogeneous neighbors and semantic incompatibility of edges are not taken into account in the above methods.

Neighbor weighting. Every node directly connects with its first-order neighbors, which influence on the node or mutual relationship is therefore stronger than those from the second and higher order neighbors. The prediction precision can be improved by strengthening the importance of first-order neighbor. To this end, Zhao et al. proposed NSHE [6], which captures neighbor information by using network pattern structure including the importance of first-order neighbors. Nevertheless, the time complexity is high and it does not solve the semantic incompatibility of edges.

Semantic incompatibility. Most of the existing methods learn node representation in the same representation space, but the edges between different nodes represent different semantics. For example, there is a collaborative relationship between authors and a writing relationship between authors and papers. Nodes and edges as different objects may not be represented accurately in a common node space. It is a semantic incompatibility problem, if edges of different types are all learned in the same representation space. Recently, approaches on embedding nodes into different representation spaces to solve semantic incompatibility are proposed, such as PME [7], HEER [8], ASPEM [9], PGRA [10], and RHINE [11]. In addition, KGE is also studied to solve the problem. TransR as a KGE approach can correctly learn the features by mapping different nodes of a KGE triple into different representation spaces. In one space only the same relationship (edge) type is specified and learnt. A limitation of TransR is it cannot extract features from high-order neighbors.

Inspired by BHin2vec and TransR, this paper proposes a node representation learning method (HNEABP), which solves the following problems simultaneously: 1) Handling unbalanced sampling through balanced walk to extract as more kinds of feature as possible. 2) Increasing importance of first-order neighbors in node representation by KGE technology. 3) Dealing with semantic incompatibility of edges using KGE method, TransR.

3   The Mechanism of HNEABP Approach

The mechanism of HNEABP is shown in Fig. 2, which contains 2 key parts. a) Balanced sampling and Skip-Gram model for comprehensive learning of neighbor information. b) Node pair and semantic information learning based on KGE for enhancing the importance of the first-order neighbor and solving semantic incompatibility.

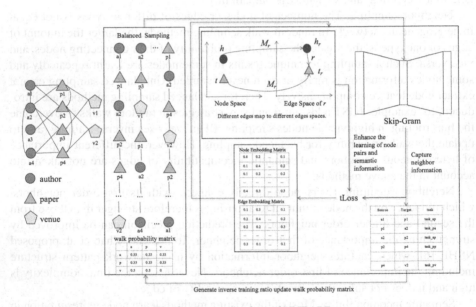

Fig. 2. The mechanism of HNEABP

3.1   Balanced Walk Method Based on Edge Number

(1) **Balanced Sampling.** BWNE, primary algorithm of HNEABP, calculates the edge number of different types during the walk and increases the sampling probability for those nodes connected by the edges which type has the minimum edge number. BWNE as balanced walk method also contains a part of random walk in order to avoid the oversampling of those nodes mentioned above. The sampling process can

not only sample neighbors uniformly, but also enhance the contribution of both low-order and high-order neighbors to the representation of the target node. Target node is the node which links with another node will be predicted. BWNE is shown in Algorithm 1.

Algorithm 1 BWNE

Input: Heterogeneous Network $G = (V, E)$, walk length L, start node number nID, random walk probability p

Output: Sampled node sequence w

initial $w[0] \leftarrow nID$

for i in 1 to L-1 do:

    pr $\leftarrow$ random(0,1) *//pr is a random number*

    if pr $< p$ { *//p probability for bias walk, 1-p probability for random walk*

        edge_num $\leftarrow$ cal(G) *//Calculate edge number of different types: ap:x_1, pp:x_2, pv:x_3, aa:x_4, av:x_5*

        edge_type $\leftarrow$ ψ (min{edge_num}) *//Obtain edge type based on the minimum number of edges*

        node_type $\leftarrow$ ϕ ($v_{w[i]}$) *//Obtain the type of current node*

        next_type $\leftarrow$ type(edge_type, node_type) *//Determine the next priority sampling node type based on the current node type and edge type.*

        if($(v_{w[i]}, v_j) \in E$ and $\phi(v_j)$==next_type) *//v_j is the neighbor of $v_{w[i]}$, the type of v_j is the type of preferential sampling (next_type).*

            $w[i+1]$ $\leftarrow$ node ID of v_j

        else $w[i+1]$ $\leftarrow$ a random node ID of $v_{w[i]}$'s neighbor   }

    else $w[i+1]$ $\leftarrow$ a random node ID of $v_{w[i]}$'s neighbor

end for

return w

Among them, *ap* denotes the number of links between authors and papers, *pv* denotes the numbers of links between papers and venues, and so on. ϕ is node type mapping function ($\phi : V \rightarrow \mathcal{A}$) and ψ represents edge type mapping function ($\psi : E \rightarrow \mathcal{R}$). $\mathcal{A}$ and $\mathcal{R}$ represent predefined types of node and edge, respectively.

(2) **Node Representation Learning Using Skip-Gram Model.** After having obtained the sampled node sequence, the features (representation) of nodes will be extracted using Skip-Gram. Node representation is the low-dimensional vector of a node. The idea of Skip-Gram is to determine the representation of neighbors by using the representation of target node. One node acts as either a target node or a neighbor node. Firstly, a target node V_i is selected from walking sequence w. V_i's k-window neighbors are the nodes within a distance of k links (including direct or indirect neighbors), which is the size of a moving window. The similarity between the embedding vector of V_i and its neighbors is then computed. Main principle is to maximize the similarity between V_i and k-window neighbors and minimize the similarity between V_i and non-neighbors. Non-neighbors are obtained via negative

sampling as shown in Eq. 1–3.

$$L_1 = -\sum_{i=1}^{l} \sum_{j=1}^{k} (L_p(v_{w[i+j]}, v_{w[i]}) + \sum_{v_0}^{N_m} L_n(v_0, v_{w[i]})) \tag{1}$$

$$L_p(v_s, v_t) = \log \sigma (f(v_s)^{\mathrm{T}} f(v_t)) \tag{2}$$

$$L_n(v_s, v_t) = \log \sigma (-f(v_s)^{\mathrm{T}} f(v_t)) \tag{3}$$

In Eq. 1–3 L_1 represents the walk loss, N_m represents negative samples, and m is the number of negative samples selected from the node set V randomly. L_p and L_n represent the losses caused by positive and negative samples respectively, σ is the sigmoid function, $f(v_s)$ and $f(v_t)$ represent the representation of source and target node.

3.2 Balanced Walk Method Based on Edge Loss

(1) **Improved Balanced Sampling**. Although BWNE can conduct a balanced sampling, it is time consuming and its adaptability is rigid to a certain degree, because the edge number of different types is calculated continually during the walk and the sampling probability must be adjusted in time according to the dynamic edge number of the minor types.

BWEL, an improved balanced walk method for HNEABP, is proposed. The key idea of BWEL is to increase the sampling of those nodes, which are connected by the edges with relatively large walk losses, instead of increasing the sampling probability of nodes which are connected by edges with the minimum number. An inverse training radio is produced by adjacency matrix and loss matrix (LM $\in \mathbb{A}^{|A| \times |A|}$). The walk probability matrix (WPM $\in \mathbb{A}^{|A| \times |A|}$) stores the jump probability of each node type and is updated by generated inverse training radio, then WPM guides the sampling of the next node. BWEL is shown in Algorithm 2. Compared with BWNE, BWEL only needs to calculate the walk loss generated by heterogeneous edges after each sampling, which reduces the time complexity.

Algorithm 2 BWEL

Input: Heterogeneous Network $G = (\mathcal{V}, E)$, walk length L, start node number nID, WPM

Output: Sampled node sequence w

initial $w[0] \leftarrow$ nID

adjacency_matrix $\leftarrow$ cal_adj(G) //*Calculate the adjacency matrix for each node*

for i in 1 to L-1 do:

    probability $\leftarrow$ WPM($\phi(v_{w[i]})$) //*Walk probability between nodes*

    weight $\leftarrow$ probability * adjacency_matrix($v_{w[i]}$) //*Calculate the probability of a node jumping to the neighbors of different types*

    next_type $\leftarrow$ max_number(weight) //*The node type with the highest jumping probability is preferentially selected as the neighbor type for the next priority sampling*

    if($(v_{w[i]}, v_j) \in E$ and $\phi(v_j)$==next_type) //*v_j is the neighbor of $v_{w[i]}$, the type of v_j is the type of preferential sampling (next_type).*

    $w[i+1] \leftarrow$ node ID of v_j

    else $w[i+1] \leftarrow$ a random node ID of $v_{w[i]}$'s neighbor

end for

return w

---

(2) **Update Strategy for WPM.** The WPM is only used for the balanced walk of the current node, so it is necessary to continuously update the WPM during the iteration process to guide balance sampling of all nodes. During the update process, multi-task learning technique is adopted. Multiple tasks are planned according to the edge of different types. $Task_{ij}$ is to predict whether a node of type $type_i$ connects to node of type $type_j$. The total loss caused by $Task_{ij}$ is divided into multiple sub-losses and saved in the loss matrix (LM $\in \mathbb{A}^{|\mathcal{A}| \times |\mathcal{A}|}$). Then we calculate the inverse training ratio for each task and save it in $S[Task_{xy}]$. Finally we use a random walk matrix (nodes are sampled with an equal probability) constrained by the inverse training ratio as the node walk probability target to update the WPM to guide the new walk to sample neighbors. The strategies are shown in Eq. 4–8.

$$LM[Task_{xy}] = -\frac{\sum_{i=1}^{l}(L_p(v_{w[i+1]}, v_{w[i]}) + \sum_{v_0}^{N_m} L_n(v_0, v_{w[i]}))}{\sum_{i=1}^{l}(I[Task_{xy}](v_{w[i+1]}, v_{w[i]}) + \sum_{v_0}^{N_m} I[Task_{xy}](v_0, v_{w[i]}))} \tag{4}$$

$$I[Task_{xy}](v_s, v_t) = \begin{cases} 1, & if\ \phi(v_s) = type_x \wedge \phi(v_t) = type_y \\ 0, & otherwise \end{cases} \tag{5}$$

$$S[Task_{xy}] = LM[Task_{xy}] / \sum_{j=0}^{|\mathcal{A}|} LM[Task_{xj}] / \sum_{j=0}^{|\mathcal{A}|} I[Task_{xj}] \tag{6}$$

$$p_{uni_{xy}} = \begin{cases} \frac{1}{degree(v_x)} & if\ (v_x, v_y) \in E \\ 0 & otherwise \end{cases} \tag{7}$$

$$L_{stochastic} = |WPM - (p_{uni_{xy}} + \delta(S[Task_{xy}] - 1))|_F^2 \tag{8}$$

In Eq. 4–8, LM[$Task_{xy}$] records the walk losses generated by edges of different types. I[$Task_{xy}$](v_s, v_t) indicates whether there is an edge between nodes of different types. $S[Task_{xy}]$ is the inverse training radio matrix storing loss ratios for multiple tasks. $L_{stochastic}$ denotes an optimization parameter for updating the walk probability matrix (WPM), which tunes the node walk probability based on the inverse training ratio. $P_{uni_{xy}}$ is a random walk probability matrix. δ is the perturbation parameter.

3.3 Node Pair and Semantic Information Learning Based on KGE

In heterogeneous network, many node representation methods specify heterogeneous nodes in the same representation space for extracting features. However, nodes and edges of different types have different semantics. The bigger the semantic difference a node pair has, the more diverse the relationship of the node pair is, thereby the longer the edge between the two nodes is. For example in Fig. 3(a), Mary likes gardening book and also the writer Hemingway, but Hemingway did not write any gardening book. The distance between Hemingway node and gardening book node is very far. During the training process, it is difficult to simultaneously obtain the similarity between Mary and the other two nodes, which is semantic incompatibility. Obviously, node representations obtained by such methods are inaccurate.

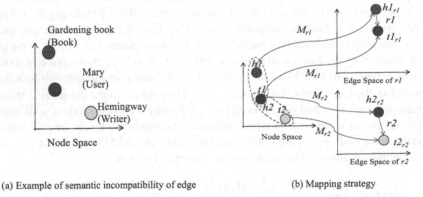

(a) Example of semantic incompatibility of edge   (b) Mapping strategy

Fig. 3. Semantic incompatibility problem and solution

In order to solve the semantic incompatibility, KGE method, TransR, is integrated in HNEABP. TransR can represent each edge as a triple, which contains node embedding set $h, t \in \mathbb{R}^d$ and edge embedding set $r \in \mathbb{R}^u$. A mapping matrix $M_r \in \mathbb{R}^{d \times u}$ is applied to map nodes from node space to the corresponding edge space based on different edge type r. The mapping method makes the distance between the head node to the tail node in the edge space as short as possible. By representing the edges of different type in different spaces, the semantic incompatibility of edges can be eliminated (see Fig. 3(b)). The margin ranking loss, as shown in Eq. 9–11, is used for computing loss.

$$h_r = hM_r, t_r = tM_r \tag{9}$$

$$f_r(h, t) = \|h_r + r - t_r\|_F^2 \tag{10}$$

$$L_2 = \sum_{(h,r,t)\in S} \sum_{(h',r,t')\in S'} \max(0, f_r(h,t) + \gamma - f_r(h',t')) \qquad (11)$$

where, h_r and t_r denote the representation of the head and tail node in the representation space of edge type r. $f_r(h,t)$ is score function. L_2 is the loss of TransR. γ is a margin. S and S' are the set of correct and incorrect triples, respectively.

3.4 Loss Function

HNEABP uses two loss matrixes, one is used to capture n-order neighbor information by Skip-Gram (see Sect. 3.1(2)), the other is used by TransR to enhance first-order neighbor information and semantic features of nodes and edges (see Sect. 3.3).

$$tLoss = \alpha L_1 + \beta L_2 \qquad (12)$$

$tLoss$ is the total loss, where L_1 and L_2 denote the Skip-Gram loss and TransR loss, respectively. α and β represent the weight coefficient.

3.5 HNEABP Algorithm

HNEABP-BWNE (Algorithm 3) and HNEABP-BWEL (Algorithm 4) adopt different balanced walk strategies and node representation techniques.

Algorithm 3 HNEABP-BWNE

Input: Heterogeneous network $G = (V, E)$, embedded dimension d, walk length L, epoch e, moving window size k, negative sample size m, learning rate l_l, triple $(h,r,t) \in S$, margin γ, random walk probability p

Output: Node embedding matrix Z, edge embedding matrix R

initial $Z \in \mathbb{R}^{|V| \times d}$, $R \in \mathbb{R}^{|R| \times d}$ //Define and initialize vectors for nodes and edges

for 1 to e:

    for each node nnID:

        $w \leftarrow$ BWNE(G, L, nnID, p)//Obtain node sampling sequence

        $L_l \leftarrow$ Skip-Gram(w, k, m, Z, R)//Extract the features (representation) of nodes

        calculate distance score for positive and negative samples with Eq. (9)-(10)

        calculate TransR loss (L_2) with Eq. (11) and the total loss ($tLoss$) with Eq. (12)

        train and minimize $tLoss$, compute the relevant parameters

    end for

end for

return Z, R

Algorithm 4 HNEABP-BWEL

Input: Heterogeneous networks $G = (\mathcal{V}, E)$, embedded dimension d, walk length L, epoch e, moving window size k, negative sample size m, learning rate l_1, l_2, triple $(h, r, t) \in S$, margin γ, WPM

Output: Node embedding matrix Z, edge embedding matrix R

initial $Z \in \mathbb{R}^{|V| \times d}$, $R \in \mathbb{R}^{|R| \times d}$, WPM=$[[0,...,0],...,[0,...,0]] \in \mathbb{R}^{|A| \times |A|}$

WPM$[i,j] \leftarrow 1$ for explicit edge;

for 1 to e:

  for each node nnID:

    $w \leftarrow$ BWEL(G, L, nnID, WPM)//*Obtain node sampling sequence*

    $L_1 \leftarrow$ Skip-Gram(w, k, m, Z, R)//*Extract the features (representation) of nodes*

    calculate distance score for positive and negative samples with Eq. (9)-(10)

    calculate TransR loss (L_2) with Eq. (11) and the total loss (*tLoss*) with Eq. (12)

    train and minimize *tLoss*, compute the relevant parameters

    compute walk loss and generate loss matrix $LM \in \mathbb{R}^{|A| \times |A|}$ using Eq. (4)-(8)

    update WPM and guide the new sampling of neighbors $\text{WPM}=\text{WPM}-l_2 \frac{\partial}{\partial \text{WPM}} L_{\text{stochastic}}$

  end for

end for

return Z, R

---

4 Verification Experiment on Node Representation Learning

4.1 Datasets and Baseline Methods

Three academic network datasets, citation network V1, V2 and ACM (ref. to Table 1), are used in experiments, which are publicly available on the AMiner platform. P-V in Table 1 means the number of edges between paper and venue nodes, and so on. HNEABP will be compared with six heterogeneous network representation methods.

Table 1. Dataset statistics

	Author	Paper	Venue	A-P	P-P	P-V	Time
V1	28646	21044	18	69311	46931	21044	2006–2015
V2	352068	315866	296	762997	59337	315866	1996–2005
ACM	485899	302395	333	957568	60462	302395	2012–2015

(a) Deepwalk [1]: use random walk for sampling and Skip-Gram for node representation.

(b) Metapath2vec [12]: adopt a specified meta-path pattern for sampling and Skip-Gram for learning node representation.

(c) JUST [3]: use a strategy by jumping to nodes of the same or different types for balanced sampling and Skip-Gram for learning node representation.

(d) BHin2vec [4]: apply inverse training ratio to update the walk probability for balanced sampling and Skip-Gram for learning node representation.

(e) NSHE [6]: preserve the neighbor information of node pairs and the structure feature of network patterns for learning node representation.

(f) TransR [5]: map nodes into different semantic spaces based on edges of different types and measure the rationality through distance between nodes in the edge space.

4.2 Experimental Setup and Evaluation Criteria

Three experiments on academic behavior prediction are designed to verify the effectiveness of heterogeneous network representation learning. The default learning rate is 0.001, the epoch is 20, the walk length L is 100, the dimension is 64, and moving window size is 7, the α is 1 and the β is 0.7. To ensure the fairness of the experiment, the datasets V1, V2 and ACM used in all mentioned methods are divided into training set and test set by a time point, e.g. the V2 data of year 1996–2002 is partitioned into training set and the part of 2003–2005 belongs to test set. Based on the same reason, the partition time point for V1 and ACM are 2011 and 2014, respectively. The evaluation indicators are F1 and AUC. The Adam optimizer is used in this paper, since it considers the first-order and the second-order moment estimation comprehensively. Besides, its parameters updating process is relatively stable.

4.3 Analysis of Experimental Results

(1) **Author's Academic Behavior Prediction.** HNEABP and six other methods accomplish firstly the representation (feature extraction) of nodes and edges from the datasets shown in Table 1. A logistic regression classifier is then applied to predict academic behaviors, namely co-authorship (A-P-A), author-paper citation (A-P-P), and author-venue participation (A-P-V). The prediction performance in terms of AUC and F1 is shown in Tables 2, 3 and 4.

Table 2. Results of co-authorship prediction (A-P-A)

Algorithm	V1		V2		ACM	
	AUC	F1	AUC	F1	AUC	F1
Deepwalk	0.794	0.272	0.814	0.483	0.821	0.471
Metapath2vec	0.77	0.679	0.745	0.652	0.727	0.649
JUST	0.857	0.415	0.834	0.561	0.795	0.627
BHin2vec	0.783	0.301	0.831	0.503	0.808	0.331
NSHE	0.784	0.60	0.78	0.601	0.794	0.645
TransR	0.854	0.689	0.828	0.589	0.835	0.579
HNEABP (BWNE)	0.853	0.735	0.858	0.639	0.852	**0.692**
HNEABP (BWEL)	**0.862**	**0.751**	**0.863**	**0.658**	**0.87**	0.657

Table 3. Results of citation prediction (A-P-P)

Algorithm	V1		V2		ACM	
	AUC	F1	AUC	F1	AUC	F1
Deepwalk	0.724	0.433	0.798	0.641	0.834	0.64
Metapath2vec	0.624	0.684	0.742	0.687	0.771	0.701
JUST	0.811	0.703	0.764	0.643	0.74	0.648
BHin2vec	**0.871**	0.63	0.842	0.636	0.857	0.61
NSHE	0.782	0.671	0.78	0.682	0.821	0.732
TransR	0.841	0.712	0.819	0.696	0.837	0.73
HNEABP (BWNE)	0.854	**0.766**	0.836	0.697	0.891	**0.749**
HNEABP (BWEL)	0.861	0.737	**0.875**	**0.709**	**0.901**	0.711

Table 4. Results of participation prediction (A-P-V)

Algorithm	V1		V2		ACM	
	AUC	F1	AUC	F1	AUC	F1
Deepwalk	0.643	0.345	0.866	0.669	0.902	0.801
Metapath2vec	0.627	0.524	0.894	0.725	0.887	0.741
JUST	0.741	0.309	0.813	0.678	0.822	0.693
BHin2vec	0.72	0.001	0.925	0.671	0.97	0.749
NSHE	0.705	0.57	0.683	0.574	0.711	0.612
TransR	0.737	0.574	0.712	0.588	0.773	0.669
HNEABP (BWNE)	0.769	0.592	0.923	0.757	0.969	0.811
HNEABP (BWEL)	**0.792**	**0.623**	**0.927**	**0.809**	**0.976**	**0.861**

The results show that HNEABP can greatly improve the prediction performance, for example, AUC increases by 0.5%–14.3% and F1 increases by 0.6%–47.9% on the co-authorship prediction. HNEABP outperforms Deepwalk because Deepwalk applies unbalanced sampling, while HNEABP samples the node information evenly. HNEABP is also better than JUST and BHin2vec in terms of the increase of AUC by an average of 5.25% and F1 by an average of 18.05%, although the latter ones use balanced sampling as well. This is because HNEABP reinforces the importance of first-order neighbors and effectively learns the semantic knowledge of nodes by mapping nodes to the corresponding edge representation space. HNEABP outperforms TransR as it conducts a balanced walk and obtains the information of mutil-order neighbor nodes effectively.

(2) **Rationality of Parameter Setting.** Parameter experiments are conducted on dataset V1. As shown in Fig. 4(a), the prediction performance in terms of the AUC and F1

is meliorated when we prolong training time from 1 to 20 epochs. The training time is determined to be 20 when F1 peak is reached. The result in Fig. 4(b) demonstrates the prediction performance reaches the peak when k is 7, so that the moving window size k is set as 7. This is because the contribution of the high-order neighbors to the target node decreases and a lot of noise disturbs the representation learning, when the neighbor order exceeds 7. In Fig. 4(c), α and β represent weight coefficients for tuning the learning focus. The purpose of HNEABP is to make the representation of adjacent nodes similar, and on this base, we strengthen the learning of first-order neighbors, i.e. β for strengthening first-order neighbors information should be smaller than α for capturing n-order neighbor information. The overall prediction performance gradually increases with the increase of β, and the most of F1 values stop increasing when β is greater than 0.7. The possible reason is most of features of first-order neighbors have been obtained at that point. After tuning α is 1 and β is 0.7. Since different datasets have different data scales and network structures, parameters need to be set separately based on datasets.

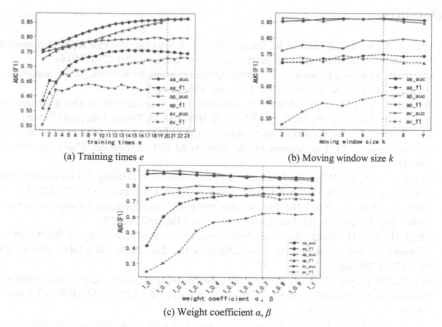

(a) Training times e (b) Moving window size k

(c) Weight coefficient α, β

Fig. 4. Hyper parameters analysis experiments

5 Conclusion

This paper proposes a heterogeneous network representation learning approach, HNE-ABP, for extracting comprehensive node features and improving academic behavior prediction. The key ideas of HNEABP are, a) develop two balanced sampling algorithms for uniformly acquiring positive and negative data samples from neighbor nodes of different types, b) strengthen the importance of the first-order neighbor and distinguish the contribution to the target node from the neighbors of different levels during the node representation (embedding), c) map heterogeneous nodes to the suitable representation spaces and solve the semantic incompatibility between edges based on KGE, d) extract structure and semantic features of nodes by learning low-order and high-order neighbor information comprehensively.

In the future, improving HNEABP by integrating multiple attributes of nodes, such as content features and community features, and extracting meta-path automatically for semantic feature learning should be studied. In addition, improving the spatiotemporal efficiency in heterogeneous network representation learning is also a challenging task.

Acknowledgments. This work is supported by the Sichuan Science and Technology Program (No 2019YFSY0032).

References

1. Perozzi B., Al-rfou R., Skiena S.: DeepWalk: online learning of social representations. In: Proceedings of the 20th ACM SIGKDD, New York, USA, pp. 701–710 (2014)
2. Grover A., Leskovec J.: Node2vec: scalable feature learning for networks. In: Proceedings of the 22nd ACM SIGKDD, New York, USA, pp. 855–864 (2017)
3. Hussein R., Yang D., Cudré-Mauroux P.: Are meta-paths necessary? Revisiting heterogeneous graph embeddings. In: Proceedings of the 27th ACM CIKM, Torino, UK, pp.437–446 (2018)
4. Lee, S., Park, C., Yu, H.: BHIN2vec: balancing the type of relation in heterogeneous information network. In: Proceedings of the 28th ACM CIKM, Beijing, China, pp.619–628 (2019)
5. Lin, Y., Liu, Z., Sun, M., et al.: Learning entity and relation embeddings for knowledge graph completion. In: Twenty-Ninth AAAI Conference, Texas, USA, pp. 2181–2187 (2015)
6. Zhao, J., Wang, X., et al.: Network schema preserved heterogeneous information network embedding. In: 29th IJCAI, Yokohama, Japan, pp. 1366–1372 (2020)
7. Chen, H., Yin, H., Wang, W., et al.: PME: projected metric embedding on heterogeneous networks for link prediction. In: Proceedings of the 24th ACM SIGKDD, London, UK, pp. 1177–1186 (2018)
8. Shi, Y., Zhu, Q., Guo, F., et al.: Easing embedding learning by comprehensive transcription of heterogeneous information networks. In: Proceedings of the 24th ACM SIGKDD, London, UK, pp. 2190–2199 (2018)
9. Shi, Y., Gui, H., Zhu, Q., et al.: Embedding learning by aspects in heterogeneous information networks. In: Proceedings of the SIAM, California, USA, pp. 144–152 (2018)
10. Chairatanakul, N., Liu, X., et al.: PGRA: projected graph relation-feature attention network for heterogeneous information network embedding. Inf. Sci. **570**, 769–794 (2021)
11. Lu, Y., Shi, C., Hu, L., et al.: Relation structure-aware heterogeneous information network embedding. In: AAAI, Hawaii, USA, pp. 4456–4463 (2019)
12. Dong, Y., Chawla, N., Swami, A.: Metapath2vec: scalable representation learning for heterogeneous networks. In: Proceedings of the 23rd ACM SIGKDD, Halifax, Canada, pp. 135–144 (2017)

A Market Segmentation Aware Retail Itemset Placement Framework

Raghav Mittal[1], Anirban Mondal[1(✉)], and P. Krishna Reddy[2]

[1] Ashoka University, Sonipat, India
raghav.mittal@alumni.ashoka.edu.in, anirban.mondal@ashoka.edu.in
[2] IIIT, Hyderabad, India
pkreddy@iiit.ac.in

Abstract. It is a well-established fact in the retail industry that the placement of products on the shelves of the retail store has a significant impact on the revenue of the retailer. Given that customers tend to purchase sets of items together (i.e., *itemsets*) instead of individual items, it becomes a necessity to *strategically* place itemsets on the shelves of the retail store for improving retailer revenue. Furthermore, in practice, customers belong to different market segments based on factors such as purchasing power, demographics and customer behaviour. Existing research efforts do not address the issue of *market segmentation* w.r.t. itemset placement in retail stores. Consequently, they fail to efficiently index, retrieve and place high-utility itemsets in the retail slots in a market segmentation aware manner. In this work, we introduce the problem of market segmentation aware itemset placement for retail stores. Moreover, we propose a market segmentation aware retail itemset placement framework, which takes high-utility itemsets as input. Our performance evaluation with two real datasets demonstrates that our proposed framework is indeed effective in improving retailer revenue w.r.t. existing schemes.

Keywords: Retail · Market segmentation · Utility mining · Indexing

1 Introduction

In brick-and-mortar retail stores, retailers seek to improve their revenue by providing customers with easy access to their desired items. Retail stores typically comprise multiple shelves (racks), which contain slots for the placement of products. These retail slots are either premium or non-premium. Premium retail slots include slots which provide high product visibility and accessibility to the consumer. These include slots near the eye or shoulder-level of the customer and the impulse-buy slots near the checkout counters; other slots are non-premium.

It is a well-established fact that strategic placement of items on the retail shelves can significantly improve the revenue of the retailer [4,8–10,16,21,23,24]. Additionally, customers tend to purchase a set of items (i.e., *itemsets* [3,15]) as opposed to individual items in order to benefit from the convenience of one-stop shopping. Therefore, there is an opportunity for the retailer to improve its revenue through strategic placement of itemsets in the premium slots. Notably, over

© The Author(s), under exclusive license to Springer Nature Switzerland AG 2022
C. Strauss et al. (Eds.): DEXA 2022, LNCS 13426, pp. 273–286, 2022.
https://doi.org/10.1007/978-3-031-12423-5_21

the past few decades, medium-to-large sized retail stores have become increasingly popular. Some of these occupy floor space upwards of a million square feet e.g., New South China Mall (Dongguan, China) and Siam Paragon Mall (Bangkok, Thailand) [2]. Given such a scale, strategic itemset placement in premium slots has become even more critical towards improving retailer revenue. Additionally, brick-and-mortar retail stores face multiple challenges such as finding a property in a good location with sustainable rent for the retail store, intense competition with online retailers and fluctuating costs of goods and labor.

In practice, customers inherently belong to different market segments based on factors such as purchasing power, demographics, ethnicity and customer behaviour [5,6,11,12,17,20,27,31]. In fact, market segmentation is a well-established area in business marketing and it has been extensively researched in retail as well as other sectors. Each market segment constitutes customers, whose purchase preferences have a high degree of similarity. For example, if market segments are based on purchasing power, affluent customers would likely buy expensive high-end items, middle-class customers would likely prefer medium-priced items and so on. Observe that if itemsets were to be placed without considering the existence of market segments, there could be a *mismatch* between the items that a given customer is exposed to and the customer's market segment [12,20,31]. For example, poorer customers being exposed to expensive items and affluent customers being targeted for buying low-end items would likely not result in sales, thereby leading to lost revenue for the retailer. *Notably, in this paper, we define users with different ranges of purchasing power as different market segments.*

Existing works focus on market segmentation [5,6,11,12,17,20,27,31], utility mining [13,29,30] and retail itemset placement [4,7–9,21–24,26]. Notably, none of the existing works consider the existence of market segments. Hence, we address the problem of *market segmentation aware* retail itemset placement for improving the retailer revenue. We consider the history of user purchase transactions on a finite set of items. Using the high-revenue itemsets extracted from these transactions as input, the problem is to (a) model the issue of itemset placement in retail stores based on market segments (b) identify high-revenue itemsets with consideration for their market segment and (c) place such itemsets in a given number of premium slots for improving the retailer revenue.

This work introduces the notion of market segmentation aware itemset placement for retail stores. In particular, we propose MATRIX, which is a **M**arket segmentation **A**ware **T**op-**R**evenue **I**temset Inde**X** for efficiently retrieving high-revenue itemsets corresponding to different market segments. MATRIX is a multi-level index, where the ith level corresponds to the top-revenue itemsets (of size i) belonging to different market segments. Notably, the number of itemsets stored for a given market segment is kept *proportional* to the number of transactions corresponding to that market segment in the transactions database. We also propose MIPs, which is a **M**arket segmentation aware **I**temset **P**lacement **S**cheme (MIPS), which exploits MATRIX, for improving retailer revenue. MIPS places high-revenue itemsets from different market segments by reserving premium slots *in proportion* to the occurrence of itemsets from each market segment in the transactions database. Our key contributions are three-fold:

1. We introduce the problem of market segmentation aware itemset placement for retail stores.
2. We propose the MATRIX index and the MIPS placement scheme, which exploits the MATRIX index for efficient retrieval and placement of high-revenue itemsets in the retail slots.
3. We conduct a performance study with two real datasets to demonstrate that MIPS is effective in improving retailer revenue w.r.t. existing schemes.

To the best of our knowledge, this is the first work to address *market segmentation aware* itemset placement in retail stores. The rest of this paper is organized as follows. Section 2 discusses related works. Section 3 details the problem framework. Section 4 presents MATRIX and MIPS. Section 5 reports the performance study. We conclude in Sect. 6 with directions for future work.

2 Related Work

Existing works can broadly be categorized into three types, namely (a) market segmentation approaches (b) utility mining approaches (c) itemset placement approaches for retail stores. We shall now discuss each of these categories.

Market Segmentation Approaches: The work in [11] proposed a normative theory of market segmentation as a mathematical model, which considered a wide gamut of possibilities for segmentation. The work in [31] discussed the advantages and disadvantages of different types of market segmentation such as geographic, demographic, psychological and behavioural. Further, the work in [12] used published data and case-studies to examine the practical implementation challenges of market segmentation, and provided insights concerning strategies for successfully implementing market segmentation in practical scenarios. Interestingly, the work in [20] reported a case-study conducted using a sample of 894 retail shoppers belonging to different age-groups in two cities in Botswana. The goal of the case-study was to identify segments of retail shoppers based on factors such as demographics, decision-making styles and overall satisfaction.

The work in [27] discussed a methodology for market segmentation in conjunction with competitive analysis in the context of supermarket retailing, and performed a large-scale study to understand the implications of the methodology for supermarket retail chains. Moreover, the work in [17] focused on retail-customer commitment as a possible criterion for market segmentation and proposed a multi-dimensional structure in this context. Furthermore, the work in [5] investigated market segmentation in the context of innovations in food retailing and determined the characteristics of market segments arising as a result of those innovations. A good survey on market segmentation can be found in [6].

Utility Mining Approaches: Utility mining approaches seek to discover high-utility itemsets (HUIs). The HUI-Miner algorithm [19] employs *utility-lists* for storing heuristic information and itemset utility values for retrieving HUIs. The Utility Pattern Growth (UP-Growth) algorithm [30] uses the Utility Pattern Tree

(UP-Tree) for identifying HUIs. The work in [28] employs the NVUV-list data structure to determine itemsets with high average utility values. Further, the MinFHM algorithm [13] uses pruning techniques for extraction of minimal HUIs i.e., the most compact itemsets with the high utility values. The CHUI-Miner algorithm [25] mines closed HUIs without generating candidates. Furthermore, incremental utility mining has been investigated in [18,32,33].

Itemset Placement Approaches for Retail Stores: The works in [7,9] proposed a framework for indexing and placement of high-utility itemsets when the physical sizes of the items can vary. Further, the work in [8,24] proposed an approach for placing high-revenue itemsets in slots with varied premiumness. The works in [22,26] proposed the kUI indexing scheme for facilitating the placement of diversified high-revenue itemsets. Moreover, the works in [21,23] proposed the (item) urgency and expiry aware URIP and PEAR itemset placement schemes. However, these approaches do not consider the issue of market segmentation.

Notably, none of the existing works address the issue of *market segmentation aware* itemset placement in retail stores for improving retailer revenue. This limits their applicability in building practical systems for retail itemset placement.

3   Proposed Framework of Itemset Placement Problem

Consider a finite set Υ of m customer transactions, where each transaction T comprises distinct and non-repetitive items. Each item i in Υ is associated with a price value ρ_i, frequency of sales σ_i and a corresponding market segment Φ. Further, assume each item i occupies a single slot in the retail store. Given N premium slots, the problem is to maximize the retailer revenue, while incorporating information about different market segments.

We now discuss some key terminology and concepts that we shall use in this paper. We define the **net revenue NR_i of an item** i as the product of its price value ρ_i and frequency of sales σ_i. We compute the frequency of sales of an item as the number of times it occurs in the transactional dataset. The NR_i of an item is computed as $NR_i = \sigma_i * \rho_i$. Further, we define the **net revenue NR_z of itemset** z as the product of its frequency of sales σ_z and the total price of all items in z. Notably, we compute the frequency of sales of an itemset z as the number of transactions that comprise all items contained in the itemset. We compute the NR_z of an itemset as $NR_z = \sigma_z * \sum_{i \in z} \rho_i$.

Let us consider Fig. 1 to further understand the context. Figure 1 provides the price values and the market segment category for items A to I. It further provides an illustration for computation of the net revenue and the market segment integer bitmap for five itemsets. Consistent with our prior discussion, observe that the net revenue of the itemset $\{A, D\}$ is $(4+3) * 6$ i.e., 42. Similarly, the net revenue of the itemset $\{A, C, G\}$ can be computed as $(4+6+5) * 3$ i.e., 45.

In practice, itemsets may contain low-end, mid-end and high-end items. Observe that while affluent customers can purchase low-priced items, low-end customers would rarely purchase high-end items. To this end, we assign a market segment value (Φ) to each item e.g., the value of Φ can be 1, 2 or 3, depending

Item	ρ	Φ
A	4	2
B	2	1
C	6	3
D	3	1
E	3	1
F	1	1
G	5	3
H	4	2
I	3	1

Items: A to I
ρ: Price
σ: Frequency of Sales
Φ: Market Segment
NR: Net Revenue

Itemset	σ	ρ	NR	Bitmap
<A, D>	6	7	42	< 0, 1, 0 >
<B, D, F, I>	3	9	27	< 0, 0, 1 >
<A, C, G>	3	15	45	< 1, 0, 0 >
<A, B, C, G, H>	1	21	21	< 1, 0, 0 >
<A, C, G, I>	3	18	54	< 1, 0, 0 >

Fig. 1. Modeling of market segmentation in itemsets

upon whether the item is a low-end, mid-end or high-end item. Next, for categorizing itemsets based on market segments, we use a bitmap array, where the first, second and third positions reflect high-end, mid-end and low-end market segments respectively. We assign the market segment of an itemset based on the item with the highest value of Φ (i.e., the item with the highest market segment) in that itemset, and set the corresponding bit to 1 only for the relevant market segment of the itemset. Itemsets belonging to the high-end, mid-end and low-end market segments are modeled as <1, 0, 0>, <0, 1, 0>, and <0, 0, 1> respectively. In Fig. 1, based on the items with the highest market segmentation value Φ, we model the integer bitmap for itemsets $\{A, D\}$ and $\{B, D, F, I\}$ as <0, 1, 0> and <0, 0, 1> respectively, as the former contains both mid-end and low-end items, but the latter contains only low-end items.

In this paper, we assume that we are given a set of high-utility itemsets (HUIs) as input. Given the history of user purchase transactions on a finite set of items, HUIs can be generated using any existing high-utility itemset mining algorithm [13, 14, 19, 29, 30]. Further, we consider revenue as a measure of utility. We use the terms *utility* and *revenue* interchangeably throughout the paper.

4   MATRIX and MIPS

This section discusses our proposed MATRIX index and the MIPS placement scheme. Figure 2 depicts the schematic diagram for our proposed framework.

The MATRIX Index: At the onset, the MATRIX index categorizes itemsets into different buckets based on the itemset size, i.e., the number of items contained in the itemset. These buckets correspond to different levels of the MATRIX index. For example, a bucket comprising itemsets that contain 3 items

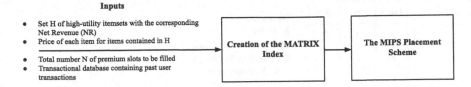

Fig. 2. Schematic diagram for our proposed framework

would constitute level 3 of MATRIX. Next, MATRIX uses the k-means cluster-ing algorithm to cluster items contained in the input itemsets on the basis of their price since we use price as a metric to segment customers. Corresponding to every bucket, i.e., every level of our proposed MATRIX index, we augment with θ hash buckets, where θ denotes the number of market segments. Figure 3 depicts an illustrative example of our proposed MATRIX index. Observe how MATRIX sorts and stores itemsets pertaining to different market segments at different levels on the basis of their net revenue. Notably, an example of how the computations in Fig. 3 are performed is shown in Fig. 1.

Notably, MATRIX is a market segmentation aware data structure, i.e., it allocates itemsets to each hash bucket proportional to the percentage of trans-actions pertaining to each market segment. Recall from our discussion in Sect. 1 how it is possible for high-end consumers to purchase low-priced items, but the converse is not true. Therefore, hash buckets of higher-end market segments may also comprise low-priced items. To ensure that items from different market seg-ments are incorporated into the MATRIX index, we examine the percentage of transactions pertaining to each market segment. Next, starting with the hash bucket corresponding to the highest market segment, we allocate itemsets to various hash buckets in the following manner.

At the onset, we consider λ, i.e., the maximum number of itemsets that can be incorporated at every level of MATRIX. In proportion to the number of transactions corresponding to each market segment, we divide λ *proportion-ally* by θ to compute the number of itemsets that are to be allocated to each hash bucket in the index. For instance, if $\lambda = 1000$, $\theta = 3$, and the percentage of high-end, mid-end, and low-end transactions in the transactional database is 20%, 30%, and 50% respectively, we allocate 200 itemsets in the high-end hash bucket, 300 itemsets in the mid-end hash bucket, and 500 itemsets in the low-end hash-bucket. Starting with the high-end market segment, we populate the corre-sponding hash-buckets with itemsets sorted in descending order of NR till they have been proportionally filled with itemsets associated with the corresponding market segment. We repeat this process for all levels of MATRIX and across all market segments until MATRIX has been fully populated.

Algorithm 1 depicts the creation of our proposed MATRIX index. In Algo-rithm 1, Line 1 categorizes the input itemsets on the basis of their size. Lines 2–4 of the algorithm cluster the items on the basis of their price and scan the transactions database to compute the percentage of transactions pertaining to

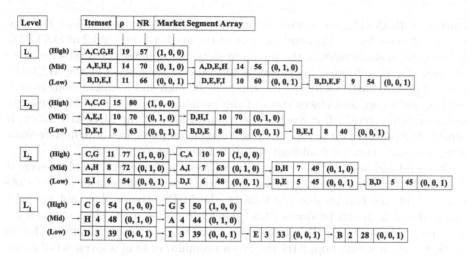

Fig. 3. Illustrative example of the MATRIX index

each market segment. In Lines 5–7, it computes the market segment bitmap corresponding to each itemset, and subsequently sorts them on the basis of their net revenue. In Lines 8–9, it populates hash buckets corresponding to every market segment based on the percentage of itemsets that are permissible, by examining the market segment bitmap corresponding to each itemset.

Algorithm 1: Creation of the MATRIX Index

Input: Set H of high-utility itemsets with item prices, λ: Maximum number of itemsets at each level, θ number of market segments, L_{max}: Highest itemset size in H

Output: The MATRIX Index

1 Categorize itemsets on the basis of size
2 **for** itemsets of size 1
3   Cluster itemsets on the basis of price
4   Scan transactional database to compute % of slots for each market segment
5 **for** itemset-sizes 2 to L_{max}
6   Compute the market segment integer bitmap for each itemset based on θ
7   Sort itemsets based on their net revenue
8   **for** each market segment Φ
9   Populate hash buckets with required % of λ number of itemsets

---

Our Proposed MIPS Placement Scheme: MIPS exploits our proposed MATRIX index for placing high-revenue itemsets in the premium slots in a market segmentation aware manner. MIPS works as follows. First, it scans the user purchase transactions database to identify the percentage of transactions associated with different market segments, and accordingly reserves the percentage of premium slots that should be allocated for each market segment. Next, starting with the highest market segment, the MIPS scheme examines the top-revenue itemset for the high-end segment of the market from level 2 of MATRIX. It then places the itemset in the premium slots and updates the number of slots available for placement. MIPS repeats the process for level 3 of the MATRIX index. Similarly, MIPS extracts the top-revenue high-end itemsets from level 4, level 5, and so on, and places them in the premium slots reserved for the high-end market segment, till it reaches the topmost level of the MATRIX index. It repeats the process for the top-2 highest-revenue itemsets, top-3 highest-revenue itemsets, and so on, until all high-end slots have been populated.

For mid-end slots, MIPS essentially follows the same process. However, it does not examine high-end and low-end itemsets for placement in the slots reserved for mid-end products. Similarly, the MIPS scheme does not consider mid-end and high-end products while placing itemsets in the slots reserved for low-end products. The MIPS placement scheme repeats the above process for all market-segments, and populates the given premium slots of a given retail store. Algorithm 2 presents the algorithm for our proposed MIPS placement scheme.

Algorithm 2: MIPS Placement Scheme

Input: MATRIX index, total slots T_S, θ market segments
Output: Placement of itemsets in the premium slots
1 Compute the % of premium slots corresponding to each market segment
2 **while** $T_S \geq 0$
3   **for** each market segment
4   **for** itemset size 2 to L_{max} in the MATRIX index
5   Retrieve itemset z with the highest net revenue
6   Place itemset z in slots for the corresponding market segment
7           $T_S = T_S$ - $|z|$
8   Remove itemset z from the MATRIX index

---

5   Performance Evaluation

This section reports our performance evaluation. We implemented our proposed placement frameworks using Python 3.8.5 on a 64-bit Intel(R) Pentium(R) 2.20 GHz processor running on Ubuntu 20.04.1 LTS with 4 GB RAM.

We performed our experiments using two real datasets, namely *Chainstore* and *Fruithut*. We obtained these datasets from the SPMF open-source data mining library [1]. *Chainstore* is a retail dataset containing 46,086 items and 1,112,949 retail customer purchase transactions from a major grocery retail outlet in California, USA. *Fruithut* contains user transactions from a major US retail outlet that emphasizes on selling fruits and fruit based products. It contains 181,970 transactions and 1265 items. Further, we divided each dataset into two parts i.e., training set and test set containing 70% and 30% of transactions respectively. We performed the placement using the training set and evaluated the performance on the test set.

Recall that our proposed framework requires a set H of high-utility itemsets as input. While any existing utility mining approach can be used to generate the set H, for our experiments, we use the kUI index [22,26] for generating the set H. Notably, the kUI index is a multi-level index, where each level corresponds to a specific itemset size. Moreover, at each level of the kUI index, the top-λ high-revenue itemsets are stored to facilitate quick retrieval of the top-revenue itemsets of any given itemset size. For our experiments, we implemented the kUI index with six levels with the value of λ being set to 4000 for each of the levels. Observe that the kUI index is oblivious to the notion of market segmentation since it maintains the top-k itemsets only on the basis of revenue.

We assign each itemset in set H to one of the market segments as discussed in Sect. 3. Recall that we use the k-means clustering algorithm to cluster items based on their respective prices. Hence, each cluster corresponds to a specific market segment with a price range. Then, for each itemset z in H, we examine the price of each item in z, and assign z to the market segment corresponding to the highest-priced item in z. Table 1 summarizes our performance study parameters.

Table 1. Parameters of the performance evaluation

Parameter	Default	Variations
Number of market segments (θ)	3	4, 5, 6, 7
Total number of slots (T_S)	6000	2000, 4000, 8000, 10000

Our performance metrics include: (a) total revenue (TR) of the retailer for the test set (b) execution time (ET) for executing the algorithm and placing itemsets in the premium slots for the training set. In the test phase, we iterate through each transaction t in the test set and add the price of the itemset to TR if the items (in t) have been placed as itemsets in the slots during the training phase. Moreover, observe that if an itemset is placed in the high-end (i.e., high-priced) section of the retail store, it would likely not be purchased by customers from low-end market segments (although customers from low-end segments could potentially purchase only some of the cheaper items in that itemset). Hence, we take a *conservative* approach by adding the price of a given itemset to TR only if the purchase transaction and the itemset are both associated with the same market segment. Notably, in practice, itemsets from different market segments would typically be placed in different sections of the retail store, thereby providing further rationale to our conservative approach in computing TR.

Recall from Sect. 2 that existing works are oblivious to the notion of market segmentation with regard to itemset placement in retail stores. Hence, for purposes of meaningful comparison, we design a two-phase reference approach, which works as follows. In the first phase, the reference approach clusters the input itemsets based on their size into k clusters, i.e., itemsets of size i belong to the ith cluster. In the second phase, for values of i ranging from 2 to k, we randomly select any itemset from i^{th} cluster and place it in the premium slots.

This process is repeated in a round-robin manner till all the premium slots are exhausted. We designate this reference approach as the **R**andmized **M**arket segment **O**blivious (**RMO**) itemset placement scheme. Observe how RMO equally prioritizes the input high-utility itemsets of different sizes towards placement, thereby making it a fairly effective itemset placement approach in itself. Notably, we do not compare the performance with a brute-force approach because RMO itself would outperform the brute-force approach.

Effect of Variations in the Total Number of Slots: Figure 4 depicts the effect of variations in the total number T_S of slots. The results in Fig. 4(a) indicate that TR increases for both RMO and MIPS with an increase in T_S. This is because more itemsets are required to fill up a larger number of slots, thereby resulting in increased TR. Our proposed MIPS approach exhibits significantly higher TR than RMO due to its ability to place itemsets in a market segment aware and revenue-conscious manner. Recall how MIPS places itemsets *in proportion* to the occurrence of itemsets from different market segments in the transactional database. In contrast, RMO selects high-utility itemsets in a randomized manner regardless of market segments, and this may possibly result in low-revenue itemsets being selected for placement in the premium slots.

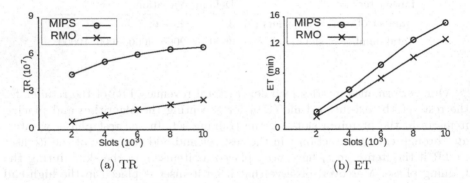

(a) TR   (b) ET

Fig. 4. Effect of variations in the total number of slots (Chainstore)

The results in Fig. 4(b) indicate that ET increases for both RMO and MIPS with an increase in the total number of slots for the Chainstore dataset. This is because a higher number of itemsets are required to be examined and processed for populating a higher number of premium slots. MIPS incurs higher ET than RMO since it meticulously places high-revenue itemsets in the retail slots in a *market segmentation* aware manner. In contrast, the RMO approach places itemsets in a randomized manner, thereby incurring lower ET. Figures 4(c)–(d) depict the results for the Fruithut dataset. Observe that the results exhibit comparable trends; actual values vary due to different dataset sizes (Fig. 5).

Effect of Variations in the Number of Market Segments: Figure 6 depicts the effect of varying the total number θ of market segments. The results in Fig. 6(a) indicate that TR remains comparable for both RMO and the MIPS

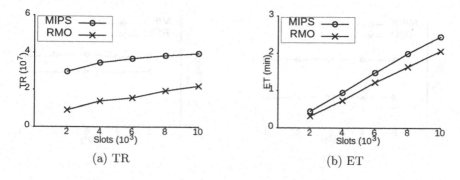

Fig. 5. Effect of variations in the total number of slots (Fruithut)

placement scheme across different values of θ. TR remains comparable for MIPS with an increase in θ due to its ability to place high-revenue itemsets corresponding to every market segment, thereby exhibiting consistent performance. RMO consistently provides lower TR to the retailer than MIPS due its inability to place high-revenue itemsets in the premium slots for different market segments (Fig. 7).

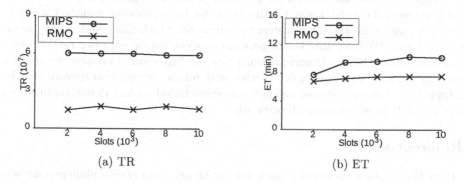

Fig. 6. Effect of variations in number (θ) of market segments (Chainstore)

Figure 6(b) depicts that ET remains comparable for both the RMO and the MIPS placement scheme with an increase in θ. This is because the number of patterns (itemsets) does not change with an increase in θ. Our proposed MIPS placement scheme incurs more ET than the RMO placement scheme on the basis of the rationale provided for results in Fig. 4(a). Furthermore, Figs. 6(c)–(d) depict the results for the Fruithut dataset. Observe that the results exhibit comparable trends w.r.t. the results in Figs. 6(a)–(b), the variations in the actual values occurring due to different dataset sizes.

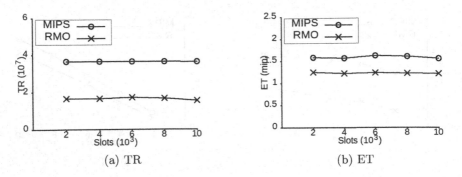

Fig. 7. Effect of variations in number (θ) of market segments (Fruithut)

6 Conclusion

Strategic placement of itemsets in retail stores can significantly impact the revenue of the retailer. While customers inherently belong to different market segments based on factors such as purchasing power and demographics, existing works have not addressed market segmentation aware retail itemset placement, thereby resulting in lost opportunities for improving retailer revenue. Hence, in this paper, we have introduced the problem of market segmentation aware itemset placement for retail stores. Furthermore, we have proposed a market segmentation aware retail itemset placement framework, which takes high-utility itemsets as input. We have also done a performance evaluation with two real datasets to demonstrate the effectiveness of our proposed framework in improving retailer revenue w.r.t. existing schemes. In the near future, we plan to investigate the integration of market segmentation approaches based on user demographics into our retail itemset placement framework.

References

1. SPMF: A Java open-source data mining library. http://www.philippe-fournier-viger.com/spmf/datasets. Accessed 1 Jun 2022
2. Largest malls in the world (2020). https://www.touropia.com/largest-malls-in-the-world/. Accessed 1 Jun 2022
3. Agrawal, R., Srikant, R.: Fast algorithms for mining association rules. In: Proceedings of the VLDB, vol. 1215, pp. 487–499 (1994)
4. Ahn, K.I.: Effective product assignment based on association rule mining in retail. Exp. Syst. Appl. **39**, 12551–12556 (2012)
5. Appel, D.L.: Market segmentation - a response to retail innovation. J. Mark. **34**(2), 64–67 (1970)
6. Beane, T., Ennis, D.: Market segmentation: a review. Eur. J. Mark. **21**(5), 20–42 (1987)
7. Chaudhary, P., Mondal, A., Reddy, P.K.: A flexible and efficient indexing scheme for placement of top-utility itemsets for different slot sizes. In: Reddy, P.K., Sureka, A., Chakravarthy, S., Bhalla, S. (eds.) BDA 2017. LNCS, vol. 10721, pp. 257–277. Springer, Cham (2017). https://doi.org/10.1007/978-3-319-72413-3_18

8. Chaudhary, P., Mondal, A., Reddy, P.K.: An efficient premiumness and utility-based itemset placement scheme for retail stores. In: Hartmann, S., Küng, J., Chakravarthy, S., Anderst-Kotsis, G., Tjoa, A.M., Khalil, I. (eds.) DEXA 2019. LNCS, vol. 11706, pp. 287–303. Springer, Cham (2019). https://doi.org/10.1007/978-3-030-27615-7_22

9. Chaudhary, P., Mondal, A., Reddy, P.K.: An improved scheme for determining top-revenue itemsets for placement in retail businesses. Int. J. Data Sci. Anal. **10**, 359–375 (2020)

10. Chen, M., Lin, C.: A data mining approach to product assortment and shelf space allocation. Exp. Syst. Appl. **32**, 976–986 (2007)

11. Claycamp, H.J., Massy, W.F.: A theory of market segmentation. J. Mark. Res. **5**(4), 388–394 (1968)

12. Dibb, S.: Market segmentation: strategies for success. Mark. Intell. Plan. **16**(7), 394–406 (1998)

13. Fournier-Viger, P., Lin, J.C.-W., Wu, C.-W., Tseng, V.S., Faghihi, U.: Mining minimal high-utility itemsets. In: Hartmann, S., Ma, H. (eds.) DEXA 2016. LNCS, vol. 9827, pp. 88–101. Springer, Cham (2016). https://doi.org/10.1007/978-3-319-44403-1_6

14. Fournier-Viger, P., Wu, C.-W., Zida, S., Tseng, V.S.: FHM: faster high-utility itemset mining using estimated utility co-occurrence pruning. In: Andreasen, T., Christiansen, H., Cubero, J.-C., Raś, Z.W. (eds.) ISMIS 2014. LNCS (LNAI), vol. 8502, pp. 83–92. Springer, Cham (2014). https://doi.org/10.1007/978-3-319-08326-1_9

15. Han, J., Pei, J., Yin, Y.: Mining frequent patterns without candidate generation. In: Proceedings of the ACM SIGMOD, vol. 29, pp. 1–12. ACM (2000)

16. Hansen, P., Heinsbroek, H.: Product selection and space allocation in supermarkets. Eur. J. Oper. Res. **3**, 474–484 (1979)

17. Iniesta, M.A., Sánchez, M.: Retail-consumer commitment and market segmentation. Int. Rev. Retail Distrib. Consum. Res. **12**(3), 261–279 (2002)

18. Lee, J., Yun, U., Lee, G., Yoon, E.: Efficient incremental high utility pattern mining based on pre-large concept. Eng. Appl. Artif. Intell. **72**, 111–123 (2018)

19. Liu, M., Qu, J.: Mining high utility itemsets without candidate generation. In: Proceedings of the CIKM, pp. 55–64. ACM (2012)

20. Makgosa, R., Sangodoyin, O.: Retail market segmentation: the use of consumer decision-making styles, overall satisfaction and demographics. Int. Rev. Retail Distrib. Consum. Res. **28**(1), 64–91 (2018)

21. Mittal, R., Mondal, A., Chaudhary, P., Reddy, P.K.: An urgency-aware and revenue-based itemset placement framework for retail stores. In: Strauss, C., Kotsis, G., Tjoa, A.M., Khalil, I. (eds.) DEXA 2021. LNCS, vol. 12924, pp. 51–57. Springer, Cham (2021). https://doi.org/10.1007/978-3-030-86475-0_5

22. Mondal, A., Mittal, R., Chaudhary, P., Reddy, P.K.: A framework for itemset placement with diversification for retail businesses. Appl. Intell., 1–19 (2022). https://doi.org/10.1007/s10489-022-03250-8

23. Mondal, A., Mittal, R., Khandelwal, V., Chaudhary, P., Reddy, P.K.: PEAR: a product expiry-aware and revenue-conscious itemset placement scheme. In: Proceedings of the DSAA. IEEE (2021)

24. Mondal, A., Saurabh, S., Chaudhary, P., Mittal, R., Reddy, P.K.: A retail itemset placement framework based on premiumness of slots and utility mining. IEEE Access **9**, 155207–155223 (2021)

25. Nguyen, L.T., et al.: An efficient method for mining high utility closed itemsets. Inf. Sci. **495**, 78–99 (2019)

26. Chaudhary, P., Mondal, A., Reddy, P.K.: A diversification-aware itemset placement framework for long-term sustainability of retail businesses. In: Hartmann, S., Ma, H., Hameurlain, A., Pernul, G., Wagner, R.R. (eds.) DEXA 2018. LNCS, vol. 11029, pp. 103–118. Springer, Cham (2018). https://doi.org/10.1007/978-3-319-98809-2_7

27. Segal, M., Giacobbe, R.: Market segmentation and competitive analysis for supermarket retailing. Int. J. Retail Distrib. Manage. **22**(1), 38–48 (1994)

28. Truong, T., Duong, H., Le, B., Fournier-Viger, P., Yun, U.: Efficient high average-utility itemset mining using novel vertical weak upper-bounds. Knowl. Based Syst. **183**, 104847 (2019)

29. Tseng, V.S., Wu, C., Fournier-Viger, P., Philip, S.Y.: Efficient algorithms for mining the concise and lossless representation of high utility itemsets. IEEE Trans. Knowl. Data Eng. **27**, 726–739 (2015)

30. Tseng, V.S., Wu, C., Shie, B., Yu, P.S.: UP-growth: an efficient algorithm for high utility itemset mining. In: Proceedings of the ACM SIGKDD, pp. 253–262. ACM (2010)

31. Tynan, A.C., Drayton, J.: Market segmentation. J. Mark. Manag. **2**(3), 301–335 (1987)

32. Vo, B., Nguyen, L.T., Nguyen, T.D., Fournier-Viger, P., Yun, U.: A multi-core approach to efficiently mining high-utility itemsets in dynamic profit databases. IEEE Access **8**, 85890–85899 (2020)

33. Wu, J.M.T., Teng, Q., Lin, J.C.W., Yun, U., Chen, H.C.: Updating high average-utility itemsets with pre-large concept. J. Intell. Fuzzy Syst. **38**, 5831–5840 (2020)

Label Selection Algorithm Based on Iteration Column Subset Selection for Multi-label Classification

Tao Peng, Jun Li, and Jianhua Xu$^{(\boxtimes)}$

School of Computer and Electronic Information, School of Artificial Intelligence,
Nanjing Normal University, Nanjing 210023, Jiangsu, China
202243029@stu.njnu.edu.cn, {lijuncst,xujianhua}@njnu.edu.cn

Abstract. In multi-label classification, each sample can be associated with a set of class labels. When the number of labels grows to the hundreds or even thousands, existing multi-label classification methods often become computationally inefficient. To this end, dimensionality reduction strategy is applied to label space via exploiting label correlation information, resulting in label embedding and label selection techniques. Compared with a lot of label embedding work, less attention has been paid to label selection techniques due to its difficulty. Therefore, it is a challenging task to design more effective label selection techniques for multi-label classification. Column subset selection is the problem of selecting a small portion of columns from a large data matrix as one form of interpretable data summarization. So, the column subset selection problem translates naturally to this purpose, as it provides simple linear models for low-rank data reconstruction. Iterative column subset selection is one of the methods to solve the problem of column subset selection, and this method can achieve a good result in the problem. In this paper, we first execute iterative column subset selection to select a small portion of columns from a large label matrix, in the prediction stage, we do some processing on the recovery matrix. So, a new method of multi-label classifier based on iterative column subset selection is proposed. The new method is tested on six publicly available datasets with varying numbers of labels. The experimental evaluation shows that the new method works particularly well on datasets with a large number of labels.

Keywords: Multi-label classification · Iterative column subset selection · Label selection · Low-rank data reconstruction · Linear models

1 Introduction

Traditional supervised learning mainly deals with a single-label (binary or multiclass) classification problem where each instance only has one of predefined

Supported by the Natural Science Foundation of China (NSFC) under grants 62076134 and 61703096.

© The Author(s), under exclusive license to Springer Nature Switzerland AG 2022
C. Strauss et al. (Eds.): DEXA 2022, LNCS 13426, pp. 287–301, 2022.
https://doi.org/10.1007/978-3-031-12423-5_22

labels. However, many real-world classification problems involve multiple label classes. In multi-label classification, each sample can be associated with multiple labels rather than only one label [9,21]. The multi-label classification has found applications in a number of statistical and machine learning tasks, such as text categorization [8,16,18], recommendation system [27], image annotation [5,15,19].

Some approaches have been proposed to address the multi-label classification problem [23] with many labels. A first attempt is by [12], which projects the n-dimensional label vector using compressed sensing, and performs training with the much lower-dimensional projected label vectors. Subsequently, many variants have been developed along this line, which use different projection mechanisms including principal component analysis [25], and other singular value decomposition [4,6]. A common characteristic is that they all reduce the possibly large number of labels to a more manageable set of transformed labels. Yet, a major limitation is that the transformed labels, though fewer in quantity, may be more difficult to learn.

The aforementioned label embedding methods have many successful applications, but their major limitation is that the transformed labels would be lack of original label real-world meanings [28], at the same time, the transformed labels is more difficult to learn.

In order to preserve label physical meanings and facilitate learning of the label, label selection methods are to choose an informative label subset, so that those unselected labels can be recovered effectively. In ML-CSSP [1], the label selection is regarded as a column subset selection problem (CSSP) [3], which is NP-complete question [22] and is solved via a randomized sampling method. Its advantage is that the recovery way is obtained directly and its disadvantage is that two high correlated labels are still selected at the same time. In [24], the authors proposed a special Boolean matrix decomposition (BMD) algorithm to approximate the original matrix exactly (EBMD), where the left low-rank matrix is a column subset of original matrix and the right low-rank matrix comes from a binary correlation matrix of original matrix. In [17], this EBMD is directly applied to label selection for multi-label classification (MLC-EBMD) to remove a few uninformative labels. However, when selecting fewer labels, although this method could be slightly modified to rank those remained labels using the number of "1" components from each label and its corresponding recovery vector in descending order, its solution is not optimal in principle. So, in LS-BaBID [14], the authors remove a few uninformative labels via EBMD and then delete some less informative labels using sequential backward selection (SBS) strategy which is widely used in feature selection field, which builds a novel label selection algorithm based on Boolean interpolative decomposition (BID) with SBS. In iterative column subset selection (IterFS) [20], this method starts from a random subset and updates feature selection taking the entirety of the subset into account, which can yield significantly excellent results. Since a straightforward implementation of this approach would be very inefficient, the author derive a series of non-trivial optimizations that make it possible to draw subsets of tens or hundreds of labels in a few seconds or milliseconds.

In this paper, we propose to use iterative column subset selection as label selection method to complete the task of multi-label classification, so as to speed up the computation and accuracy of multi-label classification task. And in the prediction stage, we do some processing on the recovery matrix, we change all the elements smaller than 0 in the Moore-Penrose generalized inverse matrix to 0, and get a surprisingly good result after processing the recovery matrix. As it turns out in experiments, our proposed method obtains better results than other methods across a wide range of datasets and performance measures.

The remainder of this paper is organized as follows. First, we give a short introduction to the algorithms used as a basis for our new approach. Building on that, the new method, called Multi-label classification using Iteration column subset selection with relu (ML-ICSSR), is described in detail. Next, performance measures are introduced. Finally, the new method is compared to and evaluated against benchmark algorithms across several standard multi-label data sets.

2   Multi-label Classification Using Iteration Column Subset Selection

The proposed method can be viewed as adding a preprocessing and a postprocessing step to the BR algorithm [2]. Using iteration column subset selection according to [20], we select k labels from the original label matrix to form a latent label matrix and a recovery matrix to represent the original label matrix. Instead of learning models for the actual labels, models are learned for the latent labels. The final labels are predicted by Boolean matrix multiplication using the recovery matrix. The advantage of this method is the introduction of a new level of abstraction, which represents the data in a more compact way than the original label space.

In the following, we will go into the details of the proposed method. In the following section, we will recall the underlying iterative column subset selection method, the training and test phases will also be explained.

2.1   Notation

Now, let's introduce some notations that will be used in this paper.

- $\mathbf{C}^+$ denotes the Moore-Penrose generalized inverse of $\mathbf{C}$.
- $\mathbf{H}_k^i$ denotes a diagonal $k \times k$ matrix whose entries are all 1, except for element $\mathbf{H}_{ii}$ which is zero.
- For a matrix $\mathbf{A}$ and a set R, $\mathbf{A}_R$ is the submatrix of $\mathbf{A}$ comprised by the columns whose indices are the elements of the set R.
- $\mathbf{A}_{i:}$ is the i-th row and $\mathbf{A}_{:i}$ is the i-th column of $\mathbf{A}$.
- $\mathbf{A}_{ij}$ is the entry in the i-th row and j-th column of matrix $\mathbf{A}$.
- Given $\mathbf{A} \in \mathbb{R}^{m \times n}$, $\mathbf{A} \setminus i$ is a $m \times (n-1)$ submatrix of $\mathbf{A}$ resulting from the removal of column i.

- In our pseudocode, we employ the function uniSampleWithoutReplacement(S,k), which returns a sample of $k \in \mathbb{N}$ elements drawn uniformly at random without replacement from the set S.
- In our pseudocode, for a set R and some $i \in \mathbb{N}$, if we employ the notation $R[i]$ we consider the set to be ordered and $R[i]$ to be its i-th element.
- Lowercase bold letters such as $\mathbf{f}$, $\boldsymbol{\delta}$ denote vectors. $\mathbf{f}_i$ is the i-th element of vector $\mathbf{f}$.
- $\mathbf{e}_i$ is the i-th vector of the canonical basis of the indicated dimensionality.
- $\circ$ denotes an element-wise vector multiplication operator.
- Given two matrices $\mathbf{A}$ and $\mathbf{B}$, $(\mathbf{A}|\mathbf{B})$ is the matrix resulting from appending the columns of $\mathbf{B}$ to $\mathbf{A}$. For example, if $\mathbf{A} \in \mathbb{R}^{m \times n}$, $\mathbf{B} \in \mathbb{R}^{m \times k}$, then $(\mathbf{A}|\mathbf{B}) \in \mathbb{R}^{m \times (n+k)}$ and consists of the column of both matrices.
- $\sigma_i(\mathbf{A})$ denotes the i-th largest singular value of $\mathbf{A}$.

2.2   Iteration Column Subset Selection

Given a matrix $\mathbf{A} \in \mathbb{R}^{m \times n}$ and a positive integer $k(k \ll n)$ smaller than the rank of $\mathbf{A}$, let $\mathbf{A}_k$ denote the set of $m \times k$ matrices comprised of k columns of $\mathbf{A}$. Find $\mathbf{X}$ such that

$$\mathbf{X} = \underset{\mathbf{X} \in \mathbf{A}_k}{\arg \min} \left\| \mathbf{A} - \mathbf{X}\mathbf{X}^+\mathbf{A} \right\|_F \tag{1}$$

where $\mathbf{X}^+$ is the Moore-Penrose pseudoinverse of $\mathbf{X}$.

For an input matrix $\mathbf{A} \in \mathbb{R}^{m \times n}$, let us consider that we want to pick k columns. First, an initial subset R of k columns is chosen uniformly at random without replacement, forming a matrix $\mathbf{C} = \mathbf{A}_R \in \mathbb{R}^{m \times k}$. Then, we can iterates until convergence as follows. For $i = 1, \ldots, k$, column i is removed from $\mathbf{C}$, forming matrix $\tilde{\mathbf{C}} \in \mathbb{R}^{m \times k-1}$, and is replaced by another column such that the objective function (1) is minimized over all possible $n - k + 1$ replacements. We do not rule out the column we removed. The algorithm will converge when no single column replacement yields an improvement in the objective function anymore.

A straightforward implementation of this approach would be very inefficient, it can be very slow when the values of k and n grow slightly. So, the authors proposed iterative column subset selection (IterFS) [20], which starts from a random subset and updates feature selection taking the entirety of the subset into account, yielding significantly better results than other state-of-the-art algorithms. And the authors derive a series of non-trivial optimizations that make it possible to draw subsets of tens or hundreds of labels in a few seconds or milliseconds. This makes the proposal comparable in speed to some of the most efficient previous proposals, and even faster in some cases, while producing better results.

We now present a series of non-trivial derivations that enable the design of Algorithm IterFS. If we have a column subset of $\mathbf{A}$, forming matrix $\mathbf{C}$, the following theorem [20] provides us with a simple criterion to identify the best single column to append to matrix $\mathbf{C}$, based on the matrix $\mathbf{E} = \mathbf{A} - \mathbf{C}\mathbf{C}^+\mathbf{A}$.

Theorem 1. Let $\mathbf{A} \in \mathbb{R}^{m \times n}$. For some $k \in \mathbb{N}$, $k < rank(\mathbf{A})$ let $\mathbf{C} \in \mathbb{R}^{m \times k}$ be a matrix comprised of a subset of the columns of $\mathbf{A}$. Let $\mathbf{E} = \mathbf{A} - \mathbf{C}\mathbf{C}^+\mathbf{A}$. Then

$$\arg\min_i \left\| \mathbf{A} - (\mathbf{C}|\mathbf{A}_{:i}(\mathbf{C}|\mathbf{A}_{:i})^+\mathbf{A} \right\|_F = \arg\min_i \frac{\left\| \mathbf{E}^T\mathbf{E}_{:i} \right\|_2^2}{\left\| \mathbf{E}_{:i} \right\|_2^2} \tag{2}$$

This means that if we have computed matrix $\mathbf{E}$, we can easily find the best column to add. If we define $\mathbf{F} = \mathbf{E}^T\mathbf{E}$, we can express this criterion as

$$t = \arg\max_{i \in [1,n]} \frac{\left\| \mathbf{F}_{:i} \right\|_2^2}{\mathbf{F}_{ii}} \tag{3}$$

In [10], efficient formulae are given for recomputing $\left\| \mathbf{F}_{:i} \right\|_2^2$ and $\mathbf{F}_{ii}$ once a column has been appended to matrix $\mathbf{C}$. Iterative column subset selection, however, does not build the column subset incrementally, but it iteratively replaces each column by another. Therefore, not only does it require to update these values when a column is added to $\mathbf{C}$, but also when it is removed (or equivalently zeroed out to be replaced by a different one). We now present a series of derivations that allow us to do this efficiently, involving fast updates of the Moore-Penrose pseudoinverse inverse.

The key of IterFS are the efficient update of the Moore-Penrose pseudoinverse of $\mathbf{C}$, the subsequent efficient update of the residual matrix $\mathbf{E} = \mathbf{A} - \mathbf{C}\mathbf{C}^+\mathbf{A}$ and the fast update of the numerator and denominator of to determine the winning column at each step of the algorithm. Firstly, the following proposition points out how to update the Moore-Penrose generalized inverse of $\mathbf{C}$.

Proposition 1. Let $\mathbf{A} \in \mathbb{R}^{m \times n}$. For some $k \in \mathbb{N}$, $k < rank(\mathbf{A})$ let $\mathbf{C} \in \mathbb{R}^{m \times k}$ be a matrix comprised of a subset of the columns of $\mathbf{A}$ such that $rank(\mathbf{C}) = k$. Let $\tilde{\mathbf{C}} \in \mathbb{R}^{m \times k}$ be the matrix resulting from zero-out column i in $\mathbf{C}$ (i.e. column i of $\tilde{\mathbf{C}}$ is comprised uniquely of zeros). Let $\rho = ((\mathbf{C}^+)_{i:})^T$ (the i-th row of $\mathbf{C}^+$ as a column vector). Then

$$\tilde{\mathbf{C}}^+ = \mathbf{C}^+ - \left\| \rho \right\|_2^{-2} \mathbf{C}^+ \rho \rho^T \tag{4}$$

In addition, the following propositions indicate how to update the residual matrix $\mathbf{E}$ when a column is removed.

Proposition 2. Let $\tilde{\mathbf{C}} \in \mathbb{R}^{m \times k} = \mathbf{C}\mathbf{H}_k^i$, $\mathbf{E} = \mathbf{A} - \mathbf{C}\mathbf{C}^+\mathbf{A}$, $\tilde{\mathbf{E}} = \mathbf{A} - \tilde{\mathbf{C}}\tilde{\mathbf{C}}^+\mathbf{A}$, $\rho = ((\mathbf{C}^+)_{i:})^T$. Then

$$\tilde{\mathbf{E}} = \mathbf{E} + \mathbf{C}_{:i}\rho^T\mathbf{A} + \left\| \rho \right\|_2^{-2} \tilde{\mathbf{C}}\mathbf{C}^+\rho\rho^T\mathbf{A} \tag{5}$$

We now provide efficient formulae to compute (3). We define $\mathbf{F} = \mathbf{E}^T\mathbf{E}$, $\tilde{\mathbf{F}} = \tilde{\mathbf{E}}^T\tilde{\mathbf{E}}$ and the vectors

$$\begin{aligned} \mathbf{f} &= (\left\| \mathbf{F}_{:1} \right\|_2^2, \ldots, \left\| \mathbf{F}_{:n} \right\|_2^2) \\ \mathbf{g} &= (\mathbf{F}_{11}, \ldots, \mathbf{F}_{nn}) \\ \tilde{\mathbf{f}} &= (\left\| \tilde{\mathbf{F}}_{:1} \right\|_2^2, \ldots, \left\| \tilde{\mathbf{F}}_{:n} \right\|_2^2) \\ \tilde{\mathbf{g}} &= (\tilde{\mathbf{F}}_{11}, \ldots, \tilde{\mathbf{F}}_{nn}) \end{aligned} \tag{6}$$

Proposition 3. Let $\boldsymbol{\delta} = (\tilde{\mathbf{E}}_{:j})^T \tilde{\mathbf{E}}$ and $\boldsymbol{\gamma} = \mathbf{E}^T \mathbf{E} \boldsymbol{\delta}$. Then

$$
\begin{aligned}
\tilde{\mathbf{f}} &= \mathbf{f} + \|\boldsymbol{\delta}\|_2^2 (\boldsymbol{\delta} \circ \boldsymbol{\delta}) \boldsymbol{\delta}_j^{-2} + 2(\boldsymbol{\gamma} \circ \boldsymbol{\delta}) \boldsymbol{\delta}_j^{-1} \\
\tilde{\mathbf{g}} &= \mathbf{g} + (\boldsymbol{\delta} \circ \boldsymbol{\delta}) \boldsymbol{\delta}_j^{-1}
\end{aligned}
\tag{7}
$$

The case of $\tilde{\mathbf{g}}$ is trivial, given that

$$
\tilde{\mathbf{g}}_k = \mathbf{F}_{kk} + \frac{\delta_k \delta_k}{\delta_j}
\tag{8}
$$

The criterion to find the current best column is

$$
t = \arg\max_{i \in [1,n]} \frac{\tilde{\mathbf{f}}_i}{\tilde{\mathbf{g}}_i}
\tag{9}
$$

Equivalent derivations yield the update formulae to use when a column is chosen and added to the subset:

$$
\begin{aligned}
\mathbf{f} &= \tilde{\mathbf{f}} + \|\boldsymbol{\delta}\|_2^2 (\boldsymbol{\delta} \circ \boldsymbol{\delta}) \boldsymbol{\delta}_j^{-2} - 2(\boldsymbol{\gamma} \circ \boldsymbol{\delta}) \boldsymbol{\delta}_j^{-1} \\
\mathbf{g} &= \tilde{\mathbf{g}} - (\boldsymbol{\delta} \circ \boldsymbol{\delta}) \boldsymbol{\delta}_j^{-1}
\end{aligned}
\tag{10}
$$

We now give an efficient formula to update $\mathbf{C}^+$ once the winning column of the current iteration is added.

Proposition 4. Let $\mathbf{C} \in \mathbb{R}^{m \times k}$ be the matrix resulting from adding column w of $\mathbf{A}$ to $\tilde{\mathbf{C}}$ at position i. Let $\mathbf{z} = \tilde{\mathbf{E}}_{:w}$ Then

$$
\mathbf{C}^+ = \tilde{\mathbf{C}}^+ - \|\mathbf{z}\|_2^{-2} (\tilde{\mathbf{C}}^+ \mathbf{A}_{:w} \mathbf{z}^T - \mathbf{e}_i \mathbf{z}^T)
\tag{11}
$$

Finally, since in this case we have added a column to the subset, we can employ the result proved in lemma 2 of [10] to update $\mathbf{E}$.

$$
\mathbf{E} = \tilde{\mathbf{E}} - \mathbf{z}\mathbf{z}^T \tilde{\mathbf{E}} \|\mathbf{z}\|_2^{-2}
\tag{12}
$$

where $\mathbf{z} = \tilde{\mathbf{E}}_{:w}$.

As described above, after a series of optimization, we can quickly determine which column should be used to replace the column in each iteration. So, the Algorithm IterFS [20] is equivalent to other greedy algorithm but much more efficient. Finally, the above procedure is summarized in Algorithm 2 (see Appendix).

2.3   Building a Recovery Matrix

The Algorithm 1 first selects $k(k \ll n)$ labels from the dataset, and the label matrix composed of the selected k labels is used as the latent label of the dataset. In order to achieve better classification results, in the stage of restoring the low-dimensional prediction label to the original label space, we do not directly use $\mathbf{X}^+$, but use (13) to deal with $\mathbf{X}^+$ first. Then, given a k-dimensional prediction vector $\mathbf{h}$, a n-dimensional label vector $\hat{\mathbf{y}}$ can be recovered as $\mathbf{h}^T \mathbf{X}^+ \mathbf{A}$.

$$
(\mathbf{X}^+)_{ij} = \begin{cases} 0 & , \quad \text{if } (\mathbf{X}^+)_{ij} \leq 0 \\ (\mathbf{X}^+)_{ij} & , \quad \text{other} \end{cases}
\tag{13}
$$

Algorithm 1. Multi-label classification using Iterative column subset selection with relu:ML-ICSSR

1: **procedure** ML-ICSSR
2:     $R \leftarrow \text{ITERFS}(\mathbf{A}, k)$
3:     $\mathbf{X} \leftarrow \mathbf{A}_R$
4:     $\mathbf{X}^+ \leftarrow$ using (13) deal with $\mathbf{X}^+$
5:     Train the classifier $f(\mathbf{x})$ from $\left\{\mathbf{x}^{(n)}, \mathbf{X}^{(n)}\right\}_{n=1}^{N}$
6:     Given a new test point $\mathbf{x}$, obtain its prediction $\mathbf{h}$ using $f(\mathbf{x})$
7:     $\hat{\mathbf{y}} \leftarrow \mathbf{h}^T \mathbf{X}^+ \mathbf{A}$
8:     return $\hat{\mathbf{y}}$
9: **end procedure**

Table 1. Statistics of six experimented multi-label data sets.

Dataset	Domain	Train	Test	Features	Labels	Cardinality	Density
Bibtex	Text	4880	2515	1836	159	2.402	0.015
Corel5k	Image	4500	500	499	374	3.522	0.009
Corel16k-s2	Image	5241	1783	500	164	2.867	0.018
Delicious	Text	12920	3185	500	983	19.020	0.019
EUR-Lex	Text	17413	1935	5000	201	2.213	0.011
Mediamill	Video	30993	12914	120	101	4.376	0.043

2.4 Learning the Model and Prediction

In the training phase, the Algorithm 1 first selects $k(k \ll n)$ labels from the dataset. Then, a binary classifier is learned for each of the k labels. In the prediction stage, we first apply the k learned classifiers on a new test sample to obtain its k-dimensional prediction vector $\mathbf{h}$. Note from (1) that $\mathbf{A} \simeq \mathbf{X}\mathbf{X}^+\mathbf{A}$. Each row of $\mathbf{A}$ (which corresponds to the n labels of a particular sample) can thus be approximated as the product of the corresponding row in $\mathbf{X}$ (which corresponds to the k selected labels of the same sample) with $\mathbf{X}^+\mathbf{A}$. So, given a new test sample $\mathbf{x}$, we can obtain its k-dimensional prediction vector $\mathbf{h}$ through the k learned classifiers. Finally, given a k-dimensional prediction vector $\mathbf{h}$, a n-dimensional label vector $\hat{\mathbf{y}}$ can be recovered as $\mathbf{h}^T\mathbf{X}^+\mathbf{A}$.

3 Experiments

In this section, we experimentally evaluate the proposed ML-ICSSR on six benchmark multi-label data sets, via comparing it with five existing methods: ML-CSSP [1], MLC-BMaD [26], MLC-EBMD [17], LS-BaBID [14] and ML-ICSS [20].

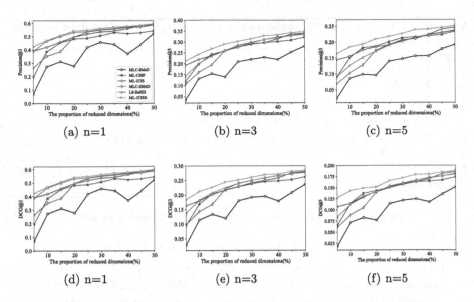

Fig. 1. Two metrics (at n = 1, 3 and 5) from six methods on Bibtex

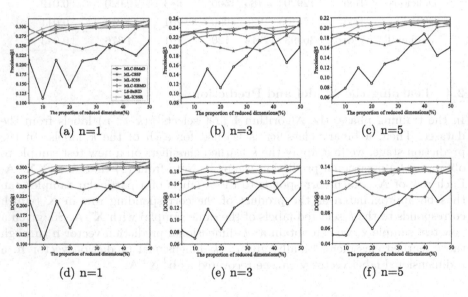

Fig. 2. Two metrics (at n = 1, 3 and 5) from six methods on Corel5k

3.1 Six Benchmark Data Sets and Two Evaluation Metrics

For the evaluation process we use six publicly available standard benchmark multi-label data sets. All of them downloaded from Mulan. Statistics on the data sets are given in Table 1.

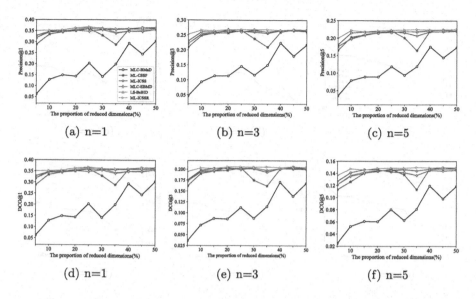

Fig. 3. Two metrics (at n = 1, 3 and 5) from six methods on Corel16k-s2

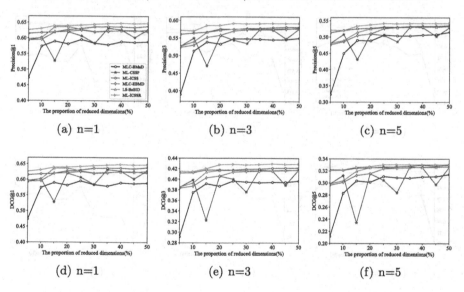

Fig. 4. Two metrics (at n = 1, 3 and 5) from six methods on Delicious

To describe the data sets, we provide two typical multi-label statistics. The first one is cardinality, which is the average number of labels per instance. This is a good measure of the dependencies between the labels. A cardinality close to one shows there are almost no dependencies in the labels which are represented in the data set, as there is only one label on average present in an instance. The label density is label cardinality divided by the total number of possible labels.

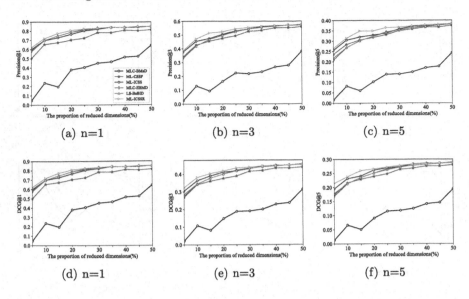

Fig. 5. Two metrics (at n = 1, 3 and 5) from six methods on EUR-Lex

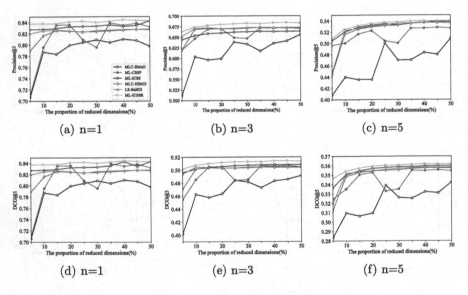

Fig. 6. Two metrics (at n = 1, 3 and 5) from six methods on Mediamill

This can be considered as the ratio of labels per instance. It is worth noting that the number of labels is more than 100 for all six data sets and three data sets belong to text categorization applications.

The traditional multi-label classification evaluation metrics are designed for low-dimensional label space [11], which are not suitable for large-scale label one [13]. For instance, the popular Hamming loss, does not prioritize predicting the few relevant labels over the millions of irrelevant ones, treats all relevant and irrelevant labels as being equally important and is biased due to missing ground truth. In our experiments, we utilize two new evaluation metrics: Precision@k, and (DisCounted Gain) DCG@k ($k = 1, 2, 3 \ldots$) [13].

For a testing instance $\mathbf{x}$, its ground label vector is $\mathbf{y} = [y_1, \ldots, y_i, \ldots, y_n]^T \in \{0,1\}^n$ and predicted function values $\hat{\mathbf{y}} = [\hat{y}_1, \ldots, \hat{y}_i, \ldots, \hat{y}_n]^T \in \{0,1\}^n$, and then such two metrics are defined as follows:

$$\text{Precision@}k = \frac{1}{n} \sum_{i \in rank_k(\hat{\mathbf{y}})} y_i \tag{14}$$

$$\text{DCG@}k = \frac{1}{n} \sum_{i \in rank_k(\hat{\mathbf{y}})} \frac{y_i}{\log_2 (i + 1)} \tag{15}$$

where $rank_k(\hat{\mathbf{y}})$ returns the top k label indexes of $\hat{\mathbf{y}}$. Finally, their average values are calculated via averaging them over all testing instances. Additionally, the higher these two metric values are, the better the label selection techniques perform.

3.2 Experimental Settings

In our experiments, we evaluate ML-ICSSR and five existing techniques (ML-CSSP, MLC-BMaD, MLC-EBMD, ML-ICSS and LS-BaBID) via training versus testing mode. To this end, we choose random forest as our base classifier, in which the number of trees is set to 100. In order to investigate how the number of selected labels (*i.e.*, k) would affect classification performance, the dimension proposition after label reduction (*i.e.*, k/n) is to set as 5% to 50% of the original label size (n) with the step of 5% [14]. In particular, the threshold for constructing the label association matrix in MLC-BMaD is tuned to be 0.7 for achieving a satisfactory comparison performance. For two metrics Precision@k and DCG@k, we set $k = 1, 3$ and 5.

3.3 Results

At first, we investigate two evaluation metrics (*i.e.*, Precision@k and DCG@k) as two functions of the different proportions of reduced labels(*i.e.*, k/n), respectively. The experimental results on six data sets in Table 2 are shown in Figs. 1, 2, 3, 4, 5 and 6.

From these six figures, we can find that ML-CSSP and MLC-BMaD are unstable, and their results fluctuate, while the experimental results of the other four techniques are almost stable. LS-BaBID can get a better result than MLC-EBMD, although both algorithms are based on exact Boolean matrix decomposition, LS-BaBID uses a better method to delete some less informative labels than MLC-EBMD. At most of label proportions, our ML-ICSSR works best, compared with five existing methods.

Table 2. The number of wins for each method and metric across six data sets.

Metric	ML-CSSP	MLC-BMaD	MLC-EBMD	LS-BaBID	ML-ICSS	ML-ICSSR
Precision@1	3	0	2	2	1	52
Precision@3	1	0	0	1	0	58
Precision@5	1	0	0	0	1	58
DCG@1	3	0	2	2	1	52
DCG@3	1	0	0	0	3	56
DCG@5	1	0	0	2	5	52
Total wins	10	0	4	7	11	328

In order to compare the six techniques more accurately, the "win" index in [7] is used in our comparison, which represents how many times each technique reaches the best metric values for all datasets and all dimension proportions of reduced labels, as shown in Table 2. ML-CSSP, MLC-BMaD, MLC-EBMD, LS-BaBID and ML-ICSS win 10, 0, 4, 7, 11 times, respectively. Our ML-ICSSR achieves the best values of 328 times for all metrics across six data sets, which is greatly than the number of wins from the total summation (32) of other four methods.

In summary, through the above experimental results, it can be concluded that our proposed method performs the best and can achieve the best results in multiple evaluation metrics, compared with five existing approaches.

4   Conclusions

The main contribution of this paper is the use of iterative column subset selection for generating latent labels in the base level step and to substitute them for the original labels. These latent labels identify and represent dependencies between the original labels in a compact manner. In the prediction stage, we do some processing on the recovery label. The experimental evaluation showed that the new method works particularly well on datasets with a large number of labels, outperforming most competing methods in that scenario. Still, there is room for further improvement. In the future work, we can explore using a more appropriate value as the threshold for modifying the generalized inverse matrix, rather than all using 0.

Appendix

The following is the detailed process of Iterative column subset selection algorithm.

Algorithm 2. Iterative column subset selection:ITERFS

1: **procedure** ITERFS($\mathbf{A}$, k)
2:　　$R \leftarrow$ uniSampleWithoutReplacement($1 \ldots n, k$)
3:　　$\mathbf{F} \leftarrow \mathbf{E}^T \mathbf{E}$
4:　　$\mathbf{f}_i \leftarrow \|\mathbf{F}_{:i}\|_2^2$; $\mathbf{g}_i \leftarrow \mathbf{F}_{ii}$ for $i = 1 \ldots n$
5:　　$\mathbf{C} \leftarrow \mathbf{A}_R$
6:　　**while** not converged **do**
7:　　　　**for** $i = 1 \ldots k$ **do**
8:　　　　　　$j \leftarrow R[i]$　　　　　　　　　　　　　▷ The i-th element in set R
9:　　　　　　$\tilde{\mathbf{C}} \leftarrow \mathbf{C}\mathbf{H}_k^i$　　　　　　　　　　　　　▷ Zero out column i
10:　　　　　$\tilde{\mathbf{C}}^+ \leftarrow \mathbf{C}^+ - \|\rho\|_2^{-2}\mathbf{C}^+\rho\rho^T$　　　　　　　　　　▷ Prop.1
11:　　　　　$\mathbf{S}_1 \leftarrow \mathbf{C}_{:i}\rho^T\mathbf{A}$
12:　　　　　$\mathbf{S}_2 \leftarrow \|\rho\|_2^{-2}\tilde{\mathbf{C}}\mathbf{C}^+\rho\rho^T\mathbf{A}$
13:　　　　　$\tilde{\mathbf{E}} \leftarrow \mathbf{E} + \mathbf{S}_1 + \mathbf{S}_2$　　　　　　　　　　　▷ Prop.2
14:　　　　　$\delta \leftarrow \tilde{\mathbf{E}}_{:j}^T\tilde{\mathbf{E}}$; $\delta \leftarrow \mathbf{E}^T\mathbf{E}\delta$
15:　　　　　$\tilde{\mathbf{f}} \leftarrow \mathbf{f} + \|\delta\|_2^2(\delta \circ \delta)\delta_j^{-2} + 2(\gamma \circ \delta)\delta_j^{-1}$
16:　　　　　$\tilde{\mathbf{g}} \leftarrow \mathbf{g} + (\delta \circ \delta)\delta_j^{-1}$　　　　　　　　　▷ Prop.3
17:　　　　　$w \leftarrow \arg\max_h \frac{\tilde{\mathbf{f}}_h}{\tilde{\mathbf{g}}_h}$
18:　　　　　$\delta \leftarrow \tilde{\mathbf{E}}_{:w}^T\tilde{\mathbf{E}}$; $\gamma \leftarrow \tilde{\mathbf{E}}^T\tilde{\mathbf{E}}\delta$
19:　　　　　$\mathbf{f} \leftarrow \tilde{\mathbf{f}} + \|\delta\|_2^2(\delta \circ \delta)\delta_w^{-2} - 2(\gamma \circ \delta)\delta_w^{-1}$
20:　　　　　$\mathbf{g} \leftarrow \tilde{\mathbf{g}} - (\delta \circ \delta)\delta_w^{-1}$
21:　　　　　$\mathbf{C}^+ \leftarrow \tilde{\mathbf{C}}^+ - \|\mathbf{z}\|_2^{-2}(\tilde{\mathbf{C}}^+\mathbf{A}_{:w}\mathbf{z}^T - \mathbf{e}_i\mathbf{z}^T)$　　▷ Prop.4
22:　　　　　$\mathbf{E} \leftarrow \tilde{\mathbf{E}} - \mathbf{z}\mathbf{z}^T\tilde{\mathbf{E}}\|\mathbf{z}\|_2^{-2}$
23:　　　　　checkConvergence()
24:　　　　　$R[i] \leftarrow w$
25:　　　　　$\mathbf{C} \leftarrow \mathbf{A}_R$
26:　　　　**end for**
27:　　**end while**
28:　　**return** R
29: **end procedure**

References

1. Bi, W., Kwok, J.: Efficient multi-label classification with many labels. In: ICML, pp. 405–413 (2013)
2. Boutell, M.R., Luo, J., Shen, X., Brown, C.M.: Learning multi-label scene classification. Pattern Recognit. **37**(9), 1757–1771 (2004)
3. Boutsidis, C., Mahoney, M.W., Drineas, P.: Unsupervised feature selection for principal components analysis. In: SIGKDD, pp. 61–69 (2008)
4. Chen, Y.N., Lin, H.T.: Feature-aware label space dimension reduction for multi-label classification. In: NIPS, vol. 25, pp. 1538–1546 (2012)
5. Chen, Z.M., Wei, X.S., Wang, P., Guo, Y.: Multi-label image recognition with graph convolutional networks. In: CVPR, pp. 5177–5186 (2019)
6. Civril, A., Magdon-Ismail, M.: Column subset selection via sparse approximation of SVD. Theor. Comput. Sci. **421**, 1–14 (2012)

7. Demsar, J.: Statistical comparisons of classifiers over multiple data sets. J. Mach. Learn. Res. **7**, 1–30 (2006)
8. Deng, X., Li, Y., Weng, J., Zhang, J.: Feature selection for text classification: a review. Multimedia Tools Appl. **78**(3), 3797–3816 (2018). https://doi.org/10.1007/s11042-018-6083-5
9. Duda, R.O., Hart, P.E., Stork, D.G.: Pattern Classification, 2nd edn. Wiley, New York (2001)
10. Farahat, A.K., Ghodsi, A., Kamel, M.S.: An efficient greedy method for unsupervised feature selection. In: ICDM, pp. 161–170 (2011)
11. Herrera, F., Charte, F., Rivera, A.J., del Jesus, M.J.: Multilabel Classification Problem Analysis, Metrics and Techniques. Springer, Cham (2016). https://doi.org/10.1007/978-3-319-41111-8
12. Hsu, D.J., Kakade, S.M., Langford, J., Zhang, T.: Multi-label prediction via compressed sensing. In: NIPS, pp. 772–780 (2009)
13. Jain, H., Prabhu, Y., Varma, M.: Extreme multi-label loss functions for recommendation, tagging, ranking & other missing label applications. In: SIGKDD, pp. 935–944 (2016)
14. Ji, T., Li, J., Xu, J.: Label selection algorithm based on Boolean interpolative decomposition with sequential backward selection for multi-label classification. In: Lladós, J., Lopresti, D., Uchida, S. (eds.) ICDAR 2021. LNCS, vol. 12822, pp. 130–144. Springer, Cham (2021). https://doi.org/10.1007/978-3-030-86331-9_9
15. Krömer, P., Platoš, J., Nowaková, J., Snášel, V.: Optimal column subset selection for image classification by genetic algorithms. Ann. Oper. Res. **265**(2), 205–222 (2018)
16. Lee, J., Yu, I., Park, J., Kim, D.W.: Memetic feature selection for multilabel text categorization using label frequency difference. Inf. Sci. **485**, 263–280 (2019)
17. Liu, L., Tang, L.: Boolean matrix decomposition for label space dimension reduction: method, framework and applications. In: CISAT, p. 052061 (2019)
18. Maltoudoglou, L., Paisios, A., Lenc, L., Martínek, J., Král, P., Papadopoulos, H.: Well-calibrated confidence measures for multi-label text classification with a large number of labels. Pattern Recognit. **122**, 108271 (2022)
19. Nowaková, J., Krömer, P., Platoš, J., Snášel, V.: Preprocessing COVID-19 radiographic images by evolutionary column subset selection. In: Barolli, L., Li, K.F., Miwa, H. (eds.) INCoS 2020. AISC, vol. 1263, pp. 425–436. Springer, Cham (2021). https://doi.org/10.1007/978-3-030-57796-4_41
20. Ordozgoiti, B., Canaval, S.G., Mozo, A.: Iterative column subset selection. Knowl. Inf. Syst. **54**(1), 65–94 (2018)
21. Rastin, N., Taheri, M., Jahromi, M.Z.: A stacking weighted k-nearest neighbour with thresholding. Inf. Sci. **571**, 605–622 (2021)
22. Shitov, Y.: Column subset selection is NP-complete. Linear Algebra Appl. **610**, 52–58 (2021)
23. Sun, S., Zong, D.: LCBM: a multi-view probabilistic model for multi-label classification. IEEE Trans. Pattern Anal. Mach. Intell. **43**(8), 2682–2696 (2020)
24. Sun, Y., Ye, S., Sun, Y., Kameda, T.: Exact and approximate Boolean matrix decomposition with column-use condition. Int. J. Data Sci. Anal. **1**(3–4), 199–214 (2016)
25. Tai, F., Lin, H.T.: Multilabel classification with principal label space transformation. Neural Comput. **24**(9), 2508–2542 (2012)
26. Wicker, J., Pfahringer, B., Kramer, S.: Multi-label classification using Boolean matrix decomposition. In: SAC, pp. 179–186 (2012)

27. Zhang, D., Zhao, S., Duan, Z., Chen, J., Zhang, Y., Tang, J.: A multi-label classification method using a hierarchical and transparent representation for paper-reviewer recommendation. ACM Trans. Inf. Syst. **38**(1), 1–20 (2020)
28. Zhang, M.L., Zhou, Z.H.: A review on multi-label learning algorithms. IEEE Trans. Knowl. Data Eng. **26**(8), 1819–1837 (2014)

Accurately Predicting User Registration in Highly Unbalanced Real-World Datasets from Online News Portals

Eva-Maria Spitzer[1], Oliver Krauss[1]([envelope]) [iD], and Andreas Stöckl[2] [iD]

[1] Advanced Information Systems and Technology, University of Applied Sciences Upper Austria, Wels, Austria
`spitzer.eva-maria@gmx.at`, `oliver.krauss@fh-hagenberg.at`
[2] Digital Media Department, University of Applied Sciences Upper Austria, Wels, Austria
`andreas.stoeckl@fh-hagenberg.at`

Abstract. Getting visitors to register is a crucial factor in marketing for online news portals. Current approaches are rule-based by awarding points for specific actions [3]. Finding efficient rules can be challenging and depends on the specific task. Registration is generally rare compared to regular visitors, leading to highly imbalanced data.

We analyze different supervised learning classification algorithms under consideration of the data imbalance. As case study, we use anonymized real-world data from an Austrian newspaper outlet containing the visitor's session behavior with around 0.1% registrations over all visits.

We identify an ensemble approach combining the Balanced Random Forest Classifier and the RUSBoost Classifier correctly identifying 76% of registrations over five independent data sets.

Keywords: Imbalanced data · Lead scoring · Label prediction

1 Introduction

Lead scoring faces highly imbalanced data sets. We compare different machine learning classifiers and sampling strategies to the problem to accurately identify leads, and apply it to real world data. The resulting ensemble classifier can be used to replace traditional lead scoring done with rule sets.

For news portals and online newspapers, getting website visitors to register and leave their email addresses and other data is essential for marketing their content. This can then be used, for example, to send email newsletters, personalize banner ads or sell digital subscription models. To optimize the registration process, it is of great interest to be able to predict whether a website visitor who has already visited a website several times will become a registered visitor or not. The process to find website visitors who are most likely to register or submit other information is commonly known as the lead scoring process. A traditional

© The Author(s), under exclusive license to Springer Nature Switzerland AG 2022
C. Strauss et al. (Eds.): DEXA 2022, LNCS 13426, pp. 302–315, 2022.
https://doi.org/10.1007/978-3-031-12423-5_23

approach to apply lead scoring works rule-based. For example, a user who opens a mail gets +10 points assigned, or in case a user does not show activity within ten days, gets −5 points. Several rules can be applied and used for lead scoring.

Rule-based scoring approaches require expert knowledge, and need to be adapted on a regular basis. This is often infeasible for smaller news agencies. To optimize the registration process a more generic and automatic process is needed. Leads could, for example, be predicted using machine learning approaches, whereby a classifier predicts whether a visitor might convert to a lead or not. The prediction can then be used to make ad placements dependent on the prediction. For example, visitors where the model predicts a possible registration can be targeted with offers to make the visitors leads.

The number of leads in such use cases usually is much smaller than the number of non-leads since, in most cases, only a few people, out of all website visitors, finally convert to a lead by signing in on an online platform or by submitting their telephone number. The huge difference in the number of leads and non-leads comes along with the problem of highly imbalanced data. Imbalanced data is a widely discussed issue through various application areas of machine learning [11,18]. The problem with data imbalance is that most conventional classifiers result in unsatisfying outputs. Working with imbalanced data lead to many misclassified samples of the minority class and to a good coverage of the majority class. Though, most of the time predicting the minority class is more relevant.

To increase the quality of classification algorithms, ensemble approaches can be used [23]. These approaches combine several classifiers to create one strong classifier that outperforms each of them.

In our work, we show algorithms that perform well on a lead prediction task using data from an online newspaper facing data imbalance. We show the performance of an ensemble approach evaluated on five different test data sets. The ensemble approach combines the Balanced Random Forest Classifier and the RUSBoost Classifier which are designed to work well with highly imbalanced data. We conduct a case study using the data of an Austrian daily online newspaper.

Comparable online offers with similar tasks have in common with our data that the data available is highly unbalanced, which needs to be considered for classification. The unbalanced nature of the classes was also the greatest challenge of our study and greatly influenced the selection of the model classes.

2 Background

The use of machine learning in digital marketing along the customer journey is widespread, as a large amount of data is generated and can be used for various marketing tasks. Artun and Levin [2], as well as Kietzmann et al. [13] give an overview of machine learning techniques used in marketing, including lead scoring. Related to our work are all techniques that deal with the prediction for optimizing the sales funnel, i.e. the process through which companies lead people when making a sale. Duncan and Elcan [8] give an overview of modeling sales funnels. A sale is not made directly in our task, but the targeted registration is a preliminary stage for the sale of digital subscriptions.

Classification algorithms can predict whether a website visitor is more or less likely to become a lead based on the visitors' behavior. Multiple classification algorithms like Random Forest [4], Logistic Regression [14], or Support Vector Machine [12] exist.

Many traditional machine learning approaches assume that the target to predict is almost balanced [10]. In many real-world scenarios the data is not balanced, negatively influencing the prediction quality. Guo et al. [10] write about the data imbalance problem and actions to handle this issue. They conclude that they detected some evidence that artificially balancing the classes does not influence the performance of the classifier's result. The reason for this is that some machine learning approaches are sensitive to data imbalance.

Kotsiantis et al. [15] write about methods to handle data imbalance. Random under-sampling intends to balance the data by removing samples of the majority class. On the contrary to under-sampling, over-sampling balances the class distribution by randomly replicating instances of the minority class.

Algorithms developed for sampling data exist. In [34], a cluster-based sampling strategy using under-sampling to achieve class balance is described. This approach intends to avoid losing relevant information from the majority class by replacing a cluster of majority samples with its centroid calculated with the KMeans algorithm.

Furthermore, machine learning algorithms for handling imbalanced data exist. A good overview of Random Forest Classifiers and unbalanced data can be found in [19]. The paper states that the Random Forest approach outperforms the SVM, DT, Functions Logistic, Naive Bayes, Adaboost, and Attribute Selected classifier. In [6], Chen et al. present two ways to use the Random Forest for imbalanced data. Weighted Random Forest is one approach that puts more weight on the less present class than the other. The Balanced Random Forest Classifier is an approach that uses a down-sampling majority voting in combination with ensemble learning for the prediction.

Another approach to handle imbalanced data is the RUSBoost algorithms proposed by Seifert et al. [26]. In their paper, the RUSBoost, Random Under-Sampling Boosting, and the SMOTEBoost algorithm outperform the AdaBoost, RUS, and SMOTE approaches. Their evaluation states that RUSBoost should be preferred over the SMOTEBoost when working with imbalanced data due to a simpler handling, speediness, and less complexity.

Polikar [22], writes about an ensemble-based system that intends to raise the prediction quality compared to a single approach. The idea behind ensemble learning is to train and weight multiple classifiers and to combine them into one strong classifier [23]. The new classifier consisting of several classifiers, then makes new predictions.

3   Collection and Overview of the Data

An Austrian daily newspaper recorded all page views and interactions on their web portal using a web tracking solution. We linked the data of the individual sessions to visitor profiles via an anonymous ID assigned by the web tracking solution. This is only possible for sessions that have given a corresponding cookie consent. This naturally limits the data quality, as sessions without consent are analyzed as different profiles. In 24% of sessions no consent is given.

Urban et al. [29] explore the impact of European legal requirements on ad tracking data. They note a 40% reduction in available tracking data. This implies that machine learning methods in this domain in the future will have to be able to work on data sets that value user privacy.

The data recorded contains more than 60 variables. These include, for example, the duration of sessions, the total number of recorded events, or the number of days since the last visit of a visitor. The daily number of sessions created on the online newspaper website is between 250,000 to 350,000. Figure 1a shows the daily number of sessions for the period 1 Nov 2020–10 Nov 2020.

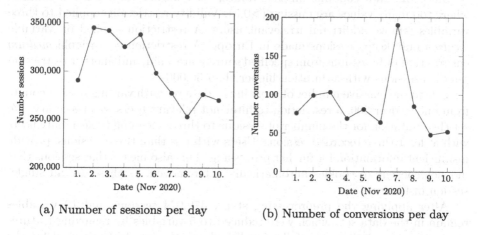

(a) Number of sessions per day   (b) Number of conversions per day

Fig. 1. Daily statistics by all visitors for the period 1 Nov 2020–10 Nov 2020.

To use a supervised learning approach, we also need the information of whether a user has made a conversion or not. In our case study, a registration, a completed subscription, and the submission of a telephone number represent a conversion goal. Figure 1b shows that the daily number of conversions in the evaluated period 1 Nov 2020–10 Nov 2020 is between 45 and 180.

To motivate visitors to register, some articles are only readable by registered users for the first few hours after publication. As shown by Fig. 1b, there is a peak in the number of conversions on 7th November. This peak might be the result of a conducted marketing campaign aiming to motivate the visitors to register. Information on campaigns is not available in the dataset however.

The small portion of users registered on the news platform leads to a meager conversion rate of around 0.1%. Meaning that only 0.1% of the total number of visitors of the online news paper website made one of the specified goals. The target used for the prediction is the information of whether a visitor converted to a lead or not. In this case study, a visitor is called a lead, if she/he achieved one of the conversion goals. The strong imbalanced between sessions (avg 301,716 per day) and conversions (avg 87 per day) represents the main challenge of our case study.

4   Data Preprocessing and Feature Extraction

Several preprocessing steps bring the raw data into the required shape to use it for the model building. A descriptive statistics analysis provide a first basic understanding of all variables in the data set.

We analyzed and evaluated each of the variables and determined if they are helpful for the prediction. We took the correlation with the target into consideration.

Since the data contains missing values, we removed variables and samples where too many values are missing (>0.5). Validity restrictions applied to three variables remove additional irrelevant users. A restriction applied on variable *location* only keeps sessions made in Europe. A restriction on variable *medium* ensures that only sessions from specified sources are valid, and another restriction removes sessions with a duration higher than 30,000 s.

Due to the massive number of users in the data set with varying session counts from one to over 200, a restriction to filter not relevant users was necessary. We set the limitation for the number of sessions to three, meaning to keep only users with at least three recorded sessions. Users with less than three sessions, provide insufficient information for further processing. This also means that sessions that did not consent to cookies automatically get removed, as these are all single-session instances.

After applying the preprocessing steps, 370,074 sessions and 12 variables remain in the data set, which were reduced to 5 variables by removing features via feature correlation. The following list shows the five features used for the prediction of the leads:

mean session unique page views shows the number of unique page views a user made on average through all sessions. Page views is a metric that represents the total number of pages loaded during a session.

mean session total time represents the time (in seconds) a user spent on their sessions on average. Whereby the duration of a session is measured from the first to the last event that happened within a session.

mean session unique custom events is the average number of custom events through all sessions of a user. Custom events are interactions made by a single user within one session such as a button click, page-scroll depth, or newsletter sign up (signing up to the newsletter is not a registration). If a user clicks on a button three times within one session, then the event is tracked only once.

max visitor days since the last session shows the maximum number of days
between the sessions of a user.

mean session total events is the average number of total events of each user
through all sessions.

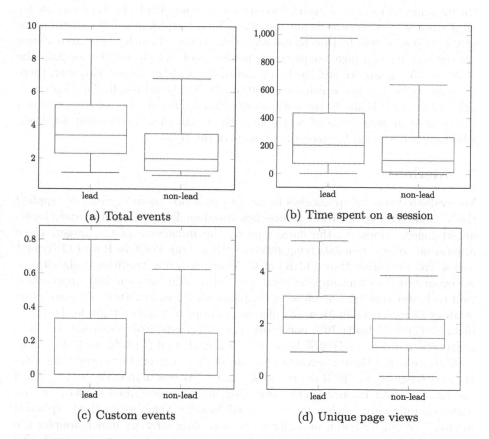

(a) Total events

(b) Time spent on a session

(c) Custom events

(d) Unique page views

Fig. 2. Average feature values separated by leads (blue) and non-leads (red) (Color
figure online)

For our use case, a very important aspect when analyzing the features are the
differences in the behavior of users who already converted to a lead and those
who did not convert. In Fig. 2a the two boxplots show that during sessions from
leads (left boxplot) more events are recorded than for sessions from non-leads
(right boxplot). Figure 2b shows that the overall time spent on a session is higher
for leads than for those who have not yet converted to a lead. Figure 2c shows
the characteristics of feature *mean session unique custom events*. Although, in
Fig. 2c, the median, the lower quantile and the lower whiskers have values of 0
there are some differences in the target visible. In the figure, the upper whiskers,
representing the largest data point excluding any outliers has a value of 0.8 for

leads and 0.625 for non-leads. This means that the number of custom events of leads is, for some users, higher than for non-leads. Furthermore, the 75-percentile has a higher value for leads than for non-leads. Figure 2d shows that most of the leads have a higher number of unique page views than most of the non-leads. Such characteristics are essential for machine learning models to learn the behavior of the users in order to distinguish between leads and non-leads. Feature *max visitor days since the last session* is not visualized by a boxplot since, the distribution of the values of this feature shows numerous zeros which leads to zero values for the metrics of which boxplots are made up of, which are the median, the 75% and 25% quantiles and the lower and upper whisker values. However, there are outliers in the data which are essential for the overall result. This feature is important, since leads, in the data, show a maximum number of days between the sessions of zero or one. Compared to the leads, there are several non-leads who have a deviation between the sessions of multiple days.

5   Methods

We evaluated several approaches to create prediction models. First, we applied the well-known classification approaches Random Forest Classifier and Generalized Linear Model to the data. Due to the imbalance of the target, these approaches return non-satisfying results with a True Positive Rate (TPR) of 1 and a True Negative Rate (TNR) of 0. After analyzing traditional algorithms, we evaluated if down-sampling strategies, also called re-sampling approaches, lead to better results. Therefore, we implemented a self-written down-sampling strategy using 5 combinations of different portions of leads and non-leads. These include: (1) 30% leads 70% non-leads, (2) 70% leads 30% non-leads, (3) 50% leads 50% non-leads, (4) 40% leads 60% non-leads and (5) 60% leads 40% non-leads. Given that these approaches lead to slightly better, but nevertheless still not satisfying scores (TPR of around 0.9 and TNR less than 0.35), we evaluated the cluster-based majority under-sampling strategy described in [34], and an under-sampling approach using the Edited Nearest Neighbor method explained in [30]. The cluster-based majority under-sampling strategy under-samples the data by replacing the original samples by the centroids of the cluster found. The Edited Nearest Neighbor under-samples the majority class data by removing samples that differ from the majority of the k nearest neighbors. The cluster-based under-sampling approach achieves a TPR of 1 and a TNR of around 0.01. On the contrary to the cluster-based approach, the Nearest Neighbor approach, achieved TPR of 0.11 and a TNR of 0.84 which are also non-satisfying.

We evaluated machine learning models developed to handle imbalanced data sets. The Python toolbox, imbalanced-learn, that intend to perform well with imbalanced data, is described in [16]. Via initial testing we discarded most algorithms known to perform well from literature since they performed poorly on our data set. We evaluated the performance of the Balanced Random Forest Classifier, as it is an implementation of random forest most applicable to imbalanced data. In addition we analyzed the RUSBoost Classifier, and the Balanced Bagging Classifier on our data set. The BRF Classifier and the RUSBoost Classifier show first promising results compared to all other considered methods. Hence, we evaluated these two approaches in more detail.

Table 1. Test data sets used for the five experiments conducted in this case study.

Experiment	From	To	Number days
1	15 Nov 2020	24 Nov 2020	10
2	22 Nov 2020	1 Dec 2020	10
3	5 Dec 2020	9 Dec 2020	5
4	6 Dec 2020	13 Dec 2020	8
5	5 Jan 2021	13 Jan 2021	9

For evaluating the two classifiers, we first condut a TrainTestSplit into one training and five test sets. The training set contains data from 1 Nov 2020–9 Nov 2020. For the test sets, we utilized different days to get a more general overview of the performance of the classifiers. Users that are already utilized in the training set are removed from the test sets, which means that the test sets only contain users that the models have not yet seen. To sum up, our test data sets vary in the number of days and time and do not include users from the training set. A random selection of the periods used for the test data sets intends to avoid possible influences and biases. The number of days in the training set ranges from five to ten. In total, we evaluated the performance of the algorithms on five experiments. They constitute all data exports available to us that were not used for validation. Table 1 gives an overview of the test data.

As shown by the table, experiments 1 and 2 contain the same number of test days (10 respectively) but differ in the date of recording. Experiments number 2, 3, and 5 vary in the number of days and dates from 5 Dec 2020 to 5 Jan 2021 with 5, 8 and 9 days respectively.

Each of the experiments varies in the number of leads and non-leads included in the test sets. Table 2, shows this difference by showing the total number of samples, the number of leads and the number of non-leads in each experiment.

Table 2. Number of samples in the training and the test sets.

Experiment	Total number samples	Number leads	Number non-leads
Training	236054	467	235587
1	214577	203	214374
2	206655	383	206272
3	134020	124	133896
4	214001	366	213635
5	212005	439	211566

The proportion of leads in the test records vary from 124 to 439. Experiment 3 and 1 have the lowest proportion of leads with 0.09% and Experiment 5 the highest with 0.21%.

6   Results

We evaluated the Balanced Random Forest Classifier and the RUSBoost Classifier by testing the models on several data sets. As described in the previous section, the conducted test sets vary in recording date and the number of days collected.

For the evaluation we calculated four error metrics and one confusion matrix [27] for each algorithm and experiment. The confusion matrix gives information about the results of the evaluated test sets. This matrix is made up of four essential parts: True Positives (TP), False Positives (FP), False Negatives (FN) and True Negatives (TN). The TP are the cases in which the model predicts non-leads, as non-leads. Compared to the TP the FP are the cases where leads are predicted as non-leads. FN are the cases where non-leads are predicted as leads and TN represent the cases where leads are predicted as leads.

Table 3a shows the performance of the Balanced Random Forest Classifier of the different data sets by exhibiting the confusion matrix values.

Table 3. Absolute evaluation values (top) and percentages (bottom) for the different classifiers on the experiment sets.

Experiment	Results			
	TP	FN	FP	TN
1	116666	97708	22	181
2	118028	88244	48	335
3	79876	54020	15	109
4	129150	84485	50	316
5	128119	83447	55	384

(a) Balanced Random Forest Classifier

Experiment	Results			
	TP	FN	FP	TN
1	185389	28985	133	70
2	162952	43320	168	215
3	108859	25037	48	766
4	173983	39652	164	202
5	168467	43099	178	261

(b) RUSBoost Classifier

Experiment	Measures			
	TPR	TNR	FNR	FPR
1	0.54	0.89	0.46	0.11
2	0.57	0.87	0.43	0.13
3	0.60	0.88	0.40	0.12
4	0.60	0.86	0.40	0.14
5	0.61	0.87	0.39	0.13

(c) Balanced Random Forest Classifier

Experiment	Measures			
	TPR	TNR	FNR	FPR
1	0.86	0.34	0.14	0.66
2	0.79	0.56	0.21	0.44
3	0.81	0.61	0.19	0.39
4	0.81	0.55	0.19	0.45
5	0.80	0.59	0.20	0.41

(d) RUSBoost Classifier

Each test set evaluated with the Balanced Random Forest Classifier achieved a True Positive score between 0.54 and 0.61 and a True Negative score between

0.86 and 0.89. The scores show that the percentage of non-leads recognized as non-leads using the Balanced Random Forest model is between 54% and 61%, and the portion of leads identified as leads is between 86% and 89%.

Table 3b shows the confusion matrix values of the RUSBoost Classifier applied to the five test sets. This algorithm results in True Positive scores between 0.79 and 0.86 and True Negative scores between 0.34 and 0.61. This states that the percentage of non-leads recognized as non-leads using the RUS-Boost Classifier is between 79% and 86%, and the portion of leads identified as leads is between 34% and 61%.

Classification metrics listed in the following give a more detailed insight into the performance of the two trained classifiers. For the evaluation of the performance of classifiers, multiple metrics such as True Positive Rate (TPR) also called recall, True Negative Rate (TNR) or inverse recall, Positive Prediction Value (PPV) also called precision, Negative Prediction Value (NPV) or inverse precision, False Negative Rate (FNR) or miss rate, or False Positive Rate (FPR) also called false alarm rate exist [28]. Due to our highly imbalanced data set, the PPV and NPV result in shallow values either close to zero or close to one. For example, the conducted experiments lead to PPV values of almost 1.0, meaning that almost 100% of the predicted non-leads were predicted correctly. It is essential to consider that the percentage of leads compared to the non-leads is at around 0.1%. Unlike the PPV the NPV shows values close to zero. This is explained by the few leads and the vast amount of non-leads in the data set. We are aware that the PPV and NPV are essential metrics for classifiers, though the other four metrics are more relevant for our specific use case facing the issue with the imbalanced data.

The TPR represents the portion of positive samples correctly predicted as positive. The TNR is the counterpart to the TPR, representing the portion of negative samples that are correctly predicted as negative. FNR is the ratio between the positive events that are wrongly classified as negative and the total number of all positive samples. The FPR metric represents the ratio between the negative events that are wrongly classified as positive and the total number of all negative samples.

Table 3c shows the achieved classification values for the test sets evaluated with the Balanced Random Forest Classifier and Table 3d for the test sets evaluated with the RUSBoost Classifier. Comparing the two result Tables 3c and 3d, each classifier shows a different strength. Whereby the Balanced Random Forest Classifier achieves a higher score for TNR than for TPR, the RUSBoost Classifier achieves a better TPR than a TNR score. When deciding which algorithm to choose, it is essential to consider the purpose of the task. If it is more critical to predict more leads truly, one should go for the Balanced Random Forest Classifier. When predicting more *non-leads* as *non-leads* is important, then the RUSBoost Classifier might be the better choice. For example, for a company that displays online advertisements with the pay-per-view price model, it is more efficient to use the RUSBoost Classifier. In this case, it is better to show fewer people the ads than showing too many false positives ads leading to higher expenditures.

As it is of great interest that the Balanced Random Forest Classifier and the RUSBoost Classifier show different strengths regarding TP and TN predictions, we combined these two classifiers into one ensemble approach.

Our approach connects the classifiers' predictions, and return only predictions where both models agree (i.e., either both predict *non-lead* or *lead*) as valid outputs. Thus, visitors, where the models disagree, are non-predictable. Although the number of predictable leads got reduced, the remaining visitors could be classified more accurately. The percentage of non-predictable non-leads is on average 24% and the percentage of non-predictable leads is on average 36%, through all experiments. Table 4 shows the results of the applied ensemble approach. Comparing the classification result scores of the ensemble approach (Table 4) with the results of the two models evaluated separately (Tables 3c and 3d), an overall improvement is visible.

Table 4. Classification measures for the experiments conducted with the ensemble approach, outperforming the Balanced Random Forest and RUSBoost Classifiers

Experiment	Measures			
	TPR	TNR	FNR	FPR
1	0.80	0.76	0.20	0.24
2	0.73	0.82	0.27	0.18
3	0.77	0.83	0.23	0.17
4	0.77	0.81	0.23	0.19
5	0.75	0.84	0.25	0.16

7 Related Work

Xie et al. [31] use an Improved Balanced Random Forests Classifier, which is a combination of the Balanced Random Forest Classifier and the Weighted Random Forest Classifier to predict churn. With data from the banking environment, the method was compared with artificial neural networks [32], decision trees [24], and class-weighted support vector machines [33], and as in our results, random forest proved to be a good classification method. Compared to Xie et al. we use an ensemble approach combining the Balanced Random Forest and the RUSBoost Classifier, with the RUSBoost classifier being more advantageous on imbalanced data.

Alshehri et al. [1] address a use case that also models the prediction of digital content purchasing behavior from website interaction data, the purchase of online courses by students. RandomForest [17], GradientBoosting [9], AdaBoost [25] and XGBoost [7] are used here as machine learning methods, and prediction accuracies between 0.82 and 0.91 are achieved when only the time spent on each step is used for modelling. They achieved a higher accuracy of 0.83 to 0.95 by adding the learner's demographic data. These accuracies are slightly higher than the prediction successes we achieved, but in the same order of magnitude. Our approach does not use demographic data, preserving user privacy. The methods examined by Alshehri et al. were not applicable to us, as they stated that they could not cope with the extremely unbalanced data.

In [20], Nygård and Mezei present a way to apply automated lead scoring using machine learning approaches. In their study they evaluated four algorithms, the Random Forest Classifier, Decision Tree, Logistic Regression, and Neural Networks. Additionally, they considered five different ways to aggregate the time-series data used, whereby one of them is not biased. The non-biased approach uses all activity data of non-leads and all except the data of the last activity date, where the purchase happened, for leads. The best model evaluated, for the mentioned aggregation approach, was the Random Forest Classifier. Nygård and Mezei do not mention whether the portion of leads and non-leads is balanced or not. As we have to face the issue with a highly unbalanced number of leads and non-leads, we can only use methods that can handle data imbalance.

8   Conclusion and Future Work

We show an approach to make the traditional lead scoring process in online environments more flexible and efficient by using machine learning methods. The mentioned algorithms lead to satisfying prediction results, correctly identifying 76% of all leads working with imbalanced anonymized data. Having evaluated several approaches to predict leads in highly imbalanced data sets, we can conclude that the RUSBoost Classifier and the Balanced Random Forest Classifier lead to good results. Our results show that the Balanced Random Forest Classifier achieves better scores when truly predicting leads as leads than the RUSBoost Classifier. Vice versa, the RUSBoost Classifier performs better when predicting non-leads truly as non-leads.

The best result in our study is achieved by an ensemble approach combining the RUSBoost Classifier and the Balanced Random Forest Classifier. This approach reaches a True Positive Rate of 0.76, and a True Negative Rate of 0.81 averaged across all experiments, meaning that the portion of non-leads/leads truly predicted as non-leads/leads are at around 76% to 81%. Although, approximately 24% of the non-leads and 35% of the leads became unpredictable using the ensemble approach, for our use case it results in the best output.

The approaches provided in this work enable the utilization of highly unbalanced data in online environments. As our data is anonymized, this goes on to show that such results can be achieved even without infringing on the privacy of online users.

More precise feature analysis and selection would enable more detailed information of the website visitors' behavior to improve the quality of the prediction, leading to more accurate results. One feature to enhance the classifier's quality could be the categorization of articles according to the user's interests. The idea behind this feature is that readers who mainly read sports articles behave differently from those who are primarily interested in politics. However, to extract the required information from the website technical changes would be necessary.

To summarize, this work shows approaches to predict leads in online environments, especially news portals, where the target is highly imbalanced. Thinking one step further, to generate value from these results, additional steps such as integrating the results into a system that automatically supports sales employees by telling if a user is likely to become a lead or not are required. Furthermore, a continuous evaluation of the models' performance enclosed with automatic retraining would support the application of this approach in a real-world production application.

In the future this work could benefit from further steps to deal with the high data imbalance, such as synthetic minority-class sample generation [5] or other methods [21], and also take into consideration more time-series related features.

References

1. Alshehri, M., Alamri, A., Cristea, A.I., Stewart, C.D.: Towards designing profitable courses: predicting student purchasing behaviour in MOOCs. Int. J. Artif. Intell. Educ. **31**, 215–233 (2021)
2. Artun, O., Levin, D.: Predictive Marketing: Easy Ways Every Marketer Can Use Customer Analytics and Big Data. Wiley Online Library (2015)
3. Benhaddou, Y., Leray, P.: Customer relationship management and small data - application of Bayesian network elicitation techniques for building a lead scoring model. In: 2017 IEEE/ACS 14th International Conference on Computer Systems and Applications (AICCSA), pp. 251–255 (2017). https://doi.org/10.1109/AICCSA.2017.51
4. Breiman, L.: Random forests. Mach. Learn. **45**(1), 5–32 (2001)
5. Chawla, N.V., Bowyer, K.W., Hall, L.O., Kegelmeyer, W.P.: SMOTE: synthetic minority over-sampling technique. J. Artif. Intell. Res. **16**, 321–357 (2002)
6. Chen, C., Liaw, A., Breiman, L., et al.: Using random forest to learn imbalanced data. Univ. Calif. Berkeley **110**(1–12), 24 (2004)
7. Chen, T., He, T., Benesty, M., Khotilovich, V., Tang, Y., Cho, H., et al.: XGBoost: extreme gradient boosting. R Package Version 0.4-2 **1**(4), 1–4 (2015)
8. Duncan, B.A., Elkan, C.P.: Probabilistic modeling of a sales funnel to prioritize leads. In: Proceedings of the 21th ACM SIGKDD International Conference on Knowledge Discovery and Data Mining, pp. 1751–1758 (2015)
9. Friedman, J.H.: Stochastic gradient boosting. Comput. Stat. Data Anal. **38**(4), 367–378 (2002)
10. Guo, X., Yin, Y., Dong, C., Yang, G., Zhou, G.: On the class imbalance problem. In: 2008 Fourth International Conference on Natural Computation, vol. 4, pp. 192–201. IEEE (2008)
11. He, H., Garcia, E.A.: Learning from imbalanced data. IEEE Trans. Knowl. Data Eng. **21**(9), 1263–1284 (2009)
12. Hearst, M.A., Dumais, S.T., Osuna, E., Platt, J., Scholkopf, B.: Support vector machines. IEEE Intell. Syst. Appl. **13**(4), 18–28 (1998)
13. Kietzmann, J., Paschen, J., Treen, E.: Artificial intelligence in advertising: how marketers can leverage artificial intelligence along the consumer journey. J. Advert. Res. **58**(3), 263–267 (2018)
14. Kleinbaum, D.G., Klein, M.: Logistic Regression. SBH, Springer, New York (2010). https://doi.org/10.1007/978-1-4419-1742-3
15. Kotsiantis, S., Kanellopoulos, D., Pintelas, P., et al.: Handling imbalanced datasets: a review. GESTS Int. Trans. Comput. Sci. Eng. **30**(1), 25–36 (2006)

16. Lemaître, G., Nogueira, F., Aridas, C.K.: Imbalanced-learn: a python toolbox to tackle the curse of imbalanced datasets in machine learning. J. Mach. Learn. Res. **18**(1), 559–563 (2017)
17. Liaw, A., Wiener, M., et al.: Classification and regression by RandomForest. R News **2**(3), 18–22 (2002)
18. López, V., Fernández, A., García, S., Palade, V., Herrera, F.: An insight into classification with imbalanced data: empirical results and current trends on using data intrinsic characteristics. Inf. Sci. **250**, 113–141 (2013)
19. More, A., Rana, D.P.: Review of random forest classification techniques to resolve data imbalance. In: 2017 1st International Conference on Intelligent Systems and Information Management (ICISIM), pp. 72–78. IEEE (2017)
20. Nygård, R., Mezei, J.: Automating lead scoring with machine learning: an experimental study. In: Proceedings of the 53rd Hawaii International Conference on System Sciences (2020)
21. Patel, D., Zhou, N., Shrivastava, S., Kalagnanam, J.: Doctor for machines: a failure pattern analysis solution for industry 4.0. In: 2020 IEEE International Conference on Big Data (Big Data), pp. 1614–1623 (2020). https://doi.org/10.1109/BigData50022.2020.9378369
22. Polikar, R.: Ensemble based systems in decision making. IEEE Circuits Syst. Mag. **6**(3), 21–45 (2006)
23. Rokach, L.: Ensemble-based classifiers. Artif. Intell. Rev. **33**(1), 1–39 (2010)
24. Rokach, L., Maimon, O.: Decision trees. In: Liu, L., Özsu, M.T. (eds.) Data Mining and Knowledge Discovery Handbook, pp. 165–192. Springer, Cham (2005). https://doi.org/10.1007/978-0-387-39940-9_2445
25. Schapire, R.E.: Explaining AdaBoost. In: Schölkopf, B., Luo, Z., Vovk, V. (eds.) Empirical Inference, pp. 37–52. Springer, Cham (2013). https://doi.org/10.1007/978-3-642-41136-6_5
26. Seiffert, C., Khoshgoftaar, T.M., Van Hulse, J., Napolitano, A.: RusBoost: a hybrid approach to alleviating class imbalance. IEEE Trans. Syst. Man. Cybern. Part A Syst. Humans **40**(1), 185–197 (2009)
27. Stehman, S.V.: Selecting and interpreting measures of thematic classification accuracy. Remote Sens. Environ. **62**(1), 77–89 (1997)
28. Tharwat, A.: Classification assessment methods. In: Applied Computing and Informatics (2020)
29. Urban, T., Tatang, D., Degeling, M., Holz, T., Pohlmann, N.: Measuring the impact of the GDPR on data sharing in ad networks. In: Proceedings of the 15th ACM Asia Conference on Computer and Communications Security, pp. 222–235, ASIA CCS 2020. Association for Computing Machinery, New York, NY, USA (2020). https://doi.org/10.1145/3320269.3372194
30. Wilson, D.L.: Asymptotic properties of nearest neighbor rules using edited data. IEEE Trans. Syst. Man Cybern. **3**, 408–421 (1972)
31. Xie, Y., Li, X., Ngai, E., Ying, W.: Customer churn prediction using improved balanced random forests. Expert Syst. Appl. **36**(3), 5445–5449 (2009)
32. Yegnanarayana, B.: Artificial Neural Networks. PHI Learning Pvt, Ltd., New Delhi (2009)
33. Ying, W.Y., Qin, Z., Zhao, Y., Li, B., Li, X.: Support vector machine and its application in customer churn prediction. Syst. Eng. Theory Pract. **7** (2007)
34. Zhang, Y.P., Zhang, L.N., Wang, Y.C.: Cluster-based majority under-sampling approaches for class imbalance learning. In: 2010 2nd IEEE International Conference on Information and Financial Engineering, pp. 400–404. IEEE (2010). https://doi.org/10.1109/ICIFE.2010.5609385

Word Alignment Based Transformer Model for XML Structured Documentation Translation

Jing An[1], Yecheng Tang[2], Yanbing Bai[2(✉)], and Jiyi Li[3]

[1] Department of Linguistics at the University of California, Santa Barbara, USA
[2] Center for Applied Statistics, School of Statistics, Renmin University of China,
Beijing, China
ybbai@ruc.edu.cn
[3] University of Yamanashi, Kofu, Japan
jyli@yamanashi.ac.jp

Abstract. In the context of globalization, the development of localized translation technology for enterprise online documents is crucial for business promotion. The enterprise online documents are represented by semi-structured text documents with markup tags, while the mainstream neural machine translation methods focus on only the plain text translation. In this research, a Word Alignment based Transformer Model was proposed for markup language translation. Experiments conducted on the Salesforce XML English-Chinese datasets, and the result demonstrated that adding a word alignment model to the translation model can improve the translation model's performance in translating text with makup tags.

Keywords: XML structured documentation · Machine translation · Word alignment · Transformer

1 Introduction

In the context of globalization, the development of localized translation technology for enterprise online documents is crucial for business promotion. The enterprise online documents are represented by semi-structured text documents with markup tags, while the mainstream neural machine translation methods [7] focus on only the plain text translation. To fill this gap, a few existing works for markup language machine translation approaches were developed[1–4]. Joanis et al. categorized the approaches on this topic into two types of methods [3]. One-stream methods [2,4] include the markup tags with the plain text to train the translation model. Two-stream methods [1,3,4] separately translate the plain text and re-insert the markup tags into the target documents. Joanis et al. [3] utilized statistical machine translation and designed markup tag transfer rules relying on word alignment to re-insert the markup tags into the translated document. Müller et al. 's approach [4] was based on statistical machine translation and

© The Author(s), under exclusive license to Springer Nature Switzerland AG 2022
C. Strauss et al. (Eds.): DEXA 2022, LNCS 13426, pp. 316–322, 2022.
https://doi.org/10.1007/978-3-031-12423-5_24

compared five variants of re-insertion-rule-based and mask-based approaches. Hanneman et al. [1] utilized neural machine translation; it spanned and unified the elements of these previous studies and compared the major markup tag representation methods. These works only separately utilized the word alignment model and translation model. Inspired that the word alignment in markups is quite related to the translation, we propose an approach that jointly learns word alignment and translation models with a multi-task loss function so that these two components can improve each other. We utilize the word alignment information inside the markups in the training set to train the component of word alignment model. The proposed approach can improve the performance on markup tag placement while preserving the performance of machine translation.

2   Methodology

As we discussed above, the major issue for the markup language translation task is finding the correct positions of tags to be inserted while maintaining the translation accuracy. Our method is a two-stream method. Our idea is using Transformer-Align model to jointly learn the word alignment information and the mapping rule between source and target languages. Using the word alignment information, the XML tags in the source language are inserted into the target language to realize the translation of markup language.

Our method is based on Transformer [7]. The attention distribution is generated by the following formulation, i.e., $A = \text{Softmax}(\frac{Q \cdot K^{\mathsf{T}}}{\sqrt{d_k}})$, where Q, K are derived from xW_q and xW_k. Here W_q and W_k are two trainable parameters, and x denotes the input sequence. In this work, we mainly utilize the distribution A to obtain the alignment information.

2.1   Transformer-Align with Fast Align Word Alignment Loss

To remedy the aforementioned issue, we replace the vanilla Transformer by Transformer-Align to jointly model the process of translation and word alignment. In order to enhance the alignment effect of Transformer-Align, we utilized the word alignment loss introduced in Transformer-Align, which is derived from Fast Align. The elaborated loss is formulated as $\mathcal{L} = \mathcal{L}_t + \lambda \mathcal{L}_a(A)$, where $\mathcal{L}_t$ denotes the standard NLL translation loss and λ is a hyper-parameter range in [0,1]. $\mathcal{L}_a(A)$ denotes the alignment loss, which can be represented as $\mathcal{L}_a(A) = -\frac{1}{I}\sum_{i=1}^{I}\sum_{j=1}^{J}\mathcal{G}_{i,j}^p log(A_{i,j})$, where $\mathcal{G}_{i,j}^p$ denote a 0-1 matrix such that $\mathcal{G}_{i,j}^p = 1$ iff Fast Align predictions include the $\langle i - j \rangle$ alignment pair else $\mathcal{G}_{i,j}^p = 0$.

2.2   Transformer-Align with XML Tag Text Alignment Loss

This approach builds on the previous approach by introducing constraints that are more closely related to the XML markup language. Make the model more focused on the alignment of text within XML tags, which helps us to precisely locate the tags in the target language.

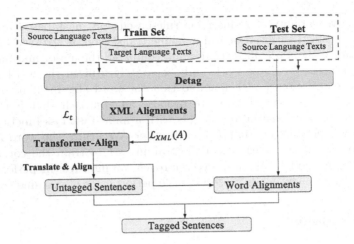

Fig. 1. The chart of Transformer-Align with XML tag text alignment loss

To simplify the relationship, we choose the XML-labeled text instead of the pure text to generate the word alignments because text within XML tags in both source and target languages can serve as the gold standard for word alignments while reducing the complex relationship from fully connected to a sparse one. Given a XML-labeled source language-target language data pairs {X,Y}, X = $(x_1, ..., x_n)$ and Y = $(y_1, ..., y_m)$, the XML alignment sets Θ_{XML} is formally defined as, $\Theta_{\mathrm{XML}} = \{\langle i - j \rangle \,|\, i \in n, j \in m\}$, where n and m are the sequence length of source language and target language, respectively. $\langle i - j \rangle$ denotes the positional pair between i-th element in the source and j-th element in the target.

Figure 2 depicts the preprocessed procedure of generating the alignment matrix $\mathcal{H}_{i,j}$. Our intention is to encourage tokens to concentrate more on other tokens which belong to the same XML tag. One can achieve this goal by incorporating the word alignment information obtained by FastAlign. Here, $\Theta_{\mathrm{XML}}^{sub}$ denotes the correct alignments. However, high-quality alignments are hard and often expensive to attain. The major challenge roots at the poor precision of Fast Align due to the limited dataset. This may bring noise signal which hinders the optimization. Also, given that the length of the text within the XML label is short, we propose a Cartesian-based alignment method to approximate the "real" alignment information. The core idea is to learn the alignments within the corresponding source and target XML tags, rather than the entire sequence.

We borrowed the merit of label-smoothing [6] that the gold alignment set is enlarged according to the Cartesian set. Through Fig. 2(b), we can see that "Encrypt" is not only aligned with "加密", but also with "字段", which belongs to the same XML tag. Based on the obtained the Θ_{XML}, the joint learning objective could be formulated as $\mathcal{L} = \mathcal{L}_t + \lambda \mathcal{L}_{\mathrm{XML}}(A)$, where $\mathcal{L}_t$ still denotes the standard NLL translation loss and λ is a hyper-parameter range in [0,1] to control the impact of XML alignment. $\mathcal{L}_{\mathrm{XML}}(A)$ denotes the newly proposed XML alignment loss, which takes the similar format of $\mathcal{L}_a(A)$ and is formally repre-

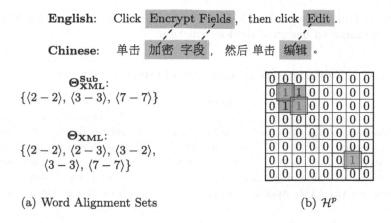

(a) Word Alignment Sets

(b) $\mathcal{H}^p$

Fig. 2. A detailed example for the alignment information.

sented as $\mathcal{L}_{\mathrm{XML}}(A) = -\frac{1}{I} \sum_{i=1}^{I} \sum_{j=1}^{J} \mathcal{H}_{i,j}^{p} log(A_{i,j})$, where $\mathcal{H}_{i,j}^{p}$ satisfies $\mathcal{H}_{i,j}^{p} = 1$ iff $\langle i - j \rangle \in \Theta_{\mathrm{XML}}$ else $\mathcal{H}_{i,j}^{p} = 0$. Figure 1 shows the structure of our XML-tag-based model and demonstrates that we use the XML alignments to assist the training of the Transformer-Align model.

3 Experiments and Result Discussion

This study uses the public English-Chinese translation datasets with XML tags provided by Salesforce [2], and the fairseq toolkit published by Facebook [5] was used as translation model. XML accuracy, matching, BLEU, named entities and numbers (NE&NUM) [2] were adopted as evaluation metrics.

Table 1 shows the experimental results. We can see that the Transformer (W/ Tag) baseline model delivers the lowest BLEU score, while the Transformer (No Tag) baseline model behaves worst in terms of XML BLEU score. Besides, all three models using word alignment achieve consistent XML BLEU improvements than Transformer (W/ Tag), and the Transformer-Align model based on XML word alignment loss delivers the best XML BLEU score, which demonstrates the effectiveness of our method.

To take a further in-depth analysis, we find that the Transformer (No Tag), as the baseline model, only focuses on the translation of unlabeled text, so its translation effect is better. However, the Transformer (No Tag) baseline model has poor predictive ability for unlabeled text. In contrast, the Transformer (W/ Tag) baseline model is trained with XML tags, which improves its ability to generate XML tags compared to the Transformer (No Tag) model, but the translation effect is sacrificed to a certain extent. The translation models based on the word alignment model have basically equivalent translation capabilities to the Transformer (No Tag) baseline model. The ability to generate XML tags is improved compared to the Transformer (W/ Tag), which is based on XML word alignment. Transformer-Align (XML Align Loss) is best at generating XML tags due to the targeted design of the loss function.

Table 1. Accuracy assessment of experimental results

Model	BLEU	XML BLEU	XML Acc.		NE & NUM	
			Struct	Match	Precision	Recall
Transformer (No Tag) [7]	63.26	32.04	1.0000	0.7410	0.9533	0.9415
Transformer (W/ Tag) [7]	49.26	49.47	0.9905	0.9890	0.6494	0.9014
Fast Align + Transformer [1]	63.26	53.16	0.9895	0.9445	0.9533	0.9415
Transformer-Align (Fast Align Loss)	62.28	52.94	0.9885	0.9450	0.9563	0.9427
Transformer-Align (XML Align Loss)	63.17	54.12	0.9900	0.9470	0.9544	0.9516

Table 2. Translation examples

Case 1	Source	Select <**uicontrol**> Rename</**uicontrol**> to rename the field's visible name
	Reference	选择 <uicontrol> 重命名 </uicontrol> ,以对字段的可见名称进行重命名。
	Transformer (W/ Tag)	选择重命名以重命名字段的可见名称。
	Transformer-Align (XML Align Loss)	选择<uicontrol> 重命名</uicontrol> 以重命名字段的可见名称。
Case 2	Source	From Setup, enter <**userinput**> Mass Delete Records </**userinput**> in the <**parmname**> Quick Find </**parmname**> box, then select <**uicontrol**> Mass Delete Records </**uicontrol**> and click the link for the type of record to delete
	Reference	从"设置"中,在 <parmname> 快速查找</parmname> 方框中输入 <userinput> 批量删除记录 </userinput> , 然后选择 <uicontrol>批量删除记录 </uicontrol> 并单击要删除记录类型的链接。
	Transformer (W/ Tag)	从"设置"中,在 <parmname> 快速查找 </parmname> 方框中输入 <userinput> 批量 </userinput> <uicontrol> 删除记录, 然后选择批量删除记录 </uicontrol> , 单击要删除的记录类型的链接。
	Transformer-Align (XML Align Loss)	从"设置"中,在<parmname> 快速查找 </parmname> 方框中输入 <userinput> 批量删除记录 </userinput> , 然后选择<uicontrol> 批量删除记录</uicontrol> , 单击要删除记录类型的链接。

Next is the comparison between the accuracy rates of different models in XML structure prediction, including structure accuracy rate and matching accuracy rate.

It shows that there are slight differences between the structural accuracy of each model, which is very close to 1. As for the matching accuracy, the Transformer (No Tag) has a particularly low matching accuracy (considering that since more than half of the data itself does not have XML tags, even if there is no prediction effect, the accuracy has a certain basic value, here about 0.74) while the Transformer (W/ Tag) delivers the highest matching accuracy. Combined with the fact that the XML BLEU score of the Transformer (W/ Tag) is lower, the Transformer (W/ Tag) has a better learning effect on XML structure, but worse learning effect of the specific insertion position of XML tags.

Finally, we compare the translation effects of different models on NE&NUM. Except the Transformer (W/ Tag), the precision and recall rates of other models on NE&NUM almost reach 1.0, which means that even without additional processing, the model has good translation capability for numbers and entities that do not need to be translated into the target language. However, the Transformer (W/ Tag) has poor translation effect on NE&NUM. Since the model regards NE&NUM and XML tags as texts that should be copied directly from the source language and the number of XML tags is large, the model is better at predicting XML tags and less effective at learning NE&NUM.

Based on the above analysis, the markup language translation model based on the word alignment model is not inferior to that of the ordinary unlabeled text translation model. For the markup language, the word alignment model can better assist the generation of XML tags and obtain good performance on markup language translation tasks. In experiments combining different word alignment models and translation models, we found that the Transformer-Align model based on XML word alignment loss performs best in XML translation. Table 2 shows the translation examples. Case 1 in Table 2 show that Transformer-Align + XML Align Loss model performs better for the translation task of tags as well as text with the tags. While case 2 in Table 2 shows that Transformer-Align + XML Align Loss model is better at recognizing NE&NUM.

4 Conclusion

This research demonstrated that adding a word alignment model to the translation model on with non-markup text can improve the translation model's performance in translating text with makup tags. This paper applies the more classic Fast Align model and Transformer-Align model in the word alignment. In the future, other word alignment models such as word alignment generated based on pre-trained models are worth to try for further improving predictive ability.

Acknowledgment. This research was supported by Public Health & Disease Control and Prevention, Major Innovation & Planning Interdisciplinary Platform for the "Double-First Class Initiative, Renmin University of China (No. 2022PDPC), fund for building world-class universities (disciplines) of Renmin University of China. Project No. KYGJA2022001. This research was supported by Public Computing Cloud, Renmin University of China.

References

1. Hanneman, G., Dinu, G.: How should markup tags be translated? In: Proceedings of the Fifth Conference on Machine Translation, pp. 1160–1173, November 2020
2. Hashimoto, K., Buschiazzo, R., Bradbury, J., Marshall, T., Socher, R., Xiong, C.: A high-quality multilingual dataset for structured documentation translation. In: Proceedings of the Fourth Conference on Machine Translation (Volume 1: Research Papers), pp. 116–127, August 2019
3. Joanis, E., Stewart, D., Larkin, S., Kuhn, R.: Transferring markup tags in statistical machine translation: a two-stream approach. In: Proceedings of the 2nd Workshop on Post-editing Technology and Practice, September 2013
4. Müller, M.: Treatment of markup in statistical machine translation. In: Proceedings of the Third Workshop on Discourse in Machine Translation, pp. 36–46, September 2017
5. Ott, M., et al.: fairseq: a fast, extensible toolkit for sequence modeling. arXiv preprint arXiv:1904.01038 (2019)
6. Szegedy, C., Vanhoucke, V., Ioffe, S., Shlens, J., Wojna, Z.: Rethinking the inception architecture for computer vision. In: 2016 IEEE Conference on Computer Vision and Pattern Recognition, pp. 2818–2826 (2016)
7. Vaswani, A., et al.: Attention is all you need. In: Advances in Neural Information Processing Systems, pp. 5998–6008 (2017)

Detecting Simpson's Paradox: A Machine Learning Perspective

Rahul Sharma[1]($\boxtimes$)(iD), Huseyn Garayev[2], Minakshi Kaushik[1](iD),
Sijo Arakkal Peious[1](iD), Prayag Tiwari[3](iD), and Dirk Draheim[1](iD)

[1] Information Systems Group, Tallinn University of Technology,
Akadeemia tee 15a, 12618 Tallinn, Estonia
{rahul.sharma,minakshi.kaushik,dirk.draheim}@taltech.ee
[2] University of Tartu, Tartu, Estonia
hugara@taltech.ee
[3] Department of Computer Science, Aalto University, Espoo, Finland
prayag.tiwari@aalto.fi

Abstract. The size of data collected around the world is growing exponentially, and it has become popular as big data. The volume and velocity of big data are facilitating the transition of machine learning (ML), deep learning (DL) and artificial intelligence (AI) from research laboratories to real life. There are numerous other claims made about Big Data. Can we, however, rely on data blindly? What happens when a dataset used to train ML models has a hidden statistical paradox? Data, like fossil fuels, is valuable, but it must be refined carefully for accurate outcomes. Statistical paradoxes are hard to observe in classical data cleaning and analysis techniques. Still, they are required to be investigated separately in training datasets. In this paper, we discuss the impact of Simpson's paradox on categorical data and demonstrate its effects on AI and ML application scenarios. Next, we provide an algorithm to automatically identify the confounding variable and detect Simpson's paradox within categorical datasets. The algorithm experiments on datasets from two real-world case studies. The outcome of the algorithm uncovers the existence of the paradox and indicates that Simpson's paradox is severely harmful in automatic data analysis, especially in AI, ML and DL.

Keywords: Big data · Artificial intelligence · Deep learning · Machine learning · Data science · Simpson's paradox · Explainable AI

1 Introduction

Human decision-making has always relied on data, but with the advancement of big data technologies, artificial intelligence (AI), data science, machine learning (ML), and deep learning (DL) have gained significant traction in artificial decision-making. These techniques are now widely used in medical sciences, social sciences, and politics, and they substantially impact human life and decisions, either directly or indirectly. In most AI use cases, ML-based trained artificial

© The Author(s), under exclusive license to Springer Nature Switzerland AG 2022
C. Strauss et al. (Eds.): DEXA 2022, LNCS 13426, pp. 323–335, 2022.
https://doi.org/10.1007/978-3-031-12423-5_25

systems are used to provide quick and precise results. Still, in some cases, the existence of statistical paradox, causal inference and uneven data distribution can mislead an AI application. Statistical paradoxes are not new to being discussed in statistics and mathematics. These terms are widely used in statistics and have been around for over a century. Expert mathematicians and statisticians adequately discussed various statistical paradoxes (e.g., Simpson's Paradox, Berkson's Paradox, Latent Variables, Law of Unintended Consequences, Tea Leaf Paradox, etc.) and addressed their severe impacts on classical data analysis. However, in modern decision support techniques, specifically AI, ML and DL, causal relationships, data fallacies and statistical paradoxes are not yet appropriately addressed.

A statistical paradox can exist in a wide variety of data. Kügelgen et al. [33] recently emphasized the importance of statistical analysis of real data and demonstrated evidence of Simpson's paradox in COVID-19 data analysis. They claim Italy's overall case fatality rate (CFR) was higher than China's. However, in every age group, China had a higher fatality rate than Italy. These observations raise numerous concerns about data accuracy and analysis. Heather et al. [20] have addressed the existence of Simpson's paradox. In psychological science, Kievit et al. [17] examined the instances of Simpson's paradox. In [14], Kaushik et al. have discussed some measures to find the impact of one numerical variable on another numerical variable. Alipourfard et al. [2] have discovered the existence of Simpson's paradox in social data and behavioural data [3]. The instances Simpson's paradox have also been discussed in various data mining techniques [10,11,13], e.g., association rule mining [1] and numerical association rule mining [15,16,31]. Therefore, understanding data, especially big data, is more critical than processing.

Most of the statistical paradoxes are fundamentally linked to various statistical challenges and mathematical logic, including causal inference [22,23], the ecological fallacy [19,26], Lord's paradox [32], propensity score matching [27], suppressor variables [8], conditional independence [9], partial correlations [12], p-technique [6], mediator variables [21], etc.

In this paper, we concentrate on a specific case of a statistical paradox called Simpson's paradox in categorical data and demonstrate its impact with some real-world case studies. Next, we provide an algorithm to detect Simpson's paradox and identify the confounding variables in categorical values. In statistics, a confounder is described as a statistic variable that influences both the dependent and independent variables, resulting in a spurious relationship. The algorithm is experimented on two datasets to detect confounder and the paradox. The paper is organized as follows.

In Sect. 2, we discuss Simpson's Paradox. In Sect. 3, we propose an algorithm for automatically detecting the Simpson's Paradox in categorical values. In Sect. 4, two real-life datasets are used to demonstrate the impact of the paradox experimentally. Finally, a discussion and conclusion is provided in Sect. 5 and Sect. 6, respectively.

2 Simpson's Paradox

In the year 1899, Karl Pearson et al. [24] demonstrated a statistical paradox in marginal and partial associations between continuous variables. Later in 1903, Udny Yule [35] explained "the theory of association of attributes in statistics" and revealed the existence of an association paradox with categorical variables. In a technical paper published in 1951 [29], Edward H. Simpson described the phenomenon of reversing results. However, in 1972, Colin R. Blyth coined the term "Simpsons Paradox" [5]. Therefore, this paradox is known by different names and is famous as the Yule-Simpson effect, amalgamation paradox, or reversal paradox [25]. Simpson's paradox can exist in any dataset irrespective of its size and type [18]. The paradox demonstrates the importance of having human experts in the loop during an automatic data analysis.

Table 1. Original Simpson's example with 2 × 2 contingency table [29]: the type of association for the entire population ($N = 52$) reverses at the level of sub-populations of men and women.

	Population $N = 52$			Men (M)= 20			Women (F) = 32		
	Success (S)	Failure ($\neg S$)	Success rate %	Success	Failure	Success Rate %	Success	Failure	Success rate %
T	20	20	50%	8	5	≈61%	12	15	≈44%
$\neg T$	6	6	50%	4	3	≈57%	2	3	≈40%

We start the discussion on the paradox by using the original example and numbers from Simpson's article [29]. In this example, analysis for medical treatment is demonstrated. Table 1 summarises the effect of the medical treatment for the entire population ($N = 52$) as well as for men and women separately in subgroups. The treatment appears effective for both men and women subgroups (Men: 61% vs 57% and Women: 44% vs 40%); however, the treatment seems ineffective at the whole population level.

We can demonstrate the above example via probability theory and conditional probabilities. Let $T = treatment$, $S = success$, $M = Men$ and $F = Women$ then,

$$\mathsf{P}(S \mid T) = \mathsf{P}(S \mid \neg T) \tag{1}$$

However, the probability for men and women is:

$$\mathsf{P}(S \mid T, M) > \mathsf{P}(S \mid \neg T, M) \tag{2}$$

$$\mathsf{P}(S \mid T, F) > \mathsf{P}(S \mid \neg T, F) \tag{3}$$

Based on Eq. 1, 2 and 3, one should use the treatment or not? As per the success rate for the men and women populations, the treatment is a success, but overall, the treatment is a failure. This reversal of results between groups population and the total population has been referred to as Simpson's Paradox. In statistics, this concept has been discussed widely and named differently by several authors [24, 35].

2.1 Impacts of Simpson's Paradox

Simpson's paradox exists in different types of data in different forms. However, classically it is expressed via 2×2 contingency tables. Let a 2×2 contingency table for treatment (T) and success (S) in the ith sub-population is represented by a four-dimensional vector of real numbers $D = (a_i, b_i, c_i, d_i)$. Then

Table 2. 2×2 Contingency table with sub population groups D1 and D2.

	Population $D = D_1 + D_2$		Sub-population D_1		Sub-population D_2	
	Success (S)	Failure $(\neg S)$	Success (S)	Failure $(\neg S)$	Success (S)	Failure $(\neg S)$
Treatment (T)	$a_1 + a_2$	$b_1 + b_2$	a_1	b_1	a_2	b_2
No-Treat. $(\neg T)$	$c_1 + c_2$	$d_1 + d_2$	c_1	d_1	c_2	d_2

$$D = \sum_{i=1}^{N} D_i = \left(\sum a_i, \sum b_i, \sum c_i, \sum d_i \right) \qquad (4)$$

is the aggregate dataset over N sub populations [30]. This can be read as given in Table 2.

Definition 1. *Consider n groups of data such that group i has A_i trials and $0 \leq X_{A_i} \leq A_i$ "successes". Similarly, consider another similar n groups of data such that group i has B_i trials and $0 \leq Y_{B_i} \leq B_i$ "successes". Then, Simpson's paradox appear if:*

$$\frac{X_{A_i}}{A_i} \leq \frac{Y_{B_i}}{B_i} \text{ for all } i = 1, 2, \ldots, n \text{ but } \frac{\sum_{i=1}^{n} X_{A_i}}{\sum_{i=1}^{n} A_i} \geq \frac{\sum_{i=1}^{n} Y_{B_i}}{\sum_{i=1}^{n} B_i} \qquad (5)$$

We could also flip the inequalities and still have the paradox since A and B are chosen arbitrarily.

$$\frac{X_{A_i}}{A_i} \geq \frac{Y_{B_i}}{B_i} \text{ for all } i = 1, 2, \ldots, n \text{ but } \frac{\sum_{i=1}^{n} X_{A_i}}{\sum_{i=1}^{n} A_i} \leq \frac{\sum_{i=1}^{n} Y_{B_i}}{\sum_{i=1}^{n} B_i} \qquad (6)$$

We use the following example to show the working of the Eqs. 5 and 6.

$$\frac{10}{20} = \frac{X_{A_1}}{A_1} > \frac{Y_{B_1}}{B_1} = \frac{30}{70} \text{ and } \frac{10}{50} = \frac{X_{A_2}}{A_2} > \frac{Y_{B_2}}{B_2} = \frac{10}{60} \text{ yet}$$

$$\frac{10+10}{20+50} = \frac{20}{70} = \frac{X_{A_1} + X_{A_2}}{A_1 + A_2} < \frac{Y_{B_1} + Y_{B_2}}{B_1 + B_2} = \frac{30+10}{70+60} = \frac{40}{130}$$

3   Detecting Simpson's Paradox

Based on the type of trends reversed in various types of data, Simpson's paradox cases are explored into two categories: classification, which involves the relative rates of binary outcomes in two groups, and regression, which involves the sign of a correlation between two variables [34]. Here, we provide an algorithm to detect the paradox in the first case, i.e. for categorical values. In the algorithm, the Pearson correlation index is used to find the relationships between two variables which allows for measuring the strength of the linear association between two variables. The output value of the Pearson correlation lies between -1 and 1. Values greater than 0 imply a positive correlation. The value 1 indicates the exact positive association, while 0 means no correlation. Values less than 0 suggest a negative association, and -1 indicates a clear negative association. The Pearson correlation coefficient is represented by r In Eq. 7. Here, x and y are input vectors, $\bar{x}$ and $\bar{y}$ are means of the variables, respectively.

$$r = \frac{\sum_{i=1}^{n}(x_i - \bar{x})(y_i - \bar{y})}{\sqrt{\sum_{i=1}^{n}(x_i - \bar{x})^2(y_i - \bar{y})^2}} \tag{7}$$

3.1   Algorithm for Detecting the Simpson's Paradox in Categorical Data (Relative Rates)

We formally describe the algorithm for detecting the Simpson's Paradox in linear trends in Algorithm 1. In the algorithm, the primary step is to convert the values of the categorical input variables to binary values. The first variable category is substituted by 0, and the second category is replaced by 1. This conversion allows the Pearson correlation index function to identify the relationship between categorical variables or between categorical and numerical (continuous) variables. We input X - categorical variable by which we condition, $X1$ - the first category of variable X, $X2$ - the second category of variable X, Y - continuous or categorical variable (with two categories) which is aggregated. Table 3 illustrates the form of an example dataset before and after the pre-processing step.

Further, the algorithm calculates the correlation index between X and Y variables with the values of the corresponding columns in the dataset. This way, we obtain information on the sign of the relationship between the variables. Next, we traverse the list of remaining categorical variables, calculate the Pearson index conditioning on each subgroup (category), count the ratio of subgroups where the correlation index reversed relative to the index in aggregated data and store the value key pairs in an array. Subsequently, we get the array element where the value (ratio) is the highest. The maximal value 1 implies the Simpson's paradox occurrence with the corresponding key of the array element being the confounding variable. Cases where the maximal ratio is less than 1 imply the absence of Simpson's paradox. However, they are also regarded as a partial occurrence of the bias and are considered in the further steps. The performance of the algorithm strongly correlates with the size of the datasets.

Algorithm 1: Identification of Simpson's Paradox in Relative Rates

---

Input: A dataset D with categorical variable x and y
Output: a pair of confounding variable and ratio of reversed association
d[x] = Preprocess(d[x]) /*conversion of categorical column to binary   */
d[y] = Preprocess(d[y])
aggreg_index = Pearson(d[x] , d[y]) /*calculate correlation index between
   columns   */
indexes = [] /*initialize index array to store key value pairs: the
   key is column and value is the number of reversed subgroups   */
cols = columns(D) /*initialize array of all columns of D   */
foreach $column \in cols$ **do**
   if *Column Is Not Categorical(column)* **then**
      | Continue
   end
   else
      subgroups = Categories(column) // get the categories of a column
      coefficients = [] // initialize empty array to store the correlation indexes
      foreach *subgroup $\in$ subgroups* **do**
         disaggreg_index = Pearson(D[x]: where D[column] = subgroup,
         D[y]: where D[column] = subgroup) *calculate corr. index between*
         columns for current subgroup
         Add index of disaggregated to correlation indexes array
      end
   end
   reversed_subgroups = RatioReversedSubgroups(aggreg_index, coefficients)
      /*calculate ratio of the correlation indexes reversed with
      respect to the correlation index for the aggregated data   */
   Add *column, reversed_subgroups* values into *indexes*
end
Store the max values of *indexes* pairs into *result*
Return *result*

Table 3. Illustration of the form of an example dataset before and after the pre-processing step.

Gender	Result	Gender	Result
Male	Success	0	1
Female	Success	1	1
Male	Failure	0	0

4 Experiments and Datasets

The Algorithm is implemented in Python on a personal computer with an Intel(R) Core(TM) i5-8265U CPU @ 1.60 GHz, 1800 Mhz, 4 Core(s), 8 Logical Processor(s), 16 GB RAM and Windows 10 × 64 operating system. We evaluate the algorithm with two real-world case studies with categorical data. The programming code, datasets, and other necessary instructions about the algorithms are available in the GitHub repository [28].

4.1 UC Berkeley Admissions Dataset Fall 1973

UC Berkeley admissions dataset [4] is a classic example of Simpson's Paradox. This dataset contained 12,763 graduate applicants (males and females) to UC-Berkeley in Fall 1973. The dataset was provided by UC-Berkeley researchers to investigate any possible cases of gender bias in the admissions. In the dataset, the admission rate for females is less than for males when data is aggregated; however, when we consider each major separately, female admission rates exceed the rates for males in most subgroups.

The aggregate data given in Table 4 demonstrate significant bias in favour of male applicants; however, data from each department given in Table 5 reveals an opposite story and bias in favour of Female applicants. Figure 1 demonstrate some hidden patterns in the dataset. As per the graph, it is clear that the overall number of women applicants is significantly less than the total men applicants. However, their rejection rate is high as compared to the male applicants. To analyze these hidden patterns and find the possible existence of Simpson's paradox in data, we use the original UC-Berkeley admission dataset having 12763 records with four attributes: *Student_id*, *Gender*, *Major* and *Admission*.

Table 4. Existence of Simpson's Paradox: a case study from UC-Berkeley admission dataset (fall 1973) [4].

	Applications	Admitted	Rejected	Admission %
Men	8442	3738	4704	44%
Women	4321	1494	2827	35%

Table 5. UC-Berkeley admission dataset (fall 1973): Percentage of acceptance rate of men and women in different departments.

Gender	Departments					
	A	B	C	D	E	F
Men	62.06%	63.04%	36.92%	33.09%	27.75%	5.90%
Women	82.41%	68%	34.06%	34.93%	23.92%	7.04%

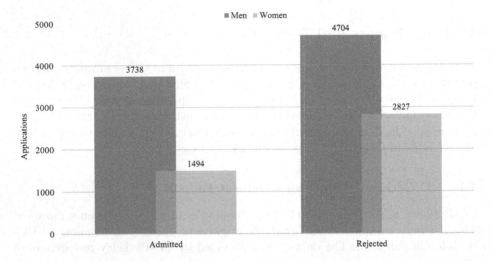

Fig. 1. Graphical representation of information in the UC-Berkeley admission dataset demonstrates hidden patterns and unbalanced data distribution.

In the algorithm, *Gender* attribute is set as X variable and *Admission* attribute is set as Y variable. To detect the paradox, the algorithm first calculates the Pearson correlation between *Gender* and *Admission* variables. In the prepossessing step, the values of *gender* variable, i.e., *Female* and *Male* are categorised by the binary values 1 and 0, similarly, the values of *admission* variable, i.e., *Failure* and *Success* are categorised by the binary values 0 and 1, respectively. Next, the algorithm traverses the complete list of variables to identify the possible confounding variable and compute the ratio of the subgroup reversals. The algorithm returns a confounder and the existence of Simpson's paradox in the dataset. As per the computation, the correlation index between the *Gender* and *Admission* variable is negative for "B, F, A, D" majors, whereas it is positive for the whole population.

4.2 Kidney Stone Treatment Dataset

We use another dataset from a real-world medical case study published by Charig et al. [7] in "The British Medical Journal" in 1986. In this study, the success rate of two different types of treatments to remove the large and small size of kidney stones are compared. In Table 6, Treatment A entails a classical open surgical procedure and treatment B entails an advanced closed surgical procedure. For both small kidney stones and large kidney stones, treatment A, i.e., open surgical procedures (*Success Rate* Small Stone Size 93%, Large Stone Size 73%) performs better than the treatment B (*Success Rate:* Small Stone Size 87%, Large Stone Size 69%), However, when the data for both the treatments is combined, the treatment B (*Success Rate:* 83%) outperforms the treatment A (*Success Rate:* 73%). Table 6 demonstrates the success rates of the treatments in detail.

Table 6. Kidney Stone Dataset: Information about the success rate of the treatments with different sizes of stones. Treatment A outperforms treatment B for large and small kidney stones, but for both kidney stones together, treatment B exceeds treatment A.

Stone size	Treatment (A) = 350			Treatment (B) = 350		
	Success (S)	Failure (F)	Success rate %	Success (S)	Failure (F)	Success rate %
Small	81	6	≈93%	234	36	≈87%
Large	192	71	≈73%	55	25	≈69%
Both	273	78	≈78%	289	61	≈83%

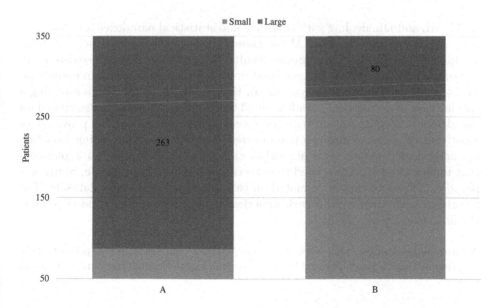

Fig. 2. Graphical representation of information in the kidney stone dataset demonstrates the hidden patterns and unbalanced data distribution for treatments A and B.

Figure 2 demonstrate the graphical representation of the hidden information in the dataset. As per the graphs, it is a perfect case of uneven distribution of sample data for both the treatments. Analyzing this dataset with the algorithm returns a confounder and the existence of Simpson's paradox. As per the computation, the correlation index between the *Treatment A* and *Treatment B* in groups is opposite to the correlation for both the treatments.

5   Discussion and Future Work

The existence of Simpson's paradoxes in real-world studies provides a direction for understanding the impact of causality in artificial decision-making. We noticed that data mining algorithms used in AI, ML and DL focus mainly on identifying the correlations in aggregate data rather than identifying the genuine causal relationships between all the data items. Therefore, understanding statistical paradoxes and evaluating causality in each combination of data items is an essential step toward fair ML models. In future, we plan to simplify the impacts of Simpson's paradox in different types of data (Continuous values) and address various other statistical paradoxes (e.g., Berkson's paradox) in datasets. Further, we intend to develop a simple framework to identify the existence of statistical paradoxes in various types of data.

6   Conclusion

In AI, ML and DL, dealing with causality and statistical paradoxes is still a challenging phenomenon. In most AI use cases, ML-based trained artificial systems are used to provide quick and precise results. Still, in some cases, the existence of statistical paradox, causal inference and uneven data distribution can easily mislead the outcome of artificial systems. In this paper, we focused on addressing a specific case of a statistical paradox called Simpson's paradox in categorical data and demonstrated its impact with some real-world case studies. We provided an algorithm to detect Simpson's paradox and identify the confounding variables in categorical datasets. This algorithm can be utilized to develop a platform that unifies most aspects related to detecting a confounding variable, Simpson's paradox. The algorithm is evaluated on two real-world case study datasets. The algorithm performed well in each experiment, and its running time is proportional to the size of a dataset.

Acknowledgements. This work has been partially conducted in the project "ICT programme" which was supported by the European Union through the European Social Fund.

References

1. Agrawal, R., Srikant, R.: Fast algorithms for mining association rules in large databases. In: Proceedings of VLDB'1994 - the 20th International Conference on Very Large Data Bases, pp. 487–499. Morgan Kaufmann (1994)
2. Alipourfard, N., Fennell, P.G., Lerman, K.: Can you trust the trend? Discovering Simpson's paradoxes in social data. In: Proceedings of the Eleventh ACM International Conference on Web Search and Data Mining, WSDM 2018, pp. 19–27. Association for Computing Machinery, New York (2018). https://doi.org/10.1145/3159652.3159684
3. Alipourfard, N., Fennell, P.G., Lerman, K.: Using Simpson's paradox to discover interesting patterns in behavioral data. In: Proceedings of the Twelfth International AAAI Conference on Web and Social Media. AAAI Publications (2018)
4. Bickel, P.J., Hammel, E.A., O'Connell, J.W.: Sex bias in graduate admissions: data from Berkeley. Science **187**(4175), 398–404 (1975). https://doi.org/10.1126/science.187.4175.398
5. Blyth, C.R.: On Simpson's paradox and the sure-thing principle. J. Am. Stat. Assoc. **67**(338), 364–366 (1972)
6. Cattell, R.B.: P-technique factorization and the determination of individual dynamic structure. J. Clin. Psychol. **8**, 5–10 (1952)
7. Charig, C.R., Webb, D.R., Payne, S.R., Wickham, J.E.: Comparison of treatment of renal calculi by open surgery, percutaneous nephrolithotomy, and extracorporeal shockwave lithotripsy. BMJ **292**(6524), 879–882 (1986). https://doi.org/10.1136/bmj.292.6524.879
8. Conger, A.J.: A revised definition for suppressor variables: a guide to their identification and interpretation. Educ. Psychol. Meas. **34**(1), 35–46 (1974)
9. Dawid, A.P.: Conditional independence in statistical theory. J. Roy. Stat. Soc. Ser. B (Methodol.) **41**(1), 1–15 (1979). https://doi.org/10.1111/j.2517-6161.1979.tb01052.x
10. Draheim, D.: DEXA'2019 keynote presentation: future perspectives of association rule mining based on partial conditionalization, Linz, Austria, August 2019. https://doi.org/10.13140/RG.2.2.17763.48163
11. Draheim, D.: Future perspectives of association rule mining based on partial conditionalization. In: Hartmann, S., Küng, J., Chakravarthy, S., Anderst-Kotsis, G., Tjoa, A.M., Khalil, I. (eds.) Proceedings of DEXA'2019 - the 30th International Conference on Database and Expert Systems Applications. LNCS, vol. 11706, p. xvi. Springer, Heidelberg (2019)
12. Fisher, R.A.: III. The influence of rainfall on the yield of wheat at Rothamsted. Philos. Trans. R. Soc. London Ser. B **213**(402–410), 89–142 (1925). Containing Papers of a Biological Character
13. Freitas, A.A., McGarry, K.J., Correa, E.S.: Integrating Bayesian networks and Simpson's paradox in data mining. In: Texts in Philosophy. College Publications (2007)
14. Kaushik, M., Sharma, R., Peious, S.A., Draheim, D.: Impact-driven discretization of numerical factors: case of two- and three-partitioning. In: Srirama, S.N., Lin, J.C.-W., Bhatnagar, R., Agarwal, S., Reddy, P.K. (eds.) BDA 2021. LNCS, vol. 13147, pp. 244–260. Springer, Cham (2021). https://doi.org/10.1007/978-3-030-93620-4_18

15. Kaushik, M., Sharma, R., Peious, S.A., Shahin, M., Ben Yahia, S., Draheim, D.: On the potential of numerical association rule mining. In: Dang, T.K., Küng, J., Takizawa, M., Chung, T.M. (eds.) FDSE 2020. CCIS, vol. 1306, pp. 3–20. Springer, Singapore (2020). https://doi.org/10.1007/978-981-33-4370-2_1
16. Kaushik, M., Sharma, R., Peious, S.A., Shahin, M., Yahia, S.B., Draheim, D.: A systematic assessment of numerical association rule mining methods. SN Comput. Sci. **2**(5), 1–13 (2021). https://doi.org/10.1007/s42979-021-00725-2
17. Kievit, R., Frankenhuis, W., Waldorp, L., Borsboom, D.: Simpson's paradox in psychological science: a practical guide. Front. Psychol. **4**, 513 (2013). https://doi.org/10.3389/fpsyg.2013.00513
18. Kim, Y.: The 9 pitfalls of data science. Am. Stat. **74**(3), 307 (2020). https://doi.org/10.1080/00031305.2020.1790216
19. King, G., Roberts, M.: EI: A(n R) program for ecological inference. Harvard University (2012)
20. Ma, H.Y., Lin, D.K.J.: Effect of Simpson's paradox on market basket analysis. J. Chin. Stat. Assoc. **42**(2), 209–221 (2004). https://doi.org/10.29973/JCSA.200406.0007
21. MacKinnon, D.P., Fairchild, A.J., Fritz, M.S.: Mediation analysis. Ann. Rev. Psychol. **58**(1), 593–614 (2007). https://doi.org/10.1146/annurev.psych.58.110405.085542. pMID: 16968208
22. Pearl, J.: Causal inference without counterfactuals: comment. J. Am. Stat. Assoc. **95**(450), 428–431 (2000)
23. Pearl, J.: Understanding Simpson's paradox. SSRN Electron. J. **68** (2013). https://doi.org/10.2139/ssrn.2343788
24. Pearson Karl, L.A., Leslie, B.M.: Genetic (reproductive) selection: inheritance of fertility in man, and of fecundity in thoroughbred racehorses. Philos. Trans. R. Soc. Lond. Ser. A **192**, 257–330 (1899)
25. Quinlan, J.: Combining instance-based and model-based learning. In: Machine Learning Proceedings 1993, pp. 236–243. Elsevier (1993). https://doi.org/10.1016/B978-1-55860-307-3.50037-X
26. Robinson, W.S.: Ecological correlations and the behavior of individuals. Am. Sociol. Rev. **15**(3), 351–357 (1950)
27. Rosenbaum, P.R., Rubin, D.B.: The central role of the propensity score in observational studies for causal effects. Biometrika **70**(1), 41–55 (1983)
28. Sharma, R., Peious, S.A.: Towards unification of decision support technologies: statistical reasoning. OLAP and Association Rule Mining. https://github.com/rahulgla/unification
29. Simpson, E.H.: The interpretation of interaction in contingency tables. J. Roy. Stat. Soc.: Ser. B (Methodol.) **13**(2), 238–241 (1951)
30. Sprenger, J., Weinberger, N.: Simpson's paradox. In: Zalta, E.N. (ed.) The Stanford Encyclopedia of Philosophy, Summer 2021 edn. Metaphysics Research Lab, Stanford University (2021)
31. Srikant, R., Agrawal, R.: Mining quantitative association rules in large relational tables. In: Proceedings of the 1996 ACM SIGMOD International Conference on Management of Data, pp. 1–12 (1996)
32. Tu, Y.K., Gunnell, D., Gilthorpe, M.S.: Simpson's Paradox, Lord's Paradox, and Suppression Effects are the same phenomenon-the reversal paradox. Emerg. Themes Epidemiol. **5**(1), 1–9 (2008)
33. Von Kugelgen, J., Gresele, L., Scholkopf, B.: Simpson's paradox in COVID-19 case fatality rates: a mediation analysis of age-related causal effects. IEEE Trans. Artif. Intell. **2**(1), 18–27 (2021). https://doi.org/10.1109/tai.2021.3073088

34. Xu, C., Brown, S.M., Grant, C.: Detecting Simpson's paradox. In: The Thirty-First International Flairs Conference (2018)
35. Yule, G.U.: Notes on the theory of association of attributes in statistics. Biometrika **2**(2), 121–134 (1903)

A Learned Prefix Bloom Filter
for Spatial Data

Beiji Zou[1,3], Meng Zeng[1,3], Chengzhang Zhu[1,2,3(✉)], Ling Xiao[1,3],
and Zhi Chen[1,3]

[1] School of Computer Science and Engineering, Central South University,
Changsha, China
anandawork@126.com, {bjzou,zengmeng,194701006,chen.zhi}@csu.edu.cn
[2] The College of Literature and Journalism, Central South University,
Changsha, China
[3] Mobile Health Ministry of Education-China Mobile Joint Laboratory,
Changsha, China

Abstract. Learned bloom filter (LBF) model has been proposed in recent work to replace the traditional bloom filter (BF). It can reduce the needed amount of memory and achieve a relatively low false positive rate (FPR). However, the LBF did not provide a good solution for multi-dimensional data, such as spatial data. In this paper, a learned prefix bloom filter (LPBF) for spatial data is presented, which supports deletion and expansion and achieves lower FPR and less memory usage than the classical BF. To our knowledge, this is the first LBF method for spatial data. Specifically, a Z-order space-filling curve is used to map the spatial data into one dimension binary code. Then, we only need to learn the suffixes of the same prefix for the corresponding sub-LBF, which reduces the learning complexity of LBF. We further use the perfect hash table to accelerate the filter and reduce the FPR. Compared with two traditional BF methods and two state-of-art LBF methods on real spatial data sets, the proposed LPBF method shows the best performance in reducing FPR, proving that the LPBF method has great potential on bloom filter for spatial data.

Keywords: Learned bloom filter · Spatial data · Machine learning · Z-order curve · Data management

1   Introduction

With the development of big data, huge amounts of geospatial data are generated from smartphones and wearable devices, which brings huge challenges in query and storage. Bloom filter (BF) [3], a classical data structure for approximate membership, can reduce the extra I/Os by checking if the requested data exists. The standard bloom filter (SBF) uses the k hash function to map each data into

© The Author(s), under exclusive license to Springer Nature Switzerland AG 2022
C. Strauss et al. (Eds.): DEXA 2022, LNCS 13426, pp. 336–350, 2022.
https://doi.org/10.1007/978-3-031-12423-5_26

k bits in the bit array, called k mapped bits [23]. Set the k mapped bits to 1 to insert an item into the database. Then, we can check if all the k mapped bits are 1 to judge the presence of the item when querying. However, SBF has two main drawbacks, that is undeletable and unextendable [23]. Therefore, many researchers focus on solving these problems with a low FPR.

To support deletion, the Counting Bloom Filter (CBF) [10,19] was designed to store counters. CBF is usually used in combination with SBF, the former is placed in slow memory to support deletion but is not expandable, and the latter is placed in fast memory for fast queries. The Cuckoo Filter [9] was presented to support both deletable and expandable dynamically while achieving the better performance of the SBF and CBF. However, there is a probability of mistaken deletion for the Cuckoo Filter, and the size of the storage space must be an exponent of 2, which increases the space overhead. For scalability, the Scalable Bloom Filter [24] and the Dynamic Bloom Filter [11] used an additional empty bloom filter at the end of the original structure for the new data. In addition, the optimized dynamic bloom filter [12] used the CBF instead of the SBF to support deletion.

In recent works, Kraska et al. [13] introduced a machine learning (ML) model for bloom filter, called Learned Bloom Filter (LBF). It required less memory than an SBF for a given FPR by using the machine learning model to learn the correlation between items in the set. The LBF has been widely concerned due to the less memory and new hope of reducing FPR beyond the theoretical. Mitzenmacher [16] further presented a Sandwiching Learned Bloom Filter Model (Sandwiching LBF), added a bloom filter before the learned oracle to reduce the FPR. Rae et al. [21] proposed a Neural Bloom Filter to learn the approximate set membership in one-shot via meta-learning, which achieved compression gains over the classical BF. Zhenwei Dai et al. [6] further used the complete spectrum of score regions to further generalized LBF. It reduces the FPR by adjusting the number of hash functions differently and allocating variable memory BF in different regions. These classical BF methods and generalize LBF methods above were only for the one-dimensional data structure.

There are some works dedicated to multi-dimensional data filters. Adina Crainiceanu et al. [4,5] designed the Bloofi for multi-dimensional data filter, which arranged the bit vectors in the form of a B+tree and exploits bit-level parallelism by packing the bloom filters. Ripon Patgiri et al. [20] proposed an r-Dimensional Bloom Filter (rDBF) for different dimensional data, and achieved better performance than Cuckoo Filter in every aspect. For LBF methods, Stephen Macke et al. [15] introduced a learned multi-dimensional data bloom filter method, which inferred the value combination connections. Then, using a classification model for multi-dimensional data filter. Angjela Davitkova et al.

[7] further optimized the LBF model for multi-dimensional data by using a compressed LBF, reducing space significantly, saving training time, and improving accuracy. However, these LBF can not fully consider the data deletable and the impact of data access frequency.

To circumvent these issues, in this paper, we proposed a novel learned prefix bloom filter (LPBF) for spatial data. It uses a learned bloom filter with a counting bloom filter to check if the items exist in a set. The model can support both deletable and expandable, and achieve lower FPR compared with the classical BF. The main contributions of this paper are as follows:

1. We proposed LPBF, a novel learned prefix bloom filter for spatial data. To our knowledge, this is the first LBF method for spatial data. To support deletion and expansion, a CBF is used as a backup bloom filter of LPBF, which achieved lower FPR and less memory usage compared with the classical BF.
2. We use a Z-order space-filling curve to map the spatial data into one dimension binary code. Then, the prefixes of the binary code are extracted to divide all data points into k clusters, and we only need to learn the suffixes of the same prefix for the corresponding sub-LBF, which can reduce the learning complexity of LBF and the FPR.
3. We use the perfect hash table to get the prefix hash codes so that the corresponding sub-LBF can be found as soon as possible. Furthermore, the negative data that its prefix code mismatch all prefixes can be filtered first when querying further reducing the FPR.

The rest of this paper is organized as follows. In Sect. 2, we briefly introduced the architecture of Bloom Filter and Learned Bloom Filter. Then, we focused on the LPBF design, introduced basic operations, and analyzed the false positive rate of the proposed method in Sect. 3, respectively. Next, we performed a series of experiments to verify the effectiveness of the LPBF method in Sect. 4. In the end, we concluded about the LPBF and future works.

2   Related Work

2.1   Bloom Filter

Bloom Filter (BF) [3] is a random data structure with high space efficiency, which uses bit arrays to express a set concisely and judge whether an item belongs to the set. Given a set $X = \{x_1, x_2, ..., x_n\}$ of n items, and using k independent hash functions to map each item in the set X to $\{1, 2, ..m\}$. Then, we can get an array of m bits for all items. For an item x_i, the location $h_i(x)$ of the ith hash function mapping is set to 1 $(1 \leq i \leq k)$. Note, once a position is set to 1, it will be fixed as 1 no matter how many times the position is mapped.

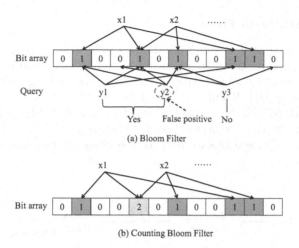

Fig. 1. Architectures of the bloom filter and counting bloom filter.

As shown in Fig. 1 (a), there is an array of bits $m = 12$, and hash functions $k = 3$, x_1 and x_2 are inserted. The values of x_1 and x_2 three hash modulus are shown in Fig. 1 (a), and the corresponding bit positions of two items are set to 1. When querying an item, the same k hash functions in the insertion process are used, then taking out the value corresponding to each bit by taking modulus. In Fig. 1 (a), y_1, y_2 and y_3 are the queried items. The results of y_1 and y_2 are the items that may exist, and y_3 does not exist in the set due to the existing 0 bit. However, the real results are that y_1 is in the set, but y_2 and y_3 are not in the set. The case that y_2 is misjudged is called false positive. It is obvious that there may be a false positive but no false negative in BF.

Let F_b be the false positive rate (FPR), that is, all k positions of an item in a bit array hashed by k functions are all 1, but it does not belong to this set. For an item i, the probability that any bit in BF is 0 after the execution of the k hash functions of the item is $(1 - \frac{1}{m})^k$. The probability that any bit in BF is 1 after all n items are inserted is $1 - (1 - \frac{1}{m})^{kn}$. Therefore, the FPR F_b can be denoted as follows:

$$F_b = [1 - (1 - \frac{1}{m})^{kn}]^k \approx (1 - e^{\frac{-kn}{m}})^k \qquad (1)$$

To solve the problem that BF can not delete items, Counting Bloom Filter (CBF) [10,19] is proposed. As shown in Fig. 1 (b), it expands each bit of the BF bit array into a small counter. The corresponding k counters are added 1 when inserting an item, and on the contrary, the corresponding k counters are deleted when deleting an item. CBF implements a delete operation by taking up several times more storage space. In this paper, CBF is used as a small backup BF to eliminate the false negatives rate (FNR) and support the deletion function.

2.2 Learned Bloom Filter

BF is usually used to determine whether an item belongs to a certain set, the process of which is regarded as a binary probability classification task. That is, we can learn a model $f(x)$ to predict an item whether belongs to the set, such as Gradient boosting [18], Recurrent Neural Network (RNN) [14] or Convolutional Neural Network (CNN) [8], called Learned Bloom Filter (LBF). However, there must be errors in the prediction results, that is $fpr \neq 0$ and $fnr \neq 0$. To ensure the $fnr = 0$, use bloom filter as a backup filter combined with binary classification.

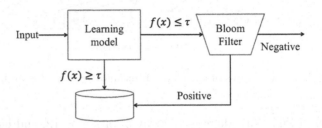

Fig. 2. Architecture of learned bloom filter.

As shown in Fig. 2, the input data first is classified by the learning model. Then, use BF to judge the left items that the output value of the classification results are less than the threshold τ. Finally, the value of the large than τ and the positive value are input to the database for the query. The BF in the model ensures the FPR is 0. Suppose the FPR of the learning model is F_p, and the FPR of backup BF is F_b as shown in Eq. 1, then, the FPR of LBF as:

$$F = F_p + (1 - F_p) \cdot F_b \approx F_p + (1 - F_p) \cdot (1 - e^{\frac{-kn}{m}})^k \qquad (2)$$

When F_p is small, the F of the Eq. 2 is approximately $F_p + F_b$.

3 The Proposed Method

3.1 Design Overview

Figure 3 shows the architecture of LPBF. The design of LPBF makes use of four key insights. First, the z-order curve is used to map the spatial data into one dimension in the data pretreatment phase. The binary value of each data can be divided into prefix and suffix, and it can be divided into N different categories according to the different prefixes, each prefix P_i contains multiple suffixes $(S_j, S_{j+1}, ..., S_t)$. Then, we use the perfect hash function to encode all prefixes and get the corresponding categories according to the hash code when looking for data. Third, N sub-LBFs are trained according to the corresponding

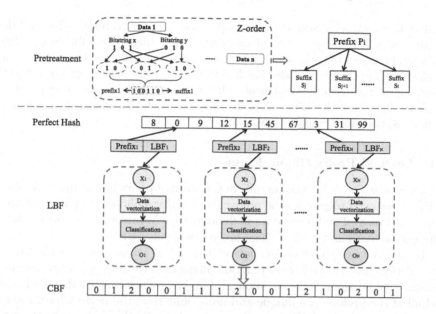

Fig. 3. Architecture of LPBF.

suffixes of the N different categories, and we can use a sub-LBF to predict an item whether belongs to the subset, as shown in the LBF part of Fig. 3. In the end, a CBF is used as the final filter to detect all negative samples from all categories that are classified by all sub-LBFs. The rest part will introduce the detials of the LPBF.

3.2 Data Pretreatment

In this part, we introduce how to encode the spatial data and the way to separate all data into different categories. We first introduce the Z-order space-filling curve, named Morton curve. It is a classical and widely used space-filling curve method that can map multi-dimensional data to one dimension due to the good locality-preserving behavior [22]. For two dimensional space, it can be mapped into one dimension by "string" all the rectangular areas of the two dimensional space with a line [17]. That is what the z-order curve does.

As the Fig. 3 pretreatment part shows, a spatial data (x, y) can be represented as a binary value by interleaving the binary representations. Then all the binary coordinate values can be expressed as a unique and non-repeated decimal value, and arranged in linear order [25] so that it can produce the Z-order curve recursively in the end. Therefore, in the data pretreatment phase, we can get unique binary values of all data to perform subsequent operations.

It is not difficult to find that each spatial data is transformed into a binary code by using the z-order curve, and each binary code can be divided into the prefix part and the suffix part. Here, both positive and negative samples are used as input samples and are mixed up when used for training. Then, all data points can be sorted into different categories according to the different prefixes. It should be noted that the number of categories we need to divide is determined by the length of the prefix code. The length of the prefix code should be set at the beginning of the data pretreatment. Therefore, we design a set of experiments in part 4.3 to test the effect of prefix lengths on FPR. Obviously, the longer prefixes is, the more categories we can get.

3.3   Learned Prefix Bloom Filter

In this part, we introduce the details of the learned prefix bloom filter. As shown in Fig. 3, we need to process the prefixes and suffixes obtained by data pretreatment separately. The binary data is divided into N categories according to the different prefixes, and each of them has its own sub-LBF.

To find the categories easier, perfect hash [2] is introduced to code the prefixes. Perfect hash, a static hash function, maps each element of a set to a series of collision-free integers. Since we know all binary prefixes that need to be hashed, and all of the prefixes are unique and non-conflicting, the perfect hash code for each category is unique. When testing a new data whether belongs to the set, it first encodes by z-order curve and gets the prefix, then uses the perfect hash to get the hash code, and further find the category it belongs to.

For the sub-LBF part, first, the data is vectorized and then gets the positive or negative results of the outputs by using the classification method. Here, we use the Gradient boosting method for classification. The loss function is $L = \sum_{i=1}^{n}[f(x_i) - y_i]^2$. Where $f(x_i)$ denotes the output of the data x_i in a sub-LBF, y_i is the truth value related to x_i. Then, we can judge an item is positive or negative according to the classification results. For N categories, there are N different learning models. In particular, each model only needs to learn the suffix binary encodings of each subset of the data, which is simpler than the original LBF model to learn the entire encoding of all data.

As is mentioned above, CBF is a BF that can support data deletion. Therefore, it is utilized as the final BF for data deletion. All negative samples identified by all sub-LBFs need to be filtered by a final CBF. Here, we only need a small CBF because of each sub-LBF filters out some positive samples. Theoretically, the size of the CBF tends to be infinitesimal when the accuracy is close to 100%.

3.4   Operations

The operations for LPBF are creation, query, insertion and deletion. For query, insertion and deletion, all of them need to convert the multi-dimensional data into one dimension at first. Then, use the perfect hash table to find the sub-LBF that each data belongs to. Next, use sub-LBFs and the CBF to finish the following operations. Therefore, we will not go into details in the data pretreatment and the choice of sub-LBF but focus on the subsequent operations.

- **Creation.** First, transform the data using the z-order curve, and get the prefixes and suffixes. Second, use the perfect hash to hash code all the prefixes, and enter the suffixes of the different prefixes into the corresponding sub-LBFs. Third, vectorize the data and train the classification model to get the classification results. Finally, use the false negative data that is output by all sub-LBFs as a positive sample to construct a CBF.

- **Query.** First, find the related sub-LBF according to the prefix. Then, use the LBF model, as shown in the LBF part of Fig. 3, to predict whether the checked data is positive. If it is determined as positive, it is considered to be present in the set. If not, we then utilize the CBF to check it is positive or negative and output the final result. Here, the LPBF focuses on judging whether an item is in the set, thus the query operations such as range queries are not within our consideration.

- **Insertion.** First, find the sub-LBF corresponding to the prefix. Then, get the classification result using the sub-LBF model after selecting the corresponding sub-LBF for the inserted data. If the predicted result is positive, we can insert the data into the set. If not, add the data into the final CBF. In particular, if no prefixes are matching the newly inserted data, add the prefix of data to the prefix array and re-encode using perfect hash function. Here, we set α as the number threshold of the new subset, and construct a new sub-LBF if the number of the new subset is large than α. Otherwise, input the new subset into final CBF directly.

- **Deletion.** We first use the corresponding sub-LBF model to get the classification result after selecting the corresponding sub-LBF for the deleted data. If the predicted result is positive, no action is required, and when looking up the deleted data from the set, it does not need to be found. If not, delete the data from the final CBF. In addition, a threshold θ is set for each sub-LBF. When the number of deleted data predicted as positive by sub-LBF reaches θ, the sub-LBF should be retrained to update the model.

3.5 Analyzing Learned Prefix Bloom Filter

We model the learned prefix bloom filter as follows. As previously introduced, all learned functions $f = \{f_1, f_2, ...f_N\}$ are used for all data keys $X = \{x_1, x_2, ...x_N\}$, and $|X| = n$, each sub-LBF function has the number of data $|x_i| = t_i$. The FPR of all learning models is represented as $F_P = \{F_p^1, F_p^2, ..., F_p^N\}$. Note, the total number of the input negative data into the learning model to calculate the F_P contains the part of negative data whose prefixes mismatch all existing prefixes, and are removed at the beginning before they input into a sub-LBF. For each learning model f_i, there are $F_n t_i$ false negatives for keys in f_i. Therefore, for the number of N sub-LBF, the backup CBF finally holds $F_N T$ keys, where $F_N = \{F_n^1, F_n^2, ..., F_n^N\}, T = \{t_1, t_2, ..., t_N\}$. Obviously, the better all learning models are trained, the smaller $F_N T$ than the number of all data keys n.

As shown in Eq. 1, the FPR of BF is calculated as F_b. For the backup CBF, the FPR F_l is expressed as:

$$F_l = [1 - (1 - \frac{1}{m})^{k \cdot F_N T}]^k \approx (1 - e^{\frac{-k \cdot F_N T}{m}})^k \tag{3}$$

Then, the overall FPR F is

$$F = \sum_{i=1}^{N} F_p^i + (1 - \sum_{i=1}^{N} F_p^i) \cdot F_l$$
$$\approx \sum_{i=1}^{N} F_p^i + (1 - \sum_{i=1}^{N} F_p^i) \cdot (1 - e^{\frac{-k \cdot F_N T}{m}})^k \tag{4}$$

According to the Eq. 4, if the number of the hash function k and the number of bits m are fixed, the overall FPR F is related to F_P and F_N. When the F_P is small enough, the Eq. 4 is approximately $\sum_{i=1}^{N} F_p^i + F_l$. Therefore, the smaller F_N, the smaller the F, which proves that as long as trains all learning models well, we can get a lower FPR. Moreover, due to the number of input data for CBF reducing significantly, the size of the bit array needed also decrease, which reduces memory usage as well. In addition, a part of negative data, whose prefixes mismatch all existing prefixes, is removed at the beginning also reduces the FPR.

4   Evaluation

4.1   Experimental Settings

In this part, four bloom filter methods are used as the comparison methods to compare with the proposed LPBF method in three spatial data sets. The comparison methods include two traditional bloom filter methods, is Standard Bloom Filter (SBF) [1] and Counting Bloom Filter (CBF) [10], and two learned bloom filter methods, is original Learned Bloom Filter (LBF) [13] and Sandwiched Bloom Filter (SandwichedBF) [16].

As for evaluation data sets, the real-world data sets Open Streets Maps (OSM)[1] are used for experiments. Specifically, the several areas of Greece, China, and Mexico data sets of OSM are used, the details of which are shown in Table 1. In addition, the synthetic 1000000 pieces of data are used as negative samples for three data sets. The operating system for all experiments executed is Windows 10 Professional Edition, and the processor is Intel(R) Core(TM) i5-9500 CPU @ 3.00 GHz 3.00 GHz, and the memory size is 24 GB. The python 3.8 is used to implement the LPBF method and all the comparison methods.

---

[1] https://download.geofabrik.de/.

Table 1. The details of data sets.

Area	Data set	Record	Size
Greece	Waterways	45640	1.28 MB
	Landuse_a	118382	3.37 MB
	Places	12763	789 KB
	Water_a	7498	204 KB
China	Landuse_a	686572	20.4 MB
	Natural	53692	890 KB
	Places	367751	14.1 MB
	Pois_a	187140	66.65 MB
Mexico	Landuse_a	166652	4.58 MB
	Natural	43939	204 KB
	Places	54046	2.1 MB
	Railways	10014	324 KB

4.2   Performance of Query

In this part, we evaluate the performance of the query by using the False Positive Rate (FPR) on three OSM data sets. We compare the FPR with different bit sizes of the bloom filter for all methods. In particular, the SandwichedBF has an Initial Filter and a Backup Filter, thus each filter uses half of the bit size. The prefix lengths of the LPBF method are fixed as 8. In addition, to ensure the fairness of the experiment, the Gradient Boosting algorithm is used in all learning models of the three learned bloom filter methods.

As shown in Fig. 4, the LPBF method shows great query performance. When the bit size is 5×10^3 to 5×10^4 in Fig. 4 (a), and 10^3 to 5×10^4 in (b) and (c), the four compared methods have high FPR, especially the two traditional BF methods, the LPBF method has maintained a low FPR. It should be noticed that the FPR of the LPBF method is slightly higher than two learned bloom filter methods when the bit size between 5×10^4 and 5×10^5, 10^4 and 1×10^5, and 10^4 and 1×10^5 in Fig. 4 (a)-(c), respectively.

The obvious advantages of LPBF on FPR may due to the following reasons. First, the FPR only correlates with bit size for traditional bloom filter methods. So, the FPR of the traditional methods are higher than the learning model because of the huge bit size correlation. Second, since the data is filtered first using the learning model, the learned bloom filter methods can use a smaller bit size to get a low FPR than traditional methods. Third, each sub-LBF of the LPBF only needs to learn the corresponding suffixes, and it uses the prefect hash to filter some negative data when querying at the beginning. Therefore, the FPR of the LPBF is lower than other learned models even if the bit size is small. However, the LPBF uses the CBF as the backup filter, which may have a slightly high FPR than the other two learned filters that use the SBF as the backup filter when the bit size is relatively small. In summary, LPBF shows the best performance overall.

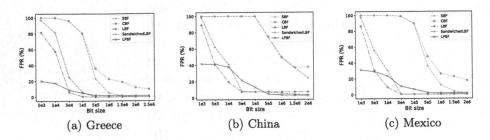

| (a) Greece | (b) China | (c) Mexico |

Fig. 4. Compare the query performance with different bit sizes on three data sets.

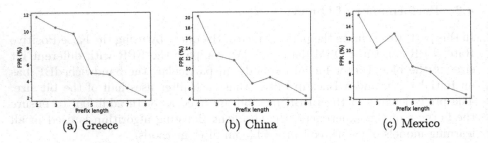

| (a) Greece | (b) China | (c) Mexico |

Fig. 5. The query performance of overall LPBF with different prefix lengths on three data sets.

4.3   Performance of Different Prefix Lengths

To explore the impact of different prefix lengths on the learning model of LPBF, we performed different prefix lengths experiments on three data sets. The bit size is fixed at 5×10^5 in all data sets. We get the FPR results of overall LPBF. The FPR results by the learning model predicted and after filtering by the entire filter are shown in Fig. 5, the FPR decreases as the prefix length increases, which means the more the prefix length, the lower the FPR. Therefore, it is obvious that the prefix length has a positive influence on the learning model.

The reason for the influence above is that in LPBF, each sub-LBF only needs to learn the corresponding suffixes. The longer the prefix, the simpler the suffix, which reduces the complicate of the learning model when the lengths of prefix increase. Moreover, a part of data that mismatches all the prefixes is filtered at the beginning of the query, which reduces the burden of the learning model.

4.4   Time Consumption

In this part, we compare the query time consumption of all methods, and the train time consumption of all learned models on three areas. Here, the prefix lengths of the LPBF method are fixed as 8. The bit size is fixed at 5×10^4.

Table 2. Compare query time of all methods on three areas. (time: s)

Area	SBF [1]	CBF [10]	LBF [13]	SandwichedBF [16]	LPBF
Greece	0.065	0.145	14.190	14.274	0.889
China	0.065	0.144	14.335	14.349	1.096
Mexico	0.064	0.144	14.321	14.242	1.135

Table 3. Time comparison of training process on three learned models. (time: s)

Area	LBF [13]	SandwichedBF [16]	LPBF
Greece	213.455	212.751	73.164
China	452.854	453.632	149.176
Mexico	225.720	225.894	77.644

As shown in Table 2, the query time of the traditional methods SBF and CBF is the smallest, and the query time of the proposed LPBF is close to the traditional methods, which is much smaller than the other two learned methods. The traditional methods only need to use the bit array for query, but the learned methods have the classification process when querying. However, the LPBF filters a part of keys by using prefix code before inputting into the sub-LBF, so it can save a lot of query time than other learned models.

We further compare the training process time on three learned models due to the traditional methods do not have the training process. As shown in Table 3, the LPBF method shows the shortest time consumption compared with other methods on three areas. This is because the LPBF method only needs to learn the suffixes, which decreases the learning complexity. Comprehensive the query and training process time consumption, and combined with the results of FPR, the proposed LPBF method shows the best performance.

5   Conclusion

In this paper, we presented a novel learned prefix bloom filter (LPBF). To our knowledge, this is the first LBF method for spatial data. It uses a CBF as a backup bloom filter of LPBF to achieve lower FPR and less memory usage compared with the classical BF. In LPBF, the Z-order curve is utilized to map the multi-dimensional spatial data into one dimension to get the binary codes of data. Then, the data can be divided into k categories according to all prefixes. All suffixes of the data are divided into different categories and input into different sub-LBFs, which reduces the learning complexity of the LBF. Moreover, we use the perfect hash table to get the prefix hash code so that the sub-LBF can be found as soon as possible and filter the negative data first when querying. The experimental results proved that the proposed LPBF method has great potential on the bloom filter for spatial data.

For future work, there are two interesting directions that can be considered. First, we can introduce the Neural Network for the learning of bloom filters and adaptive selection of the number of networks. Second, the weight of data can be considered for model learning to help reduce FPR.

Acknowledgements. This work is supported in part by the National Key R&D Program of China (2018AAA0102100), the Scientific and Technological Innovation Leading Plan of High-tech Industry of Hunan Province (2020GK2021), the National Natural Science Foundation of China (61902434), the International Science and Technology Innovation Joint Base of Machine Vision and Medical Image Processing in Hunan Province (2021CB1013).

References

1. Alexiou, K., Kossmann, D., Larson, P.: Adaptive range filters for cold data: avoiding trips to siberia. Proc. VLDB Endow. **6**(14), 1714–1725 (2013)
2. Belazzougui, D., Boldi, P., Pagh, R., Vigna, S.: Theory and practice of monotone minimal perfect hashing. ACM J. Exp. Algorithmics **16** (2011)
3. Bloom, B.H.: Space/time trade-offs in hash coding with allowable errors. Commun. ACM **13**(7), 422–426 (1970)
4. Crainiceanu, A.: Bloofi: a hierarchical bloom filter index with applications to distributed data provenance. In: 2nd International Workshop on Cloud Intelligence, ACM VLDB 2013, pp. 4:1–4:8 (2013)
5. Crainiceanu, A., Lemire, D.: Bloofi: multidimensional bloom filters. Inf. Syst. **54**, 311–324 (2015)
6. Dai, Z., Shrivastava, A.: Adaptive learned bloom filter (ada-bf): efficient utilization of the classifier with application to real-time information filtering on the web. In: Advances in Neural Information Processing Systems 33: Annual Conference on Neural Information Processing Systems 2020, NeurIPS 2020 (2020)

7. Davitkova, A., Gjurovski, D., Michel, S.: Compressing (multidimensional) learned bloom filters. In: Workshop on ML for Systems at NeurIPS 2021 (2021)
8. Ding, Y., Ma, Z., Wen, S., Xie, J., Chang, D., Si, Z., Wu, M., Ling, H.: AP-CNN: weakly supervised attention pyramid convolutional neural network for fine-grained visual classification. IEEE Trans. Image Process. **30**, 2826–2836 (2021)
9. Fan, B., Andersen, D.G., Kaminsky, M., Mitzenmacher, M.: Cuckoo filter: practically better than bloom. In: Proceedings of the 10th ACM International on Conference on emerging Networking Experiments and Technologies, CoNEXT 2014, pp. 75–88. ACM (2014)
10. Fan, L., Cao, P., Almeida, J.M., Broder, A.Z.: Summary cache: a scalable wide-area web cache sharing protocol. IEEE/ACM Trans. Netw. **8**(3), 281–293 (2000)
11. Guo, D., Wu, J., Chen, H., Luo, X.: Theory and network applications of dynamic bloom filters. In: 25th IEEE International Conference on Computer Communications, Joint Conference of the IEEE Computer and Communications Societies, IEEE INFOCOM 2006 (2006)
12. Guo, D., Wu, J., Chen, H., Yuan, Y., Luo, X.: The dynamic bloom filters. IEEE Trans. Knowl. Data Eng. **22**(1), 120–133 (2010)
13. Kraska, T., Beutel, A., Chi, E.H., Dean, J., Polyzotis, N.: The case for learned index structures. In: Proceedings of the 2018 International Conference on Management of Data, SIGMOD 2018, pp. 489–504. ACM (2018)
14. Li, S., Li, W., Cook, C., Zhu, C., Gao, Y.: Independently recurrent neural network (indrnn): building a longer and deeper RNN. In: 2018 IEEE Conference on Computer Vision and Pattern Recognition, CVPR 2018, pp. 5457–5466 (2018)
15. Macke, S., Beutel, A., Kraska, T., Sathiamoorthy, M., Cheng, D.Z., Chi, E.H.: Lifting the curse of multidimensional data with learned existence indexes. In: Workshop on ML for Systems at NeurIPS 2018 (2018)
16. Mitzenmacher, M.: A model for learned bloom filters and optimizing by sandwiching. In: Advances in Neural Information Processing Systems 31: Annual Conference on Neural Information Processing Systems 2018, NeurIPS 2018, pp. 462–471 (2018)
17. Mokbel, M.F., Aref, W.G.: Space-Filling Curves, Encyclopedia of GIS, pp. 1068–1072. (2008)
18. Natekin, A., Knoll, A.: Gradient boosting machines, a tutorial. Front. Neurorobotics **7** (2013)
19. Nayak, S., Patgiri, R.: countbf: a general-purpose high accuracy and space efficient counting bloom filter. In: 17th International Conference on Network and Service Management, CNSM 2021, Izmir, pp. 355–359. IEEE (2021)
20. Patgiri, R., Nayak, S., Borgohain, S.K.: RDBF: a r-dimensional bloom filter for massive scale membership query. J. Netw. Comput. Appl. **136**, 100–113 (2019)
21. Rae, J.W., Bartunov, S., Lillicrap, T.P.: Meta-learning neural bloom filters. In: Proceedings of the 36th International Conference on Machine Learning, ICML 2019, vol. 97, pp. 5271–5280 (2019)
22. Ramsak, F., Markl, V., Fenk, R., Zirkel, M., Elhardt, K., Bayer, R.: Integrating the UB-tree into a database system kernel. In: Proceedings of 26th International Conference on Very Large Data Bases, VLDB 2000, pp. 263–272 (2000)
23. Wu, Y., et al.: Elastic bloom filter: deletable and expandablefilter using elastic fingerprints. IEEE Trans. Comput. **71**, 1 (2021)

24. Xie, K., Min, Y., Zhang, D., Wen, J., Xie, G.: A scalable bloom filter for membership queries. In: Proceedings of the Global Communications Conference, GLOBECOM 2007, pp. 543–547. IEEE (2007)
25. Zhang, S., Ray, S., Lu, R., Zheng, Y.: SPRIG: a learned spatial index for range and KNN queries. In: Proceedings of the 17th International Symposium on Spatial and Temporal Databases, ACM SSTD 2021, pp. 96–105 (2021)

ReferEmo: A Referential Quasi-multimodal Model for Multilabel Emotion Classification

Alvar Esperanca and Xiao Luo[✉] [iD]

Purdue School of Engineering and Technology, Indiana University-Purdue University
Indianapolis, Indianapolis, IN 46202, USA
`aesperan@indiana.edu, luo25@iupui.edu`

Abstract. Textual emotion classification is a task in affective AI that branches from sentiment analysis and focuses on identifying emotions expressed in a given text excerpt. It has a wide variety of applications that improve human-computer interactions, particularly to empower computers to understand subjective human language better. Significant research has been done on this task, but very little of that research leverages one of the most emotion-bearing symbols we have used in modern communication: Emojis. In this research, we propose ReferEmo, a model that processes Emojis as textual inputs and leverages DeepMoji to generate affective feature vectors used as reference when aggregating different modalities of text encoding. To evaluate ReferEmo, we experimented on two benchmark datasets: SemEval'18 and GoEmotions for emotion classification, and achieved competitive performance compared to state-of-the-art models tested on these datasets. Notably, our model performs better on the underrepresented classes of each dataset. The source code of ReferEmo is available on Github (https://github.com/alvarosness/ReferEmo).

Keywords: Emotion classification · Multimodal model · Natural language processing · AI

1 Introduction

Sentiment analysis is the branch of affective AI consisting of various methodologies for identifying emotional valence expressed in text. Over the past years, there has been an increase in the prevalence of sentiment analysis in research and industry. Use cases include identifying customer satisfaction by inferring the sentiment being expressed in product reviews and determining job satisfaction from Voice of Employee surveys, and monitoring the emotional state of a large population by inferring the sentiment expressed in public communication platforms, particularly in the case of significant events such as political elections, and major health crises. Identifying early signs of mental health conditions by identifying the sentiment expressed in published online content. Due to the increasing popularity of Emojis, there is great interest in analyzing and studying their usage in text content for sentiment analysis [18,20,22].

© The Author(s), under exclusive license to Springer Nature Switzerland AG 2022
C. Strauss et al. (Eds.): DEXA 2022, LNCS 13426, pp. 351–366, 2022.
https://doi.org/10.1007/978-3-031-12423-5_27

The textual emotion or sentiment classification falls into two categories: lexicon-based approaches and machine learning-based approaches. The lexicon-based approaches utilize curated words and their associations to classify a text, whereas the machine learning-based approaches train a model to classify text. The defined lexicons are either used in rule-based models that rely on keyword frequency count [30] or keyword search [6] or as input features to machine learning-based models [21]. The machine learning-based approaches include those using either traditional or deep learning models. The traditional machine learning approach is similar to text classification, which includes steps such as first assigning Unicodes to the emojis, then applying feature selection, and classification algorithms, such as multinomial Naïve Bayes [19] for emotion classification or sentiment detection. With the advance of deep learning, the recent literature on emotion classification investigates various deep language models. NTUA-SLP [5] utilized the BiLSTM with a multi-layer self attention mechanism to predict affective content in tweets. They converted the emojis into word-level expressions to include emojis in the system. BERT based models has been applied to emotion classification [3,9]. The autoencoder-based approach has also been used to construct a latent variable representation from the latent emotion module to guide the prediction [12].

We propose the ReferEmo, a referential emotion encoder with three main components: a reference encoder, a BiLSTM and BERT text encoder, and an attention-based feature aggregation Layer. The reference encoder generates affective feature vectors aimed to enrich the text encodings with more affect value. The BiLSTM and BERT text encoders generate contextual token embeddings from the input sequence, including emojis. The BERT encoder encapsulates much knowledge given that it has been pretrained on a very large corpus allowing for quick learning of the task. The BiLSTM encoder, however, supplements the BERT encoder allowing the entire model to achieve higher sensitivity. The attention-based feature aggregation combines the text encodings and the reference vector by having the text encodings attend to the feature vector, thus giving more weight to tokens whose affective value is highest.

The main contributions of this paper are summarized as follows: (1) We proposed a novel emotion classification model - ReferEmo, (2) demonstrated its competitive performance compared to state-of-the-art models in multilabel emotion classification, and (3) showed that the ReferEmo performs better on underrepresented classes.

2 Related Work

2.1 Lexicon Based Emotion Classification

Affective lexicons have been extensively used since the early stages of automatic affect detection research and still provide useful linguistic features that aid in more contemporary methodologies. These lexicons consist of curated sets of words and their associated set of affect scores.

Lexicons like the Liu Lexicon [14] consists of a set of words with either a positive or a negative label. There are lexicons such as AFINN [25], and Senti-WordNet [29] that do quantify how positive or how negative the word is, allowing for better distinction of expressed effect. Emotion classification models have used these lexicons, particularly TCS Research [21] and SeerNet [11], with some significant results. EmoLex [24] is a lexicon that builds on top of the ones stated prior and is better suited for the emotion classification task. Much like Liu Lexicon, EmoLex does not quantify the association between a word and each emotion. These lexicons have also been used in emotion classification models such as NELEC [1], and SINAI [4]. Despite their prevalence, these lexicons can present some limitations. Namely the fact that most of these lexicons are a form of local representations of emotion.

2.2 Deep Learning Based Emotion Classification

NTUA-SLP [5] was the best performing model submitted to the SemEval'18 Task 1 [23]. The authors developed a recurrent neural network consisting of BiLSTMs with deep attention where the input embeddings were word2vec word vectors whose dimensions are augmented with hand-picked affective features. The augmentation of the word vectors did not improve their performance on the multilabel classification task though it did improve their performance on regression tasks.

Seq2Emo [16] is another most recent relevant deep learning-based model for emotion classification. Its architecture mimics a sequence to sequence model where an encoder BiLSTM network transforms a sequence of tokens into a sequence of encodings for each emotion class decoded by another BiLSTM network. Their model does not use any other emotion or sentiment information such as lexicons or pretrained affective embeddings. Their performance is akin to that of NTUA-SLP despite the little additional information that it uses.

GoEmotions is the most recent benchmark dataset for emotion classification. The authors of the GoEmotions [9] have applied BERT [10] on to the dataset with some significant results. Notably, a standard pretrained BERT had already embedded much affective information, leading to quick learning and improved performance on other tasks when pretrained on the GoEmotions dataset.

2.3 Emojis in Emotion Classification

Emojis are a pictorial representation of various concepts, including emotions, objects, and activities. Since their dawn in the early 2010s, emojis have become increasingly commonplace in our modern forms of electronic communication.

Hu et al. [15] conducted a study on the usage of emojis while focusing on the intent behind their widespread use [15] and found that emojis are used to express positive or negative sentiment, further increase the amount of sentiment expressed in a text excerpt, and adjust the tone of a message to convey sarcasm, irony, humor, or closeness. Ai et al. [2] conducted a similar study with the focus

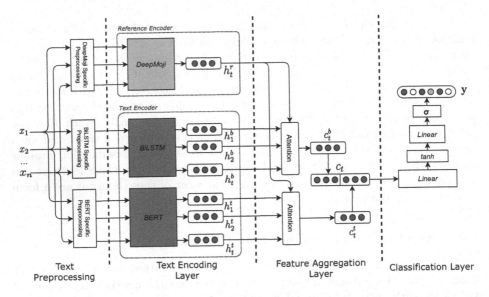

Fig. 1. ReferEmo architecture.

on understanding what leads an emoji to be more popular than others by analyzing the relationship between emojis and the context in which they are presented [2]. They found that emoji usage is characterized into two distinct functions: complementary and supplementary. The most popular emojis are also the ones that convey the most sentiment. Delobelle et al. [8] argued that emojis were not used enough in NLP models. They found that in most NLP research, emojis have either been underutilized or not utilized at all. The proper use of emojis can increase the performance of contextual models, with an observed increase of 5.85% in performance once emojis were used in their conversational model.

3   ReferEmo: Referential Emotion Encoder

The proposed model, as illustrated in Fig. 1, encodes a sequence of tokens, including words and emojis, defined as $\mathbf{x} = (x_1, x_2, \cdots, x_T)$, and generates a feature vector c_t used by a classification model to predict the target labels defined as $\mathbf{y} = (y_1, y_2, \cdots, y_K) \in \{0, 1\}^K$, where $y_i = 1$ indicates that the ith emotion is being expressed in the input sequence. The architecture consists of three distinct layers: encoding layer, feature aggregation layer, and classification layer, which are preceded by a module-specific preprocessing step. The following sections explain in detail the components of each of the layers.

3.1   Preprocessing

Each of the modules of the text encoding layer has different text preprocessing requirements. The BERT Encoder requires the text to be processed as SentencePiece tokens and uses BERT built-in embedding to generate the input token embeddings. The BiLSTM Encoder requires the text to be processed as word tokens with the addition of some preprocessing steps that preserve hashtags (e.g., #blessed), mentions (e.g., @BarackObama), and emojis. These tokens are then vectorized using GloVe [27] word vectors extended with emojis. The DeepMoji [13] model requires the text to be processed as word tokens and transformed into input embeddings using its own pretrained word vectors. These word vectors are pretrained on an emoji prediction task.

3.2 Encoding Layer

The encoding layer is responsible for creating the various feature vectors from the input sequence used by the upper layers. This layer consists of two types of encoders.

Reference Encoder. The reference encoder is a model that generates an affective feature vector h_t^r, shown as Eq. 1. As the name suggests, this affective feature vector serves as a reference to enrich the word embeddings of the input sequence with affect knowledge. We use a pretrained DeepMoji [13] model as the reference encoder.

$$h_t^r = DeepMoji(\mathbf{x}) \tag{1}$$

DeepMoji is a 2-layer BiLSTM with Attention. It has been pretrained on the task of predicting the occurrence of emojis in an input text. This model has been shown to perform well on tasks such as emoji prediction and sarcasm detection achieving state-of-the-art performance. Furthermore, other models have leveraged the knowledge learned by DeepMoji in their models with significant improvement to their performance. These facts suggest that DeepMoji generates good affective feature vectors and that our model can benefit from using its feature vectors as reference. Note that the use of the pretrained DeepMoji model is a deliberate design choice. Any model that can generate an affective feature vector can be used.

Text Encoder. This encoder generates a set of contextual token embeddings from the input sequence. These token embeddings are later enriched with the affective knowledge from the reference vector generated by the reference encoder. BiLSTM and BERT are two state-of-the-art network architectures that take sequences of words or tokens to encode the text content. The recent research on text classification shows that they work better than other architectures.

We use a 2-layer BiLSTM that encodes an input sequence $\mathbf{x} = (x_1, \cdots, x_n)$ into the contextual token embeddings $h_1^b, \cdots, h_t^b$, shown as Eq. 2. These token embeddings are the intermediate outputs of the BiLSTM at timestep i and are defined as

$$h_i^b = BiLSTM_i(x_i) \tag{2}$$

where h_i^b is the hidden state of the BiLSTM at timestep i. It summarizes all of the sequence information up to x_i from both the forward and backward directions.

In addition to the BiLSTM, we also use BERT as an additional text encoder. BERT is a transformer-based model that is pretrained on a large corpus for a masked language modeling task. BERT has shown state-of-the-art performance on many NLP tasks ranging from machine translation to sentiment analysis. Particularly, its performance on sentiment-related tasks suggests that BERT can assimilate how affectiveness is expressed in text. Preliminary experimental results have shown that the BiLSTM encoder yielded a better precision while the BERT encoder yielded a better recall. We use the two encoders to improve the sensitivity of the model while still maintaining high specificity. We define the token embeddings generated by BERT as the sequence $h_1^t, \cdots, h_t^t$ where h_i^t represents the embedding of the ith token in the sequence within the context of the entire input sequence.

3.3   Attention-Based Feature Aggregation Layer

This layer receives as inputs the reference feature vector h_t^r and the token embeddings $[h_1^t, \cdots, h_t^t]$ and $[h_1^b, \cdots, h_t^b]$ from the previous layer and generates a context vector that aggregates the values of the sequence embeddings and the affective value of the reference vector.

In our architecture, we use attention as the aggregation mechanism. Given that there are two sets of token embeddings, two sets of attention scores are computed. One set of attention scores is between the reference vector and the BiLSTM token embeddings, and the other set is between the reference vector and the BERT token embeddings. Using attention not only generates a feature vector that better encodes longer sequences with long dependencies but also allows us to visualize the alignment scores between the reference feature vector and the token embeddings, thus providing information regarding which tokens carry the most affect value that allows us to assess the relationship between these input tokens and the classification label.

The attention scores with respect to the BiLSTM embeddings are defined as Eqs. 3 and 4

$$s_{r,i}^b = V^\top \tanh(W_q h_t^r + W_v h_i^b) \tag{3}$$

$$\alpha_{r,i}^b = \frac{\exp(s_{r,i}^b)}{\sum_{j=0}^{T} \exp(s_{r,j}^b)} \tag{4}$$

where V, W_q, and W_v are learned parameters, $s_{r,i}^b$ is the attention score for the ith token embedding, and $\alpha_{r,i}^b$ is the normalized attention score. The attention scores with respect to the BERT embeddings are defined in an identical manner.

The context vector for each of the embeddings is computed as the attention-weighted sum of the token embeddings, shown as Eqs. 5 and 6.

$$c_t^b = \sum_{i=0}^{T} \alpha_{r,i}^b h_i^b \qquad (5)$$

$$c_t^t = \sum_{i=0}^{T} \alpha_{r,i}^t h_i^t \qquad (6)$$

The final context vector is defined as the concatenation of c_t^b and c_t^t, shown as Eq. 7.

$$c_t = [c_t^b; c_t^t] \qquad (7)$$

3.4 Classification Layer

The last layer consists of a two-layer fully connected neural network with a *tanh* activation between the two layers and a sigmoid activation at the output layer, shown as Eq. 8. Since the model architecture is designed for multilabel classification tasks, using sigmoid as the activation of the output layer is the most suitable option.

$$\mathbf{y} = \sigma(W_2 \tanh(W_1 c_t + b_1) + b_2) \qquad (8)$$

4 Experiments and Results

4.1 Experimental Settings

Datasets. To evaluate our proposed model, we trained it on two recent benchmark datasets for emotion classification: SemEval'18 Affect in Tweets [23] and the GoEmotions [9] datasets. SemEval'18 consists of tweets, whereas GoEmotions consists of Reddit posts and comments. The SemEval'18 corpus has more emojis than the GoEmotions counterpart. However, GoEmotions is the most recent dataset and the largest manually annotated dataset with the most extensive set of emotion classes. Both of these datasets were provided with the splits for training, validating, and testing. We also observe that the distribution of labels in the datasets differs significantly. For instance, GoEmotions has many more examples with only one label compared to SemEval'18. These details can be observed in Table 1.

Baselines. We compare the performance of the ReferEmo model with the top-performing model submitted to the SemEval'18 Task 1 and a most recent state-of-the-art method.

- **NTUA-SLP** [5], the top-ranked model in the SemEval'18 Task 1 competition.

Table 1. Summary statistics of the datasets

	SemEval'18	GoEmotions
Train size	6,838	43,410
Valid size	886	5,426
Test size	3,259	5,427
Total size	10,983	54,263
Number of emotions	11	27 + Neutral
Labels per example		
1	13.48%	83.75%
2	40.89%	14.97%
3	31.49%	1.21%
4+	14.13%	0.07%

- **Seq2Emo** [16], the most current state-of-the-art model for the emotion classification task.

Both models are compared on the SemEval'18 and the GoEmotions datasets through implementing the code published on GitHub. Since SemEval'18 is a popular benchmark dataset, we also included the published results of other methods.

Hyperparameter Tuning. In this research, we used the pretrained BERT and DeepMoji models and tuned the hyperparameters for the BiLSTM encoder, including the number of layers (n_layers) and the number of hidden units (hdim) per layer. In addition to those, the dropout probability (dropout_p) and the learning rate (lr) are also tuned. Table 2 reports the optimal hyperparameters and the search space for each of them.

Table 2. Hyperparameters and search space.

	SemEval'18	GoEmotions	Range
hdim	833	938	[128, 1024]
n_layers	3	1	[1, 3]
dropout_p	0.42	0.47	[0.2, 0.7]
lr	2.02e−05	2.07e−05	[1e−5, 1e−1]

Evaluation Metrics. To evaluate the performance of our model, we use the three metrics [23] used in the SemEval'18 competition. These metrics are Jaccard Index [28], macro- and micro-averaged F1 scores [7]. The Jaccard Index is

Table 3. Summarized performance on the SemEval'18

	Jaccard	Micro F1	Macro F1
Random	0.19	0.31	0.29
SVM-Unigrams	0.44	0.57	0.44
PlusEmo2Vec	0.58	0.69	0.50
TCS research	0.58	0.69	0.53
LEM	–	0.67	0.56
BNet	0.59	0.69	0.56
NTUA-SLP	**0.59**	0.70	0.53
Seq2Emo	0.59	0.70	0.52
ReferEmo (AVG)	0.58	**0.71**	**0.57**
ReferEmo (STD)	0.01	0.01	0.01

Table 4. Summarized performance on the GoEmotions

	Jaccard	Micro F1	Macro F1
BERT	0.53	0.59	0.46
NTUA-SLP	0.48	0.54	0.44
Seq2Emo	**0.54**	**0.60**	0.47
ReferEmo (AVG)	0.53	0.56	**0.48**
ReferEmo (STD)	0.01	0.01	0.00

commonly used as a multilabel classification accuracy measure. It measures the overlap between ground truth and predicted labels. It is defined as:

$$J(G, P) = \frac{|G \cap P|}{|G \cup P|}$$

where G is the ground truth label, and P is the predicted label. The final Jaccard Index is computed by averaging the Jaccard indices of all the documents in the dataset.

$$Jaccard = \frac{1}{|D|} \sum_{d \in D} J(G_d, P_d) \tag{9}$$

We also report the precision, recall, and F1 scores at the class level to thoroughly exam the performance of our model and the baseline models.

4.2 Performance Comparison

Table 3 shows the performance of our proposed model on the SemEval'18 dataset. In addition to the performance of Seq2Emo and NTUA-SLP, TCS Research [21], PlusEmo2Vec [26], LEM [12], BNet [17], as well as the random and SVM-unigram baselines of the competition. Table 4 shows the performance of our proposed model on the GoEmotions dataset in addition to the performance of BERT, NTUA-SLP, and Seq2Emo.

Compared to NTUA-SLP and Seq2Emo, our Jaccard and Micro F1 scores are very similar. Our Macro F1 score is slightly better than the baselines as shown in Tables 3 and 4. Further inspection of the class level performance suggests we have improved the macro measure while still performing the same on the other measures.

On the SemEval'18 dataset, through further investigation, we found that our performance on the precision metric is not much different from the other baselines, and in some instances, some baseline performs slightly better (i.e., anticipation, joy, love, and pessimism). However, in terms of recall, our model performs much better, leading to more balanced F1 scores as shown in Table 5 in Appendix. The improvement on the recall metric indicates that, despite not being as precise as the other baseline models, our model makes up for it by being more sensitive to the correct examples, particularly in the anticipation, fear, love, and pessimism classes which are underrepresented in the training dataset. On the GoEmotions dataset, our performance on the precision metric is better on the relief and remorse classes. On the other hand, the recall is significantly better than the baselines, like the SemEval'18 dataset. Again, the higher recall is due to the higher sensitivity of our model to the correct classes. These results are shown in Table 6 in Appendix. Our model tends to perform better on the underrepresented classes. The better performance is more prevalent on the GoEmotions dataset as opposed to the SemEval'18 dataset. Grief and relief are the two least represented classes in the GoEmotions dataset. Our model could not identify grief despite the other baselines having identified some of it, whereas our model was able to identify relief much better than the other baseline models.

4.3   Ablation Study

We perform an ablation study to examine the effects of the different ReferEmo modules on its performance. We trained three variants of our proposed model. The first and second variants have the BERT and the BiLSTM Text Encoders removed, respectively. These first two variants are similar in architecture, given that the proposed model is designed to have swappable and removable encoders. The third variant has the DeepMoji Reference Encoder removed, which requires a change in the architecture. Specifically, the attention-based feature aggregation is replaced by averaging the feature vectors of the BERT and BiLSTM encoders, and those averaged vectors are then concatenated to form the final context vector.

These variants were all trained with the optimal hyperparameters described prior and the same number of epochs, including the very shallow network for the BiLSTM when training on the GoEmotions dataset. The results are summarized in Figs. 2 and 3. As the results show, the model that has the BERT encoder removed suffers significantly in performance. The contribution of DeepMoji is not high as expected. Through investigating the performance of individual emotions

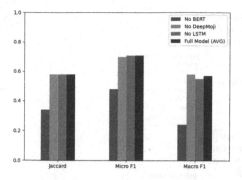
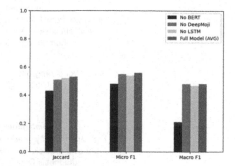

Fig. 2. Ablation study on SemEval'18. **Fig. 3.** Ablation study on GoEmotions.

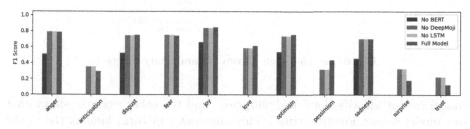

Fig. 4. Ablation study for individual emotions in the SemEval'18 dataset.

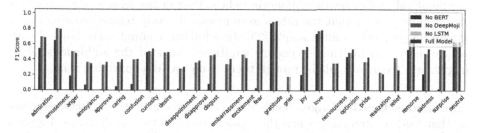

Fig. 5. Ablation study for individual emotions in the GoEmotions dataset.

on both datasets (shown in Figs. 4 and 5), we found that the BERT encoder has the most impact on the system performance. The contribution of DeepMoji and BiLSTM is similar. For some categories, such as 'surprise', 'trust' in SemEval'18, and 'grief' and 'relief' in GoEmotions, using either DeepMoji or BiLSTM can produce better results.

Label Ambiguity. Many examples were labeled ambiguously, making it difficult for a model to properly learn the relationships between the input tokens and the emotion classes. This ambiguity is even more prevalent when not even human annotators can agree with the gold standard label. The ambiguous labeling in Fig. 6 shows some examples of these cases. One example shows that our

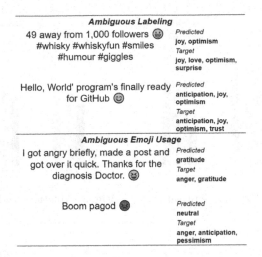

Fig. 6. Sample tweets shown the ambiguity inherent

model cannot identify 'love' and 'surprise', and the other example shows that our model cannot identify 'trust'. This ambiguity, in turn, hinders the model performance, especially when it is a rather sensitive model like ReferEmo. These ambiguous labels are more present in the 'anticipation' and 'trust' classes of the SemEval'18 dataset than any other class. Part of the reason why the labels are so ambiguous is that the raters were given relatively relaxed conditions for attributing a label to an example [23]. In addition, examples are labeled with one primary and multiple secondary emotions. Some of the ambiguous labels might have secondary labels, though there is no way of confirming that.

Ambiguous Emoji Usage. Some of the examples in both datasets use emojis that lead to ambiguity when the model attempts to classify them. Figure 6 provides two examples to demonstrate the cases of ambiguous emoji usage. The first example in the Ambiguous Emoji Usage has a happy emoji, but it is labeled as 'anger' emotion. In the GoEmotions dataset, the happy emojis are commonly associated with more positive emotions such as love and joy. Similarly, some examples of the anticipation emotion in the SemEval'18 dataset use more negative emotions, such as the 'anger' emoji.

5   Conclusion

In this research, we proposed and evaluated a referential quasi-multimodal model for emotion classification, ReferEmo, which combines encodings from different modalities of text encoding and enriches these encodings with an emotional reference vector. In addition to that, this model processes emojis as input tokens. We evaluated our proposed model on two benchmark multilabel emotion classification datasets and achieved competitive performance compared to state-of-the-art baselines. Notably, our proposed model is more sensitive to true labels and performs significantly better on underrepresented classes. We concluded that the BERT text encoder has a significant impact on the performance of our model. The future work includes better utilization of the DeepMoji reference encoder and BiLSTM text encoder to maximize the advantages of each when integrating with the BERT text encoder.

Appendix

Table 5. Performance emotions of the SemEval'18 dataset. * and † denote $p < .05$

	Example Count	Precision			Recall			F1		
		Our model	NTUA-SLP	Seq2Emo	Our model	NTUA-SLP	Seq2Emo	Our model	NTUA-SLP	Seq2Emo
anger	2,544	**0.79 ± 0.03**	0.77 ± 0.04	0.76 ± 0.04	0.78 ± 0.04	0.77 ± 0.05	0.79 ± 0.05	**0.79 ± 0.01†***	0.77 ± 0.00	0.77 ± 0.01
anticipation	978	0.37 ± 0.06	0.39 ± 0.05	0.38 ± 0.10	**0.28 ± 0.15***	0.14 ± 0.02	0.12 ± 0.08	**0.29 ± 0.07 †***	0.20 ± 0.03	0.17 ± 0.09
disgust	2,602	**0.73 ± 0.02***	0.69 ± 0.02	0.68 ± 0.02	**0.78 ± 0.05**	0.77 ± 0.04	0.76 ± 0.04	**0.75 ± 0.01†***	0.73 ± 0.01	0.72 ± 0.01
fear	1,242	0.72 ± 0.09	0.82 ± 0.03	0.78 ± 0.05	**0.77 ± 0.05†***	0.67 ± 0.04	0.65 ± 0.06	0.74 ± 0.04	0.74 ± 0.02	0.71 ± 0.02
joy	2,477	0.81 ± 0.02	0.86 ± 0.01	0.85 ± 0.03	**0.89 ± 0.03†***	0.83 ± 0.02	0.83 ± 0.04	0.85 ± 0.00	0.84 ± 0.01	0.84 ± 0.01
love	700	0.52 ± 0.06	0.71 ± 0.04	0.67 ± 0.07	**0.76 ± 0.08†***	0.47 ± 0.07	0.50 ± 0.13	**0.61 ± 0.02†***	0.56 ± 0.03	0.56 ± 0.05
optimism	1,984	0.67 ± 0.02	0.71 ± 0.02	0.70 ± 0.03	**0.87 ± 0.03†***	0.67 ± 0.04	0.71 ± 0.07	**0.75 ± 0.00†***	0.69 ± 0.01	0.70 ± 0.02
pessimism	795	0.41 ± 0.03	0.48 ± 0.05	0.48 ± 0.05	**0.46 ± 0.06†***	0.20 ± 0.02	0.17 ± 0.07	**0.43 ± 0.01†***	0.29 ± 0.02	0.25 ± 0.07
sadness	2,008	0.67 ± 0.05	0.77 ± 0.06	0.75 ± 0.04	**0.75 ± 0.06†***	0.63 ± 0.05	0.63 ± 0.04	**0.70 ± 0.01†***	0.68 ± 0.02	0.69 ± 0.01
surprise	361	**0.59 ± 0.10†**	0.43 ± 0.02	0.56 ± 0.25	0.11 ± 0.05*	0.11 ± 0.02	0.07 ± 0.04	0.18 ± 0.06*	0.17 ± 0.02	0.12 ± 0.07
trust	357	0.29 ± 0.05	0.24 ± 0.06	0.31 ± 0.39	0.09 ± 0.07*	0.07 ± 0.04	0.03 ± 0.03	0.13 ± 0.08	0.10 ± 0.06	0.04 ± 0.04

Table 6. Performance emotions of the GoEmotions dataset. * and † denote $p < .05$

	Example count	Precision			Recall			F1		
		ReferEmo	NTUA-SLP	Seq2Emo	ReferEmo	NTUA-SLP	Seq2Emo	ReferEmo	NTUA-SLP	Seq2Emo
admiration	4,130	0.62 ± 0.03	0.63 ± 0.01	0.66 ± 0.04	**0.75 ± 0.02†***	0.67 ± 0.01	0.65 ± 0.03	**0.68 ± 0.01**	0.65 ± 0.00	0.65 ± 0.01
amusement	2,328	0.72 ± 0.04	0.75 ± 0.02	0.78 ± 0.01	**0.87 ± 0.03†**	0.82 ± 0.02	0.87 ± 0.02	0.79 ± 0.01	0.78 ± 0.02	0.83 ± 0.01
anger	1,567	0.46 ± 0.03	0.45 ± 0.04	0.61 ± 0.09	**0.50 ± 0.02†***	0.36 ± 0.03	0.32 ± 0.02	**0.48 ± 0.01**	0.40 ± 0.02	0.41 ± 0.02
annoyance	2,470	0.31 ± 0.03	0.36 ± 0.04	0.49 ± 0.13	**0.40 ± 0.07†***	0.21 ± 0.04	0.19 ± 0.08	**0.34 ± 0.02**	0.26 ± 0.03	0.26 ± 0.09
approval	2,939	0.32 ± 0.04	0.40 ± 0.02	0.46 ± 0.06	**0.42 ± 0.05†**	0.25 ± 0.02	0.24 ± 0.03	**0.36 ± 0.01**	0.31 ± 0.02	0.31 ± 0.02
caring	1,087	0.37 ± 0.04	0.34 ± 0.02	0.46 ± 0.08	**0.44 ± 0.05**	0.27 ± 0.04	0.27 ± 0.05	**0.40 ± 0.02**	0.30 ± 0.03	0.34 ± 0.03
confusion	1,368	0.32 ± 0.03	0.38 ± 0.06	0.52 ± 0.06	**0.54 ± 0.07**	0.36 ± 0.03	0.29 ± 0.06	**0.40 ± 0.03**	0.37 ± 0.03	0.37 ± 0.05
curiosity	2,191	0.46 ± 0.01	0.48 ± 0.02	0.50 ± 0.05	**0.64 ± 0.06**	0.41 ± 0.06	0.43 ± 0.10	**0.54 ± 0.02**	0.44 ± 0.02	0.46 ± 0.06
desire	641	0.53 ± 0.02	0.52 ± 0.07	0.61 ± 0.10	**0.46 ± 0.05**	0.34 ± 0.04	0.33 ± 0.10	**0.49 ± 0.02**	0.41 ± 0.03	0.42 ± 0.09
disappointment	1,269	0.30 ± 0.04	0.27 ± 0.05	0.42 ± 0.06	**0.31 ± 0.06**	0.16 ± 0.04	0.16 ± 0.04	**0.30 ± 0.04**	0.20 ± 0.04	0.23 ± 0.04
disapproval	2,022	0.33 ± 0.05	0.32 ± 0.03	0.43 ± 0.02	**0.46 ± 0.08**	0.26 ± 0.04	0.24 ± 0.03	**0.38 ± 0.01**	0.29 ± 0.03	0.31 ± 0.03
disgust	793	0.40 ± 0.06	0.51 ± 0.05	0.60 ± 0.02	**0.56 ± 0.05**	0.41 ± 0.04	0.38 ± 0.05	0.46 ± 0.03	0.45 ± 0.03	0.46 ± 0.04
embarrassment	303	0.42 ± 0.06	0.42 ± 0.07	0.60 ± 0.06	**0.40 ± 0.02**	0.32 ± 0.09	0.31 ± 0.09	0.41 ± 0.02	0.35 ± 0.07	0.41 ± 0.08
excitement	853	0.36 ± 0.05	0.43 ± 0.03	0.51 ± 0.10	**0.51 ± 0.05**	0.33 ± 0.02	0.31 ± 0.05	**0.42 ± 0.03**	0.37 ± 0.02	0.38 ± 0.03
fear	596	0.56 ± 0.05	0.60 ± 0.04	0.64 ± 0.05	**0.76 ± 0.07**	0.68 ± 0.04	0.63 ± 0.06	0.64 ± 0.01	0.64 ± 0.04	0.63 ± 0.01
gratitude	2,662	0.91 ± 0.03	0.93 ± 0.01	0.94 ± 0.02	0.89 ± 0.01	0.88 ± 0.01	0.89 ± 0.01	0.90 ± 0.01	0.90 ± 0.01	0.91 ± 0.01
grief	77	0.00 ± 0.00	0.32 ± 0.12	0.23 ± 0.33	0.00 ± 0.00	0.23 ± 0.09	0.10 ± 0.15	0.00 ± 0.00	0.26 ± 0.09	0.14 ± 0.20
joy	1,452	0.51 ± 0.04	0.55 ± 0.03	0.60 ± 0.08	**0.63 ± 0.01**	0.51 ± 0.05	0.52 ± 0.09	**0.57 ± 0.02**	0.53 ± 0.03	0.55 ± 0.04
love	2,086	0.72 ± 0.01	0.74 ± 0.01	0.77 ± 0.05	**0.85 ± 0.01**	0.80 ± 0.02	0.83 ± 0.05	0.78 ± 0.01	0.77 ± 0.01	0.80 ± 0.01
nervousness	164	0.32 ± 0.06	0.38 ± 0.11	0.63 ± 0.26	**0.38 ± 0.08**	0.23 ± 0.07	0.17 ± 0.08	**0.35 ± 0.06**	0.28 ± 0.08	0.25 ± 0.09
optimism	1,581	0.49 ± 0.04	0.59 ± 0.05	0.60 ± 0.05	**0.58 ± 0.01**	0.47 ± 0.03	0.45 ± 0.05	**0.53 ± 0.02**	0.52 ± 0.01	0.51 ± 0.03
pride	111	0.58 ± 0.10	0.46 ± 0.07	0.69 ± 0.11	0.34 ± 0.03	0.32 ± 0.07	0.30 ± 0.07	**0.43 ± 0.05**	0.36 ± 0.05	0.41 ± 0.07
realization	1,110	0.21 ± 0.02	0.36 ± 0.03	0.47 ± 0.14	**0.21 ± 0.03**	0.16 ± 0.03	0.15 ± 0.04	0.21 ± 0.02	0.22 ± 0.03	0.22 ± 0.05
relief	153	**0.28 ± 0.19**	0.15 ± 0.04	0.08 ± 0.17	**0.25 ± 0.20**	0.11 ± 0.04	0.05 ± 0.12	**0.26 ± 0.19**	0.12 ± 0.03	0.06 ± 0.14
remorse	545	**0.58 ± 0.04**	0.53 ± 0.04	0.57 ± 0.03	**0.83 ± 0.03**	0.64 ± 0.07	0.53 ± 0.22	**0.68 ± 0.03**	0.58 ± 0.05	0.52 ± 0.12
sadness	1,326	0.50 ± 0.08	0.54 ± 0.02	0.60 ± 0.05	**0.59 ± 0.03**	0.51 ± 0.04	0.46 ± 0.06	**0.54 ± 0.04**	0.52 ± 0.02	0.52 ± 0.04
surprise	1,060	0.50 ± 0.03	0.56 ± 0.04	0.62 ± 0.08	**0.55 ± 0.06**	0.46 ± 0.04	0.39 ± 0.12	**0.52 ± 0.04**	0.51 ± 0.03	0.46 ± 0.07
neutral	14,219	0.62 ± 0.01	0.65 ± 0.01	0.62 ± 0.02	0.64 ± 0.02	0.56 ± 0.02	0.69 ± 0.02	0.63 ± 0.01	0.60 ± 0.01	0.65 ± 0.00

References

1. Agrawal, P., Suri, A.: NELEC at SemEval-2019 Task 3: think twice before going deep. arXiv arXiv:1904.03223 [cs] (April 2019)
2. Ai, W., Lu, X., Liu, X., Wang, N., Huang, G., Mei, Q.: Untangling emoji popularity through semantic embeddings. In: Proceedings of the International AAAI Conference on Web and Social Media, vol. 11, no. 1 (April 2017). https://ojs.aaai.org/index.php/ICWSM/article/view/14903
3. Alhuzali, H., Ananiadou, S.: SpanEmo: casting multi-label emotion classification as span-prediction. arXiv preprint arXiv:2101.10038 (2021)
4. Plaza-del Arco, F.M., Molina-González, M.D., Martin, M., Ureña-López, L.A.: SINAI at SemEval-2019 Task 3: using affective features for emotion classification in textual conversations. In: Proceedings of the 13th International Workshop on Semantic Evaluation, Minneapolis, Minnesota, USA, pp. 307–311. Association for Computational Linguistics (June 2019). https://doi.org/10.18653/v1/S19-2053. https://aclanthology.org/S19-2053
5. Baziotis, C., et al.: NTUA-SLP at SemEval-2018 Task 1: predicting affective content in tweets with deep attentive RNNs and transfer learning. arXiv arXiv:1804.06658 [cs] (April 2018)
6. van den Broek-Altenburg, E.M., Atherly, A.J.: Using social media to identify consumers' sentiments towards attributes of health insurance during enrollment season. Appl. Sci. 9(10), 2035 (2019). https://doi.org/10.3390/app9102035. https://www.mdpi.com/2076-3417/9/10/2035
7. Chinchor, N.: MUC-4 evaluation metrics. In: Proceedings of the 4th Conference on Message Understanding, MUC4 1992, pp. 22–29. Association for Computational Linguistics, USA (June 1992). https://doi.org/10.3115/1072064.1072067
8. Delobelle, P., Berendt, B.: Time to take emoji seriously: they vastly improve casual conversational models. arXiv arXiv:1910.13793 [cs] (October 2019)
9. Demszky, D., Movshovitz-Attias, D., Ko, J., Cowen, A., Nemade, G., Ravi, S.: GoEmotions: a dataset of fine-grained emotions. arXiv arXiv:2005.00547 [cs] (June 2020)
10. Devlin, J., Chang, M.W., Lee, K., Toutanova, K.: BERT: pre-training of deep bidirectional transformers for language understanding. arXiv:1810.04805 [cs] (May 2019)
11. Duppada, V., Jain, R., Hiray, S.: SeerNet at SemEval-2018 Task 1: domain adaptation for affect in tweets. arXiv:1804.06137 [cs] (April 2018)
12. Fei, H., Zhang, Y., Ren, Y., Ji, D.: Latent emotion memory for multi-label emotion classification. In: Proceedings of the AAAI Conference on Artificial Intelligence, vol. 34, pp. 7692–7699 (2020)
13. Felbo, B., Mislove, A., Søgaard, A., Rahwan, I., Lehmann, S.: Using millions of emoji occurrences to learn any-domain representations for detecting sentiment, emotion and sarcasm. In: Proceedings of the 2017 Conference on Empirical Methods in Natural Language Processing, pp. 1615–1625 (2017). https://doi.org/10.18653/v1/D17-1169. arXiv:1708.00524
14. Hu, M., Liu, B.: Mining and summarizing customer reviews. In: Proceedings of the 10th ACM SIGKDD International Conference on Knowledge Discovery and Data Mining, pp. 168–177 (2004)

15. Hu, T., Guo, H., Sun, H., Nguyen, T.T., Luo, J.: Spice up your chat: the intentions and sentiment effects of using emoji. arXiv arXiv:1703.02860 [cs] (March 2017)

16. Huang, C., Trabelsi, A., Qin, X., Farruque, N., Mou, L., Zaïane, O.: Seq2Emo: a sequence to multi-label emotion classification model. In: Proceedings of the 2021 Conference of the North American Chapter of the Association for Computational Linguistics: Human Language Technologies, Online, pp. 4717–4724. Association for Computational Linguistics (2021). https://doi.org/10.18653/v1/2021.naacl-main.375. https://www.aclweb.org/anthology/2021.naacl-main.375

17. Jabreel, M., Moreno, A.: A deep learning-based approach for multi-label emotion classification in tweets. Appl. Sci. **9**(6), 1123 (2019)

18. Kelly, R., Watts, L.: Characterising the inventive appropriation of emoji as relationally meaningful in mediated close personal relationships. In: Experiences of Technology Appropriation: Unanticipated Users, Usage, Circumstances, and Design, vol. 2 (2015)

19. LeCompte, T., Chen, J.: Sentiment analysis of tweets including emoji data. In: 2017 International Conference on Computational Science and Computational Intelligence (CSCI), pp. 793–798. IEEE (2017)

20. Ljubešić, N., Fišer, D.: A global analysis of emoji usage. In: Proceedings of the 10th Web as Corpus Workshop, pp. 82–89 (2016)

21. Meisheri, H., Dey, L.: TCS research at SemEval-2018 Task 1: learning robust representations using multi-attention architecture. In: Proceedings of the 12th International Workshop on Semantic Evaluation, New Orleans, Louisiana, pp. 291–299. Association for Computational Linguistics (2018). https://doi.org/10.18653/v1/S18-1043. http://aclweb.org/anthology/S18-1043

22. Miller, H., Kluver, D., Thebault-Spieker, J., Terveen, L., Hecht, B.: Understanding emoji ambiguity in context: the role of text in emoji-related miscommunication. In: Proceedings of the International AAAI Conference on Web and Social Media, vol. 11 (2017)

23. Mohammad, S., Bravo-Marquez, F., Salameh, M., Kiritchenko, S.: SemEval-2018 Task 1: affect in tweets. In: Proceedings of the 12th International Workshop on Semantic Evaluation, New Orleans, Louisiana, pp. 1–17. Association for Computational Linguistics (2018). https://doi.org/10.18653/v1/S18-1001. http://aclweb.org/anthology/S18-1001

24. Mohammad, S., Turney, P.: Emotions evoked by common words and phrases: using mechanical turk to create an emotion lexicon. In: Proceedings of the NAACL HLT 2010 Workshop on Computational Approaches to Analysis and Generation of Emotion in Text, Los Angeles, CA, pp. 26–34. Association for Computational Linguistics (June 2010). https://www.aclweb.org/anthology/W10-0204

25. Nielsen, F.Å.: A new ANEW: evaluation of a word list for sentiment analysis in microblogs. arXiv preprint arXiv:1103.2903 (2011)

26. Park, J.H., Xu, P., Fung, P.: PlusEmo2Vec at SemEval-2018 Task 1: exploiting emotion knowledge from emoji and #hashtags. arXiv arXiv:1804.08280 [cs] (April 2018)

27. Pennington, J., Socher, R., Manning, C.: GloVe: global Vectors for Word Representation. In: Proceedings of the 2014 Conference on Empirical Methods in Natural Language Processing (EMNLP), Doha, Qatar, pp. 1532–1543. Association for Computational Linguistics (2014). https://doi.org/10.3115/v1/D14-1162. http://aclweb.org/anthology/D14-1162

28. Rogers, D.J., Tanimoto, T.T.: A computer program for classifying plants. science (1960). https://www.science.org/doi/abs/10.1126/science.132.3434.1115

29. Sebastiani, F., Esuli, A.: SENTIWORDNET: a publicly available lexical resource for opinion mining. In: Proceedings of the 5th International Conference on Language Resources and Evaluation, pp. 417–422 (2006)
30. Udochukwu, O., He, Y.: A rule-based approach to implicit emotion detection in text. In: Biemann, C., Handschuh, S., Freitas, A., Meziane, F., Métais, E. (eds.) Natural Language Processing and Information Systems. Lecture Notes in Computer Science, vol. 9103, pp. 197–203. Springer, Cham (2015). https://doi. org/10.1007/978-3-319-19581-0_17. http://link.springer.com/10.1007/978-3-319-19581-0_17

Community Detection in Attributed Networks via Kernel-Based Effective Resistance and Attribute Similarity

Clara Pizzuti⬡ and Annalisa Socievole(✉)⬡

National Research Council of Italy (CNR), Institute for High Performance
Computing and Networking (ICAR), Via Pietro Bucci, 8-9C, 87036 Rende, CS, Italy
{clara.pizzuti,annalisa.socievole}@icar.cnr.it

Abstract. In attributed graphs, community detection methods group
nodes by considering structural closeness and attribute similarity. In this
work, we investigate the simultaneous use of kernels on the adjacency
matrix of the graph and the node attribute similarity matrix. First,
we weight the input adjacency matrix of the graph with the *effective
resistance* between nodes, a Euclidean distance metric derived from the
field of electric circuits that takes into account all the alternative paths
between two nodes. Then we apply kernels for computing the similarity
between nodes both in terms of structure and attributes. Simulations on
synthetic networks show that kernels effectively improve the quality of
the obtained partitions that better fit with the ground-truth.

Keywords: Attributed networks · Community detection · Effective
resistance · Kernel on graphs

1 Introduction

Real-world complex networks such as social, telecommunication, biological,
medicine, or economy networks are usually characterized by a *structure* representing the interconnections between the nodes, and *attributes* describing the
features the nodes or the edges have [5]. In a mobile social network, for example,
each mobile user is connected through Bluetooth or Wi-Fi links to other users,
and is characterized by a profile specifying attributes like user ID, hometown,
checkins, interests, and so on. Such kind of networks graphs where a *structural*
and a *compositional* dimension coexist are referred to as *attributed networks* [5].

For clustering these networks, both the structural and the compositional
dimensions are usually considered since the analysis of the two aspects provides
a deeper understanding of the underlying communities [5]. Two approaches are
mainly exploited to balance structure and attributes. The community detection algorithms belonging to the first category optimize a single objective that
combines the structural and the compositional quality functions, which usually
correspond to high intra-community (a) edge density and (b) attribute node similarity, respectively. The other approach tries to jointly optimize the two functions. This work focuses on *node*-attributed graphs, where attributes are only on

© The Author(s), under exclusive license to Springer Nature Switzerland AG 2022
C. Strauss et al. (Eds.): DEXA 2022, LNCS 13426, pp. 367–372, 2022.
https://doi.org/10.1007/978-3-031-12423-5_28

nodes, and on the latter strategy in order to avoid one dimension would prevail over the other one during the computation. In some datasets, for example, there may be highly distant nodes with similar attributes where a single-objective method would produce sparse and eventually unconnected communities. On the contrary, the joint optimization of the two objectives would allow a simultaneous evaluation of both perspectives resulting in a proper weight of structure and attributes. To this end, *MultiObjective Evolutionary Algorithms* (MOEAs) can offer a valid solution to the community detection problem when competing objectives are present [14].

Intra-community edge density and attribute homogeneity are typically computed through *similarity* or dissimilarity (i.e. distance) measures able to quantify the closeness between nodes both from a structural and a compositional point of view. The *effective resistance* [11], for example, a measure derived from the electrical network theory, has shown to be an effective distance measure to weight the original graph by improving the solutions found by the community detection methods [9,13]. The distance between two nodes in a graph has been classically measured as the length of the *shortest path* between them. Differently from this type of distance, however, the effective resistance includes more information since it considers every path between any two nodes, even longer than the shortest one, weighting them as parallel paths. Although the effective resistance distance has been widely used for network analysis showing interesting graph aspects, its use for community detection is still limited.

When measuring similarity on graphs, *kernels* functions [1] are often used since they are able to transform the input data into a form facilitating data analysis. Given a graph G, kernels produce a symmetric positive semidefinite matrix K, which is a Gram matrix of some vectors in a Euclidean space quantifying the closeness between vertices in a meaningful manner. It follows that distance measures derived from a kernel are Euclidean. Different kernels have been defined so far, the most used in clustering tasks include *Communicability*, *Heat*, *Personalized Page Rank*, *Free Energy*. They differ in the input matrix on which the kernel is applied, the adjacency matrix A, the Laplacian L, the Markov matrix P, and the parameters used. Usually, the choice of a particular kernel is application dependent.

Kernel-based clustering has been mainly applied to networks without attributes showing its ability in obtaining better partitions [2,16]. More recently, this strategy has been extended to attributed networks and evaluated in [3]. Here, the adjacency matrix of the graph is updated by summing attribute similarity to the existing edge weights. Hence, network structure and attribute similarity are combined. Then, a set of top-performing kernels are applied to this matrix and used to cluster nodes with two classic clustering algorithms. The two main drawbacks of such methodology, however, are (1) how to weight structure and attributes by choosing proper weight coefficients and (2) the supervised nature of the clustering methods adopted where the resulting number of communities needs to be specified in advance.

In this work, we study the problem of community in attributed networks by exploiting kernels as in [3], but in a way that overcomes the outlined drawbacks. In particular, we consider an unsupervised community detection method for attributed networks, namely *MOGA@Net* [15], a multi-objective optimization method that jointly optimizes the structural quality of the communities as first objective, and the attribute similarity as the second objective, and we investigate two novel aspects. We study the effect of kernel similarity on *MOGA@Net* by weighting the adjacency matrix A of the network with the effective resistance between nodes and using kernels for measuring node similarity. In such a way, we aim to combine the ability of effective resistance with that of kernels in detecting better communities. More specifically, both objective functions, the one dealing with the structure and based on effective resistance, and the one dealing with attributes are evaluated on similarity matrices derived by applying kernels. We call this approach KERGA-@Net, *Kernel and Effective Resistance based bi-objective Genetic Algorithm for @tributed Networks*. Experimenting on synthetically generated attributed networks, we demonstrate that by applying the Free Energy kernel, which has been shown to be one the most effective kernels [3], our unsupervised method is able to outperform other benchmark community detection schemes.

2   KERGA-@Net

KERGA-@Net is built on our previous work MOGA-@Net [15], a MOEA able to uncover communities over attributed networks by contemporarily optimizing two functions: (1) the *structural quality of the partition* f_S and (2) the *intra-community homogeneity of node attributes* f_T. MOGA-@Net works on two input matrices: the adjacency matrix A of the graph G representing the network, containing the edges between node couples (i,j), and the similarity matrix sim_T, containing for each element (i, j) the similarity in terms of attributes between the nodes i and j according to a given metric. For each tested solution (i.e. a particular partitioning), the method computes f_S, by using the information on edges contained in A, and f_T by exploiting the similarity between nodes contained in sim_T. More specifically, MOGA-@Net evolves a population of possible solutions satisfying the two objectives f_S and f_T for a fixed number of generations and finally chooses a candidate solution using the framework of Pareto optimality [7]. To represent the solutions, the *locus-based adjacency representation* [12] is used, while as genetic operators, the *uniform crossover* and *random mutation* are adopted.

Differently from MOGA-@Net, KERGA-@Net weighs edges through effective resistance and improves both attribute and structural similarity computation through kernels. In particular, the method computes the effective resistance matrix Ω from A, containing for each element (i,j) the effective resistance distance between the two nodes, and sim_T from the attribute data. Since Ω is a distance matrix and basically, a dissimilarity matrix, it is turned into a similarity matrix and each value is normalized in the range $[0, 1]$. Over these two resulting similarity matrices, a kernel k_{id} is chosen and applied in order to obtain the kernel-based matrices K_S and K_T. Until a termination condition is not satisfied (i.e. the computed solution is not better than the previous one or the maximum number of generations is achieved), the method jointly evaluates the two objective functions f_S and f_T computed on K_S and K_T, respectively, over an individual and assigns a rank to the solution based on Pareto dominance. Once the termination condition is satisfied, a solution is chosen from the Pareto front by taking the one with the highest structural quality function.

3   Results

We performed a number of experiments on synthetic attributed networks. KERGA-@Net has been written in MATLAB R2020a, by using the Global Optimization Toolbox, which implements the NSGA-II framework of Deb et al. [6].

For generating the networks we used the LFR-EA benchmark proposed by Elhadi and Agam [8]. The generator creates community networks by using the *mixing parameter* μ and the noise parameter ν: the first determines the rate of intra- and inter- communities edges while the second manages the noise of attributes (i.e. the degree of common features between nodes). More specifically, low values of the mixing parameter and of the attribute noise produce graphs with a dense and clear community structure with similar attributes between the nodes belonging to the same community. Viceversa, high values of μ and ν generate networks with a confused community organization with not dense communities and different features between nodes.

For our experiments, we generated a set of networks with 1000 nodes and two numerical attributes with the same parameters used in [15], by varying the mixing parameter in the range $[0.1; 0.9]$ and the attribute noise in the range $[0; 0.9]$. For each network type, we generated 10 network samples for a given couple (μ, ν) for a total of 100 synthetic LFR-EA networks for each mixing parameter value. Overall, we tested our method on 900 synthetic networks.

To assess the effectiveness of our approach, we selected the Free Energy kernel, and then compared KERGA-@Net to 4 community detection methods: the Louvain method [4] which works only on the network structure and does not consider the attributes, the attribute-only method k-means [10], the multi-objective community detection method for attributed networks MOGA-@Net which integrates attributes and network structure, and the kernel-based spectral community detection method by Aynulin and Chebotarev [3] also designed for attributed networks. As performance metric, we computed the Cumulative NMI (CNMI) [8], which allows the integration of the well-known Normalized Mutual Information (NMI) over different settings of the structure mixing parameter μ and attribute noise ν as

$$CNMI = \frac{\sum^\mu \sum^\nu NMI}{S} \tag{1}$$

where S is the number of samples of network graphs considered. According to the definition of these quality indexes, the larger they are, the better the found communities match the ground truth. If the NMI or the CNMI value is 1, the detected communities are equal to the real ones.

For KERGA-@Net and MOGA-@Net, the population size and the number of generations have been set to 100 and 100, respectively, mutation rate to 0.4 and crossover fraction to 0.8. In addition, we considered *weighted modularity* as structure fitness function f_S, and similarity based on *Euclidean distance* for the attribute fitness function f_T since the attributes are numerical. For the kernel-based algorithm by Aynulin and Chebotarev, we equally weighted structure and attributes by setting $\beta = 0.5$ and considered Free Energy as suggested by their work [3]. Table 1 shows the results obtained by KERGA-@Net and the other algorithms. Our method has the best performance on all the datasets. The second best is the other multi-objective algorithm MOGA-@Net: this result demonstrates our initial intuition that kernels together with the effective resistance are useful for improving the clustering quality. The kernel-based method by Aynulin and Chebotarev achieves lower CNMI thus showing the limit of combining structure and attribute similarity metrics. Finally, the results achieved by Louvain and k-means confirm that structure-only or attribute-only methods are not enough for properly detecting communities in attributed networks, when networks are synthetically generated.

Table 1. Methods comparison in terms of CNMI.

Community detection	Dataset
Louvain	0.6993
k-means	0.3544
MOGA-@Net	0.8783
Aynulin & Chebotarev	0.8253
KERGA-@Net	**0.9109**

4   Conclusions

We studied the role of effective graph resistance and kernel functions on MOGA-@Net, a multiobjective genetic community detection algorithm for attributed networks that contemporarily optimizes two objectives: the structural quality and the attribute homogeneity of the communities. The input network was initially weighted through the effective resistance. The first objective was hence measured in terms of densely connected communities over the effective resistance matrix where a kernel was also applied. The attribute homogeneity was measured through a classic similarity metric and transformed with the same kernel adopted for the structure. Over synthetically generated networks, we found that the Free Energy kernel is able to improve the quality of the communities compared to other state-of-the-art methods. In future work, we are planning to apply mathematical functions (i.e. transformations) to each element of the kernel matrices.

References

1. Avrachenkov, K., Chebotarev, P., Rubanov, D.: Similarities on graphs: Kernels versus proximity measures. Eur. J. Comb. **80**, 47–56 (2019)
2. Aynulin, R.: Impact of network topology on efficiency of proximity measures for community detection. In: Cherifi, H., Gaito, S., Mendes, J.F., Moro, E., Rocha, L.M. (eds.) COMPLEX NETWORKS 2019. SCI, vol. 881, pp. 188–197. Springer, Cham (2020). https://doi.org/10.1007/978-3-030-36687-2_16
3. Aynulin, R., Chebotarev, P.: Measuring proximity in attributed networks for community detection. In: Benito, R.M., Cherifi, C., Cherifi, H., Moro, E., Rocha, L.M., Sales-Pardo, M. (eds.) Complex Networks & Their Applications IX. COMPLEX NETWORKS 2020. Studies in Computational Intelligence, vol. 943, pp. 27–37. Springer, Cham (2020). https://doi.org/10.1007/978-3-030-65347-7_3
4. Blondel, V.D., Guillaume, J.L., Lambiotte, R., Lefebvre, E.: Fast unfolding of communities in large networks. J. Stat. Mech: Theory Exp. **2008**(10), P10008 (2008)
5. Bothorel, C., Cruz, J.D., Magnani, M., Micenkova, B.: Clustering attributed graphs: models, measures and methods. Netw. Sci. **3**(3), 408–444 (2015)
6. Deb, K., Pratap, A., Agarwal, S., Meyarivan, T.: A fast and elitist multiobjective genetic algorithm: NSGA-II. IEEE Trans. Evol. Comput. **6**(2), 182–197 (2002)
7. Ehrgott, M.: Multicriteria Optimization, vol. 491. Springer, New York (2005). https://doi.org/10.1007/3-540-27659-9
8. Elhadi, H., Agam, G.: Structure and attributes community detection: comparative analysis of composite, ensemble and selection methods. In: Proceedings of the 7th Workshop on Social Network Mining and Analysis, pp. 1–7 (2013)
9. Gancio, J., Rubido, N.: Community detection by resistance distance: automation and benchmark testing. In: Benito, R.M., Cherifi, C., Cherifi, H., Moro, E., Rocha, L.M., Sales-Pardo, M. (eds.) Complex Networks & Their Applications X. COMPLEX NETWORKS 2021. Studies in Computational Intelligence, vol. 1015, pp. 309–320. Springer, Cham (2021). https://doi.org/10.1007/978-3-030-93409-5_26
10. Hartigan, J.A., Wong, M.A.: Algorithm as 136: a k-means clustering algorithm. J. R. Stat. Soc. Ser. C (Appl. Stat.) **28**(1), 100–108 (1979)
11. Klein, D.J., Randić, M.: Resistance distance. J. Math. Chem. **12**(1), 81–95 (1993)
12. Park, Y., Song, M.: A genetic algorithm for clustering problems. In: Proceedings of the Third Annual Conference on Genetic Programming, vol. 1998, pp. 568–575 (1998)
13. Pizzuti, C., Socievole, A.: An effective resistance based genetic algorithm for community detection. In: Proceedings of the 13th International Conference on Evolutionary Computation Theory and Applications - ECTA, pp. 28–36. INSTICC, SciTePress (2021)
14. Pizzuti, C., Socievole, A.: A genetic algorithm for community detection in attributed graphs. In: Sim, K., Kaufmann, P. (eds.) EvoApplications 2018. LNCS, vol. 10784, pp. 159–170. Springer, Cham (2018). https://doi.org/10.1007/978-3-319-77538-8_12
15. Pizzuti, C., Socievole, A.: Multiobjective optimization and local merge for clustering attributed graphs. IEEE Trans. Cybern. **50**(12), 4997–5009 (2019)
16. Sommer, F., Fouss, F., Saerens, M.: Comparison of graph node distances on clustering tasks. In: Villa, A.E.P., Masulli, P., Pons Rivero, A.J. (eds.) ICANN 2016. LNCS, vol. 9886, pp. 192–201. Springer, Cham (2016). https://doi.org/10.1007/978-3-319-44778-0_23

Self-supervised Learning for Building Damage Assessment from Large-Scale xBD Satellite Imagery Benchmark Datasets

Zaishuo Xia[1], Zelin Li[1], Yanbing Bai[1(✉)], Jinze Yu[2], and Bruno Adriano[3]

[1] Center for Applied Statistics, School of Statistics, Renmin University of China, Beijing, China
ybbai@ruc.edu.cn
[2] Department of Artificial Intelligent, Connected Robotics Inc., Tokyo, Japan
[3] RIKEN Center for Advanced Intelligence Project (AIP), Tokyo, Japan

Abstract. In the field of post-disaster assessment, for timely and accurate rescue and localization after a disaster, people need to know the location of damaged buildings. In deep learning, some scholars have proposed methods to make automatic and highly accurate building damage assessments by remote sensing images, which are proved to be more efficient than assessment by domain experts. However, due to the lack of a large amount of labeled data, these kinds of tasks can suffer from being able to do an accurate assessment, as the efficiency of deep learning models relies highly on labeled data. Although existing semi-supervised and unsupervised studies have made breakthroughs in this area, none of them has completely solved this problem. Therefore, we propose adopting a self-supervised comparative learning approach to address the task without the requirement of labeled data. We constructed a novel asymmetric twin network architecture and tested its performance on the xBD dataset. Experiment results of our model show the improvement compared to baseline and commonly used methods. We also demonstrated the potential of self-supervised methods for building damage recognition awareness.

Keywords: Self-supervised learning · Building damage assessment · Satellite imagery · xBD dataset

1 Introduction

When disasters, such as hurricanes, tsunamis, and earthquakes occur, people need to know the locations of affected residents, which is crucial for rescue and localization. Easily accessible remote sensing images enable us to identify building damage more accurately and to locate damaged buildings easier. However, this method was based on comparing images before and after the disaster manually by experts, which was time-consuming.

© The Author(s), under exclusive license to Springer Nature Switzerland AG 2022
C. Strauss et al. (Eds.): DEXA 2022, LNCS 13426, pp. 373–386, 2022.
https://doi.org/10.1007/978-3-031-12423-5_29

The introduction of deep learning methods gives another direction for automatic damage assessment, where models can classify building damage after training on labeled pre-and post-disaster remote sensing images [1–5]. Some scholars take the difficulty of obtaining large amounts of labeled data of disaster into consideration, and disaster types and datasets require different labeling methods. They adopted a semi-supervised or unsupervised approach, using only small amounts of labeled or/and unlabeled data for research [6–10].

However, building an accurate model for new disasters is still far from enough. First, existing semi-supervised learning requires a certain amount of labeled data and does not fully address the problem of the complex labeling of disaster datasets [6–8]. Second, current unsupervised learning generally relies on pre-training on labeled data [9,10], which has not good enough for portability. Third, the interpretability of deep learning is poor, so it cannot be reliably used in our daily life.

To solve these problems, we generate supervision from entirely unlabeled datasets to clarify the features in a self-supervised setting. Also, it turns out that the new self-supervised learning using entirely unlabeled datasets can achieve similar results as supervised learning [16,19]. In the pre-training stage, a dataset containing images before and after the disaster is used, and both types of images are passed through the model to show the implied features. The model has two main parts, using contrast learning and image reconstruction which learn global information as well as local information, respectively. In the second stage, we use the weights of the pre-trained model to pass the images before and after the disaster through our model, and then join them together and pass them through a semantic segmentation head to compare with the labeled images. Finally, the effect of our model is reflected by the $F1$-score. The main contributions of this paper are as follows.

1 We propose a novel model for building damage assessment. The model utilizes the transformer encoder structure internally, which has an excellent performance in remote sensing images. This structure is applied to build damage assessment and has achieved ideal results.

2 The proposed model does not require labeled data. It can learn the features of images from unlabeled data only using a contrast learning self-supervised approach. Without labeled data, our result is very close to labels.

2 Related Work

Deep Learning Achievements in Building Damage Recognition. Gupta et al. [11] provided a labeled dataset including nineteen different events and more than 20,000 images, which is one of the largest and highest-quality datasets. Supervised methods have yielded excellent results with this boost [1–5].

Supervised methods require labeled images, but the disaster domain has less labeled data than the traditional application domain of deep learning. In addition, supervised learning relies on manual labeling for training, which often leads to costly models. Lee et al. [6] used two semi-supervised methods, MixMatch and

FixMatch, to train models by fusing features from labeled and unlabeled data (containing damaged and undamaged areas), obtaining good results. Later, Xia et al. [7] used less labeled data and trained the model with only a tiny number of positive examples (images of damaged areas) and the rest unlabeled, achieving results comparable to supervised learning by combining positive examples and unlabeled data. Ghosh et al. [8] employed a novel two-part graph neural network-based framework with only a small amount of labeled data.

However, semi-supervised methods do not fully address the problem of complex labeling; moreover, most of them have demanding requirements on labeled data. Therefore, some scholars have averted their eyes to unsupervised methods. Li et al. [9] used post-hurricane disaster data and transfer learning. They first completed pre-training on the labeled source domain, aligning the features in the source and target domains by maximum mean discrepancy (MMD), and later transferred to two new types of hurricane data to complete the classification. They [10] later proposed a new generative adversarial network to align the source and target domains in an unsupervised approach to achieve better results on both transferred tasks.

The unsupervised methods above require labeled data for pre-training and later transfer to new tasks, making the model more complex and possibly requiring repeated pre-training for different tasks, which is time-consuming and labor-intensive. Akiva et al. [21] used a self-supervised approach to identify flood trajectories, and their method outperformed current semantic segmentation methods for the flood trajectory segmentation task, demonstrating that self-supervised learning can perform well in disaster identification.

In addition, Chowdhury et al. [12] proposed a high-resolution UAV dataset HRUD, incorporating images after Hurricane Michael, on which they wished to evaluate the performance of semantic segmentation models. They [13] later gave ReDNet, which used a self-attention mechanism to achieve high accuracy on HRUD, unveiling the potential of the self-attention mechanism.

Development of Self-supervised Learning. Self-supervised learning is a category of machine learning that requires no labeled data or any pre-training on labeled data to learn features. An early self-supervised learning training model, by making features as close as possible between positive samples and as far as possible between negative samples [14,15], is known as contrastive learning. However, such methods need to compare features between a large number of images and often require a large batch_size [14] or memory_bank [15]. The BYOL [16] model proposed later shows that self-supervised learning can work without differentiating the classes of images, i.e., without negative samples, but the model performance has decreased. The recently proposed DINO [17] model uses the Transformer encoder and BYOL-based structure and solves the problem of model instability.

Above research shows that self-supervised learning without any labeled data or pre-training can perform no less well than supervised learning.

Transformer Structure. The Transformer [18] structure used a self-attention mechanism and was first proposed by the Google team in 2017, bringing significant results and breakthroughs in NLP. Inspired by the Transformer structure, the Google team proposed Vision Transformer [19] and achieved better results on image datasets. The model of MoCo v3 [20] was the first to combine the ViT structure with self-supervised contrast learning. However, the instability generated by the combination of the two affected the results to a certain extent. DINO [17] applied a method to solve the instability problem.

The Transformer structure exceeds other methods in all areas and shows promising results even when confronted with instability problems by combining with contrast learning.

Inspired by the research mentioned, we conducted this study on a dataset containing pre- and post-disaster images. For the first time in the field, we employed a contrastive learning approach that combines a self-supervised method with a Transformer structure, yielding positive results.

3   Data Description

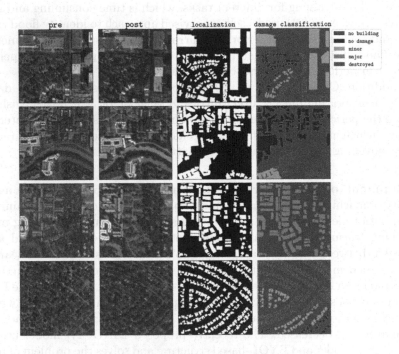

Fig. 1. Examples of xBD dataset

Inspired by the literature mentioned above, we conducted our study on a dataset containing pre-and post-disaster images. For the first time in the field, we employ a contrastive learning approach that combines a self-supervised method with a Transformer structure. Combining these methods, our model yields positive results.

We conduct experiments on the xBD dataset, the largest and highest-quality satellite remote sensing image dataset for natural disasters. It contains 22,068 remote sensing images of 19 different disaster types, such as earthquakes, floods, wildfires, volcanic eruptions, and car accidents. Since there are pre-and post-disaster remote sensing images, they can be used to construct the tasks of localization and damage classification. The dataset comes from a 5,000 km^2 area in 15 countries with high-resolution images. Two annotations are provided for a pair of images(pre and post), respectively, for localization problem and damage classification, as shown in the Fig. 1.

4   Model

4.1   Overview

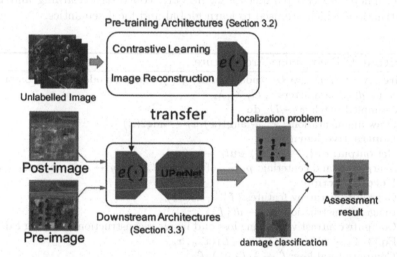

Fig. 2. Overview of the proposed model. The first stage utilizes pre-training architecture, while the second stage utilizes downstream architecture. Localization problem and damage classification are solved uniformly in the second stage.

Problem Definition: Building damage classification usually consists of two subtasks: localization and damage classification. For the former one, we classified each pixel in the pre-disaster image as "building" or "no building". When it comes to damage classification, for the corresponding pixels in the post-disaster image, we assign values from 0 to 4 depending on the damage degree, where 0 means "no building", 1 as "no damage", 2 as "minor damage", 3 as "major damage", and 4 as "being destroyed".

The whole process consists of two main stages. In the first stage or pre-training stage, we pre-train encoder $e(\cdot)$ on a large amount of unlabeled data using a self-supervised contrastive learning approach. Then we transfer the pre-training encoder $e(\cdot)$ to the downstream task for building damage classification (localization problem and damage classification) during the second stage.

The two stages are summarized in detail in the Fig. 2. Localization problem and damage classification are solved uniformly in the second stage. We show the architecture of the two stages below in detail.

4.2   Pre-training Architecture

A self-supervised contrastive learning framework was used to pre-train encoder $e(\cdot)$ [17] a novel style with no negative samples (DINO), which achieved good performance in transfer learning. However, DINO was designed based on image classification and could not learn the location information of remote sensing images well. With reference to [22], we improved Dino so that it can be suitable for remote sensing images. Figure 3 shows an overview of our pre-training architecture. There are two parallel tasks, namely, contrastive learning and image reconstruction, which are used to learn global and local semantics.

Algorithm 1. Pre-training Architecture

Require: A set of images D; student model $g_s(\cdot)$, teacher model $g_t(\cdot)$, reconstruction decoder $d(\cdot)$; parameters τ, C, λ_1, λ_2

1: **for** sampled batch $x \leftarrow D$ **do**
2:    Draw augmentations: $x_s \leftarrow \text{aug1}(x), x_t \leftarrow \text{aug2}(x)$
3:    **Contrastive learning:**
4:    Get output: $out_s \leftarrow g_s(x_s), out_t \leftarrow g_t(x_t)$
5:    Sharpening and centering: $p_s \leftarrow \text{softmax}(out_s/\tau), p_t \leftarrow \text{softmax}(out_t - C)$
6:    **Reconstruction:**
7:    Get multi-hierarchy features:$\{f\} \leftarrow e_s(x_s)$
8:    Image reconstruction:$x_{re} \leftarrow d(\{f\})$
9:    Compute contrastive learning loss and image reconstruction loss using Eq. 2 and Eq. 3: $\mathcal{L}_1 \leftarrow CE(p_s, p_t), \mathcal{L}_2 \leftarrow L1(x_{re}, x_s)$
10:    Compute total loss: $\mathcal{L} \leftarrow \lambda_1 \mathcal{L}_1 + \lambda_2 \mathcal{L}_2$
11:    Update student model $g_s(\cdot), \lambda_1, \lambda_2$ to minimize $\mathcal{L}$
12:    Update teacher model $g_t(\cdot)$ and C by EMA
13: **end for**

4.2.1   Contrastive Learning

Contrastive learning is used to learn the global semantics of images, which is instrumental for coarse-grain task. Asymmetric twin network architecture was used to learn two neural networks simultaneously, called the teacher network $g_t(\cdot)$ and the student network $g_s(\cdot)$. The network architecture is shown in the upper part of the Fig. 3. It is widely used in contrastive learning [16,17]. Both

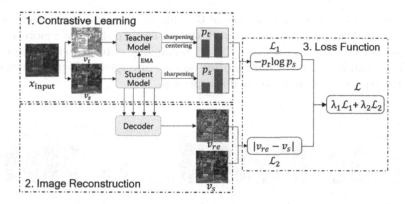

Fig. 3. Pre-training architecture. Contrastive learning is displayed in the upper part, while image reconstruction is displayed in the lower part.

the teacher $g_t(\cdot)$ and the student network $g_s(\cdot)$ consists of transformer encoder $e(\cdot)$ and projector, both of which are initialized with same given weights.

Suppose the set of images is labeled as D. An image $x \in D$ is randomly selected, with different image augmentation transforms aug1, aug2 to result in v_s and v_t which are different views of x. We input v_s and v_t into the student network $g_s(\cdot)$ and teacher network $g_t(\cdot)$ respectively, and the two outputs out_s and out_t are returned via encoder $e(\cdot)$ and projector. The corresponding probability distributions p_s and p_t are obtained by applying a softmax function to the outputs. The views of any randomly augmented version of a sample image should have similar feature representations. The cross-entropy loss between p_s and p_t is calculated for updating the parameters of the student network θ_s.

The asymmetry of twin networks is mainly presented in the following two points.

- **The two networks utilize different parameter updating methods.** The parameters of the student network θ_s were updated by minimizing the loss function $\mathcal{L}$, while those of the teacher network θ_t by exponential moving average (EMA). Specifically, the teacher network parameter θ_t update rule is $\theta_t \leftarrow \lambda\theta_t + (1-\lambda)\theta_s$, where the hyperparameter λ follows the cosine schedule. Referring to [17], such a momentum encoder update can result in better convergence of the model.
- **Sharpening and centering.** To prevent the model from collapsing, two operations are added to the model output, sharpening and centering. However, only sharpening is applied to the student network $g_s(\cdot)$ and both operations are applied to the teacher network $g_t(\cdot)$. Both operations take place for the softmax function. Sharpening is achieved by adding a temperature parameter τ to the softmax function, as shown in Eq. 1.

$$P(x) = \frac{\exp\left(g_\theta(x)/\tau\right)}{\sum_{k=1}^{K} \exp\left(g_\theta(x)^{(k)}/\tau\right)} \tag{1}$$

Sharpening enhances the variance of the softmax output. As for centering, it is only applied to the teacher network where the output is made more average by subtracting the vector $C \in \mathbb{R}^{out}$. By balancing centering with sharpening, collapse can be avoided while ensuring model convergence. Figure 4 shows the feature space after centering and sharpening.

Fig. 4. Example of sharpening and centering. A simulated feature space has been used for an intuitive representation in low dimensions.

4.2.2   Image Reconstruction

Compared with natural images, remote sensing images variate in scale and are small in geographical size. Image reconstruction is introduced to learn better the local semantics in remote sensing images concerning [22]. As shown in the lower part of Fig. 3, the reconstruction of the image is carried out by a lightweight decoder $d(\cdot)$ using the multi-layer feature maps $\{f\}$ output by encoder $e(\cdot)$. Since the student network $g_s(\cdot)$ retains the gradient backpropagation, the encoder of the student network $e_s(\cdot)$ is used in this task. For details, the input 3-channel RGB images x are fed to the encoder of the student network $e_s(\cdot)$, and the output multi-level sequence features are fed to a lightweight decoder $d(\cdot)$. The lightweight decoder $d(\cdot)$ learns to recover the original image x from the multi-layer feature maps $\{f\}$ by multi-stage feature fusion.

- **Multi-layer feature maps $\{f\}$.** The output of each stage of Swin Transformer is selected as multi-layer feature maps $\{f\}$. Swin Transformer[23] is designed with reference to the layered feature representation of convolutional neural networks, where the whole model is grouped into different layers, with each downsampling the feature maps output from the previous layer, in which the layer features are calculated by moving windows. Therefore, multi-layer and multi-scale, the feature maps output by each stage of Swin Transformer have different resolutions.
- **Lightweight decoder $d(\cdot)$.** A lightweight decoder $d(\cdot)$ is used to recover the original image x from multi-layer feature maps $\{f\}$, inspired by [22]. It is necessary to ensure that the decoder $d(\cdot)$ is lightweight, so that not only the computation can be reduced, but also the encoder $e(\cdot)$ can be better trained. The decoder $d(\cdot)$ we use consists of several fusion layers, each containing a 3×3 convolutional layer and a ReLU layer. The number of fusion layers corresponds to the number of feature maps, and the final output is restored to the same resolution as the original image x by a 1×1 convolutional layer.

4.2.3   Loss Function $\mathcal{L}$

Corresponding to contrastive learning and image reconstruction, our total loss function $\mathcal{L}$ consists of two parts. For contrastive learning, the loss of the probability distribution of the student network output p_s and of the teacher network output p_t is calculated by cross-entropy loss (Eq. 2).

$$\mathcal{L}_1(p_s, p_t) := -p_t \log p_s \tag{2}$$

As for the image reconstruction, $L1$ loss function is used to calculate the loss of the original image x versus the decoder output x_{re}, as shown as Eq. 3.

$$\mathcal{L}_2(x, x_{\mathrm{re}}) := |x - x_{\mathrm{re}}| \tag{3}$$

Since our loss are applied to multiple tasks, we add learnable weights λ_1, λ_2 to the final loss function $\mathcal{L}$, which could be described as $\mathcal{L} := \lambda_1 \mathcal{L}_1 + \lambda_2 \mathcal{L}_2$, where λ_1, λ_2 are the learnable parameter and updated with gradient descent.

4.3   Architectures for Downstream Tasks

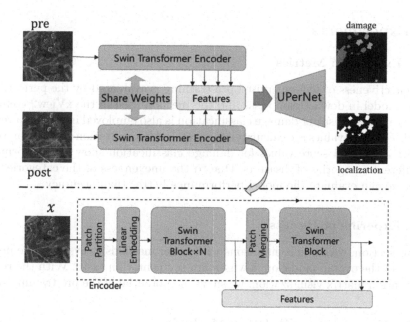

Fig. 5. Architectures for downstream tasks. This is an end-to-end model, using the common encoder-decoder architecture. The internal structure of the encoder is shown below the dotted line. Swin Transformer encoder is used.

In the second stage, our goal is to fine-tune the encoder, which has been pre-trained in the first stage, to be applied for building damage classification. For the downstream task, we use the Swin Transformer [23] pre-trained by contrastive learning as an encoder, with a semantic segmentation head, to handle localization and damage classification consistently, as shown in Fig. 5. Figure 5 also shows feature maps output from inside of the encoder. Regarding the multi-temporal characteristics of disaster images, our model is a siamese network. Pre- and post-disaster images are fed into two Swin Transformers, which share the weights. The output of each layer of the Swin Transformer is a 2D feature map. We concatenate the 2D feature maps corresponding to the pre-and post-disaster images along the feature dimension. These 2D feature maps are fed into the UPerNet [24] semantic segmentation head, which undergoes multi-layer feature fusion and upsampling to finally obtain a prediction map of the original image resolution size. There are 5 channels in the feature map, and the elements of different channels represent the probability that the corresponding pixel point is predicted to be of that class. For damage classification, the prediction map is applied to the argmax function along the channel dimension to obtain a 1-channel mask. The value of each pixel in the mask is the category with the highest probability in the prediction map. For localization, referring to [11], we classify the points with predicted values greater than or equal to 1 as buildings. In this way, we apply the model to both the localization and damage classification.

5   Experiments

5.1   Evaluation Metrics

The effectiveness of self-supervised pre-training is evaluated by the performance of the model in downstream tasks and the method used in the xView2 competition to evaluate disaster damage classification is also employed in this research. In detail, $F1$-score values are calculated separately for localization and damage classification, and $F1$-score values for damage classification show the performance in different categories of disasters. Due to the unevenness of the categories, the $F1$-score can evaluate the model performance better than accuracy.

5.2   Experiment Results

In this section, we made comparisons with other methods to evaluate the performance of the overall model on downstream segmentation tasks. With the results of attention matrix visualization, we show the potential of the pre-training stage.

5.2.1   Comparison with Other Methods

To evaluate the effectiveness of self-supervised contrast learning pre-training, we compared it with other pre-training methods (DINO [17], MoCo v3 [20]). Pre-training with ImageNet is a currently widespread initialization method, which was used for comparison. The entire training data is used for self-supervised pre-training.

- **$F1$-score under limited annotation.** For fine-tuning on the downstream task (the second stage), we controlled the amount of labeled data and fine-tuned using 1% and 20% of the labeled data, respectively. We used different initialization methods with the same downstream task architecture, controlled learning rate of 1e−4, and loss function of Dice-loss with CE-loss. The results are shown in Table 1. Our method has better performance in both cases.
- **10% amount of annotated data with training process.** We explored the performance of the pre-training model on limited labeled data by training it longer. Using 1% labeled data with a learning rate of 1e−4, we trained more epochs while fine-tuning (the second stage). Using our pre-training model, and ImageNet pre-training model with random initialization method respectively, 500 epochs have been tested to compare the training process. The results are shown in Fig. 6. It turns out that our model loss converges faster and has a higher $F1$-score.

Table 1. $F1$-score under limited labeled data. Nan means that the model has no valid output for that category, which is a result of insufficient samples.

Amount	Method	Localization	Damage	No damage	Minor	Major	Destroyed
1%	ImageNet	0.461	0.321	0.387	0.136	0.234	NaN
	DINO	0.522	0.366	0.480	0.157	0.439	NaN
	MoCo v3	0.550	0.379	0.425	0.124	0.337	NaN
	Ours	0.539	0.390	0.486	0.261	0.345	NaN
20%	ImageNet	0.661	0.587	0.604	0.278	0.471	0.456
	DINO	0.714	0.601	0.667	0.229	0.384	0.447
	MoCo v3	0.650	0.639	0.562	0.230	0.392	0.400
	Ours	0.678	0.636	0.646	0.314	0.480	0.380

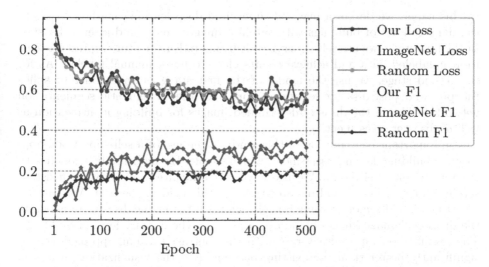

Fig. 6. 10% labeled data with training process. The line graph was eventually plotted as we recorded every 5 epochs.

5.2.2   Results of Attention Matrix Visualization

We visualized the attention matrix of encoder in the pre-training stage. Though the pre-training does not have labeled data involved, the visualization results of the attention matrix are rich in semantic information. Figure 7 shows the visualization results of the attention matrix in the transformer, where our model learns the approximation without any labeled information.

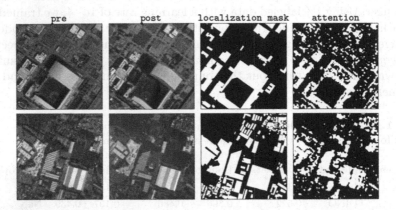

Fig. 7. Pre, post, mask, and attention matrix. The result is obtained from the first stage(pre-training), and we don't use any annotations in the first stage. Attention matrix, which was trained without labeled data, is very close to the man-made mask.

6   Conclusion

In this paper, we design a new self-supervised deep learning method to assess the damage level of buildings with satellite images before and after a disaster. In the first stage, we use self-supervised learning to learn feature representations from unlabeled data and achieve results close to mask manual labeled. As for the second stage, we use the model of the first stage as an encoder to splice the pre- and post-disaster features before and after a semantic segmentation network to obtain the disaster-determined images for building identification as well as disaster assessment.

An important contribution of this study is the use of a self-supervised app-roach in building damage assessment, which enables the model to be trained with less reliance on labeled data. We also solved the difficulties caused by inconsistent satellite image labeling and disaster types in this field in the past. We can use only the original remote sensing images before and after the disaster to generate the disaster damage images of an area and assess the damage level of buildings. The results show an accuracy rate with the contrast learning approach that is significantly better than the existing baseline, and the visualization results in the unlabeled case are close to the manual labeling. The self-supervised app-roach has been proved to have outperforming results and considerable potential for building damage assessment. In the future, we plan to explore the combination of self-supervised models deeper with building damage identification using multi-temporal information and improve the model architecture for better performance.

Acknowledgements. This research was supported by Public Health & Disease Control and Prevention, Major Innovation & Planning Interdisciplinary Platform for the "Double-First Class" Initiative, "Renmin University of China" (No. 2022PDPC), fund for building world-class universities (disciplines) of Renmin University of China. Project No. KYGJA2022001, fund for building world-class universities (disciplines) of Renmin University of China. Project No. KYGJF2021001, Beijing Golden Bridge Project seed fund (No. ZZ21021). This research was supported by Public Computing Cloud, Renmin University of China.

References

1. Rudner, T.G., et al.: Multi3Net: segmenting flooded buildings via fusion of multiresolution, multisensor, and multitemporal satellite imagery. Proc. AAAI Conf. Artif. Intell. **33**(01), 702–709 (2019)
2. Kyrkou, C., Theocharides, T.: EmergencyNet: efficient aerial image classification for drone-based emergency monitoring using atrous convolutional feature fusion. IEEE J. Sel. Topics Appl. Earth Observ. Remote Sens. **13**, 1687–1699 (2020)
3. Zheng, Z., Zhong, Y., Wang, J., Ma, A., Zhang, L.: Building damage assessment for rapid disaster response with a deep object-based semantic change detection framework: from natural disasters to man-made disasters. Remote Sens. Environ. **265**, 112636 (2021)
4. Gupta, R., Shah, M.: RescueNet: joint building segmentation and damage assessment from satellite imagery. In: 2020 25th International Conference on Pattern Recognition (ICPR). IEEE, pp. 4405–4411 (2021)
5. Adriano, B., et al.: Learning from multimodal and multitemporal earth observation data for building damage mapping. ISPRS J. Photogramm. Remote. Sens. **175**, 132–143 (2021)
6. Lee, J., et al.: Assessing post-disaster damage from satellite imagery using semi-supervised learning techniques. arXiv preprint arXiv:2011.14004 (2020)
7. Xia, J., Yokoya, N., Adriano, B.: Building damage mapping with self-positive unlabeled learning. arXiv preprint arXiv:2111.02586 (2021)
8. Ghosh, S., Maji, S., Desarkar, M.S.: Unsupervised domain adaptation with global and local graph neural networks in limited labeled data scenario: application to disaster management. arXiv preprint arXiv:2104.01436 (2021)
9. Li, Y., Hu, W., Li, H., Dong, H., Zhang, B., Tian, Q.: Aligning discriminative and representative features: an unsupervised domain adaptation method for building damage assessment. IEEE Trans. Image Process. **29**, 6110–6122 (2020)
10. Lin, C., Li, Y., Liu, Y., Wang, X., Geng, S.: Building damage assessment from post-hurricane imageries using unsupervised domain adaptation with enhanced feature discrimination. IEEE Trans. Geosci. Remote Sens. **60**, 1–10 (2021)
11. Gupta, R., et al.: XBD: a dataset for assessing building damage from satellite imagery. arXiv preprint arXiv:1911.09296 (2019)

12. Chowdhury, T., Rahnemoonfar, M., Murphy, R., Fernandes, O.: Comprehensive semantic segmentation on high resolution UAV imagery for natural disaster damage assessment. In: 2020 IEEE International Conference on Big Data (Big Data), pp. 3904–3913. IEEE (2020)

13. Chowdhury, T., Rahnemoonfar, M.: Attention based semantic segmentation on UAV dataset for natural disaster damage assessment. In: 2021 IEEE International Geoscience and Remote Sensing Symposium IGARSS, pp. 2325–2328. IEEE (2021)

14. Chen, T., Kornblith, S., Norouzi, M., Hinton, G.: A simple framework for contrastive learning of visual representations. In: International Conference on Machine Learning. PMLR, pp. 1597–1607 (2020)

15. He, K., Fan, H., Wu, Y., Xie, S., Girshick, R.: Momentum contrast for unsupervised visual representation learning. In: Proceedings of the IEEE/CVF Conference on Computer Vision and Pattern Recognition, pp. 9729–9738 (2020)

16. Grill, J.-B., et al.: Bootstrap your own latent-a new approach to self-supervised learning. In: Advances in Neural Information Processing Systems, vol. 33, pp. 21 271–21 284 (2020)

17. Caron, M., et al.: Emerging properties in self-supervised vision transformers. In: Proceedings of the IEEE/CVF International Conference on Computer Vision, pp. 9650–9660 (2021)

18. Vaswani, et al.: Attention is all you need. In: Advances in Neural Information Processing Systems, vol. 30 (2017)

19. Dosovitskiy, A., et al.: An image is worth 16 × 16 words: transformers for image recognition at scale. arXiv preprint arXiv:2010.11929 (2020)

20. Chen, X., Xie, S., He, K.: An empirical study of training self-supervised visual transformers. arXiv e-prints, pp. arXiv-2104 (2021)

21. Akiva, P., Purri, M., Dana, K., Tellman, B., Anderson, T.: H2O-Net: self-supervised flood segmentation via adversarial domain adaptation and label refinement. In: Proceedings of the IEEE/CVF Winter Conference on Applications of Computer Vision, pp. 111–122 (2021)

22. Wang, L., Liang, F., Li, Y., Ouyang, W., Zhang, H., Shao, J.: RePre: improving self-supervised vision transformer with reconstructive pre-training. arXiv preprint arXiv:2201.06857 (2022)

23. Liu, Z., et al.: Swin transformer: hierarchical vision transformer using shifted windows. In: Proceedings of the IEEE/CVF International Conference on Computer Vision, pp. 10 012–10 022 (2021)

24. Xiao, T., Liu, Y., Zhou, B., Jiang, Y., Sun, J.: Unified perceptual parsing for scene understanding. In: Proceedings of the European Conference on Computer Vision (ECCV), pp. 418–434 (2018)

Warehousing Methodologies

What Logical Model Is Suitable for Relational Trajectory Data Warehouses?

Application to Agricultural Autonomous Robots

Konstantinos Oikonomou[1], Georgia Garani[1], Sandro Bimonte[2]([⊠]),
and Robert Wrembel[3] [iD]

[1] University of Thessaly, Volos, Greece
garani@uth.gr
[2] Universitye Clermont Auvergne, TSCF, INRAE, Clermont-Ferrand, France
sandro.bimonte@inrae.fr
[3] Poznan University of Technology, Poznan, Poland
robert.wrembel@cs.put.poznan.pl

Abstract. The ability to query vast amount of historical data for statistical analysis and reporting is provided by Data Warehouses. They facilitate Business Intelligence for effective decision-making significantly. In recent years, great progress has been made in movement monitoring devices, such as smart phones and GPSs. The storing and managing of spatio-temporal data related to the trajectories of moving objects in a data warehouse is called Trajectory Data Warehouse (TDW). The relational approach is adopted widely for the logical representation of TDWs, since it is based on the classic database approach where data representation and processing are handled on structured data. In this paper, the key idea is to consider different logical relational TDW models, i.e. flat, segment and complex, which are compared and evaluated. The study is based on a novel classification of OLAP queries, the cardinality of facts and the resolution of each trajectory in segments. Real data provided by agricultural autonomous robots is used, where experiments on size and time performances are conducted and discussed.

Keywords: Data warehouse · Trajectory data · OLAP

1 Introduction

Data Warehouse (DW) and OLAP systems are first citizens of Business Intelligence tools [9]. Warehoused data are stored according to the multidimensional model, which organizes data into analysis subjects (i.e. facts), and analysis axes (i.e. dimensions). These data are explored and analyzed by means of OLAP systems that allow to navigate into warehoused data. Despite of the proliferation of new NoSQL Database Management Systems (DBMS), existing OLAP servers are

© The Author(s), under exclusive license to Springer Nature Switzerland AG 2022
C. Strauss et al. (Eds.): DEXA 2022, LNCS 13426, pp. 389–403, 2022.
https://doi.org/10.1007/978-3-031-12423-5_30

essentially based on the relational model, and more precisely on the classical *star schema* logical representation of the multidimensional model. Nowadays, in the Internet of Things era, DW and OLAP systems are confronted with trajectory data. Indeed, more and more data representing the movement in time and space of humans, animals, vehicles and objects are produced by means of new sensors, smartphones, and other connected objects equipped with GPS systems. Trajectory data is complex data. Commonly, a trajectory is defined as set of ordered tuples $(pi, ti, ai1, ..., ain)$ where: pi is of the form (x;y) representing geographical coordinates, ti is the timestamp, and $ai1, ..., ain$ are numerical attributes. The introduction of these data into DW has lead to the concept of *Trajectory DW* (TDW). Several works have been proposed in the last decade to study the conceptual, logical and physical design of TDWs [1,2]. According to [1], two main approaches have been proposed to store trajectory data: (i) *cell*, and (ii) *segment*. The cell approach divides the space into a regular grid, associates all points of the trajectory to a set of cells, and stores the numerical aggregated measures for each cell into the fact table of star schema. The segment approach stores the trajectory's start and end points with their timestamps as measures. These approaches provide an aggregated/approximated representation of the trajectory (i.e. by cell and segment). Thus, they are not well suited when decision-makers need a complete representation of the trajectory with all set of $(pi, ti, ai1, ..., ain)$ elements, such as in the case of monitoring the mechanical data of vehicles [3]. Therefore, two other approaches have been proposed that entirely store the trajectory as a measure: the *flat* approach, which consists of storing each $(pi, ti, ai1, ..., ain)$ element as a fact tuple [3], and the *complex* approach where one fact tuple contains all the points (the set of (pi) sub-elements) of the trajectory [21]. However, the complex approach proposed by [21] has limited potential for two main reasons. Firstly, because numerical attributes of the trajectory are not used as measures, and secondly, it is implemented in the PostgreSQL-based extension MobilityDB, which cannot be therefore applied in other DBMSs. To conclude, there is not a standard logical model for relational TDWs, and the complex approach has not been yet fully implemented. Moreover, this issue implies also a lack of an existing well-defined benchmark for relational TDWs. Indeed, all existing works define a set of ad-hoc OLAP queries over their logical schemata.

Motivated by these issues, in this work we (i) provide a new classification of trajectory OLAP queries (ii) present **two** new logical relational models for the complex approach allowing to represent $(pi, ti, ai1, ..., ain)$ element as a single measure (iii) compare the three approaches using a real case study concerning the analysis of agricultural autonomous robots odometry data. The experiments we conduct take into account the number of trajectories (i.e. cardinality of facts), and the resolution of the trajectory (i.e. the number of $(pi, ti, ai1, ..., ain)$ elements composing the trajectory).

The paper is organized in the following way: Sect. 2 presents related work, the agricultural autonomous robots data analysis scenario is shown in Sect. 3; Sect. 4 presents the three logical models. Experiments and results are described in Sect. 5 and Sect. 6 concludes the paper.

2 Related Work

Trajectory Data Warehouse has received a lot of attention in the last years, as shown in [2] and [1]. As described in Sect. 1, [1] groups existing works in two main classes: cell and segment approaches. Among cell based approaches, it is worth reporting [10,11], which present a formal framework and a system for visual analysis, and [5] that studies approximated aggregations.

The segment approach, which is more similar to our requirement for a complete trajectory elements data representation, has been studied in several works, as described in what follows. [4] presents a TDW for the analysis of animals' movement. The fact table stores the paths (i.e. segments) of the animals' movements according to the star schema. Each path is associated with numerical measures such as the speed, which are also preaggregated with min and max. The spatial and temporal dimensions contain the start and end spatial and temporal elements as most detailed levels, respectively. However, this work does not present any OLAP queries and performance experiments. In the same line, [17] describes a TDW for the analysis of hospital patients trajectories, where only start and stop points are stored with their duration as measure. [16] proposes a conceptual and logical model for TDWs using data collected by humans about Point of Interests (POIs). The logical model is based on the snowflake schema, where only the coordinates of the POIs are stored as dimensions. However, this work does not present performance experiments. [18] proposes the usage of TDWs to analyze athletes' measurements along their training states. Observations about athletes are stored in an ad hoc dimension. Therefore, no real measures have been associated with the fact table. Time and location for each observation are stored as dimensions. Authors implement their model in a MDX-based OLAP Server, and compare SQL and MDX queries' time performance. [6] shows a tool for the visualization and analysis of TDW. However, no details about the logical model has been provided. Authors present the conceptual model where the *episod* is a measure. An episod is a relevant element of a trajectory, and it is described by the location, the time and other numerical attributes. [19] combines cell and segment approaches. In this work, authors propose a conceptual model to represent the semantics of the trajectory. Although no logical model is shown, authors state that points, timestamps, and some aggregated numerical measures associated with the segment are stored in the fact table. The representation of the semantic features of the trajectories has been also studied by means of the usage of ontologies [13]. Finally, [14] proposes storing the distance among the points of the segment to allow answering ad hoc OLAP queries and improve time and storage performance.

The flat approach has been proposed in few works. [3] presents an implementation of TDW for storing GPS and odometry data of vehicles. Authors use the flat storage approach, where points, timestamps and other numerical attributes of the vehicles' trajectories are stored as measures in the fact table. By means of some physical tuning methods of PostgreSQL, they are able to store and analyze some billions of data. However, performance of queries are not reported, and their logical schema is not compared to others. The same approach is used by

[20], which adds a supplementary fact table to provide decision-makers with an aggregated view of the trajectory data. A flat approach using multiple fact tables (constellation schema) for a TDW is proposed in [8], where the logical model stores segments of trajectories in a fact table. A case study is presented where a semantic TDW is applied to human resources management for organizational analysis. Existing works that use the flat approach to store trajectory data are not able to store numerical attributes associated with each point-timestamp as measures. Moreover, no work exists that compare performance (storage and time computation of OLAP queries) of the different logical models. Finally, OLAP queries provided on those existing logical models are not generic, but guided from the case study. This means that a generic set of OLAP queries over TDWs has not been identified yet in order to be able to define a TDW benchmark, contrary to the classical DW [15].

Table 1 reports main works related to logical models and compares them to our proposal.

Table 1. Existing work study

Proposal	Approach	Performance study	Trajectory OLAP queries
[4]	Segment	No	No
[19]	Segment	No	No
[18]	Flat	No	No
[20]	Flat with no measures	No	No
Our proposal	Flat, complex and segment	Yes	Yes

3 Case Study: Agricultural Autonomous Robots Case Study

Agroecology is the new paradigm of agriculture production that aims to protect biodiversity, environment, and sustainable production. In this context, the usage of autonomous robots, coupled with IoT devices, becomes mandatory since: (i) they are lightweight, which implies less soil compaction, (ii) they are autonomous, which permits to avoid long and tedious agricultural practices to the farmers, and (iii) they are embedded with sensors, which allows precise agricultural tasks. In order to effectively use autonomous agricultural robots, their behaviour in the field must be precisely set, since in a uncontrolled environment (such as field), different events (meteorological, mechanical, etc.) can disturb the planned work of the robots. Real-time analysis of autonomous agricultural robots and their evolving environment is indeed a mandatory issue, but not sufficient. Indeed, real-time analysis can be used to detect some possible anomalies, but only the usage of historical data can allow decision-makers to find out the right action to these anomalies.

To analyze these historical data, we propose a multidimensional model composed of the following dimensions: `Alert` represents the mechanical faults identified by the robot, such as blocking wheel. It is organized in two levels: alert and type of alert; `Action` defines the response action provided by the robot to a mechanical fault, such as stop action. Actions are also grouped by their types; `Crop` represents the crop present in the field (e.g. wheat); `Campaign` represents the agronomic year associated with a particular crop, for example September 2019 - September 2020; `Equipment` is the tool associated with the robot for a particular agricultural task such as spraying; `Robot` has two levels. The less detailed represents the robot used for the task and the type of robot according to its size; `Scheduling` designs a set of planned trajectories to achieve the agricultural task, for example: first spray and then return to the warehouse; `Location` is the spatial dimension that groups plots, where each robot evolves, in farm and city; `Time` is the temporal dimension. The minimum granularity chosen by the decision-makers is minute, where hour and day are also included. The fact contains the measure that represents the trajectory made by a robot. The trajectory is characterized by the point, the timestamp and the speed of the robot. Using this multidimensional model it is possible to answer OLAP queries like this: *What is the max of the internal average speed of each trajectory, for each robot, equipment, crop type, and plot?* (Fig. 1d).

4   TDW OLAP Queries and Logical Multidimensional Models

In this section, firstly we introduce a classification of TDW OLAP queries, and then we present three logical models: the flat one, and our new complex and segment models.

4.1   TDW OLAP Queries Classification

In what follows, we present a classification of the OLAP queries over trajectories data, where the criteria are detailed. Queries' examples are shown in Table 1.

Numerical Aggregation Type: *Distributive* (e.g. sum, min and max), *Algebraic* (such as average) and *Holistic* (for example distinct count and median). This classification is important since it allows the usage of materialized views to speed-up queries. As described next, it could be used for improving the complex and segment models performance.

Table 2. TDW OLAP queries examples

Aggregation	Granularities		
	Point	Trajectory	Inter-Trajectories
Distributive	Max speed for each robot and plot	Min speed for each trajectory	Min speed of the internal max delay of each trajectory, for each robot, equipment, crop type, and plot
Algebraic	Average speed for each robot and plot	Average speed for each trajectory	Average speed of the internal max delay of each trajectory, for each robot, equipment, crop type, and plot
Holistic	Median speed for each robot and plot	Median speed for each trajectory	Median speed of the internal max delay of each trajectory, for each robot, equipment, crop type, and plot

Granularity: *Point*, *Trajectory* and *Inter-Trajectories*. *Trajectory* queries return an aggregated numerical value for each trajectory. *Inter-Trajectories* queries compute an OLAP Trajectory query for each trajectory, and then aggregate all these numerical aggregated values in a single one. *Point* queries consider all elements of all trajectories involved in the query independently of the trajectories they belong.

To better explain our classification, we propose an example with a cartographic representation in Fig. 1. Figure 1a shows the points and the numerical value (i.e. speed) for two trajectories provided by two different robots. We omit the timestamp for readability sake. Figure 1b shows a *Point Distributive* query. All points of the two trajectories are considered as a single input. Then, the min is applied, and one numerical value (i.e. 8 km/h) is returned. Figure 1c shows a *Trajectory Distributive* query, where each trajectory is considered as a different input, and for each one the max speed is returned. Figure 1d is a *Inter-Trajectories Distributive* query. In this query, at first the average is computed for each trajectory: 19.5 Km/h for Robot A and 9.5 Km/h for Robot B. Then, the max of these two numerical values is returned.

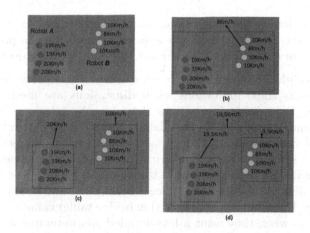

Fig. 1. Queries examples: a) input trajectories data, b) *Point Distributive* query - min, c) *Trajectory Distributive* query - max, d) *Inter-Trajectories Distributive* query - max of average

4.2   TDW Logical Models

Flat Model. The generic representation of this approach is shown in Fig. 2a. It is based on the star schema, where dimensions are denormalized. For each non spatial and temporal dimension (i.e. thematic dimension), there is a table with all the hierarchy levels. For sake of readability, Fig. 2a presents only one thematic dimension. The spatial dimension contains all spatial levels (`SpatialLevel`i) and a particular most detailed level **Point**. This level represents one point of the trajectory and it has a spatial type. Then, according to [12], a spatial topological relationship of inclusion must be defined between the level **Point** and the level `SpatialLevel1`. For example in our case study, a point of the robot's trajectory must be contained in a plot. The temporal dimension is defined in the same way, where the **Timestamp** level is the temporal value of a trajectory point, and the remaining levels are classical temporal levels. The fact table contains the foreign keys to dimensions tables and all numerical attributes of the trajectory as measures. This approach splits the data of the trajectory among fact and dimensions Tables. For each element of the trajectory, a tuple of the fact table is created. The spatial and temporal dimensions tables are forced to contain also spatial and temporal data for each trajectory's element. Therefore, a fact does not correspond to the analysis subject representing the overall trajectory, but it represents a single element of the trajectory. An example using our case study is shown in Fig. 6a, where only the fact table, the spatial and temporal dimensions are depicted for the sake of readability.

Complex Model. This model is shown in Fig. 2b. The main goal of this approach is to provide a logical representation that is as close as possible to the conceptual multidimensional representation. Therefore, the spatial, the temporal and the thematic dimensions present only their common levels. The fact table comes with n vectors. A vector is used to represent all the points of the trajectory, one for all the timestamps, and the other ones for each numerical attribute of the trajectory. Using this approach, the fact table presents a tuple for each trajectory, and dimensions are not affected by the trajectory data.

Segment Model. The complex approach storing all trajectory data in a unique measure presents some limitations when dimensions are used for describing episodic facts. Episodic facts are facts that describe relationships among the dimensions, and no measure is calculated. Examples are Alert and Action dimensions of our case study. Indeed, alerts and actions characterize the trajectory in the same way as its numerical attributes. In this case, a many to many association (implemented with a bridge table) between the fact table and each of these dimensions is needed, since several alerts and actions can occur in one trajectory. However, this approach (i.e. bridge table) cannot be sufficient for decision-makers when they want a less detailed spatio-temporal granularity to identify the parts of the trajectory. For example, decision-makers want to analyze alerts and actions per minute. Therefore, there is a need to split the trajectory into segments when these particular dimensions are involved in the multidimensional application. Splitting the trajectory into segments is also necessary when the most detailed levels of the spatial and temporal dimensions are on scales smaller than those of the trajectory. The segment model is shown in Fig. 2c. For example, in our case study, if the trajectory evolves over 2 plots, then the trajectory must be split into 2 segments. This constraint has been described in [12]. The same issue is valid for the temporal dimension, as we have described above. Therefore, in some particular cases a particular version of the complex approach is needed, which stores in the fact table a segment of the trajectory with all its elements. An example of the segment approach using our case study is shown in Fig. 6b[1], where only the fact table, the spatial and temporal dimensions are depicted, for the sake of readability. The fact table shown in Fig. 6b presents 3 vectors of 60 elements, which means that the trajectory is split in segments containing 60 trajectory elements. We use this size since in this scenario robots have been set to collect data each second. Therefore, the size of the segment is compliant with a less detailed level of the temporal dimension which is the minute. However, it is very important to note that for other kinds of analysis, decision-makers could need robots' data at high resolution (for example 1 datum each 10 ms). Finally, it is essential to mention that the complex and segment approaches have an important advantage compared to the flat one concerning the materialization of preaggregated measures (i.e. materialized views) [9]. Indeed, as we have described above aggregation functions can be distributive and algebraic, meaning that their result can be reused for the computation of queries using coarser levels. Therefore, by adding a preaggregated measure in the fact table, it is possible to improve query performance and storage size in respect to the flat approach. This is because the materialization of the flat approach needs to create a new fact table with all foreign keys and aggregated measures. For example, it will be possible to answer the query shown in Fig. 1b with a simple SELECT query, by simply adding a new measure to the fact table containing the max speed. This tuning technique can be also applied to the segment approach.

---

[1] Due to pages limit, this figure is presented in the Appendix.

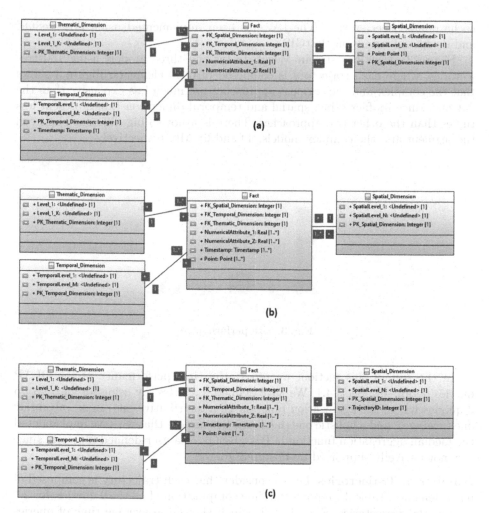

Fig. 2. Logical multidimensional models: a) flat, b) complex, c) segment

5   Experiments

In this section, we detail some experiments we conducted to evaluate the storage and time performance of TDW OLAP queries defined in Sect. 4. In particular, for time performance we analyze how the number of trajectories, and the resolution of each trajectory varies according the flat, segment and complex models. We use these queries since they are the most common in our case study.

The experiments were performed on a machine running Windows 10 64-bit, using pgAdmin4, both installed on a NVMe SSD, 8 GB of RAM and an AMD RYZEN 5 5600 H @3.3 GHz 64-bit Six core processor. The method used to retrieve the time was to take the average time of 10 runs per query and each time before the query was rerun the server restarted so that we could avoid potential cache related acceleration. The DW has been implemented using PostgreSQL and its spatial extension PostGIS.

Figure 3 shows the size in Megabyte for the three models using 160 trajectories, where each trajectory is composed of 6640 elements. Figure 3 shows that the complex and the segment approaches are much less expensive than the flat one, since its fact table, spatial and temporal dimension tables present less tuples than the other two approaches. There is a not so big difference between the segment and the complex models, 49 and 68 Mb, respectively.

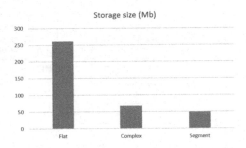

Fig. 3. Size performance

In the rest of this section, we report the time performance of the OLAP queries detailed in Sect. 4.1. We have defined 9 queries, by using for each group 3 queries where the aggregation function is: a distributive one (max), algebraic one (average), and a holistic one (median). Let us note that we have implemented the median aggregation function in PostgreSQL as a user-defined function, since it is not natively supported by PostgreSQL.

Number of Trajectories. Let us consider that each trajectory is composed of 6640 elements. Table 3 [2] reports the time computation of all the 9 queries, for 20, 60 and 100 trajectories. From Fig. 4, which shows the average time of queries belonging to the same group, it is possible to conclude that: (i) the complex and segment approaches perform better for *Point* queries, (ii) all approaches are well suited for Trajectory queries (under 1 s), and (iii) complex approach is the best solution for *Inter-trajectories* queries. From a more global point of view, it obviously appears that by increasing the number of trajectories, the time performance of queries also increases.

Resolution of Trajectories. In this subsection, we compare the performance of the three approaches with 160 trajectories stored, and vary the number of elements in each trajectory. The results are shown in Table 4 [3] using 360, 900 and 6640 elements. The resolution 360 has been tested since it represents the number of elements commonly used in moving objects benchmarks (i.e. [7]). The

---

[2] Due to pages limit, this Table is presented in the Appendix.
[3] Due to pages limit, this Table is presented in the Appendix.

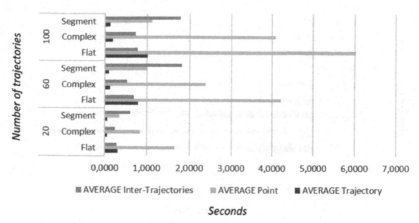

Fig. 4. Average time queries

other two resolutions have been tested to simulate the cases where ms frequency is used for data acquisition.

Figure 5 shows the average time for each group of queries. Obviously, a high resolution corresponds to high time results. All approaches have feasible results with 360 resolution. The *Flat* approach is the worst for *Point* queries, and the *Complex* one is the best when a high resolution is used.

To conclude, the complex and segment approaches perform well with all the different types of tested queries and require less storage space than the flat approach, when the resolution is not too much high (i.e. up to 900 elements). In the case of high resolution (i.e. 6640 elements) the flat model is also interesting, but the complex and segment approaches can radically improve their performance by adding a preaggregated value as discussed in the previous section.

We have also evaluated the previous described queries using a WHERE statement (i.e. OLAP Slice operator) on the thematic dimension Crop. The above described results have not been significantly impacted.

Finally, it is important to note that in order to achieve generic conclusions about the performance of the proposed logical models, a real benchmark for TDW should be provided which should take into account selectivity and scale factor, as well as spatial selection queries.

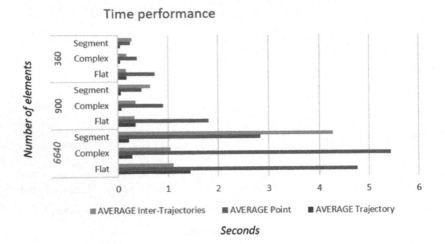

Fig. 5. Average time queries

6 Conclusion and Future Work

TDWs are DWs used for mobility data analysis. Mobility data are defined on the spatio-temporal domain considering that moving objects are changing their locations as time progresses. The applications of such approaches are increased significantly, particularly in the field of GIS. The main contribution of this work is to compare different logical models and conclude which one is more preferable for trajectory data querying. Therefore, in this paper the problem of performance is addressed. For this reason, initially, relational OLAP queries have been classified into two different groups. In particular, according to their numerical aggregation type, the queries are distinguished as distributive, algebraic or holistic. In terms of granularity, queries are identified as point, trajectory, and inter-trajectories. Based on these taxonomies, three logical models, i.e. flat, complex and segment, are presented and compared. The main feature of the complex model is a vector which represents the points of a trajectory. In the segment model, trajectories are partitioned into segments which facilitates the analysis in low hierarchical levels for in-depth querying and analysis. One main advantage of complex and segment approaches, is the materialization of preaggregated measures. The proposals of this research are validated through an implementation for a case study dealing with mobility and other semantical data provided by agricultural autonomous robots. The results of the experiments conducted prove that overall, the complex and segment approaches use less storage space and achieve better performance than the flat one. In combination with the inclusion of a preaggregated value, the results lean even further towards the complex and segment models. Future works include to extend the experiments to other logical TDW models using new OLAP queries' classifications, different DBMSs and platforms for explaining better the behavior of trajectory data.

Acknowledgement. This work is supported by the French National Research Agency project IDEX-ISITE initiative 16-IDEX-0001 (CAP 20-25).

7 Appendix

Table 3. Time performance of queries according to number of trajectories; Traj.Distr. - Trajectory Distributive, Traj.Alg. - Trajectory Algebraic, Trj.Hol. - Trajectory Holistic, Pt.Distr. - Point Distributive, Pt.Alg. - Point Algebraic, Pt.Hol. - Point Holistic, ITraj.Dist. - Inter-Trajectories Distributive, ITraj.Alg. - Inter-Trajectories Algebraic, ITraj.Hol. - Inter-Trajectories Holistic

Group	Query	20			60			100		
		Flat	*Segment*	*Complex*	*Flat*	*Segment*	*Complex*	*Flat*	*Segment*	*Complex*
Traj.Distr.	Q1	0,202	0,053	0,062	0,492	0,092	0,117	0,693	0,101	0,184
Traj.Alg.	Q2	0,2	0,056	0,061	0,061	0,095	0,125	0,691	0,105	0,185
Traj.Hol.	Q3	0,521	0,06	0,062	1,463	0,11	0,131	1,678	0,146	0,187
Pt.Distr.	Q4	0,993	0,312	0,821	0,821	0,974	2,307	3,723	1,017	3,723
Pt.Alg.	Q5	1,04	0,315	0,817	2,552	0,986	2,373	3,824	1,021	3,893
Pt.Hol.	Q6	2,9	0,411	0,867	7,589	1,1	2,453	10,523	1,312	4,52
ITraj.Distr.	Q7	0,201	0,593	0,215	0,42	1,772	0,502	0,686	1,787	0,689
ITraj.Alg.	Q8	0,203	0,599	0,213	0,419	1,778	0,486	0,778	1,774	0,702
ITraj.Hol.	Q9	0,457	0,603	0,319	1,218	1,873	0,602	0,875	1,779	0,781

Table 4. Time queries performance according to resolution of trajectories

Group	Query	360			900			6640		
		Flat	*Segment*	*Complex*	*Flat*	*Segment*	*Complex*	*Flat*	*Segment*	*Complex*
Traj.Distr	Q1	0,168	0,044	0,047	0,22	0,067	0,066	1,23	0,218	0,27
Traj.Alg	Q2	0,116	0,045	0,045	0,213	0,057	0,06	0,992	0,199	0,249
Traj.Hol	Q3	0,24	0,048	0,047	0,609	0,062	0,062	2,098	0,209	0,286
Pt.Distr	Q4	0,492	0,201	0,368	1,076	0,401	0,913	3,945	2,647	5,057
Pt.Alg	Q5	0,491	0,206	0,38	1,089	0,412	0,88	3,938	2,658	5,037
Pt.Hol	Q6	1,221	0,326	0,381	3,277	0,586	0,917	6,422	3,186	6,225
ITraj.Distr	Q7	0,116	0,269	0,163	0,212	0,639	0,31	0,99	4,245	1,012
ITraj.Alg	Q8	0,166	0,27	0,162	0,218	0,645	0,321	0,997	4,311	1,015
ITraj.Hol	Q9	0,204	0,279	0,199	0,535	0,646	0,396	1,324	4,249	1,108

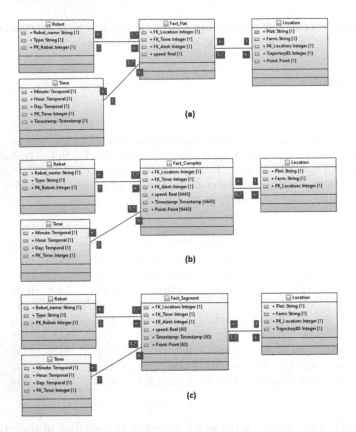

Fig. 6. Logical multidimensional examples: a) flat, b) complex, c) segment

References

1. Ribeiro de Almeida, D., de Souza Baptista, C., Gomes de Andrade, F., Soares, A.: A survey on big data for trajectory analytics. ISPRS Int. J. Geo-Information **9**(2), 88 (2020)
2. Alsahfi, T., Almotairi, M., Elmasri, R.: A survey on trajectory data warehouse. Spat. Inf. Res. **28**(1), 53–66 (2020)
3. Andersen, O., Krogh, B.B., Thomsen, C., Torp, K.: An advanced data warehouse for integrating large sets of GPS data. In: Int. Workshop on Data Warehousing and OLAP (DOLAP), pp. 13–22. ACM (2014)
4. Arfaoui, N., Akaichi, J.: Modeling herd trajectory data warehouse. Int. J. Eng. Trends Technol. **6**, 1–9 (2011)
5. Braz, F., Orlando, S., Orsini, R., Raffaeta, A., Roncato, A., Silvestri, C.: Approximate aggregations in trajectory data warehouses. In: International Conference on Data Engineering (ICDE) Workshop, pp. 536–545. IEEE (2007)
6. Campora, S., de Macedo, J.A.F., Spinsanti, L.: St-toolkit: a framework for trajectory data warehousing. In: AGILE Conference on Geographic Information Science (2011)

7. Düntgen, C., Behr, T., Güting, R.H.: Berlinmod: a benchmark for moving object databases. VLDB J. **18**(6), 1335–1368 (2009)
8. Garani, G., Adam, G.K.: A semantic trajectory data warehouse for improving nursing productivity. Health Inf. Sci. Syst. **8**(1), 25 (2020). https://doi.org/10. 1007/s13755-020-00117-5, https://doi.org/10.1007/s13755-020-00117-5
9. Kimball, R., Ross, M.: The data warehouse toolkit: the complete guide to dimensional modeling, 2nd edn. Wiley (2002)
10. Leonardi, L., et al.: T-warehouse: Visual olap analysis on trajectory data. In: International Conference on Data Engineering (ICDE), pp. 1141–1144. IEEE (2010)
11. Leonardi, L., Orlando, S., Raffaetà, A., Roncato, A., Silvestri, C., Andrienko, G., Andrienko, N.: A general framework for trajectory data warehousing and visual OLAP. GeoInformatica **18**(2), 273–312 (2013). https://doi.org/10.1007/s10707-013-0181-3
12. Malinowski, E., Zimányi, E.: Spatial hierarchies and topological relationships in the spatial MultiDimER model. In: Jackson, M., Nelson, D., Stirk, S. (eds.) BNCOD 2005. LNCS, vol. 3567, pp. 17–28. Springer, Heidelberg (2005). https://doi.org/10. 1007/11511854_2
13. Manaa, M., Akaichi, J.: Ontology-based trajectory data warehouse conceptual model. In: Madria, S., Hara, T. (eds.) DaWaK 2016. LNCS, vol. 9829, pp. 329–342. Springer, Cham (2016). https://doi.org/10.1007/978-3-319-43946-4_22
14. Marketos, G., Theodoridis, Y.: Ad-hoc olap on trajectory data. In: International Conference on Mobile Data Management, pp. 189–198. IEEE (2010)
15. Nambiar, R.O., Poess, M.: The making of tpc-ds. VLDB **6**, 1049–1058 (2006)
16. Oueslati, W., Akaichi, J.: Mobile information collectors trajectory data warehouse design. Int. J. Manag. Inf. Technol. (IJMIT) **2** (2010)
17. Oueslati, W., Hamdi, H., Akaichi, J.: A mobile hospital trajectory data warehouse modeling and querying to detect the breast cancer disease. In: International Conference on Intelligent Information Processing, Security and Advanced Communication (IPAC), pp. 93:1–93:5. ACM (2015)
18. Porto, F., et al.: A metaphoric trajectory data warehouse for olympic athlete follow-up. Concurrency Comput. Pract. Exp. **24**(13), 1497–1512 (2012)
19. Wagner, R., de Macedo, J.A.F., Raffaetà, A., Renso, C., Roncato, A., Trasarti, R.: Mob-warehouse: a semantic approach for mobility analysis with a trajectory data warehouse. In: Parsons, J., Chiu, D. (eds.) ER 2013. LNCS, vol. 8697, pp. 127–136. Springer, Cham (2014). https://doi.org/10.1007/978-3-319-14139-8_15
20. Tang, B., Shen, G., Zhang, C.: Data warehousing of vehicle trajectory. In: Int. Conf. on Software Engineering and Service Science (ICSESS), pp. 935–938 (2015)
21. Vaisman, A., Zimányi, E.: Mobility data warehouses. ISPRS Int. J. Geo-Information **8**(4), 170 (2019)

BPF: An Effective Cluster Boundary Points Detection Technique

Vijdan Khalique[1]([⊠]) [iD] and Hiroyuki Kitagawa[2] [iD]

[1] Graduate School of Systems and Information Engineering, University of Tsukuba, Tsukuba, Japan
khalique.vijdan@kde.cs.tsukuba.ac.jp
[2] International Institute for Integrative Sleep Medicine, University of Tsukuba, Tsukuba, Japan
kitagawa@cs.tsukuba.ac.jp

Abstract. In a dataset, *boundary points* are located at the extremes of the clusters. Detecting such boundary points may provide useful information about the process and it can have many real-world applications. Existing methods are sensitive to outliers, clusters of varying densities and require tuning more than one parameter. This paper proposes a boundary point detection method called *Boundary Point Factor (BPF)* based on the outlier detection algorithm known as Local Outlier Factor (LOF). *BPF* calculates Gravity values and BPF scores by combining original LOF scores of all points in the dataset. Boundary points can be effectively detected by using BPF scores of all points where boundary points tend to have larger BPF scores than other points. *BPF* requires tuning of one parameter and it can be used with LOF to output outliers and boundary points separately. Experimental evaluation on synthetic and real datasets showed the effectiveness of our method in comparison with existing boundary points detection methods.

Keywords: Boundary points detection · Cluster boundary · Data mining

1 Introduction

Data mining comprises a multitude of techniques to extract knowledge from the data. Clustering is one such technique that focuses on dividing a dataset into subsets such that data belonging to a subset are similar in some way [9]. In cluster analysis, it is desirable to extract useful features about the data by clustering the data objects as the grouping of data may represent a phenomenon. For example, a cluster of data objects may represent a specific behavior of customers in a customer dataset or, in an image dataset, a cluster of images may share similar properties. On the other hand, outlier detection is a data mining task which is related to clustering. An outlier is defined as a data object that deviates from the majority of the data objects [6]. The majority of the data objects share common

© The Author(s), under exclusive license to Springer Nature Switzerland AG 2022
C. Strauss et al. (Eds.): DEXA 2022, LNCS 13426, pp. 404–416, 2022.
https://doi.org/10.1007/978-3-031-12423-5_31

features and may form one or more clusters. Hence, an outlier is a data object that is isolated and does not belong to any cluster. In the past, many techniques have been proposed for detecting outliers based on distance [2,8], density [3,7] and angle [10,14]. Importantly, we do not make any distinction between noisy points and outliers and use the term outliers to refer to the points which do not belong to any cluster.

Unlike clustering and outlier detection, a limited research has been dedicated to the *boundary points detection*. In [13], border or boundary points are defined as points which are located at the extremes of a class region or near free pattern space. In other words, boundary points are located at the border of a cluster i.e. boundary points are forming the boundary of the cluster. Hence, boundary points detection can be defined as the task of detecting the points which are situated at the boundary of a cluster [17]. Consequently, detecting boundary points may provide useful information. For example, in a disease detection system, the normal data objects may represent healthy patients, the outliers may represent patients who have contracted a certain disease and the boundary points may represent normal patients with high likelihood of developing the disease where they may show many symptoms but somehow have not yet developed the disease. As a result, such boundary cases should be monitored closely as they may reveal interesting information about the disease. Similar motivating examples are presented in the related papers [5,12,16,17].

Outlier detection is a well studied problem. Several techniques have been proposed to solve outlier detection and one such technique is Local Outlier Factor (LOF) [3]. It is one of the most popular and competitive density-based outlier detection method [19] that detects the outliers based on the relative density of a target object according to its local neighborhood. The points located in the inner region of a cluster are referred to as the *core* points, while the outliers are isolated points in the less dense regions. Moreover, the neighborhood of the core points is in all directions, while the boundary points are in the dense region and they may have their neighborhood in one direction. If LOF is applied on such dataset and top-m data objects are considered as output, the list of top-m outliers may contain outliers and boundary points. However, the problem of boundary points detection requires the detection of boundary points only while ignoring the core points and outliers. This paper proposes to use the average unit vector between a target point and its k-nearest neighbors pointing towards the dense region. We call the scalar value obtained by taking the norm of this average unit vector as *Gravity* of the target point. The ratio of the Gravity and LOF values are used for calculating the *Boundary Point Factor (BPF) scores* of all the data points. The data points with larger BPF scores can be considered as boundary points and they are ranked at higher ranks w.r.t. BPF scores than cores and outlier points. We have shown experimentally that our proposed method is robust to the presence of outliers and it can deal with datasets having multiple clusters of various densities. Hence, following are the advantages of our proposed method:

- *BPF* can be used in conjunction with LOF. Hence, one can obtain outliers and boundary points using LOF and *BPF* respectively as our method efficiently shares the computation to calculate the k-nearest neighbor and LOF values.

- *BPF* has one tune-able parameter k (number of nearest neighbors) which can be tuned in the same way as tuning k for LOF.
- *BPF* is robust to the presence of outliers and clusters of different densities.

This paper makes the following key contributions: i) proposes a boundary points detection method *BPF* based on the ratio of LOF and Gravity values, and ii) experiments on synthetic and real datasets to demonstrate the effectiveness of the proposed method.

2   Related Work

The boundary of a cluster is composed of points which are at the decision region of the cluster. Different from outliers, boundary points have clear class labels like core points but they can offer a different and interesting information about the process which the data is representing (as explained in the examples in the previous section). One of the initial works in this direction was [17]. Xia et al. in [17] proposed BORDER algorithm which exploited the property that boundary points have smaller number of reverse k-nearest neighbors (RkNN) than core points and therefore, boundary points can be identified based on this property. Since, the computation of RkNN is expensive, they proposed to use G-ordering kNN join [18] method to speed up the computation of RkNN. BORDER was found to be effective in identifying boundary points in the dataset without outliers. In case of datasets with many outliers, BORDER cannot distinguish between outliers and boundary points as both of them tend to have smaller number of RkNN. In order to solve the problem in BORDER, Qiu et al. in [15] proposed BRIM which was successful in detecting the boundary points in the datasets with many outliers. Given a distance *eps*, BRIM uses the observation that the *eps*-neighborhood of a boundary point can be distributed in either positive or negative direction based on the diameter line which divides its *eps*-neighborhood into two parts. Also, based on *eps*-neighborhood, the densities of boundary points are greater than outliers. BRIM uses these observations to detect boundary points effectively. The major drawback of BRIM is that it uses *eps*-neighborhood to calculate boundary degree of each point, and therefore it does not perform well in datasets with clusters of different densities and scales.

More recently, Li et al. in [12] proposed a method of detecting outliers and boundary points based on the geometrical measures. The proposed algorithm BPDAD combined the two observations that outliers and boundary points have lower local densities and smaller variance of angles than their neighbors. However, this work does not directly solve the problem of detecting boundary points, rather its output is outliers and boundary points together. In [5], BorderShift algorithm was proposed that uses similar observations regarding the densities of outliers, boundary and core points. BorderShift uses Parzen Window (kernel density estimation) to estimate the local density of a point and MeanShift vector to determine the direction of the dense region. It requires tuning of three parameters k, λ_1 and λ_2 to achieve good results, where tuning λ_1 and λ_2 are particularly difficult as they are highly dependent on the number of outliers in the dataset.

Furthermore, methods proposed in [4] and [16] are for high-dimensional data and they perform the projection of high-dimensional data points to lower dimensions for boundary points detection. However, it is challenging to tune their parameters as they require prior knowledge of the number of outliers in the dataset. It is more appropriate that the boundary points detection algorithms should have minimum number of tune-able parameters and the user should not have to assume ratio of outliers in the dataset. The boundary points detection algorithm should take a dataset with or without outliers as input and output the top-m boundary points. This research proposes such an algorithm with only one parameter and does not require any prior knowledge of the dataset.

3 Preliminaries

In this section, we briefly review the definitions related to Local Outlier Factor. We recommend the readers to refer to [3] for further details.

Given a point p in the dataset D, let k represents the number of nearest neighbors. The LOF score of p ($LOF(p)$) can be calculated using two key concepts: Reachability Distance $reach\text{-}dist_k(p, o)$ and Local Reachability Density $lrd_k(p)$. The following definitions present these concepts followed by the definition of LOF.

Definition 1. *Reachability distance of a point p w.r.t. point o is defined as:*

$$reach\text{-}dist_k(p, o) = max\{k\text{-}distance(o), dist_k(p, o)\},$$

where, $k\text{-}distance(o)$ is the Euclidean distance of o with its kth neighbor and $dist_k(p, o)$ is the Euclidean distance between point p and o.

Definition 2. *The Local Reachability Density of a point p denoted as $lrd_k(p)$ is defined as:*

$$lrd_k(p) = 1/\left(\frac{\sum_{o \in N_k(p)} reach\text{-}dist_k(p, o)}{|N_k(p)|}\right),$$

where, $N_k(p)$ is the set of k-nearest neighbors of the point p and $|N_k(p)|$ represents the cardinality of $N_k(p)$.

Intuitively, the local reachability density is the estimation of the density of p w.r.t. its neighbors $o \in N_k(p)$. More concretely, $lrd_k(p)$ is the reciprocal of average reachability distance of p with its k-nearest neighbors. Therefore, larger the reachability distances of p, smaller is $lrd_k(p)$. Based on Definitions 1 and 2, we can define the local outlier factor (LOF).

Definition 3. *The local outlier factor of p is defined as:*

$$LOF(p) = \frac{\sum_{o \in N_k(p)} \frac{lrd_k(o)}{lrd_k(p)}}{|N_k(p)|}.$$

$LOF(p)$ is the outlier factor of the point p which indicates its degree of outlierness. If p is a core then $LOF(p)$ is close to 1 and if p is a boundary point, then $LOF(p)$ is greater than core points but still close to 1. In case p is an outlier, $LOF(p)$ is greater than 1. The details about the range of LOF score are explained in [3]. The next section explains the proposed method based on the definition of LOF.

4   Proposed Method

4.1   Definitions and Observations

This section introduces the definitions related to the proposed method and explains the basic observations. The following definition explains the Gravity $G(p)$ of a given point.

Definition 4. *Given a point $p \in D$, the set of k-nearest neighbors of p $N_k(p)$, a point $o \in N_k(p)$ and the norm $\|.\|$, the unit vector of p to o denoted as $\hat{p}o$ can be given as $\hat{p}o = \frac{\vec{po}}{\|\vec{po}\|}$. Hence, the Gravity of p can be defined as:*

$$G(p) = \frac{1}{|N_k(p)|} \left\| \sum_{o \in N_k(p)} \frac{\vec{po}}{\|\vec{po}\|} \right\|.$$

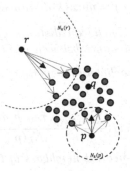

Fig. 1. Observation.

Consider the boundary point p shown in Fig. 1. Taking the sum of the unit vectors originating from p to its k-nearest neighbors will result in a single vector (solid arrow). The value of $G(p)$ will be larger for p than the core point q because q is surrounded by many points in all directions resulting in a smaller $G(q)$. However, Gravity value of outlier depends on the data distribution. Hence, the following may hold for boundary and core points:

$$G(q) < G(p).$$

On the other hand, the LOF scores of data points w.r.t. their k-neighborhood can be calculated using Definition 3. As shown in [3], the LOF values of core and boundary points are close to 1 ($LOF(q), LOF(p) \approx 1.$) and the LOF scores of outliers are greater than 1. Consequently, for the points p, q and r shown in Fig. 1, the following may hold:

$$LOF(q), LOF(p) < LOF(r).$$

Based on these observations, Boundary Point Factor (BPF) score can be defined in the following definition.

Definition 5. *Given a point $p \in D$, the $LOF(p)$ and the Gravity $G(p)$, the Boundary Point Factor score of p $BPF(p)$ can be calculated as follows:*

$$BPF(p) = \frac{G(p)}{LOF(p)}.$$

From Definition 5, the following may hold:

$$BPF(q), BPF(r) < BPF(p).$$

By calculating the BPF scores of all data point in a dataset, boundary points are more likely to have larger BPF scores than core points and outliers. In this way, our proposed method can output top-m boundary points in the dataset.

4.2   Algorithm

The main idea of *BPF* algorithm is to calculate the BPF scores of all the points in the dataset and sort them in descending order w.r.t. the scores. Given the number of points to consider as boundary points m, *BPF* will output the list $\mathbb{B}$ of top-m boundary points in the dataset. The steps are given in Algorithm 1.

4.3   Runtime Complexity

Let n represent number of points in a d-dimensional dataset. The most time consuming operation for *BPF* and its competitors like BORDER [17], BRIM [15], BPDAD [12] and BorderShift [5] is the k-nearest neighbors search for each point which have the complexity of $O(n^2 d)$. But the runtime complexity can be improved to $O(n \log n)$ by using a suitable indexing technique. Furthermore, BPF score is calculated based on the Gravity values and LOF scores of n points. The complexity of calculating Gravity values and LOF scores can be given as $O(nkd)$ and $O(nk^2)$ respectively. Hence, the overall runtime complexity of *BPF* algorithm is $O(n^2 d + nkd + nk^2)$.

Algorithm 1: *BPF*

    Input : D, k
    Output: $\mathbb{B}$
1 **for** $p \in D$ **do**
2 | $N_k(p) \leftarrow k\text{-}NN$ of p
3 | **for** $o \in N_k(p)$ **do**
4 | | calculate $BPF(p)$ using Definition 5 store in $\mathbb{C}$
5 | **end**
6 **end**
7 $\mathbb{B} \leftarrow$ Sort $\mathbb{C}$ in descending order w.r.t. BPF scores and get top-m points.
8 **return** $\mathbb{B}$

---

5 Experiments and Results

This section presents the experimental setup for evaluating the effectiveness of the proposed method on synthetic and real datasets.

5.1 Synthetic Data

We compared *BPF*'s accuracy on synthetic datasets with BorderShift [5], BPDAD [12], BRIM [15] and BORDER [17]. The parameter of these methods are tuned according to the suggested range in the respective papers. Particularly for BPDAD, it uses fixed parameter values (details in [12]) and automatically outputs the boundary points. Therefore, we use the default parameter values given in the original paper for BPDAD and output detected boundary points. For *BPF*, the value of k is tuned in the range [30,60]. The parameter values are mentioned along with the results. Furthermore, the evaluation on synthetic datasets are of two types: 1) demonstrating the accuracy by showing the boundary points detected by all methods like [5,12,15–17], and 2) quantitatively showing the accuracy of all the methods in terms of precision, recall and F1 score like [5,16]. In addition to these metrics, we show Area Under ROC Curve (AUC ROC) and Area Under Precision-Recall Curve (AUC PR).

    Figures 2, 3, 4 and 5 show the results and their captions show the description of the synthetic datasets (where n is the dataset size and m is top-m boundary points detected) and the parameter values. The dark points indicate the top-m boundary points detected by each method. Diamonds, Rings, Mix1 and Mix2 datasets are used to illustrate the boundaries detected by *BPF* while, Mix3 is used to evaluate the accuracy quantitatively.

    On diamonds [5,15,17] and rings [16] dataset, *BPF* performs well by detecting the boundary points clearly in comparison with other methods. However, a few outliers are detected but the results are still comparable or better than other methods specially from BPDAD and BORDER which detected outliers along with boundary points. BRIM does not perform well particularly on rings dataset due to the different densities of the rings. Furthermore, its is hard to tune an appropriate *eps* value for BRIM. In these experiments, the number of

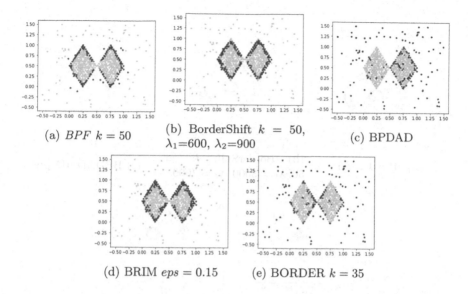

(a) *BPF* $k = 50$

(b) BorderShift $k = 50$, $\lambda_1 = 600$, $\lambda_2 = 900$

(c) BPDAD

(d) BRIM *eps* $= 0.15$

(e) BORDER $k = 35$

Fig. 2. Diamonds dataset: $n = 1000$, $\#outliers = 100$, $m = 300$.

outliers are known beforehand, and therefore the best results for BorderShift are shown. However, the tuning of λ_1 and λ_2 is a difficult task if prior knowledge about the number of outliers in the dataset is unknown.

The robustness of our method is demonstrated on Mix1 and Mix2 datasets that contain clusters of different shapes and densities. Mix2 is similar to the dataset used in LOF paper [3] with clusters having points from uniform and normal distribution. It can be seen that *BPF* performs well on both datasets.

Next, in Table 1, we quantitatively show the effectiveness of *BPF* on Mix3 dataset (n = 2400, #outliers = 200) by obtaining ground truth about the boundary points. Mix3 dataset contains two clusters of uniformly distributed points with different radii and number of points. The small and large clusters have radius $r_1 = 1$ and $r_2 = 2$ with 1400 and 800 points respectively and 290 boundary points in the ground truth. To obtain the ground truth, points within $r_1 - (r_1 - 0.05)$ in the small cluster and $r_2 - (r_2 - 0.2)$ in the large cluster are considered as the boundary points. The precision, recall and F1 scores are given in Table 1 at different parameters by considering top 350 points returned by each method. The number of boundary points returned by BPDAD are in the parenthesis. Overall, *BPF* outperformed all methods and showed almost consistent performance on different parameter values. Similarly, BRIM and BORDER did not show significant change in the accuracy, while BorderShift's accuracy changed with the change in λ_1 and λ_2 values at fixed k. λ_1 and λ_2 cover the top 350 points and their values are changed with the interval of 200 points. Hence, the results suggest that tuning λ_1 and λ_2 is difficult when number of outliers are not known. In addition, the outlier detection accuracy (precision, recall and F1)

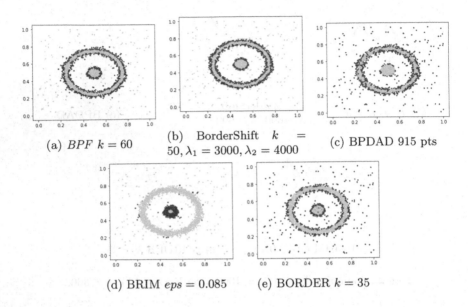

(a) *BPF* $k = 60$

(b) BorderShift $k = 50, \lambda_1 = 3000, \lambda_2 = 4000$

(c) BPDAD 915 pts

(d) BRIM *eps* $= 0.085$

(e) BORDER $k = 35$

Fig. 3. Rings dataset: $n = 4200, \#outliers = 200, m = 1000$.

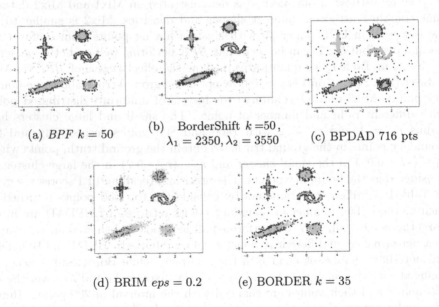

(a) *BPF* $k = 50$

(b) BorderShift $k = 50, \lambda_1 = 2350, \lambda_2 = 3550$

(c) BPDAD 716 pts

(d) BRIM *eps* $= 0.2$

(e) BORDER $k = 35$

Fig. 4. Mix1 dataset: $n = 3800, \#outliers = 150, m = 1200$.

of LOF is 0.94 for detecting top 200 outliers on k = 50,75,100. This shows that LOF and *BPF* can work with the same *k* values.

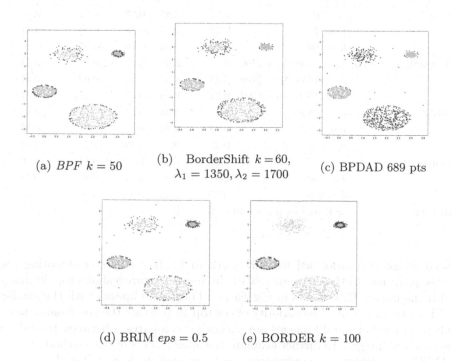

(a) *BPF* $k = 50$

(b)   BorderShift $k = 60$,
$\lambda_1 = 1350, \lambda_2 = 1700$

(c) BPDAD 689 pts

(d) BRIM $eps = 0.5$

(e) BORDER $k = 100$

Fig. 5. Mix2 dataset $n = 1710, \#outliers = 10, m = 350$.

Futhermore, we report the AUC ROC and AUC PR in Table 1. The AUC ROC and AUC PR consider the ranking of ground truth. Therefore, we did not include the results of BPDAD as the output is not based on the ranking. For BorderShift, we sorted the output in ascending order w.r.t. the scores and considered the ranking from λ_2 until the start of the list (backwards) to calculate AUC ROC and AUC PR. For *BPF*, BORDER and BRIM, we considered the ranking of points according to the calculated scores. The results show that *BPF* is more effective in detecting boundary points than other methods.

5.2   Real Data

Since, there are no benchmark real datasets available for verifying the performance of boundary point detection, we use the same datasets used in the related work for evaluation. Similar to [5], we used Olivetti Research Laboratory (ORL) face dataset [1]. The dataset consists of 400 images of 92×112 pixels of frontal faces of 40 people where each pixel represents the gray value in the range of 0–255. The images of frontal faces are considered as normal images, while the images with left and right profile faces are considered as the boundary images.

Table 1. Comparison of accuracy of *BPF* and other methods on Mix3 dataset

Methods	Parameter	Precision	Recall	F1 score	AUC ROC	AUC PR
BPF	$k = 50$	0.71	0.85	0.77	0.97	0.8
	$k = 75$	0.72	0.87	0.78	0.98	0.81
	$k = 100$	0.69	0.83	0.76	0.97	0.8
BorderShift ($k = 50$)	$\lambda_1 = 1950, \lambda_2 = 2300$	0.5	0.61	0.55	0.9	0.44
	$\lambda_1 = 1750, \lambda_2 = 2100$	0.45	0.55	0.49	0.62	0.31
	$\lambda_1 = 1550, \lambda_2 = 1900$	0.21	0.26	0.23	0.34	0.21
BRIM	$eps = 0.2$	0.4	0.48	0.44	0.67	0.48
	$eps = 0.3$	0.43	0.52	0.47	0.77	0.55
	$eps = 0.4$	0.39	0.47	0.43	0.75	0.52
BORDER	$k = 100$	0.39	0.47	0.425	0.81	0.29
	$k = 150$	0.41	0.49	0.44	0.85	0.31
	$k = 200$	0.39	0.47	0.43	0.86	0.31
BPDAD	# boundary pts = 740	0.16	0.42	0.23	–	–

Each image is transformed from 92×112 to 1×10304 by concatenating each subsequent row to the previous row of the image. Figure 6 shows top 40 images with the largest BPF scores in Fig. 6a and bottom 40 images with the smallest BPF scores in Fig. 6b. The majority of the top 40 faces are the non-frontal images whereas, the bottom 40 images are shown for comparison between frontal and non-frontal images. To further evaluate the effectiveness of our method, we used MNIST [11] dataset of handwritten digits as used in [5] and [16]. The dataset contains 60000 images in training and 10000 images in testing set of 28×28 pixels. We selected digit '3' from testing set of MNIST dataset. It contains 1010 images of digit '3' which are considered as core and boundary images. The easily recognizable 3s can be regarded as cores and distorted 3s can be boundary images. We also selected 100 random images from digits 0,2,4,6,7 and 9 from testing set which may be considered outliers. The preprocessing is performed

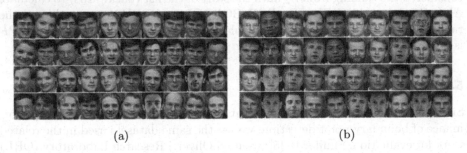

(a)   (b)

Fig. 6. Top (a) and bottom (b) 40 faces detected by *BPF* ($k = 50$)

in the same way as done for ORL dataset. Figure 7a shows the top 50 boundary points detected by *BPF* with the larger BPF scores. The boundary images are distorted 3s and they are ranked higher, while core and outlier images are ignored. Figure 7b shows core and outliers images with smaller and similar BPF scores therefore, a mixture of cores and outliers can be seen. We further show the LOF scores as labels of each digit to show that LOF scores of core points are close to 1 and outliers have LOF scores > 1.

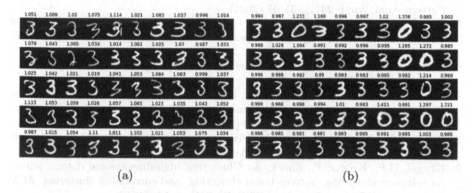

(a)   (b)

Fig. 7. Top (a) and bottom (b) 50 digits detected by *BPF* ($k = 50$). The LOF scores are labeled at the top of each digit.

6 Conclusion

This paper proposes an effective boundary points detection technique called *BPF* based on LOF and Gravity values. According to the proposed formulation, boundary points tend to have larger BPF scores than cores and outliers. Therefore, *BPF* can detect boundary points more effectively while ignoring the core points and outliers. We experimentally demonstrated the effectiveness of *BPF* where it showed comparable or better results compared with other methods on the synthetic datasets with multiple clusters of different shapes and densities, and outliers. Furthermore, our method has one parameter which can be tuned in the same range as LOF. In addition, *BPF* can be used in conjunction with LOF as they share k-nearest neighbor and LOF computation. Hence, *BPF* can be integrated with LOF to output outliers and boundary points separately. In the future, we intend to work on detecting boundary points in streaming data where the boundaries of the clusters may change with time.

Acknowledgement. This work was partly supported by JSPS KAKENHI Grant Number JP19H04114 and AMED Grant Number JP21zf0127005.

References

1. The ORL database of faces. https://cam-orl.co.uk/facedatabase.html
2. Angiulli, F., Pizzuti, C.: Outlier mining in large high-dimensional data sets. IEEE Trans. Knowl. Data Eng. **17**(2), 203–215 (2005)
3. Breunig, M.M., Kriegel, H.P., Ng, R.T., Sander, J.: LOF: identifying density-based local outliers. In: Proceedings of 2000 ACM SIGMOD International Conference on Management of Data, pp. 93–104 (2000)
4. Cao, X.: High-dimensional cluster boundary detection using directed Markov tree. Pattern Anal. Appl. **24**(1), 35–47 (2021)
5. Cao, X., Qiu, B., Xu, G.: BorderShift: toward optimal MeanShift vector for cluster boundary detection in high-dimensional data. Pattern Anal. Appl. **22**(3), 1015–1027 (2019)
6. Hawkins, D.M.: Identification of Outliers, vol. 11. Springer, Cham (1980). https://doi.org/10.1007/978-94-015-3994-4
7. Jin, W., Tung, A.K., Han, J.: Mining top-n local outliers in large databases. In: Proceedings of 7th ACM SIGKDD International Conference on Knowledge Discovery and Data Mining, pp. 293–298 (2001)
8. Knorr, E.M., Ng, R.T., Tucakov, V.: Distance-based outliers: algorithms and applications. Very Large Data Bases J. **8**(3–4), 237–253 (2000)
9. Kriegel, H.P., Kröger, P., Zimek, A.: Clustering high-dimensional data: a survey on subspace clustering, pattern-based clustering, and correlation clustering. ACM Trans. Knowl. Discov. from Data (TKDD) **3**(1), 1–58 (2009)
10. Kriegel, H.P., Schubert, M., Zimek, A.: Angle-based outlier detection in high-dimensional data. In: Proceedings of 14th ACM SIGKDD International Conference on Knowledge Discovery and Data Mining, pp. 444–452 (2008)
11. LeCun, Y., Cortes, C.: MNIST. http://yann.lecun.com/exdb/mnist/
12. Li, X., Wu, X., Lv, J., He, J., Gou, J., Li, M.: Automatic detection of boundary points based on local geometrical measures. Soft Comput. **22**(11), 3663–3674 (2017). https://doi.org/10.1007/s00500-017-2817-y
13. Li, Y., Maguire, L.: Selecting critical patterns based on local geometrical and statistical information. IEEE Trans. Pattern Anal. Mach. Intell. **33**(6), 1189–1201 (2010)
14. Pham, N., Pagh, R.: A near-linear time approximation algorithm for angle-based outlier detection in high-dimensional data. In: Proceedings of 18th ACM SIGKDD International Conference on Knowledge Discovery and Data Mining, pp. 877–885 (2012)
15. Qiu, B.-Z., Yue, F., Shen, J.-Y.: BRIM: an efficient boundary points detecting algorithm. In: Zhou, Z.-H., Li, H., Yang, Q. (eds.) PAKDD 2007. LNCS (LNAI), vol. 4426, pp. 761–768. Springer, Heidelberg (2007). https://doi.org/10.1007/978-3-540-71701-0_83
16. Qiu, B., Cao, X.: Clustering boundary detection for high dimensional space based on space inversion and Hopkins statistics. Knowl.-Based Syst. **98**, 216–225 (2016)
17. Xia, C., Hsu, W., Lee, M.L., Ooi, B.C.: Border: efficient computation of boundary points. IEEE Trans. Knowl. Data Eng. **18**(3), 289–303 (2006)
18. Xia, C., Lu, H., Ooi, B.C., Hu, J.: GORDER: an efficient method for KNN join processing. In: Proceedings of 30th International Conference on Very Large Data Bases, vol. 30, pp. 756–767 (2004)
19. Zimek, A., Schubert, E., Kriegel, H.P.: A survey on unsupervised outlier detection in high-dimensional numerical data. Stat. Anal. Data Mining ASA Data Sci. J. **5**(5), 363–387 (2012)

Hypergraphs as Conflict-Free Partially Replicated Data Types

Aruna Bansal[✉]

SIT, Indian Institute of Technology Delhi, Delhi, India
aruna.bansal@cse.iitd.ac.in

Abstract. Hypergraphs provide a natural mathematical way to accommodate hierarchical, relational, navigational, semi-structured, complex & higher-order relationships in real-world settings. Applications such as social media (that have their data centers in different geographical locations) and bibliographic paper publication portals (that deal with a diverse group of authors and reviewers from a wide geographical area) require data availability while coping with high network latencies. Replication is the commonly used approach for achieving a high degree of availability while facilitating local query processing, but it requires expensive (and often infeasible) concurrency control to ensure consistency. Moreover, administrative and security policies may prohibit certain parts of the database from being fully replicated at certain sites. This paper proposes a solution to partially replicate hypergraphs in conformance with distribution policies while ensuring data availability and network latency. Our approach leverages the novelty of Conflict-free Replicated Data Types (CRDTs) (that guarantee strong eventual consistency without requiring complex concurrency control mechanisms) with hypergraph structures and semantics to consistently update and propagate (fully or partly) hypergraphical information across multiple replicas. We also show how concurrent processes meet convergence conditions, proving the soundness of our approach.

Keywords: Hypergraphs · Complex relationships · Higher-ordered relationships · Graph database · Consistency · Data replication · CRDTs

1 Introduction

Hypergraphs are interesting data structures that can offer useful properties of various data models (such as schema of relational databases, and navigations of graph databases) to cope with the semi-structured, hierarchical, navigational, complex, higher-order relationships [2, 9–11, 19, 20, 25]. *Complex relationships* link and represent multiple entities (that exist independently of a relationship, and are self-contained) and/or relations (for building higher-order relations), describing either a collection or a structure. Hypergraphs depicting complex relationships are visible in a monotonically growing bibliographic dataset where a journal article is

© The Author(s), under exclusive license to Springer Nature Switzerland AG 2022
C. Strauss et al. (Eds.): DEXA 2022, LNCS 13426, pp. 417–432, 2022.
https://doi.org/10.1007/978-3-031-12423-5_32

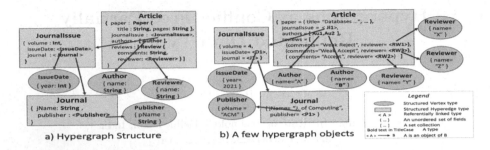

Fig. 1. Example showing a hypergraph structure to capture article relationship (part a), and hypergraph objects storing an article record (part b).

represented as a *structured higher-order* relationship between its authors, a collection of reviews, and a journal issue. A publisher publishes many journals, each of which often has many *issues*. The italicized *issue* refers to a journal issue. An *issue* is tied to its journal by its publication year and volume. Figure 1 depicts relevant entities and relationships as structured vertices (i.e., `Author`, `Reviewer`, `IssueDate`, and `Publisher`) and hyperedges (i.e., `Article`, `JournalIssue`, and `Journal`) of a hypergraph, each with a set of attributes.

In real-world, publishing journal articles/conference papers involves paper submission, review, and publication. These steps are often carried out at distant domain sites, with each site notifying the other sites of the article's revised status. These sites may span a wide geographical area and have diverse network connections. Thus, these systems need to be distributed with a high requirement of information availability (for efficient local query processing), along with network latencies. Since replication offers *availability* at the expense of *strong consistency* between the copies, that further requires synchronization [12]. Here, a weaker notion of consistency is required to ensure the consistency of replicated copies. *Conflict-free Replicated Data Types (CRDTs)* is a reasonable choice for maintaining consistency in highly dynamic environments [22,23]. CRDTs address the twin requirements of availability of data and operation under network partitioning without requiring complicated concurrency control mechanisms.

Requirements for Partial Replication: Furthermore, due to several constraints, including administrative policies (e.g., ownership of data), security and privacy concerns, and lack of physical resources, the entire data may not be required to distribute to all the domain replicas. A double-blind review, for instance, may require a limited exchange of in-process article information, and reviewers may be provided limited details about the journal issue. *Sharding* is a commonly used approach for determining data replication. However, distribution and replication of these partial chunks of data, called *shards*, needs attention.

Our Approach: In [3], we have leveraged the hypergraph semantics with CRDT approach to provide consistent updating and propagation of hypergraphical information across multiple replicas in the distributed environment. We introduced the first instance of a well-formed higher-order hypergraph CRDT called HgCRDT for full replication [3]. Now, we propose a modified version of HgCRDT: **H**ypergraph as **C**onflict-free **P**artially **R**eplicated **D**ata **T**ypes (**HgCPRDTs**), emphasizing the distribution of *shards* of the well-formed higher-order hypergraphs to multiple sites, ensuring the availability of data. The partial distribution of hypergraphical shards is performed according to distribution constraints/policies (hereafter DP). The hypergraphs are *well-formed* built using *schema* and *types*, which further brings *acyclicity* [9] into the structure. We exclude specifics about our ongoing implementation to emphasize the suitability of our proposed hypergraphs to be used with conventional CRDTs. However, the implementation introduces a database paradigm for managing our envisioned hypergraphs in a distributed (currently in a multi-threaded) environment.

Our Contribution: The major contributions of this paper are as follows:

1. We introduce and define *well-formed higher-order recursively-defined mutable partially-replicated hypergraph* termed as **Hypergraph as Conflict-free Partially Replicated Data Types (HgCPRDTs)**.
2. We formally specify HgCPRDTs to incorporate query, add, remove and modify operations on hypergraphical *atoms*.
3. We also give a proof-of-correctness showing how concurrent processes meet convergence conditions (essential for eventual consistency) in HgCPRDT.

2   Conflict-Free Replicated Data Types (CRDTs)

CRDTs [22,23] are data structures that guarantee *strong eventual consistency* [13] with minimal synchronization. The CRDT applies to data type representations in which the operations performed are conflict-free, allowing local modification to the data, replicating the data/operations asynchronously at the distributed locations, and then immediately returning to computation.

We are interested in a Set-based CRDT: **2P-Sets** [22] that uses two sets (say, A and R) as its payload (i.e., an internal data structure representation), for adding and removing elements. The set elements are interrogated using *query* operations. The *update operations* (that change the internal state of the data object) include *add* and *remove* operations that act as two phases of a sequential set. In the 2P-Set, once the elements are removed from A, and added to the remove (or *tombstone*) set R, cannot be reintroduced. The 2P-set uses "remove-wins" semantics, so $remove(e)$ takes precedence over $add(e)$. The 2P-set is a good data structure for storing shared session data, such as shopping carts, shared documents, or spreadsheets.

3   Partially Replicated Hypergraphs as a CRDT

Definition 1 (Hypergraph). *A hypergraph defined as:* $\mathbb{G}: (V, H)$, *where* $V = \{v_1, v_2, ..., v_n\}$ *is a set of vertex objects, and* $H = \{he_1, he_2, ..., he_m\}$ *a set of hyperedge objects representing entities and relationships, respectively.* $n, m \geq 0$. *Each* $v_i \in V$ *is an instance of a vertex type,* v_T, *and has a non-empty finite set of distinct attributes* F, *each with one or more scalar values. Each* $he \in H$ *is an instance of a hyperedge type,* he_T, *and includes* A, *a finite set of distinct intrinsic (i.e.,* $A' \in A$*) and extrinsic attributes (i.e.,* $A'' \in A$*).* $|A'| \geq 0, |A''| > 0$. *An attribute* $a \in A' \cup A''$ *may be a collection.*

Definition 2 (Intrinsic attribute). *An intrinsic attribute* $a \in he.A'$ *defines a property for storing a scalar (e.g., String, Int, Float, Boolean) or sub-structure value.*

Definition 3 (Sub-structure). *A sub-structure value* a_S *of a sub-structure type* a_{S_T} *includes a set of attributes* B, *where each* $b \in B$ *has either a scalar value or an atom object* $x \in (V \cup H)$. *The attribute* b *may be a collection too.*

Definition 4 (Extrinsic attribute). *An extrinsic attribute* $b \in he.A''$ *referentially*[1] *links one or a more atom objects* $y \in (V \cup H)$ *that exists outside* he.

We use well-formed higher-order *hypergraphs* in HgCPRDT as a collection of *schematic & typed vertex* and *directed hyperedge* objects, so are different than the traditional hypergraphs as given in [9]. We use the term hypergraph *atom* to abstractly refer to the schematic typed vertices and hyperedges. The atom *types* are defined in an object-oriented framework, and are derived from user-defined *schema* that provides domain-specific structures of different kinds of vertex and hyperedges. Note, vertex is the basic unit of information. Also, in a hyperedge *he*, an extrinsic attribute helps build the associated relationship.

Definition 5 (Well-formed hyperedge). *A hyperedge he is well-formed when added to* H *on satisfying the following constraints, avoiding any cycle and self-loop for every* $u_i \in U$, *where* $U \in (he.A'' \cup he.a_S)$, *and* $u_i \in H$:

1. $u_i \neq he$, *and*   ▷ *no self-loops*
2. $\forall\, he' \in H$:   ▷ *no cyclic dependency*
   (a) $u_i \notin he'.U$,
   (b) $\forall\, he'' \in he'.U : he'' \notin he$

Definition 6 (Well-founded hypergraphs). *A hypergraph* $\mathbb{G}$ *is well-founded if for every new hyperedge he to be added in* H, $\forall u \in (he.A'' \cup he.a_S)$, *the atom* u *must exist in* $(V \cup H)$.

---

[1] Referential linking means an object is linked via its implicit id.

For our HgCPRDT approach, we ignore intrinsic properties and consider hyperedges of the form $he(\textbf{mutable } atom \ set \ U)$, which connects a *set* of *atoms* U irrespective of being in a sub-structure of he. By insisting that the set of atoms U already exist when creating a hyperedge $he(U)$, we ensure that our hypergraphs are **well-founded**. As a consequence, a hyperedge cannot appear within its own set. Hyperedges are **mutable**, in that we permit the set of atoms to be modified. Notably, we treat atoms as **typed objects** with a unique *implicit* identity, avoiding the need to store the entire hyperedge where hyperedge members (particularly, the referentially-linked atoms) are themselves (independent) objects. Such hyperedges are **well-formed** and **acyclic**. Every hyperedge built as per the definition of 5 yields a well-formed hypergraph. Furthermore, two hyperedges with the same set are *not* the same. The implicit object ids, the hyperedge type, and the internal attributes make the hyperedges different.

The `Article(paper, reviews, authors, journalIssue)` in Fig. 1(b) is a well-formed acyclic typed hyperedge built atop existing (relevant) reviewers, authors, and journal issue objects. The `paper` (with `title` & `pages` attributes) and `reviews` (with `comments` & `reviewer` attributes) are inherent and essential parts of a published article and are thus used as *intrinsic sub-structure attributes* of the `Article`. The `reviews` is a collection, and its `reviewer` attribute belongs to `Reviewer` vertex type. The *journalIssue* and *authors* contain independent objects of `JournalIssue` hyperedge and `Author` vertex types, respectively, and are thus considered *extrinsic attributes* of the `Article`. The `Article` and `JournalIssue` are higher-ordered hyperedges due to the complex relationships between hyperedges (i.e., $Article \rightarrow JournalIssue \rightarrow Journal$; and $JournalIssue \rightarrow Journal$). The `Article(reviews.reviewer, authors, journalIssue)` is considered an illustration of the form $he(mutable \ atom \ set \ U)$.

Traditional CRDTs use state-based and operation-based replication, yielding in *Convergent Replicated Data Types (CvRDTs)* and *Commutative Replicated Data Types (CmRDTs)*, respectively [22,23]. A *state* captures the *update* information and is transmitted to all the replicas in CvRDTs; whereas, only the *update operations* are replicated to all the replicas in CmRDTs. Since the eventual transmission of the entire state in CvRDTs may be costly in terms of memory and computation for large data structures, we prefer **operation-based CmRDT replication** in our work to make HgCPRDT suitable for large hypergraphs.[2] The **communication model** of the HgCPRDT is similar to that of the CRDTs [22], that works with an underlying *causally-ordered broadcast communication protocol* [16] via a reliable delivery mechanism that helps to further

---

[2] A conference may include sub-conferences, workshops, and journals; and a journal or a conference may receive a large volume of submissions as shown in [1].

reduce inconsistencies between replicas by restricting the operations seen in possibly different orders at the replicas to only *concurrent operations*. Therefore, the same replica can simultaneously send and receive different or the same messages. We assume that the communication infrastructure notifies the replicas about the sender with update information.

To facilitate the adoption of our technique, we use the template provided by the *operation-based 2P2P-Graph specification* [22,23] for hypergraphs. Our proposed hypergraph specification uses two tombstone sets (VR, HR) to represent the 2P-Sets, which relaxes in some instances the requirement for a causal order of delivery; an example includes multiple removals of the same atom in a set; and thus permits some additional asynchrony. Other variants of CRDT Sets are possible, such as OR-Sets, though the commutativity properties need to be carefully verified for each such choice.

Note that, vertices are the base case for atoms (which also include hyperedges) and that hyperedges relate the atoms of a *set* to each other. The novelty of this work lies in this treatment of such well-founded recursive hypergraphical structures. Another novelty is that the set incident on a hyperedge is itself *mutable* 2P-Set. Hyperedges are built on object references to store the atom set rather than the complete hyperedge itself (in the implementation). Consequently, hyperedges are mutable, as we may add and remove atoms incident on the hyperedge. The use of tombstone sets allows deletion of an atom from a set; however, since the atoms are implemented as typed objects having their implicit identity, the atoms persist across such modifications. The usage of implicit object identities explains why traditional CRDT models like Key-Value pairs and maps are not suitable for encoding hypergraphs, even after some transformation.

Hypergraphs are particularly well-suited for **partial replication** since a hyperedge's projection containing only some of its set's atoms is still a hyperedge. In other words, hyperedges are closed under projections to a subset of atoms. The HgCPRDT specification is designed to support sharding with full or partial replication of shards under specific user-defined DPs, considering the number of active threads/systems in a domain according to the availability requirements. Under the presumption of a static number of domain replicas, a DP (that may be a string) must specify which atom type may be shared with which replica.

4   Specification of HgCPRDTs

The HgCPRDT specifies the payload as four sets: VA, VR, HA, and HR for adding and removing vertices and hyperedges (initialized as empty); and comprises its three types of interface operations: *auxiliary*, *query*, and *update* operations (i.e., add, remove, and modify). The applicability of HgCPRDT operations in distributing in-process article objects is shown in Fig. 2.

Table 1. HgCPRDT: payload and auxiliary operations

▷ VA : *vertex add set*, VR : *vertex remove set*,
▷ HA : *hyperedge add set*, HR : *hyperedge remove set*

payload set VA, VR, HA, HR
▷ *internal data structure of a replica*
  initial ϕ, ϕ, ϕ, ϕ ▷ *"initial" specifies initial values of payload sets at every replicas*

query *inShard* (atom a, location l) : boolean b   ▷ *"let" marks non − mutating statements*
  let $b = \begin{cases} true & if\ a\ is\ to\ be\ replicated\ to\ location\ l\ as\ per\ sharing\ policy \\ false & otherwise \end{cases}$

shardAtom (atom a): atom set array   ▷ *atom set array A*
  if a is a vertex: $A := shardVertex(a)$
  otherwise: $A := shardHyperedge(a)$
  return A

shardAtomset (atom set X): atom set array
▷ *atom set array T, S; location set L*
  $\forall(l \in L):\ T[l] := \emptyset$
  $\forall(x \in X):\ S := shardAtom(x)$
  $\forall(l \in L):\ T[l] := T[l] \cup S[l]$
  return T

shardVertex(vertex v): vertex set array   ▷ *vertex set array S*
  $\forall(l \in L):\ S[l] := \{v \mid inShard(v, l)\}$
  return S

shardHyperedge(hyperedge $he(U)$) : hyperedge set array   ▷ *he(mutable atom set U)*
  $U' := shardAtomset(U)$
  $\forall(l \in L):\quad S[l] := \{he(U'[l]) \mid inShard(he(U'[l]), l)\}$ ▷ *hyperedge set array S*
  return S

Auxiliary operations (Table 1) help in preparing shards to be used by query and update operations. These are non-mutable operations that are performed locally at each replica. *inshard* embodies the DP for each atom- given an atom and a location, it returns true if and only if the atom is (to be) present at that location. Given an atom a, *shardAtom* computes a location-indexed array of atom sets, which are either the singleton $\{a\}$ or the empty set $\emptyset$ according to the DP. In our implementation, this is efficiently represented as a bit-vector. *shardAtom* is defined in terms of *shardVertex* and *shardHyperedge*. *shardAtomset* computes a location-indexed array of atom sets, each component of which is the subset of the given atom set X that is to be replicated at that location. Since hyperedges have a set of atoms, *shardHyperedge* uses *shardAtomset* to project those atoms of the given hyperedge that are replicated at each location. Note that if the hyperedge itself is not replicated at a particular location, we need not worry about replicating the node's incident on it. Here too, we use bit-vectors for efficient representation of the location-indexed arrays.

Table 2. HgCPRDT: query operations

---

query *lookupAtom* (atom a) : boolean b   ▷ *"query" indicates a non − mutable operation*
  if a is a vertex: *lookupVertex* (a)
  otherwise: *lookupHyperedge* (a)

query *lookupAtomset* (atom set S) : boolean
  let $b = \left(\bigwedge\limits_{\forall u \in\ S} lookupAtom(u) \right)$

query *lookupVertex* (vertex v) : boolean
  let $b = (v \in (VA \setminus VR))$

query *lookupHyperedge* (hyperedge $he(U)$) : boolean
  let $b = (lookupAtomset(U) \wedge\ he(U) \in (HA \setminus HR))$

query *within* (atom a, hyperedge $he(U)$) : boolean b   ▷ *checking for acyclic hyperedges*
  let $b = \begin{cases} true & \text{if } a = he(U) \vee a \in U \vee \exists he(U') \text{ s.t.} \\ & lookupHyperedge(he(U')) \wedge a \in U' \wedge within(he(U'), he(U)) \\ false & \text{otherwise} \end{cases}$

Query operations (Table 2) are immutable operations that interrogate the state of the mutable object (hypergraphs in this case) locally. Unlike full replication, partial replication limits the queries and answers at a location to those that are permitted to be asked and those that are permitted to be present in that replica. Accordingly, the correctness criterion for the eventual consistency of the partial replicas is that the answer to a permitted query at any location is the *projection*, as per the DP, of the answer to the query asked on the (perhaps hypothetical) complete hypergraph object.

The *lookupAtom* checks for the presence of an atom, whether a vertex (*lookupVertex*) or a hyperedge (*lookupHyperedge*), as the case may be, in the partial replica of the hypergraph present at that location. The *lookupAtom* operation is lifted to sets of atoms using a conjunction. The precondition of the *lookupHyperedge* query checks for the existence of all atoms in the set. Since we permit the set to be mutable, the payload sets HA, HR only contain reference-based structures for the hyperedges, and the set is accessed by dereferencing. *within* operation checks that the given hyperedge should be acyclic. Therefore, it recursively checks if a given atom a appears within a given hyperedge.

Update Operations are mutable global operations executed by *source* replicas (the site that send their update information to other replicas), and *downstream* replicas (i.e., a recipient site) in two phases- *prepare at source*, and *effect at downstream*. A *source replica* initiates an update operation and tailors an argument (i.e., *location-specific update shards*) locally in *prepare at source* phase according to a DP. The location-specific shards are prepared so that all necessary data are sent to the *downstream replicas* so that any query operation permitted by the DP at any particular location can be performed *strictly locally*. After receiving the argument, each downstream replica processes it in the later phase atomically and asynchronously using a causal delivery order. The causal

Table 3. HgCPRDT: add operations

---

update *addAtom* (atom *a*)   ▷ *"update" indicates a mutable operations*
  if *a* is a vertex: *addVertex(a)*
  otherwise: *addHyperedge(a)*

update *addVertex* (vertex *v*)
  ▷ *A source replica prepares location-specific shards for downstreams in this phase.*
  prepare at source
    parameter *v*   ▷ *"parameter" explicitly describes the arguments used by a replica*
    let $S = shardVertex(v)$
  ▷ *Each downstream uses "effect at downstream" phase to process the received shards.*
  effect at downstream *l* s.t. $S[l] \neq \emptyset$ ▷ *l is the location of a downstream*
    parameter $S[l]$
    ▷ *"pre" specifies preconditions that must be satisfied for an operation to be invoked*
    pre $S[l] = \{v\}$
    $VA := VA \cup \{v\}$

update *addVertexSet* (vertex set *X*)
  prepare at source
    parameter *X*
    let $S = shardAtomset(X)$
  effect at downstream *l* s.t. $S[l] \neq \emptyset$
    parameter $S[l]$
    $\forall(v \in S[l]): \; VA := VA \cup \{v\}$

update *addHyperedge* (hyperedge *he(U)*)   ▷ *he(atom set U)*
  prepare at source
    parameter *he(U)*
    pre *lookupAtomset(U)*
    let $S = shardHyperedge(he(U))$
  effect at downstream *l* s.t. $S[l] \neq \emptyset$
    parameter $S[l]$
    pre $(S[l] = \{he(U')\}) \wedge lookupAtomset(U')$
    $HA := HA \cup S[l]$

---

delivery order of asynchronous update messages from the source to downstream sites ensures that any local query will result in a legitimate answer. Note that the notion of sources and downstream sites is not statically fixed. Nor do we assume that any single site has a copy of the complete hypergraph in our framework.

Add Operations (Table 3): The *addAtom* operation adds a vertex or a hyperedge, depending on the kind of atom specified. In these **update** operations, the arguments are prepared at the source location, then the partial replicas (or shards) for each location is computed. The operation is then *effected* immediately at the source, and if the parameter is non-trivial, also sent asynchronously but reliably to the downstream locations, where it is affected atomically. If the shard for a downstream location is empty, then we do not send it any message: this is not only for efficiency but also to prevent information leakage. Adding a set of hyperedges can be realized by iterating the *addHyperedge* operation. Note that when adding a hyperedge, all atoms in its set must exist, and thus the corresponding add operations for all these atoms must have been delivered earlier. The sharding operations take care to ensure the necessary invariant properties.

Table 4. HgCPRDT: remove operations

---

update *removeAtom* (atom a)
  if a is a vertex: *removeVertex*(a)
  otherwise: *removeHyperedge*(a)

update *removeVertex* (vertex v)
  prepare at source
    parameter v
    pre $lookupVertex(v) \land \forall(he\{U\} \in (HA \setminus HR)) : \neg U.lookupVertex(v)$
    let $S = shardVertex(v)$
  effect at downstream l s.t. $S[l] \neq \emptyset$
    parameter $S[l]$
    pre $(S[l] = \{v\}) \land addVertex(v)$ delivered
    $VR := VR \cup \{v\}$

update *removeHyperedge* (hyperedge $he(U)$)
  prepare at source
    parameter $he(U)$
    pre $lookupAtom(he(U)) \land \forall(he\{U'\} \in (HA \setminus HR)) : \neg U'.lookupAtom(he(U))$
    let $S = shardHyperedge(he(U))$
  effect at downstream l s.t. $S[l] \neq \emptyset$
    parameter $S[l]$
    pre $(S[l] = \{he'\}) \land addHyperedge(he')$ delivered
      $\land \forall(he\{U'\} \in (HA \setminus HR)) : \neg he.U'.lookupAtom(he')$
    $HR := HR \cup \{he'\}$

Table 5. HgCPRDT: modify operation

---

update *changeHyperedge* (hyperedge $he(U)$, S^+, S^-)   ▷ $he($**mutable** *atom set U*$)$
  prepare at source   ▷ $S^+ =$ *atoms to be added,* $S^- =$ *atoms to be removed at source*
    parameter $he(U)$, *atom set* S^+, S^-
    pre $lookupAtomset(S^+) \land U.lookupAtomset(S^-) \land lookupHyperedge(he(U)$
      $\land \forall(x \in S^+) : \neg within(x, he(U))$
    let $D = shardHyperedge(he(U))$
    let $DS^+ = shardAtomset(S^+)$ ▷ $DS^+ =$ *atoms to be added at a downstream*
    let $DS^- = shardAtomset(S^-)$ ▷ $DS^- =$ *atoms to be removed at a downstream*
  effect at downstream l s.t. $D[l] \neq \emptyset$
    parameter $D[l]$, $DS^+[l]$, $DS^-[l]$
    pre $(D[l] = \{he'\}) \land addHyperedge(he')$ delivered $\land lookupAtomset(DS^+[l])$
      $\land \forall(x \in DS^+) : \neg within(x, he')$
    $\forall(x \in DS^-[l]) : U.removeAtom(x);$
    $\forall(x \in DS^+[l]) : U.addAtom(x);$

Remove Operations (Table 4): Again, the source location prepares the shards to be removed from each location, sending the parameter asynchronously to only those downstreams where some non-trivial action needs to be performed. An atom incident on a hyperedge can only be deleted after the hyperedge itself has been removed. Note that deleting an atom requires that it should not be incident on *any* hyperedge (should not be in the set of any hyperedge). Thus the precondition ensures that it cannot possibly appear within any higher-order hyperedge. We do not present here the *remove* operations lifted to a set of atoms.

Modify Operations (Table 5)**:** It is always possible to modify a hyperedge in a hypergraph by deleting the existing edge and replacing it with the modified edge. It entails ensuring that any atoms present (recursively) within the new set of the new hyperedge must already exist (and must not be the hyperedge itself). However, since hyperedges are complex structures, their modification is expensive. Instead, we specify the modification of a hyperedge by the addition or removal of atoms in a set via *changeHyperedge* operation. Note that we now require a set to *itself* be *mutable* 2P2P-Set: the vertices and hyperedges in the sets $U.V, U.H$ are respectively subsets of the two 2P-Sets V, H of the global hypergraph object, represented by payload VA, VR and HA, HR in the tombstone implementation. By global hypergraph, we mean the replica's state consisting of payload sets, irrespective of any particular hyperedges.

The *changeHyperedge* operation takes an existing hyperedge $he(U)$, and the atom sets S^+, S^- that are to be added to and removed from the set U. For simplicity, assume that $S^+ \cap S^- = \emptyset$, $S^+ \cap U = \emptyset$ and $S^- \subseteq U$. For readability, we use the set operations of intersection and subset. These conditions can be expressed in terms of the query operations. Note that in the precondition of *changeHyperedge*, we need to check that the set S^+ being added should exist in the (global) hypergraph, whereas the set being deleted S^- should already be in the *mutable* source and target sets of the given hyperedge. Note also that in the **effect** phase, the atoms from the various sets are removed/added to the set of the hypergraph. Observe that the atoms are only removed from the set of the hyperedge, but *not* from the (global) hypergraph because hyperedges are formed using references of existing other atoms. Vertex modify is a trivial operation, and therefore we ignore it in this paper.

4.1   Example with HgCPRDT Operations

Figure 2 demonstrates the partial distribution of an in-process article record in the HgCPRDT framework on the three sites- S1, S2, and S3, each with initial empty payload sets. Vertices and hyperedges are added to the system in case of no earlier existence. Initially, S1 acts as a *source* and adds two vertices A and B for authors, and a hyperedge Ar for an article. Meanwhile, S2 initiates *add* vertices for reviewers X, Y, and Z. Note that a *source* site prepares shards for every downstream but transmits only the non-empty shards. Thus, ensuring a double-blind review, neither the authors nor the reviewers are shared with S2, S1, respectively. However, the *source* sites broadcast the shards of the updated articles with *null* objects for the non-shared objects. Since each review has two fields: *comments* and *reviewers*, the updated article at S1 only includes the comments, whereas S3 contains the both. Now, S3 initiates *add* operations for a publisher, P and a journal, J. Due to the causal sequence, the arrival of the *add P* and *add J* operations at $S1$ does not result in a conflict, although both operations came from the same source. This is followed by the inclusion of an *issue*, $JI2$ (having *volume* 3) at S3, which requires a revision to the article. In case there is a need to change the *issue* (having another *vol.*) of the article, the deletion of $JI2$ is required; however, the *removeHyperedge(JI2)* operation fails due to the hyperedge Ar being dependent on it. Thus, the article is updated with a newly added *issue*

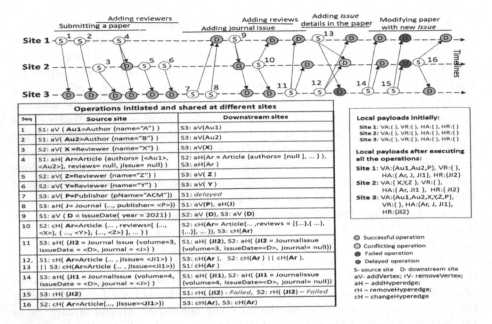

Fig. 2. The partial distribution of an in-process article using HgCPRDT operations.

hyperedge $JI1$ (having *vol.* 4), which may follow the deletion of $JI2$ to add $JI2$ in the tombstone set (HR). Due to a delay in the network, concurrent operations may reach downstream. Some concurrent operations on the same object do not or minimally affect the payload sets. On the other hand, the concurrent operations on different objects that affect the payload sets are handled via the causal delivery order of operations. Examples include: $addVertex(P) \parallel addHyperedge(J)$ at S1, and $changeHyperedge(Ar) \parallel changeHyperedge(Ar)$ at S2.

4.2   Proof of Correctness

Most of the arguments related to 2P-Sets and 2P2P-Graphs [22] carry over in the proof that this specification implements a CRDT. It is easy to show that *add* operations or *remove* operations on unrelated atoms naturally commute. If, however, an atom appears (recursively) within the set of another atom, then adding the second atom must causally follow the addition of the first atom. The delivery order ensures it. The reverse holds for remove operations. Concurrent add and remove operations on the same atom are resolved by the 2P-Set condition [22]. The cases of concurrent $remove(u)$ and $add(w)$ [or $add(u)$ and $remove(w)$] operations where there is a some relationship between u, w, are dealt with using the 2P-Set conditions, the conditions on adding or removing atoms, and transitivity.

Operations other than the *remove* are independent of the *changeHyperedge*. The tombstone set will ensure that removal prevails over modifications. Modifications to different hyperedges commute. Consider two concurrent modifications to the same hyperedge with changes S_1^+, S_1^- and S_2^+, S_2^- respectively. We claim that the operations can safely commute (refer to the Lemma 1), resulting in set $(U \cup S_1^+ \cup S_2^+) \setminus (S_1^- \cup S_2^-)$. Atoms appearing in the corresponding add set (or removal set) pose no problem. The assumptions about the sets of atoms being added or removed from a given set within each operation allow the commutation.

Lemma 1. *Concurrent* changeHyperedge(he, S_1^+, S_1^-) *and* changeHyperedge(he, S_2^+, S_2^-) *commute. (For the proof, see the Appendix A.)*

5 Related Work

The basic CRDTs include counters, registers, sets, and ordered lists [21,23,24]. A complex CRDT includes *Graph* [22,23] that uses two 2P-Sets for adding and removing vertices and edges. Existing work on higher-order CRDT includes JSON data structure [17] (composes lists, maps, and registers to embed JSON types); Riak [5] (defines maps as a CvRDT); Causal trees [14] (represents ordered trees into graphs); Logoot [24] (uses a sparse non-mutable n-ary tree to nest ordered lists); higher-order patterns [18]. The CRDT-based partial replication includes [?]- uses delta mutations over CvRDTs; [8]- computes digests for partial replication; [4]- defines a partial order over operations; [?]- performs sharding on a set of *particle*; [6]- introduces a non-uniform replication on *top-K* without removals and histograms, and *top-Sum*; and [7]- clubs replica sharding with a delta-based CRDT approach. An instance of CRDTs employed in databases includes *SU_Sets* [15] for RDF graphs and the SPARQL update operations. Our work differs in that it deals with hypergraphs, a more complex, higher-order data type, and contains modifiable mutable structures. We use hyperedges of well-formed schematic hypergraphs to represent complex relationships, allowing the nesting of hyperedges built on references that make their members independent. Additionally, sharding in HgCPRDT is performed at source, as dictated by schema according to administrative policies.

6 Conclusion

We have introduced a framework for partial replication of the commutative conflict-free hypergraphs. We have proposed and specified a new CRDT- a *well-formed higher-order recursively-defined mutable partially-replicated hypergraph* named HgCPRDT as an operation-based specification of 2P2P that works with tombstone sets. In HgCPRDT, hyperedges are modifiable and closed under projections to a subset of atoms. We have proved its correctness by showing how concurrent processes meet convergence conditions and guarantee strong eventual consistency in HgCPRDT. Future work involves studying the performance and scalability of partially replicated hypergraphs and developing systems dealing with various large hypergraphs on real data. Changes in the properties of vertices and hyperedges must also be accounted for in addition to changes in the structure.

Acknowledgements. I wish to acknowledge the fruitful discussions and collaborative work of Sanjiva Prasad (IIT Delhi, India), the peer reviews of Madhulika Mohanty (Inria, Saclay) and Himanshu Gandhi (IIT Delhi, India), and the anonymous reviewer's comments.

A   Proof of Lemma 1

Proof. According to the *changeHyperedge* operation, a set of atoms S^+ are added to, and a set of atoms S^- are removed from a hyperedge. Therefore:

$$changeHyperedge\ (he,\ S_1^+,\ S_1^-) = U \cup S_1^+ \text{ and } U \setminus S_1^- = (U \cup S_1^+) \setminus S_1^-$$

Similarly,

$$changeHyperedge\ (he,\ S_2^+,\ S_2^-) = U \cup S_2^+ \text{ and } U \setminus S_2^- = (U \cup S_2^+) \setminus S_2^-$$

The concurrent execution of both the change operations on the same hyperedge results on each of the replicas:

$$changeHyperedge\ (he,\ S_1^+,\ S_1^-) \parallel changeHyperedge\ (he,\ S_2^+,\ S_2^-)$$
$$= (U \cup S_1^+ \cup S_2^+) \setminus (S_1^- \cup S_2^-) \parallel (U \cup S_2^+ \cup S_1^+) \setminus (S_2^- \cup S_1^-)$$

Further, the commutative *set-union* operation makes the results equivalent:

$$(U \cup S_1^+ \cup S_2^+) \setminus (S_1^- \cup S_2^-) \equiv (U \cup S_2^+ \cup S_1^+) \setminus (S_2^- \cup S_1^-)$$

Therefore, modification of concurrent operations to the same hyperedge commute.

References

1. ACL2020: General conference statistics (2008). https://acl2020.org/blog/general-conference-statistics/. Accessed 06 Apr 2022
2. Angles, R., Gutierrez, C.: Survey of graph database models. ACM Comput. Surv. (CSUR) **40**(1), 1 (2008)
3. Bansal, A.: Conflict-free replicated hypergraphs. In: The Fourteenth International Conference on Advances in Databases, Knowledge, and Data Applications (DBKDA), p. 8 (2022)
4. Baquero, C., Almeida, P.S., Shoker, A.: Making operation-based CRDTs operation-based. In: Magoutis, K., Pietzuch, P. (eds.) DAIS 2014. LNCS, vol. 8460, pp. 126–140. Springer, Heidelberg (2014). https://doi.org/10.1007/978-3-662-43352-2_11
5. Brown, R., Lakhani, Z., Place, P.: Big(ger) Sets: decomposed delta CRDT Sets in Riak. In: Proceedings of the 2nd Workshop on the Principles and Practice of Consistency for Distributed Data, p. 5. ACM (2016)

6. Cabrita, G., Preguiça, N.: Non-uniform replication. In: 21st International Conference on Principles of Distributed Systems (OPODIS 2017). Leibniz International Proceedings in Informatics (LIPIcs), vol. 95, pp. 24:1–24:19 (2018)

7. Deftu, A., Griebsch, J.: A scalable conflict-free replicated set data type. In: 2013 IEEE 33rd International Conference on Distributed Computing Systems (ICDCS), pp. 186–195. IEEE (2013)

8. Enes, V., Almeida, P.S., Baquero, C., Leitão, J.: Efficient synchronization of state-based CRDTs. In: 2019 IEEE 35th International Conference on Data Engineering (ICDE), pp. 148–159. IEEE (2019)

9. Fagin, R.: Degrees of acyclicity for hypergraphs and relational database schemes. J. ACM **30**(3), 514–550 (1983)

10. Gallo, G., Longo, G., Pallottino, S., Nguyen, S.: Directed hypergraphs and applications. Discret. Appl. Math. **42**(2–3), 177–201 (1993)

11. Ghaleb, F., Taha, A.A., Hazman, M., ElLatif, M., Abbass, M.: RDF-BF-hypergraph representation for relational database. Int. J. Math. Comput. Sci. **15**(1), 41–64 (2020)

12. Gilbert, S., Lynch, N.: Perspectives on the CAP theorem. Computer **45**(2), 30–36 (2012)

13. Gomes, V.B., Kleppmann, M., Mulligan, D.P., Beresford, A.R.: Verifying strong eventual consistency in distributed systems. In: Proceedings of the ACM on Programming Languages vol. 1, no. OOPSLA, p. 109 (2017)

14. Hall, A., Nelson, G., Thiesen, M., Woods, N.: The causal graph CRDT for complex document structure. In: Proceedings of the ACM Symposium on Document Engineering (2018)

15. Ibáñez, L.D., Skaf-Molli, H., Molli, P., Corby, O.: Live linked data: synchronising semantic stores with commutative replicated data types. Int. J. Metadata Semant. Ontol. **8**(2), 119–133 (2013)

16. de Juan-Marín, R., Decker, H., Armendáriz-Íñigo, J.E., Bernabéu-Aubán, J.M., Muñoz-Escoí, F.D.: Scalability approaches for causal multicast: a survey. Computing **98**(9), 923–947 (2015). https://doi.org/10.1007/s00607-015-0479-0

17. Kleppmann, M., Beresford, A.R.: A conflict-free replicated JSON datatype. IEEE Trans. Parallel Distrib. Syst. **28**(10), 2733–2746 (2017)

18. Leijnse, A., Almeida, P.S., Baquero, C.: Higher-order patterns in replicated data types. In: Proceedings of the 6th Workshop on Principles and Practice of Consistency for Distributed Data. ACM (2019)

19. Levene, M., Poulovassilis, A.: An object-oriented data model formalised through hypergraphs. Data Knowl. Eng. **6**(3), 205–224 (1991)

20. Prasad, S.: Designing for scalability and trustworthiness in mHealth systems. In: Natarajan, R., Barua, G., Patra, M.R. (eds.) ICDCIT 2015. LNCS, vol. 8956, pp. 114–133. Springer, Cham (2015). https://doi.org/10.1007/978-3-319-14977-6_7

21. Preguiça, N., Marquès, J.M., Shapiro, M., Letia, M.: A commutative replicated data type for cooperative editing. In: 2009 29th IEEE International Conference on Distributed Computing Systems, pp. 395–403. IEEE (2009)

22. Shapiro, M., Preguiça, N., Baquero, C., Zawirski, M.: Conflict-free replicated data types. In: Défago, X., Petit, F., Villain, V. (eds.) SSS 2011. LNCS, vol. 6976, pp. 386–400. Springer, Heidelberg (2011). https://doi.org/10.1007/978-3-642-24550-3_29

23. Shapiro, M., Preguiça, N., Baquero, C., Zawirski, M.: A comprehensive study of Convergent and Commutative Replicated Data Types. Research report RR-7506, INRIA Centre Paris-Rocquencourt (2011)

24. Weiss, S., Urso, P., Molli, P.: Logoot-undo: distributed collaborative editing system on P2P networks. IEEE Trans. Parallel Distrib. Syst. **21**(8), 1162–1174 (2010)
25. Wolf, M.M., Klinvex, A.M., Dunlavy, D.M.: Advantages to modeling relational data using hypergraphs versus graphs. In: 2016 IEEE High Performance Extreme Computing Conference (HPEC), pp. 1–7. IEEE (2016)

A Method for Summarizing Trajectories with Multiple Aspects

Vanessa Lago Machado[1,2](✉) , Ronaldo dos Santos Mello[1] ,
and Vania Bogorny[1]

[1] PPGCC, Universidade Federal de Santa Catarina, Florianópolis, Brazil
vanessalagomachado@gmail.com,
{r.mello,vania.bogorny}@ufsc.br
[2] Instituto Federal Sul-Rio-Grandense, Passo Fundo, Brazil

Abstract. Trajectory data mining and analysis have been largely studied in the past years. These tasks are complex and non-trivial due to the data volume and heterogeneity. One solution for these problems is data summarization in order to generate representative data. Few works in the literature address this solution, and none of them consider space, time, and unlimited semantic dimensions and their data type details. This paper proposes a grid-based method for multiple aspects trajectory data summarization named MAT-SG. It brings several contributions: (i) trajectory segmentation into a spatial grid according to data point dispersion; (ii) it expresses a set of trajectory data by a sequence of representative points with representative values for each dimension, considering their data type particularities. We evaluate MAT-SG over two datasets to assess volume reduction and accuracy.

Keywords: Trajectory summarization · Multiple aspects trajectory · Representative trajectory

1 Introduction

With the explosion of the *Internet of Things*, many technologies have emerged, such as portable devices, embedded computing and location-based social networks, which provide data about the movement of objects. These collected data are called *moving object trajectories*, and they are often used in data analysis activities by many application domains, such as traffic control [1,8,17], animals migration [4,8], hurricane prediction [10,16,17] and vessel monitoring [6].

Trajectory data have quickly evolved over time. A *raw trajectory* is a sequence of trajectory points over the geographic space in the time *(x, y, t)* [5]. When a raw trajectory is enriched with semantic information, such as a *point of interest (PoI)* the object had visited, this trajectory is known as *semantic trajectory*. At last, when an entire trajectory, or some of its points, is associated with many semantic contexts, we have the recent concept of *multiple aspects trajectory (MAT)* [12].

© The Author(s), under exclusive license to Springer Nature Switzerland AG 2022
C. Strauss et al. (Eds.): DEXA 2022, LNCS 13426, pp. 433–446, 2022.
https://doi.org/10.1007/978-3-031-12423-5_33

Fig. 1. An example of MAT [12]

Figure 1 shows a MAT of an individual during a day. In this example, the raw trajectory (spatial and temporal dimensions) is enriched with semantic information such as PoIs, means of transportation, weather and health information.

Understanding patterns in trajectories can help data analysts make better decisions. Recommendation systems, for example, deal with the analysis of users' behaviors to find products of interest or suggest actions that will let he/she more healthy/satisfied. Figure 2 (left) shows the MATs of a woman. On the right side, her MATs are summarized in a representative MAT that presents the actions that she frequently does. From that, we see that she goes to work on weekdays and lunch at a vegetarian restaurant between 0:30 pm and 1:30 pm.

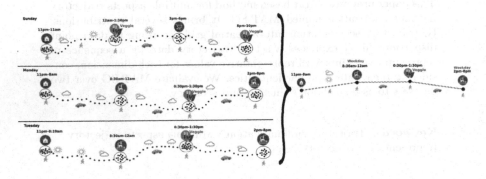

Fig. 2. Example of MATs (left) and a representative MAT for them (right)

Based on the knowledge provided by the representative MAT of Fig. 2 (right), a recommendation system can learn about her habits and make recommendations. Suppose the same woman is in a different location and she decides to move around at 1 pm. Suppose also that the system identifies a vegetarian restaurant on her trajectory. Given this context, it recommends this restaurant to her as 1 pm is the time that she usually has lunch and she prefers vegetarian restaurants.

From this example, we see that *trajectory data summarization* is helpful to reduce the complexity of the data to be processed for further analysis. Some surveys point out that semantic trajectory data summarization is an open issue [7,18]. This lack of works is probably due to the complexity of these data, as different semantic contexts may coexist and be related to parts of a trajectory, which makes data summarization tasks more challenging.

This paper proposes a novel method for summarizing MAT data: *MAT-SG (Multiple Aspect Trajectory Summarization based on a spatial Grid)*. It is based on a spatial grid that covers a set of input MATs. For all points in a same grid cell we generate a representative point. In turn, a *representative MAT* is generated from the sequence of representative points. Our data summarization aims to reduce the volume of MATs data with low accuracy loss.

The main contribution of this paper is the detailed treatment of all MAT dimensions (*spatial, temporal* and *semantic*) for summarization purposes. Regarding spatial summarization, we segment the input MATs in a grid of cells by dimensioning the cell size according to the dispersion of the MATs points. Regarding temporal summarization, we discover and rank the most significant time intervals in a cell. Regarding semantic summarization, we rank the semantic values that best represent the behaviour of the cell points. We evaluate our method over two datasets (Foursquare and a synthetic one), with promising results.

The rest of this paper is organized as follows. Section 2 presents the basic concepts associated with MAT-SG. Section 3 is dedicated to related work. Section 4 describes the proposed method. Section 5 presents an evaluation and Sect. 6 concludes the paper and outlines future works.

2   Fundamentals

This section defines basic concepts that are relevant to this paper [12,13,15].

Definition 1 *Aspect (asp).* *Let asp = (desc, asp_type, ATT) a relevant real-world fact, where desc is its description, asp_type is the aspect type that characterizes it, and ATT is a set of attribute-value pairs that describe its properties.*

An aspect type is a categorization of a real-world fact. It is essentially any information that can be annotated on a trajectory. For instance, it is possible to define aspect types such as *social media post, weather condition, PoI,* and *mean of transportation.* Each one can contain attributes with different types of categorical and numerical data. A hotel, for example, can be described by a *name, address* and *stars.* A MAT can hold several aspects, as specified next.

Definition 2 *Multiple Aspect Trajectory (MAT).* *A MAT is a sequence of points $(p_1, p_2, ..., p_n)$, with $p_i = (x, y, t, A)$ being the i-th point of the trajectory generated in the location (x,y) at timestamp t, and described by the set $A = \{a_1 : v_1, a_2 : v_2, ..., a_r : v_r\}$ of r attributes of the related aspects.*

A MAT holds a set of points, and a MAT point is a complex element with many aspects' attributes (attribute-value pairs) besides space and time data.

This work is also related to *data summarization.* It aims to reduce the size of data to obtain approximate data in comparison with the original contents in a larger dataset so most of them satisfy the user requirements [9]. It can be defined in different ways, according to the considered focus, like interest, density, frequency, and pairwise distance [14]. We understand *trajectory summarization* as a process that abstracts data from a set of MATs and generates a *representative MAT* without necessarily hold a 100% similarity with all the individual MATs.

3   Related Work

Most of works that generates representative trajectories are limited to raw trajectories [3,4,6,8,10]. Table 1 shows the related work.

Table 1. Related work comparison

Study	Summarized dimensions			Activities	Mapping data specification
	Space	Time	Semantic		
Lee et al. (2007) [10]	X			– Partitioning & cluster (subtrajectory) – Fusion of each group of points	
Buchin et al. (2013) [3]	X			– Partitioning into subtrajectories – Median trajectory computation	
Ayhan and Samet (2015) [2]	X	X		– Partitioning & Fusion (subtrajectory) – Clustering & Fusion (points)	
Etienne et al. (2016) [6]	X	X		– Median trajectory computation	
Agarwal et al. (2018) [1]	X	X		– Partitioning & cluster (subtrajectory)	
Buchin et al. (2019) [4]	X	X		– Partitioning & cluster (subtrajectory) – Minimal GD computation	
Gao et al. (2019) [8]	X	X		- Clustering points — ROIs	
Seep and Vahrenhold (2019) [17]	X	X	X	– Finite State Machine	
Rodriguez and Ortiz (2020) [16]	X	X		– Partitioning & cluster (subtrajectory) – Fusion of each group of points	
MAT-SG	X	X	X	– Partitioning in cell grid – Clustering points – Representative Point Computation	X

Partitioning and clustering are the main activities performed by the related work. They usually partition the trajectories into subtrajectories or points and cluster the partitioned data. Then, data summarization occurs at each cluster.

Only one method considers the semantic dimension like MAT-SG [17]. However, all attributes of the points are treated as spatial or non-spatial data, i.e., semantic data are not analyzed individually as categorical or numeric data. The work adopts an approach that identifies a sequence of transitions common to most of the movements using a finite state machine. Each state denotes a common point, and a sequence of states generates the representative trajectory. A strong limitation of this work is the lack of method details, as it is a short paper.

Different from related work, MAT-SG is a detailed data summarization method that generates a representative MAT for a set of MATs, i.e., a trajectory enriched with unlimited semantic information treated as categorical or numerical data. Our approach treats semantic data individually, allowing us to understand the patterns and influence of each data in the representative trajectory. It also holds mapping data between the input MATs and the representative MAT.

4   The MAT-SG Method

MAT-SG is inspired by the literature gap regarding MAT summarization. We assume the input MATs was already filtered by some criterion, so the representative MAT denotes the main behavior of these input MATs considering spatial density and frequency of each aspect attribute value.

Figure 3 gives an overview of MAT-SG. The trajectory data previously filtered is given as input. The method holds two internal steps: *(i) spatial segmentation*; and *(ii) data summarization*. In order to identify patterns by spatial density, we chose to segment MAT points into a grid in the first step. Clusters of nearby points in the same cell are then generated. The second step, in turn, generates the representative trajectory (*rt*). It computes a representative point (p_r) for each relevant cell summarizing each dimension. Then, the *rt* is given as output.

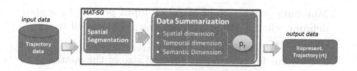

Fig. 3. MAT-SG overview.

The data considered by MAT-SG is based on the data model shown in Fig. 4. It standardizes the input data representation, and maintains the representative MAT points as well as their mappings to the input MAT points. As shown in the data model, a MAT can contain many points. Each point, in turn, holds information about all dimensions: *space* (x and y coordinates), *time* (a timestamp), and *semantic* (a set of the attributes with their corresponding values). Each attribute belongs to a categorical or numerical data type.

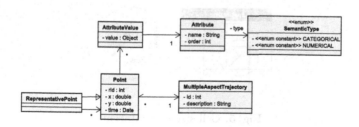

Fig. 4. The conceptual model for MAT-SG

Another concept is the *representative point* p_r. It is a point generated by MAT-SG, and a sequence of p_r's composes the representative MAT. A p_r is generated from many MAT points, and a relationship between p_r and the corresponding MAT points is modeled for maintaining mapping data. We also consider p_r as a MAT point specialization, so it can also hold attributes.

MAT-SG takes as input a set of parameters besides the input MATs, as detailed in Table 2. Only rc and τ_{rv} can be optionally set by the analyst, otherwise default values are considered. MAT-SG starts by calculating $\tau_{rc} = |T.points| * rc$, which is based on a proportion rc. For example, given $rc = 1\%$ and $|T.points| = 200$, then $\tau_{rc} = 2$, i.e., only cells with at least 2 points are considered relevant to hold a p_r. Next, MAT-SG performs its steps, as detailed in the following.

Table 2. Parameters of MAT-SG

Parameter	Explanation	Default
T	Set of previously filtered input MATs	–
rc	Minimum proportion of all MAT input points $|T.points|$, defining when a cell is considered a relevant cell to compute rt	$\tau_{rc} = 2$
z	A constant for calculating the cell size of the spatial grid	≥ 2
τ_{rv}	Rate of representativeness in the temporal ranking of rt	10%

4.1   Spatial Segmentation

The first MAT-SG step segments the points of the input MATs over a grid of squared cells. Figure 5 shows a spatial grid with a highlighted cell. The cell size is based on a threshold (τ_s) that specifies the maximum spatial distance between two points in the cell, i.e., the diagonal length of the cell.

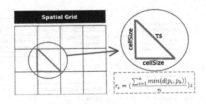

Fig. 5. Cell size computation

We automatically compute τ_s (Equation in Fig. 5) based on the average of the minimum spatial distance of the input MATs points to provide dynamic space segmentation for clustering these input points. Given T with n points, we compute the Euclidean distance $d()$ for each point $p_i \in T$ with the nearest point $p_k \in T$. Then, τ_s is computed as z times the average value, where z is based on preliminary experiments over a training MAT dataset to identify the better value for spatial segmentation. More details are given in Sect. 5.

Algorithm 1 details the spatial segmentation step. Its advantage is that only cells with points are generated. It allocates T points on a spatial grid, which is implemented as an inverted index (or *inverted list*), where the *key* is the identity of the cell position and the *value* is a list of the T points allocated in the interval.

Algorithm 1: *segmentIntoSpatialGrid*

---

input : T, z
output: *spatialCellGrid*   /* inverted list */
1 *spatialCellGrid* $\leftarrow \emptyset$;
2 $\tau_s \leftarrow compute\tau_s(z)$;
3 *cellSize* $\leftarrow computeCellSize(\tau_s)$;
4 **foreach** $t \in T$ **do**
5   **foreach** $p \in t$ **do**
6   *key* $\leftarrow getCellPosition(p_x, p_y, cellSize)$;
7   **if** *spatialCellGrid.get(key)* $= \emptyset$ **then**
8   *spatialCellGrid.new(key)*;
9   *spatialCellGrid.get(key).put(p)*
10   **else**
11   *spatialCellGrid.get(key).append(p)*;
12   **end**
13   **end**
14 **end**
15 **return** *spatialCellGrid*

First, the cell size of the spatial grid is computed (lines 2–3). In order to define the cell where each point $p \in T$ will be allocated, we obtain the grid position key of p (line 6), considering the cell size given by the *getCellPosition* function: $(\frac{p_x}{cellSize}, \frac{p_y}{cellSize})$. Then, p is allocated into the grid cell of this position (lines 6–12), if it exists. Otherwise, a new key is created to insert it (lines 8–9).

4.2 Data Summarization

The second MAT-SG step receives *spatialCellGrid* as input and summarizes points in the same cell to generate a p_r. Cells with less than τ_{rc} points are discarded, as they are assumed as weak representative. The p_r generation takes into account the analysis of the three MAT dimensions for all points in the cell (*cell.P*). The summarization of each dimension is added to p_r.

For *spatial dimension*, the *centroid* point is computed [19], i.e., the average of the (x,y) coordinates in *cell.P*. For the *temporal dimension*, we compute the *significant temporal intervals* in which all timestamps in *cell.P* fit, as follows.

Definition 3 *Significant Temporal Intervals (STI)*. *An STI is a set of time intervals* $\{[ts_f - ts_i], ..., [ts_k - ts_m]\}$ *that contains all* $ts \in cell.P$.

MAT-SG defines an STI rank that refers to all intervals $t_i \in STI$ and their tendency. We use the predefined threshold τ_{rv} to define which t_i are considered

Algorithm 2: *computeTemporalDimension*

```
input  : cell.P, τrv                          /* cell.P is a set of points in a cell */
output: rankSTI                               /* ranking of representative STIs for cell.P */
1  foreach p ∈ cell.P do
2  |   Time.add(p.time);
3  end
4  Time.sort();
5  ΔTime ← computeTimesDifference(Time);
6  VΔTime ← computeValidValues(ΔTime);
7  τt ← computeTimeThreshold(VΔTime);
8  STIaux ← ∅;
9  rankSTI ← ∅;
10 foreach tsi ∈ Time do
11 |   STIaux.append(tsi);
12 |   if δi > τt and (|STIaux|/|Time|) ≥ τrv then
13 |   |   rankSTI.new(STIaux);
14 |   |   rankSTI.get(STIaux).put(|STIaux|/|Time|);
15 |   |   STIaux ← ∅;
16 end
17 return rankSTI
```

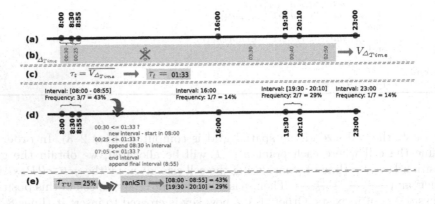

Fig. 6. An example of temporal dimension summarization in a grid cell

representative for p_r, i.e., the t_i's with a frequency rate $\geq \tau_{rv}$. Algorithm 2 computes the ranking of representative STIs and Fig. 6 exemplifies this process.

First, a *Time* list is generated to hold all $ts \in cell.P$. It is sorted for better analyzing the time intervals (lines 1 to 4), as shown in Fig. 6 (a). Then, we consider a computed threshold (τ_t) to define when a $ts \in Time$ is close to another, and aggregate ts's to generate an *STI*, as explained in the following.

Consider δ_i a time difference of two consecutive timestamps ($\delta_i = ts_{i+1} - ts_i$), and $\Delta_{Time} = \{\delta_1; \delta_2; ...; \delta_{n-1}\}$ a set of δ_i's for all $ts \in Time$ (line 5). In line 6, we set the *valid time interval set* $V_{\Delta_{Time}}$ as all $\delta_i \in \Delta_{Time}$ that fit into the average $\overline{\Delta_{Time}}$ plus or minus the standard deviation $\sigma\Delta_{Time}$, as defined by Eq. 1.

$$V_{\Delta_{Time}} = \{\delta_i \in \Delta_{Time}, 1 \leq i \leq (n-1) \mid (\overline{\Delta_{Time}} - \sigma\Delta_{Time}) \leq \delta_i \leq (\overline{\Delta_{Time}} + \sigma\Delta_{Time})\} \quad (1)$$

In line 7, we define τ_t as the average of $V_{\Delta_{Time}}$ ($\overline{V_{\Delta_{Time}}}$). Its purpose is to eliminate all $\delta_i \in \Delta_{Time}$ that represent outliers. Figure 6 (b) shows all $\delta_i \in \Delta_{Time}$ and $V_{\Delta_{Time}}$. In this case, 07:05 is an outlier. Figure 6 (c) computes τ_t.

The STI is built based on τ_t (lines 10 to 16). We initially append to STI_{aux} the next $ts_i \in Time$ (line 11), and while δ_i is less than τ_t, we consider ts_i part of an STI and continue to append subsequent timestamps. When δ_i becomes higher than τ_t and its frequency is considered representative (line 12), we assume that an $sti \in STI$ is discovered, and we insert STI_{aux} as a new key into a $rankSTI$ inverted list (line 13), and its frequency as the value of this key (line 14). An sti may also be a punctual ts_i when it is very distant from its neighbors, i.e., when δ_i to its neighbors are higher than τ_t. Figure 6 (d) shows the STI generation. Each $ts \in Time$ is analyzed to verify whether ts is an isolated timestamp or part of a time interval. In the example, we have a first $sti_1 = \{08:00, 08:30, 08:55\}$ as all their $\delta_i <= \tau_t$. An $sti_2 = \{16:00\}$ holds a single ts as the time differences to its neighbors exceed τ_t. This process is repeated to all the remaining $ts \in Time$. According Fig. 6 (e) and $\tau_{rv} = 25\%$, $STI = \{[08:00–08:55], [19:30–20:10]\}$.

At last, we summarize the *semantic dimension*. As it can be composed of multiple aspects, we divide it into two types: *(i)* categorical (*e.g.*, mean of transportation and weather condition) and *(ii)* numerical (*e.g.*, temperature and air humidity). For categorical types, as well as the temporal dimension, we rank the representative *mode* values, i.e., the most commonly observed values for each aspect in the cell. For numerical types, we compute the *median* value[1].

Our summarization approach is based on spatial segmentation, i.e., this dimension has priority. So, if all points in the same cell are semantically different, at least one representative point considering the spatial dimension is computed. That shows the representativeness of this location in input MATs.

4.3 Running Example

We now exemplify the application of MAT-SG. Let $T = \langle q, r, s \rangle$, where $q = \langle p_{q_1}, p_{q_2}, ..., p_{q_n} \rangle$, $r = \langle p_{r_1}, p_{r_2}, ..., p_{r_m} \rangle$ and $s = \langle p_{s_1}, p_{s_2}, ..., p_{s_t} \rangle$ are the input MATs of some individuals. Figure 7 presents them and some related aspects: *price* they spend in a PoI, the *PoI* itself, *weather* condition and *rain* precipitation.

We consider $rc = 10\%$, $z = 6$ and $\tau_{rv} = 30\%$ as input values. As $|T.points| = 15$, a relevant cell must contain at least 2 points. Figure 8 (a) shows T segmentation into a grid of cells, and Fig. 8 (b) shows the resulting $rt = \langle p_{rt_1}, p_{rt_2}, ..., p_{rt_k} \rangle$ (yellow line - spatial dimension summarization). Detailed output is illustrated in Fig. 8 (c). Data summarization occurs at cells containing more than one point.

For the temporal dimension, we find some relevant $sti \in STI$. This is the case of p_{rt_1}, where only one sti is identified, considering $\tau_t = 15$ min. It represents 75% of the cell points. For the cases we have two or three points in a cell, the analyst set $\tau_t = 100$ min. For p_{rt_2}, for example, we have two punctual occurrences, since

---

[1] We prefer the median value instead of the mean value when the data are not symmetrically distributed since it is less sensitive to the influence of outliers [11].

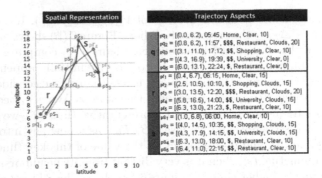

Fig. 7. Sample data with point aspects information for trajectories q, r and s.

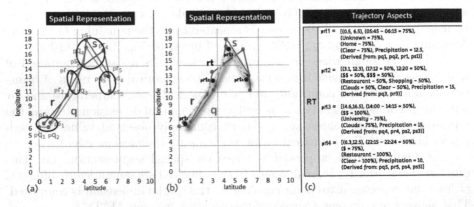

Fig. 8. A spatial segmentation (a) and the resulting representative trajectory (b, c)).

the time difference between p_{q_3} and p_{r_3} is higher than 100 min. These ts's do not generate an STI.

For the semantic dimension, we compute the median of *rain precipitation*, which is a numeric data. For the categorical data, we define a frequency ranking considering representativeness values. This is the case for *price, POI*, and *weather*. These rankings can be useful to the analyst. For p_{rt_1}, for example, we see that, in most of the cases, the location refers to *Home*, so we can strongly suppose that this is a residential area. We also set categorical data as *Unknown* when the aspect has no value, as this information can also be relevant. We realize, for example, that p_{rt_1} has no evaluated *price*. It occurs because the most common PoI is *Home*, and an individual does not spend money at home. It also highlights a relationship between these aspects.

5 Experimental Evaluation

This section presents a first MAT-SG evaluation. Our method is implemented in Java, and all experiments ran on a Dell Inspiron laptop, with an Intel Core i5 processor and 16 GB memory.

We evaluate our method on two datasets: (i) *Running Example* (Sect. 4.3), with $rc = 10\%$ and $\tau_{rv} = 30\%$; and (ii) *Foursquare*, a publicly real-world dataset with MATs from anonymous users in New York City [20]. It contains the following aspects: (i) *latitude* and *longitude*; (ii) *time of the day*; (iii) *day of the week*, (iv) *PoI description*, (v) *PoI category*, (vi) *PoI subcategory*, (vii) *rating level* of the PoI (a numeric classification), (viii) *price level* of the PoI (a numeric classification), and (ix) *weather condition*. All aspects are categorical. We focus on *User 6* that has 13 MATs with 225 points, and we set $rc = 1\%$ and $\tau_{rv} = 20\%$.

In this experiment, we evaluate *data reduction* and the *accuracy* compared to the input data. We use this information to analyze and define the best cell size to segment input MATs and generate rt. We analyze $|rt|$ for evaluating rate reduction, i.e., the number of rt points. In terms of accuracy, we analyze how much rt represents the input MATs by evaluating two criteria: (i) covered MAT points (T^c); and (ii) information on the covered MATs (R_M).

We define a *proximity* metric for evaluating the accuracy that informs how much each $t \in T$ is close to rt in all dimensions. We base this metric on the *match* function for MATs called *MUITAS* [15]. MUITAS is the state-of-the-art w.r.t. MAT similarity measure. MUITAS measures the similarity between two MATs quantifying the distance between their points, and it only considers complete attribute matching. However, as rt follows a different structure, given by simple or rank values, we consider partial matches and compute a score based on the number of matching attributes. For each $p_t \in t$, we compute the closest point in rt and the sum of the scores of the best matches (the $parity(t, rt)$ function). From this parity, it is possible to know how much t data are captured by rt. For computing the proximity of each $t \in T$ to rt, we define the function described in Eq. 2. Finally, we compute our R_M metric as a median of proximity scores for $t \in T$. The results for each dataset are presented in Table 3.

$$proximity(t, rt) = \frac{parity(t, rt)}{|t|} \qquad (2)$$

Regarding the gain value of R_M for each z value parameter, the highest values are associated with $z = 6$ (65%) for the Running Example and $z = 9$ for Foursquare (47,38%). As R_M compares rt against all input MATs, our first conclusion is that MAT-SG shows promising results since R_M depicts how much information rt captures from the input MATs at a point granularity. This is the finest granularity for trajectory comparison purposes, so we may have heterogeneous similarity scores when the input MATs and the MAT lengths are large, i.e., the scores tend not to be very high for all point-level comparisons and, consequently, the final R_M value tends to be directly proportional.

Table 3. R_M results for different cell sizes according to z value.

Dataset × Z value	Running example					Foursquare - User 6				
	R_M	Gain	$\|rt\|$	T^c	$\%T^c$	R_M	Gain	$\|rt\|$	T^c	$\%T^c$
2	47,50%		3	8	53%	32,93%		15	138	61%
3	51,17%	8%	3	9	60%	35,23%	7%	15	149	66%
4	51,17%	0%	4	11	73%	35,95%	2%	17	155	69%
5	56,17%	10%	4	11	73%	39,28%	9%	19	169	75%
6	**65,00%**	**16%**	**4**	**14**	**93%**	39,67%	1%	19	175	78%
7	68,00%	5%	4	14	93%	42,66%	8%	18	183	81%
8	66,00%	−3%	4	14	93%	41,85%	−2%	19	180	80%
9	68,70%	4%	4	14	93%	**47,38%**	**13%**	**22**	**198**	**88%**
10	58,67%	−15%	3	14	93%	43,48%	−8%	19	193	86%

We also realize that, for both datasets, as the value of z increases, the covered MAT points also increase. However, the covered information given by R_M tends to increase to a maximum point after decreasing or having a weak increase. From this trend, we see that, although the covered MAT points increase, these points could not be semantically similar. Then, the z value given by this maximal point of R_M can be considered to generate rt with the maximum possible coverage of both MAT points and information. Nevertheless, we must execute more experiments with different MAT datasets to prove this hypothesis better.

From this reasoning, we consider rt generated by the highest R_M value. For the Running Example, rt covers 93% of all input MAT points ($T^c = 14$). W.r.t. covered information, rt captures 65% of all input MATs considering all aspects. In terms of volume, rt represents a reduction of $73,33\%$ of all input MATs ($|rt| = 4$ and $|T.points| = 15$). As the average size of all $t \in T$ is 5, $|rt|$ is close to the size of each $t \in T$. For Foursquare, rt covers 88% of the MAT points ($T^c = 198$). rt achieves a score of $47,38\%$ covered information, and reduces $90,22\%$ of all input MATs. The average size of all $t \in T$ is $17,31$, close to $|rt|$.

Regarding Foursquare, an rt analysis indicate several user patterns: (i) usage of train as transport on Tuesdays; (ii) travels by train on Thursday and Friday mornings before 6 am, and on those same days, travels between 5:35 pm and 5:52 pm by bus and/or plane; and (iii) goes out to eat on Sundays.

6   Conclusion

This paper presents a pioneer method to summarize MATs named MAT-SG[2]. MAT-SG considers spatial, temporal and different semantic attributes that characterize MATs, abstracting each one of these dimensions according to their singularities. Another differential is the mapping between input MATs and the

---

[2] Avalaible in https://github.com/vanessalagomachado/MAT-SG/tree/master.

representative MAT through a data model. It allows persistence and querying of representative MATs, as well as their origins, and it also allows the analyst to identify patterns on the data, and the representativeness of some MAT points.

W.r.t. evaluation, we propose an evaluation metric based on the state-of-art on MATs similarity that can be considered to improve MAT-SG by finding the best z parameter and also be used to compare the summarization quality of related work proposals. From the results, we see that MAT-SG is a promising method. It was not possible to compare MAT-SG against the unique close baseline [17] as its source code was not available and the presented evaluation did not show the output data to allow us a comparison with our output.

As future works, we aim to reduce the complexity of our method. MAT-SG currently has a quadratic complexity $O(n^2)$ w.r.t. the number n of points of all input MATs, which is dominated by the *computeCellSize* function in Algorithm 1. We also intend to improve MAT-SG by considering dependencies between aspects, like *price* depending on *PoI* in our running example. Experiments involving the baseline and larger MAT datasets are also expected.

Acknowledgements. This work has been co-funded by the Brazilian agencies CAPES (Finance code 001) and FAPESC (Project Match - Grant 2018TR 1266), as well as the European Union's Horizon 2020 research and innovation programme under GA N. 777695 (MASTER).

References

1. Agarwal, P.K., Fox, K., Munagala, K., Nath, A., Pan, J., Taylor, E.: Subtrajectory clustering: models and algorithms. In: Proceedings of the 37th ACM SIGMOD-SIGACT-SIGAI Symposium on Principles of Database Systems, pp. 75–87 (2018)
2. Ayhan, S., Samet, H.: DICLERGE: divide-cluster-merge framework for clustering aircraft trajectories. In: Proceedings of the 8th ACM SIGSPATIAL International Workshop on Computational Transportation Science, pp. 7–14 (2015)
3. Buchin, K., et al.: Median trajectories. Algorithmica **66**(3), 595–614 (2013)
4. Buchin, M., Kilgus, B., Kölzsch, A.: Group diagrams for representing trajectories. Int. J. Geogr. Inf. Sci. **34**(12), 2401–2433 (2019)
5. Erwig, M., Schneider, M., Vazirgiannis, M., et al.: Spatio-temporal data types: an approach to modeling and querying moving objects in databases. GeoInformatica **3**(3), 269–296 (1999)
6. Etienne, L., Devogele, T., Buchin, M., McArdle, G.: Trajectory box plot: a new pattern to summarize movements. Int. J. Geogr. Inf. Sci. **30**(5), 835–853 (2016)
7. Fiore, M., et al.: Privacy in trajectory micro-data publishing: a survey. Trans. Data Privacy **13**, 91–149 (2020)
8. Gao, C., Zhao, Y., Wu, R., Yang, Q., Shao, J.: Semantic trajectory compression via multi-resolution synchronization-based clustering. Knowl.-Based Syst. **174**, 177–193 (2019)
9. Hesabi, Z.R., Tari, Z., Goscinski, A., Fahad, A., Khalil, I., Queiroz, C.: Data summarization techniques for big data—a survey. In: Khan, S.U., Zomaya, A.Y. (eds.) Handbook on Data Centers, pp. 1109–1152. Springer, New York (2015). https://doi.org/10.1007/978-1-4939-2092-1_38

10. Lee, J.G., Han, J., Whang, K.Y.: Trajectory clustering: a partition-and-group framework. In: SIGMOD, pp. 593–604. ACM (2007)
11. McCluskey, A., Lalkhen, A.G.: Statistics ii: central tendency and spread of data. Continuing Educ. Anaesth. Crit. Care Pain **7**(4), 127–130 (2007)
12. dos Santos Mello, R., et al.: MASTER: a multiple aspect view on trajectories. Trans. GIS **23**(4), 805–822 (2019)
13. Mello, R.S., Schreiner, G.A., Alchini, C.A., Santos, G.G., Bogorny, V., Renso, C.: Dependency rule modeling for multiple aspects trajectories. In: Ghose, A., Horkoff, J., Silva Souza, V.E., Parsons, J., Evermann, J. (eds.) ER 2021. LNCS, vol. 13011, pp. 123–132. Springer, Cham (2021). https://doi.org/10.1007/978-3-030-89022-3_11
14. Panagiotakis, C., Pelekis, N., Kopanakis, I., Ramasso, E., Theodoridis, Y.: Segmentation and sampling of moving object trajectories based on representativeness. IEEE Trans. Knowl. Data Eng. **24**(7), 1328–1343 (2012)
15. Petry, L.M., Ferrero, C.A., Alvares, L.O., Renso, C., Bogorny, V.: Towards semantic-aware multiple-aspect trajectory similarity measuring. Trans. GIS **23**(5), 960–975 (2019)
16. Rodriguez, D.F., Ortiz, A.E.: Detecting representative trajectories in moving objects databases from clusters. In: Rocha, Á., Ferrás, C., Montenegro Marin, C.E., Medina García, V.H. (eds.) ICITS 2020. AISC, vol. 1137, pp. 141–151. Springer, Cham (2020). https://doi.org/10.1007/978-3-030-40690-5_14
17. Seep, J., Vahrenhold, J.: Inferring semantically enriched representative trajectories. In: 1st ACM SIGSPATIAL International Workshop on Computing with Multifaceted Movement Data, MOVE 2019, pp. 1–4. ACM (2019)
18. Wang, S., Bao, Z., Culpepper, J.S., Cong, G.: A survey on trajectory data management, analytics, and learning. ACM Comput. Surv. **54**(2), 1–36 (2021)
19. Wood, G.B., Wiant, H.V., Jr., Loy, R.J., Miles, J.A.: Centroid sampling: a variant of importance sampling for estimating the volume of sample trees of radiata pine. Forest Ecol. Manag. **36**(2–4), 233–243 (1990)
20. Yang, D., Zhang, D., Zheng, V.W., Yu, Z.: Modeling user activity preference by leveraging user spatial temporal characteristics in LBSNs. IEEE Trans. Syst. Man Cybern. Syst. **45**(1), 129–142 (2015)

Quality Versus Speed in Energy Demand Prediction
Experience Report from an R&D project

Witold Andrzejewski[1], Jędrzej Potoniec[1], Maciej Drozdowski[1]([⊠]),
Jerzy Stefanowski[1], Robert Wrembel[1], and Paweł Stapf[2]

[1] Poznan University of Technology, Poznań, Poland
{Witold.Andrzejewski,Jedrzej.Potoniec,Maciej.Drozdowski,
Jerzy.Stefanowski,Robert.Wrembel}@cs.put.poznan.pl
[2] Kogeneracja Zachód S.A., Poznań, Poland
p.stapf@kogeneracjazachod.pl

Abstract. Effective heat energy demand prediction is essential in combined heat power systems. The algorithms considered so far do not sufficiently take into account the computational costs and ease of implementation in industrial systems. However, computational cost is of key importance in edge and IoT systems, where prediction algorithms are constantly updated with new arriving data. In this paper, we propose two types of algorithms for heat demands prediction: (1) novel extensions to the algorithm originally proposed by E. Dotzauer and (2) based on a kind of autoregressive predictor. They were developed within an R&D project for a company operating a cogeneration system and for their real dataset. We evaluate the algorithms experimentally focusing on prediction quality and computational cost. The algorithms are compared against two state-of-the art artificial neural networks.

Keywords: Time-series analysis · Energy demand forecasting · Artificial neural networks · Time-quality trade-off

1 Introduction

District heating systems (DHS) are widely used in North-Western Europe to deliver heat and hot water to households. A DHS is often a cogeneration system or combined heat power (CHP), if heat is also used to generate electric energy. The amount of produced electric energy is directly related to the amount of produced heat. Hence, **for CHP systems, prediction of heat demand is essential** to submit bids for electric energy in the priciest hours in an energy market. Constructing computationally efficient intelligent systems faces a number of challenges in industrial applications where data is collected in extensive sensor networks. Often, such raw data is incomplete and of low quality. In this paper, we report on our experience from an R&D project in designing algorithms for heat demand prediction for Kogeneracja Zachód, a company running a cogeneration system in Poland. We evaluate the algorithms w.r.t. prediction quality

© The Author(s), under exclusive license to Springer Nature Switzerland AG 2022
C. Strauss et al. (Eds.): DEXA 2022, LNCS 13426, pp. 447–452, 2022.
https://doi.org/10.1007/978-3-031-12423-5_34

and computational cost, both in the training and testing phases. The trade-off between prediction error and computational cost is essential in the edge/fog systems, if IoT devices (e.g. energy counters) with low computational resources were supposed to not only measure but also predict energy consumption and incrementally train models with new arriving data.

This paper is organized as follows. In Sect. 2 the goals in energy prediction of the R&D project are formulated. The prediction algorithms we contribute are presented in Sect. 3. The test dataset is outlined and the prediction algorithms are evaluated in Sect. 4. Section 5 concludes the paper. Further details on the research described here and the context of related works can be found in [1].

2   Problem Formulation

The studied DHS has a tree structure. In selected nodes, energy counters are installed. There are two types of energy counters: (1) hot water energy (HW) and (2) heat energy (HE) counters. Past energy readings, exogenous variables: atmospheric temperature, humidity, wind speed, sky overcast and a three-day weather forecast are known with 1 h resolution. It is required to forecast any counter readings in the next three days. Prediction of the energy for the next 72 h, especially for the "sum" in the tree root, is required to plan selling electricity on day-ahead markets. Formally, given past energy readings of a certain counter $\mathcal{Y} = (Y_1, \ldots, Y_t)$ (where Y_i are scalars), past values of exogenous variable values $\mathcal{W} = (W_1, \ldots, W_t)$ (W_i are vectors), and their forecast $\mathcal{W}' = (W_{t+1}, \ldots, W_{t+72})$ it is required to forecast energy demands $\mathcal{F} = (Y_{t+1}, \ldots, Y_{t+72})$.

3   Examined Algorithms

3.1   Dotzauer Method and Extensions

In [2] heat demand Y_i for hour i is modeled as a sum of two components:

$$Y_i = f(T_i) + g(i), \tag{1}$$

where $f(T_i)$ is a function of the atmospheric temperature T_i, g_i is an array of corrections calculated for hours in a week i. $f(T_i)$ is a piecewise linear function with five segments. A shorthand notation DPLW will be used to refer to this algorithm.

Extensions. The following versions of this Dotzauer method were examined in our studies:

- Linear temperature model Weekly/Yearly corrections (denoted DLW/DLY),
- Piecewise Linear temperature Weekly/Yearly corrections (DPLW/DPLY),
- Spline temperature model Weekly/Yearly corrections (DSW/DSY),
- Isotonic regression temperature model Weekly/Yearly corrections (DIW/DIY),
- Multivariate temperature model Weekly/Yearly corrections (DMW/DMY).

Table 1. Exogenous variables. Notation: T - atmospheric temperature, DL - day length in hours, DT - day type (1: Monday–Thursday, 2: Friday, 3: Saturday, 4: Sunday), V - wind speed in m/s, $\sqrt{V}$ - square root of wind speed, TV - product of temperature T and wind speed V, $T\sqrt{V}$ - product of temperature T and $\sqrt{V}$, pY - season of the year (spring,...,winter), Oc - overcast in oktas, H - humidity.

Algorithm version	T	DL	DT	V	$\sqrt{V}$	TV	$T\sqrt{V}$	pY	Oc	H
DMW/DMY	*	*	*	*	*	*	*	*	*	*
WRNH0/WRWH0	*	*								
WRNH1/WRWH1	*	*			*		*			
WRNH2/WRWH2	*	*		*	*	*	*	*		
WRNH3/WRWH3	*	*	*	*	*	*	*	*	*	*
WRNH4/WRWH4										

3.2 W-Regressors

WRNH builds independent prediction models for each hour of the week. For each hour i of the week a moving average of the last week energy consumption is calculated (168 samples). In order to construct a linear regression fit for hour i tuples (a_j, W_j, Y_j) are used as input data points, where: a_j is energy consumption 168-h moving average at hour j, W_j are weather conditions at hour j, Y_j is the actual energy consumption at hour j, for hours $j : j \bmod 168 = i$ in the past. We applied various combinations of atmospheric conditions in vector W_j resulting in five versions WRNH0,...,WRNH4 of this algorithm (cf. Table 1). Tuples (a_j, W_j, Y_j) are used to fit linear regression $Y_i = \overline{k_i} \times [a, W]^T + l_i$, where a is a moving average, W is a vector of weather conditions, $[a, W]$ is a vector of independent variables, $\overline{k_i}$ is a vector of directional coefficients, and Y_i is the modeled energy consumption. When predicting energy consumption for future hour $t + p$, for $p = 1,\ldots,72$, at the current hour t, the energy prediction is calculated as $Y_{t+p} = \overline{k}_{(t+p) \bmod 168} \times [a_t, W_{t+p}]^T + l_{(t+p) \bmod 168}$, where a_t is the moving average of energy consumption at prediction moment t, W_{t+p} is a forecast of weather conditions for hour $t + p$.

WRWH. The WRWH regressors build $168 \times 72 = 12096$ linear models. In order to develop a model for future hour $i + p$, where $i = t \bmod 168$ were the hours of the week of the current moment t, tuples (a_j, W_j, Y_q) for $j : j \bmod 168 = i$ and $q : q = j + p$, were used. Future hour $t + p$ energy consumption is calculated as $Y_{t+p} = \overline{k}_{(t \bmod 168),p} \times [a_t, W_{t+p}]^T + l_{(t \bmod 168),p}$, where $\overline{k}_{i,p}$ is a vector of directional coefficients for the current hour of the week i and energy consumption shifted p hours into the future. There are five versions WRWH0, ..., WRWH4 of this method (cf. Table 1).

3.3   Neural Networks

Feed-Forward Neural Network (FFNN). We used an FFNN implemented in PyTorch. Its best topology consisted of 99 inputs, two hidden layers with 10 and 46 neurons, and 72 outputs to forecast for the 72 h (with ReLU functions). Beyond the past energy readings all weather attributes mentioned in Table 1 for the current time t were the FFNN inputs. Optuna, a hyperparameter optimization framework, was applied to tune hyperparameters of this model. Separate neural networks with the same architecture were trained for each counter in DHS.

Radial Basis Function Neural Network (RBFNN). We used a single hidden layer consisting solely of RBF neurons with Gaussian functions. To decide on the number of hidden, output neurons and the input features we used Optuna, and arrived at 16 hidden RBF neurons and 72 linear output neurons. Further details on arriving at our neural network architectures, training them can be found in [1].

Furthermore, the moving average of the last 100 h was used as a reference prediction algorithm. It is denoted as algorithm C-100.

4   Experimental Analysis of Prediction Algorithms

The test dataset comprises heat measurements from 28 HW and 83 HE counters in the DHS serving a town with about 30000 inhabitants. The records cover the period since the 1st of September 2015 until 28th of February 2019. The dataset includes: measurement timestamp, measured energy consumption, weather conditions, type of day, season. The measurements were collected for billing purposes, rather than for heat demand prediction and optimization, so they have several drawbacks from the datamining perspective. Information on the artifacts found in the test data and on the procedures applied to clean the data are presented in [1]. This test dataset has been shared for research purposes, see [3].

Data from interval 2016-01-01 until 2017-12-31 was used as the training dataset. Interval from 2019-01-02 until 2019-02-25 was used for testing. Each of the algorithms calculated for each energy counter a 72-h energy consumption forecast starting at each hour of the testing interval (1320×72 prediction points). All the codes were written in Python 3.7.4 and tested on a PC with Windows 10 and Intel i7-8550U CPU @ 1.80 GHz (no GPU acceleration).

4.1   Basic Evaluation

Quality of the predictions of 8 best algorithms aggregated over all 1320×72 prediction points are presented quantitatively in Table 2 for the "sum" counter with respect to Mean Absolute Percentage Error (MAPE) and Mean Squared Error (MSE) measures. It can be seen that RBFNN and simple Dotzauer model variants with weekly corrections provided the best predictions of the "sum" of consumed energy for MAPE. For MSE quality measure RBFNN, FFNN are the best algorithm whereas C-100 and DLW (which is the best Dotzauer-like algorithm) have MSE three times worse than FFNN (not shown in Table 2). Thus, RBFNN, FFNN are the best for "sum" counter. More details are given in [1].

Table 2. 8 best algorithms on "Sum" counter

MAPE								
Method	RBFNN	DLW	DMW	DPLW	DIW	WRWH0	FFNN	DSW
MAPE [%]	16.45	16.49	16.49	16.96	17.1	17.7	18.43	18.45
MSE								
Algo.	FFNN	RBFNN	WRWH0	WRWH1	WRNH0	WRNH1	WRNH2	WRWH2
MSE	217884	237987	437482	452190	463599	477858	568191	575916

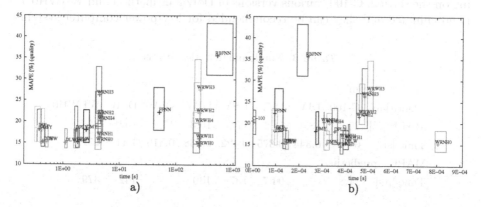

Fig. 1. Prediction quality (MAPE) vs computational cost. a) Prediction quality vs training time. b) Prediction quality vs predicting time.

4.2   Time-Quality Trade-Off

Time performance of the prediction algorithms is important when these algorithms are supposed to be used in low-power IoT or embedded devices. In Fig. 1 run-time vs quality trade-off is shown (run time on the horizontal axis while prediction quality (MAPE) on the vertical axis). Results for each algorithm are presented as interquartile boxes aggregated over all counters. That is, a box for each algorithm spans between Q1 and Q3 in time and prediction quality. Median of quality and run-time is also marked. This way of visualizing mutual algorithm performance has three-fold advantages: 1) it is possible to recognize algorithm differences with respect to prediction quality, 2) differences in time efficiency are visible, 3) it is possible to analyze how these algorithms trade run-time for prediction error. In Fig. 1a training time is shown, in Fig. 1b time of calculating a single 72-h forecast is shown.

As far as training these algorithms in low-power computers is considered, Dotzauer-like and WRNH algorithms are feasible choices, whereas using RBFNN or WRWH algorithms seems to be less convenient. Conversely, calculating one 72-h prediction is far less costly computationally and all the considered algorithms managed it in less than 1 ms. Considering prediction quality, most of the interquartile ranges overlap, so it is rather hard to draw sharp conclusions on algorithm superiority, but still, some tendencies can be observed. Algorithms

WRNH0, WRWH0 using only atmospheric temperature and day length are the best among WRNH/WRWH methods. Thus, an extensive set of weather attributes, impedes rather than helps obtaining good energy predictions.

Figure 1 allows also to identify algorithms that are nondominated. We used median run-time and median accuracy as indicators of algorithm position on the run-time vs quality space. The set of nondominated methods comprises the algorithms for which no other algorithm has both better accuracy and run-time. For better exposition, the nondominated algorithms are collected in Table 3. Depending on the setting, C-100, various versions of Dotzauer method and WRWH0 are nondominated and they can be recommended for low-power computer systems.

Table 3. Nondominated algorithms

MAPE - training							
Algorithm	C-100	DIY	DLY	DSY	DIW	DSW	DLW	WRWH0
MAPE [%]	20.5	18.4	18.1	17.9	15.3	15.2	15.0	14.5
Time [s]	0	0.344	0.375	0.422	0.438	0.516	1.11	251.2
MAPE - predicting							
Time [μs]	11.8	–	94.7	107	130	–	142	379

5   Conclusions

In this paper several algorithms for DHS energy consumption prediction were analyzed in order to verify their utility in predicting energy consumption every hour, for 72-h intervals, both in total and for each energy counter separately. Another aspect, important in the use on low-power devices, was the computational complexity of the algorithms. It turned out that the simple methods offer the best run-time quality trade-off. More advanced neural networks have low potential for training on low-power devices. However, a more detailed inspection revealed that accuracy of energy consumption prediction depends very much on the energy counter and accuracy measure. For example, RBFNN, FFNN networks are the best for the total energy prediction.

Acknowledgments. Research partially supported by Kogeneracja Zachód S.A.

References

1. Andrzejewski, W., Potoniec, J., Drozdowski, M., Stefanowski, J., Wrembel, R., Stapf, P.: Quality versus speed in energy demand prediction for district heating systems (2022). https://doi.org/10.48550/ARXIV.2205.07863
2. Dotzauer, E.: Simple model for prediction of loads in district-heating systems. Appl. Energy **73**(3–4), 277–284 (2002). https://doi.org/10.1016/S0306-2619(02)00078-8
3. Test dataset: real data sets on district-heating systems energy consumption (2022). https://www.cs.put.poznan.pl/rwrembel/energy-cons-data.html

Author Index

Printed in the United States
by Baker & Taylor Publisher Services

Printed in the United States
by Baker & Taylor Publisher Services